Lithium-Ion Batteries

Masaki Yoshio • Ralph J. Brodd • Akiya Kozawa
Editors

Lithium-Ion Batteries

Science and Technologies

Springer

Editors
Prof. Em. Masaki Yoshio
Advanced Research Center, Saga University
1341 Yoga-machi, 840-0047, Japan
yoshio@cc.saga-u.ac.jp

Ralph J. Brodd
Broddarp of Nevada, Inc.,
2161 Fountain Springs Dr.
Henderson, NV 89074, USA
ralph.brodd@earthlink.net

Akiya Kozawa
ITE Aichi Office, 2-15-19 Kamejima
Nakamura-ku, Nagoya 453-0013
Japan

ISBN: 978-0-387-34444-7 e-ISBN: 978-0-387-34445-4
DOI: 10.1007/978-0-387-34445-4

Library of Congress Control Number: 2008940833

Printed on acid-free paper

springer.com

Foreword

My Way to Lithium-Ion Batteries

Yoshio Nishi

I have been engaged in research and development (R&D) on novel materials for electronic appliances for 40 years since I joined Sony Corporation in 1966. I started my scientific career in Sony as a researcher of zinc-air batteries. After 8 years in R&D on electrochemistry, my research field was shifted against my will to electroacoustic materials, specifically diaphragm materials for electroacoustic transducers including loudspeakers, headphones, and microphones. My R&D work also extended to cabinet materials for speaker systems. This about-face was uncomfortable for me at first, but it forced me to devote myself to the investigation of various classes of materials unfamiliar to me, covering pulp and paper, metals (i.e., Ti, Al, Be), ceramics (B4C, TiN, BN, SiC), carbonaceous materials (carbon fibers, intrinsic carbon, artificial diamond), reinforcing fibers for FRP (carbon fibers, aromatic polyamide fibers, glass fibers, SiC fibers, superdrawn polyethylene fibers), organic polymers (polyamides, polyethylene, polypropylene, polymethylpentene, polyimides, polysulfones, polyetherimides, polyethersulfones, PET), boards (plywood, particle board), resin composites (bulk molding compounds, resin concretes, artificial marble), and so on. I also was engaged in development of piezoelectric loudspeakers employing poly(vinylidene difluoride) (PVdF).The remarkably successful output from my R&D activities in those days were organic polymer whiskers and bacterial cellulose. The former was the first organic whisker in the world discovered by M. Iguchi,[1] which is composed of polyoxymethylene (POM). Organic whiskers have favorable properties for speaker diaphragms, namely, a high Young's modulus and low density. Sony made the joint developmental work with Dr. Iguchi on POM whiskers and succeeded in their mass production on a small scale and in their application to speaker diaphragms through the development of composite materials with the whisker and polyethylene.[2]

Bacterial cellulose is biosynthesized by the help of *Acetobacter xylinum* cultivated in media containing monosaccharides or disaccharides. We developed headphone diaphragms composed of bacterial cellulose sheets that had a high specific elastic modulus comparable to aluminum and titanium foils with low specific gravity.[3] The Technical Award was given for this work from the Agrochemical Society of Japan.

After 12 years of work on electroacoustic materials, I resumed investigating novel electrochemical cells in 1986. My efforts were focused on cells with nonaqueous electrolytes, especially on those making use of carbon-lithium alloy anodes. In 1990, Sony announced that novel high-powered rechargeable cells were completed. $LiCoO_2$ was used as a cathode-active material and a tailor-made carbonaceous material was developed as an anode. Lithium was inserted in this carbon anode when a cell was charged and lithium was extracted from the anode during discharge. We gave the name of *lithium-ion battery* or LIB to this battery system. My accumulated experience in a wide range of advanced materials described above was greatly helpful in my new R&D activities, because LIB required various sophisticated materials including ceramics (i.e., $LiCoO_2$), carbonaceous materials (i.e., anodes), polymer films (i.e., separators), adhesives (i.e., binders for cathode and anode materials), and organic solvents (i.e., electrolytes).

In the case of POM synthesis, for example, it was necessary to control the moisture content in the solution of raw materials to a very low level (a few ppm)2 and I could turn this technique to advantage when I made up nonaqueous electrolytes in which water content must be as low as possible. Biaxially drawn polyethylene microporous film was adopted as a separator and this material was analogous to superdrawn polyethylene fibers described above. As a binder for active electrode materials, PVdF was used, which was familiar to me as a piezoelectric speaker material.

The anode-active material is one of the most important items in LIB. The oldest description about carbon/lithium-negative electrodes appeared, as far as I know, in the German Patent filed in August 7, 1978,4 in which it was described that "Bei derartigen galvanischen Elementen kann das Leichmetall der negativen Elektrode aus Be, Mg, Ca, B, Al, Sc, C, Si, As oder aus einer überwiegend aus diesen Leichtmetallen aufgebauten Leichtmetallegierung oder aber aus Lithiumlegierungen der genannten Stoff bestehen." Since then, considerable patents and papers referred to carbon/lithium electrodes.5–8

In regard to positive electrodes, we had developed $AgNiO_2$ as a cathode-active material in silver oxide cells.9 An Ag insertion-extraction was observed in NiO2 layers during charge-discharge reactions. By analogy with this phenomenon, a bright idea occurred to us that $LiMO_2$ (M = Ni, Co, etc.) could be used as cathode-active materials in lithium cells. J. B. Goodenough et al., however, previously showed us that $LiCoO_2$ and $LiNiO_2$ cathodes can be reversibly charged and discharged in aprotic solvents in metallic lithium/$LiMO_2$ cells.10 From these facts, it might be said that the LIB system itself was not an original invention but a novel combination of already known technologies.

In my opinion, the most important thing in developing electrochemical cells is to create a means for confining all the materials (cathode and anode active materials, electrolytes, separators, current collectors, etc.) in a limited and enclosed space, and for enabling energy as large as possible to be brought out without any safety problems. We completed this goal and succeeded in introducing LIB to practical use in 1991.12

The first-generation cells had energy densities of 200 Wh dm^{-3} and 80 Wh kg^{-1}, only slightly greater than those of nickel-metal hydride (NiMH) cells of those days.13 LIB performance has been improved continuously since then, and now energy densities reach 560 Wh dm^{-3} and 210 Wh kg^{-1} or more.

We started R&D on LIB with gel electrolytes in 1988 prior to the commercialization of LIB and put them into the market in 1998. They have comparable performances to conventional LIB and gravimetric energy densities are higher than those of LIB because polymer films can be utilized as enclosure materials instead of metallic cans.14 I believe that the advent of LIB has realized the so-called ubiquitous era. LIB has made it possible for people to utilize mobile equipment for a longer time out of doors. And in Japan, almost a billion LIBs are manufactured annually at the present time.

Since I started my R&D activities in Sony, it has been my eager dream to use and enjoy a new product in which a novel material or a device that I developed myself is made use of. At last in 1976, a new hi-fi loudspeaker system with a novel diaphragm which I developed was put on the market and I wanted to buy and use it. To my regret, however, the price was too high for me: 2-million yen! I succeeded in introducing audio equipment constantly into the market in which my novel materials were adopted including organic polymer whiskers, superdrawn polyethylene, artificial diamonds, and so on. These products also were too expensive for me. In 1988, a headphone was commercialized in which bacterial cellulose was used as diaphragms and I expected that I could buy this because a headset usually was much cheaper than a speaker. But I was at a loss for words to hear that the price of this headset was 360-thousand yen.

From the price point of view, LIB is much cheaper than a loudspeaker and the price is acceptable to me. Unfortunately, however, independent batteries are of no use. I must purchase expensive mobile gear such as notebook computers, cellular phones, digital cameras, and so forth.

References

1. M. Iguchi, *Br. Polymer J.*, 5 (1973) 195
2. Y. Nishi, M. Uryu, *New Mater. New Process.*, 3 (1985) 102
3. M. Iguchi, T. Suehito, Y. Watanabe, Y. Nishi, M. Uryu, *J. Mater. Sci.*, 17 (1982) 1632
4. Y. Nishi, M. Uryu, S. Yamanaka, K. Watanabe, K. Kitamura, M. Iguchi, N. Mitsuhashi, *J. Mater. Sci.*, 25 (1990) 2997
5. F. Heinz, German Patent, DE 2834485 C2, filed in 1978
6. A. G. MacDiarmid, A. Heager, P. J. Nigrey, U.S. Patent 4,442,125, filed in 1980
7. R. Yazami, Ph. Touzain, Abstract No. 23, International Meeting on Lithium Batteries, 1982
8. S. Basu, U.S. Patent 4,423,125, filed in 1982
9. F. P. McCullough, Jr., U.S. Patent, filed in 1984
10. T. Nagaura and T. Aida, Japanese Patent S62-11460, filed in 1980
11. J. B. Goodenough, M. Miaushima, P. J. Wiseman, British Patent, filed in 1979
12. Y. Nishi, *Chem. Rec.*, 1 (2001) 406
13. Y. Nishi, Lithium Ion Batteries, M. Wakihara, O. Yamamoto, Eds., Kodansha/Wiley-VCH, Tokyo (Japan), 1998, p. 181
14. Y. Nishi, Advances in Lithium-Ion Batteries, W. A. van Schalkwijk, B. Scrosati, Eds., Kluwer/Plenum, New York, 2002, p. 233

Contents

Contributors

Hidekazu Awano
Nippon Chemical Industrial Co., Ltd., 9-11-1, Kameido, Koto-ku,
Tokyo 136-8515, Japan, hidekazu.awano@nippon-chem.co.jp

Ralph J. Brodd
Broddarp of Nevada, Inc., 2161 Fountain Springs Dr., Henderson, NV 89074,
USA, ralph.brodd@earthlink.net

Jaephil Cho
Department of Applied Chemistry, Hanyang University,
Ansan 426-791, South Korea, jpcho@hanyang.ac.kr

Nikolay Dimov
Department of Applied Chemistry, Saga University, 1 Honjo, Saga 840-8502,
Japan, nikjapan@yahoo.com

Kazuma Gotoh
Department of Chemistry, Faculty of Science, Okayama University,
Okayama 700-8530, Japan

Tatsuo Horiba
Shin-Kobe Electric Machinery Co., Ltd., 2200 Oka, Fukaya, Saitama 369-0297,
Japan, t.horiba@shinkobe-denki.co.jp

Kazuhiko Kikuya
Toda Material Corp. Kyushu, 1-26 Hibiki-machi, Wakamatsu, Kitakyushu,
Kazuhiko_Kikuya@todakogyo.co.jp

Michiyuki Kono
R&D Division, Dai-Ichi Kogyo Seiyaku Co., Ltd., 55, Nishi-Shichijo,
Higashikubo-Cho, Shimogyo-Ku, Kyoto 600-8873, Japan,
m-kono@dks-web.co.jp

Akiya Kozawa
ITE Aichi Office, 2-15-19 Kamejima, Nakamura-ku, Nagoya 453-0013, Japan

Mariko Maeda
Research Center, Kureha Corporation, 16, Ochiai, Nishiki, Iwaki, Fukushima 974-8232, Japan

Hidekazu Mori
Zeon Co., Yako 1-2-1, Kawasaki 210-9507, Japan, h.mori@zeon.co.jp

Aisaku Nagai
Research & Technology Division, Kureha Corporation, 3-3-2, Nihonbashi-Hamacho, Chuo-ku, Tokyo 103-8552, Japan, a-nagai@kureha.co.jp

Yoshio Nishi
1-7-302 Wakabadai, Asahi-ku, Yokohama 241-0801, Japan, west24440@mmm-keio.net

Tatsuya Nishida
Hitachi Chemical Co. Ltd., Japan, 13-1, Higashi-Cho 4-chome, Hitachi City 317-8555, Japan, t-nishida@hitachi-chem.co.jp

Hideyuki Noguchi
Department of Applied Chemistry, Saga University, 1 Honjyo, Saga 840-8502, Japan

Koichi Numata
Mitsui Mining & Smelting Co, Ltd., Corporate R&D Center, 1333-2 Haraichi, Ageo, Saitama 362-0021, Japan, k_numata@mitsui-kinzoku.co.jp

Takao Ogino
Research and Development Division, Bridgestone Corporation, 3-1-1 Ogawa-higashi, Kodaira 187-8531 Tokyo, Japan

Zempachi Ogumi
Department of Energy & Hydrocarbon Chemistry, Graduate School of Engineering, Kyoto University, Kyoto 606-8501, Japan, ogumi@scl.kyoto-u.ac.jp

Masaki Okada
Nanyo Research Laboratory, TOSOH Corporation, 4560, Kaisei-cho, Shin-nanyo 746-8501, Japan, okada_m@tosoh.co.jp

Shigeto Okada
Institute for Materials Chemistry and Engineering, Kyushu University, 6-1 Kasuga Koen, Kasuga-shi, Fukuoka 816-8580, Japan, s-okada@cm.kyushu-u.ac.jp

Toshiyuki Osawa
Kanagawa Industry Technology Center, 705-1, Shimoizumi, Ebina, 243-0435, Japan, tohsawa@kanagawa-iri.go.jp

Masashi Otsuki
Central Research, Bridgestone Corporation, 3-1-1 Ogawa-higashi, Kodaira, Tokyo 187-8531, Japan, masashi.otsuki@bridgestone.co.jp

Byungwoo Park
School of Materials Science and Engineering and Research Center for Energy Conversion & Storage, Seoul National University, Seoul, Korea

Premanand Ramadass
Celgard, LLC, 13800 South Lakes Dr., Charlotte, NC 28273, USA

Kazuhiko Shimizu
Research Center, Kureha Corporation, 16, Ochiai, Nishiki, Iwaki, Fukushima 974-8232, Japan

Michael E. Spahr
TIMCAL Ltd., Strada Industriale, CH-6743 Bodio, Switzerland, m.spahr@ch.timcal.com

Yang-kook Sun
Department of Chemical Engineering, Hanyang University, Seoul, Korea

Kazuo Tagawa
Hohsen Corporation, 10-4-601 Minami Semba 4-chome, Chuo-ku, Osaka 542-0081, Japan

Makoto Ue
Mitsubishi Chemical Group, Science and Technology Research Center, Inc., Battery Materials Laboratory, 8-3-1 Chuo, Ami, Inashiki, Ibaraki 300-0332, Japan, 3707052@cc.m-kagaku.co.jp

Masami Ueda
Toda Kogyo Corp., 1-26 Hibiki-machi, Wakamatsu, Kitakyushu, Fukuoka 808-0021, Japan

Hongyu Wang
Advanced Research center, Saga university, 1341 Yoga-machi, Saga 840-0047, Japan, wanghongyu@hotmail.com

Jun-ichi Yamaki
Institute for Materials Chemistry and Engineering, Kyushu University, 6-1 Kasuga Koen, Kasuga-shi, Fukuoka 816-8580, Japan, yamaki@cm.kyushu-u.ac.jp

Haruhisa Yamamoto
ZEON Corporation, Yako 1-2-1, Kawasaki 210-9507, Japan
Hiroshi Yamamoto Toda Kogyo Corp., 1-1-1 Shinoki, Onoda, 756-0847, Japan

Masaki Yoshio
Advanced Research center, Saga university,
1341 Yoga-machi, Saga 840-0047, Japan,
yoshio@cc.saga-u.ac.jp

Hideya Yoshitake
Specialty Chemicals & Products Company, Ube Industries, Ltd., 1-2-1 Shibaura, Minato-ku, Tokyo 105-8449, Japan, 27452u@ube-ind.co.jp

Zhengming (John) Zhang
Celgard, LLC, 13800 South Lakes Dr., Charlotte, NC 28273, USA,
johnzhang@celgard.co

Introduction: Development of Lithium-Ion Batteries

Masaki Yoshio, Akiya Kozawa, and Ralph J. Brodd

1 Introduction

In response to the need for better batteries, the lithium-ion battery (LIB) was conceived and developed in Japan by Asahi Kasei Co.[1] and first commercialized by Sony Co.[2] in 1991, followed by A&T Battery Co. (a joint company of Toshiba Battery and Asahi Kasei Co.) in 1992. The LIB was accepted immediately because of its high-energy density, good performance, and no memory effect as occurred with nickel–cadmium (Ni–Cd) or nickel-hydride (Ni–MH) batteries. LIBs have been used mainly for portable electronics, especially cellular phones and notebook computers. Recently, the application area has been extended to power tools and battery-assisted electric bicycles. Several companies are working to adapt the lithium-ion system for use in hybrid electric vehicles to replace the Ni–MH.

2 History of Lithium Batteries

Today, the major battery systems are the rechargeable lead acid and the primary manganese dioxide-zinc. Both have a long history and are at an advanced state of technical maturity. The LIB is poised to challenge these established systems as the demand for higher-performance battery systems continues. Lithium has a low atomic number and a high electrode potential that results in significantly high-energy density for the LIB compared to lead and zinc in the traditional batteries. However, the development of new high-energy lithium systems has been neither simple nor easy. It has required a total system approach and the development of breakthrough technologies based on new anodes, cathodes, and nonaqueous electrolytes to continue the steady improvement of high-energy lithium battery systems.

Masaki Yoshio
Advanced Research center, Saga University, 1341 Yoga-machi, Saga 840-0047, Japan
E-mail: yoshio@cc.saga-u.ac.jp, GBE03502@nifty.com

Lithium-metal anode primary batteries based on nonaqueous electrolytes such as propylene carbonate-lithium perchlorate and lithium negative electrodes were developed in the early 1970s, with Matsushita introducing a lithium-carbon monofluoride (Li-CFx) primary cell in 1973, followed by Sanyo which commercialized primary lithium-manganese dioxide primary cells (Li-MnO_2) in 1975. These cells were used for LED fishing floats, cameras, and memory backup applications. A strong research effort then was mounted to convert lithium primary cells into rechargeable cells with high energy density. Table 1 documents the various research efforts. In the 1970s and 1980s most efforts concentrated on inorganic cathode compounds. Conducting polymer materials such as polyacetylene were developed as possible negative and positive electrode materials. However, these polymer materials have less density than water, and the batteries made with these materials offered no competitive advantage when it is enlarged, except for the polyacene (PAS) battery. The low-density conducting polymer cathodes have found a use only for coin cells for memory backup.

The early rechargeable lithium cells were plagued with safety problems caused by the tendency of lithium-metal anodes to form dendrites and powder deposits on recharging. The use of the high-performance perchlorate electrolytes was discontinued for safety reasons due to dendrite formation and very reactive fine powder deposits during recharge. In 1989, Moli Energy had the heat generation related to lithium metal in an AA-size cell. There was a shift to Li-Al alloy anode for greater safety in coin cells. However, the metallurgy of the alloy proved unacceptable for wound AA-size cells. Tadiran developed a dioxolane-based electrolyte that spontaneously polymerized at temperatures above 110°C.[3] The polymerized electrolyte had high resistance and shut down cell operation to provide a safety measure for the cells. The lithium-metal rechargeable cells now are restricted mainly to small-capacity coin cells. An early blend of lithium battery that still is in use is based on the electrochemical system Li-Al-PAS. In conjunction with a small solar cell this type of battery provides a convenient and compact power sources which now is used widely for road sign nighttime illumination or similar applications in remote

Table 1 Various rechargeable lithium metal battery systems developed

System	Voltage	Wh/kg	Wh/l	Company
Li/TiS2	2.1	130	280	'78 Exxon
LiAl/TiS2				'79 Hitachi
Li/$LiAlCl_4$-SO_2/C	3.2	63	208	'81–85 Duracell
Li/V_2O_5	1.5	10	40	'89 Tohsiba
Li/$NbSe_3$	2.0	95	250	'83–86 Bell Lab
LiAl/Polyaniline	3.0	–	180	'87 Bridgestone
LiAl/Polypyrolle	3.0	–	180	'89 Kanebo
Li/Al/Polyacene	3.0	–	–	'91 Kanebo/Seiko
Li/MoS_2	1.8	52	140	'87 MoLi
Li/CDMO(Li_xMnO_2)	3.0	–	–	'89Sanyo
Li/$Li_{0.3}MnO_2$	3.0	50	140	'89 Tadiran
Li/VO_x	3.2	200	300	'90 HydroQuebec

areas as a capacitor PAS-PAS or Li-doped PAHs-PAS hybrid battery, in which PAHs (polycyclic aromatic hydrocarbons) is the reformed structure from PAS.[4,5] Some other lithium alloys also are being developed for use as active materials for LIB and will be considered later in this book.

3 History of LIB and Patents on Lithium-Ion Cells

Since lithium metal constituted a safety problem, attention shifted to the use of a lithium-intercalation material as an anode. H. Ikeda of Sanyo was the first to patent an intercalation material in an organic solvent such as graphite in his June 1981 Japanese Patent No. 1769661.[6] One year before the Ikeda patent on graphite, Goodenough filed his $LiCoO_2$ patent for an intercalation cathode material.[7] S. Basu of Bell Laboratories filed U.S. Patent 4,423,125, in 1982, based on his finding of lithium intercalation in graphite at room temperature.[8] Previously, Basu had found lithium intercalation into graphite in molten salt electrolytes at high temperatures (U.S. Patent 4,304,8259). I. Kuribayashi and A. Yoshino developed a new cell design using an intercalation carbon anode and a $LiCoO_2$ cathode and filed patents worldwide.1 Using a pilot plant developed for rechargeable Li-MnO_2 cells, Sony Energytec Inc. began to produce commercial cells (called the Li-Ion Battery) based on the Asahi patents in 1991.2 They also introduced electronic circuitry to control the charge-discharge, the use of a current interrupt device to interrupt current flow on buildup of excessive internal cell pressure, and the use of a "shut-down" polymer separator.

The name "lithium-ion" now is accepted by the battery community worldwide, although there is no lithium metal in the cell. However, very often lithium-metal deposition occurs during charging with the graphite anode and it may cause the many troubles on the LIB. Both electrodes operate by intercalation of lithium ions into the structure of the active materials. AT Battery Co., a joint venture of Toshiba Battery Co. and Asahi Chemical Co., was the second to commercialize the technology using Asahi patent portfolio. Table 2 shows the prominent patents in the lithium-ion battery field.

4 Electrolyte Additives: A Means for Increasing the Energy Density and Safety of the LIB

Figure 1 depicts the increase in capacity of the cylindrical 18,650 cell (18 mm in diameter and 65.0 mm long) from 1992 to 2006. The early LIBs had a capacity of 800 mAh and an end-of-charge voltage of 4.1 V. The initial cells used hard-carbon anode materials which had a capacity of about 200 mAh/g, and the $LiCoO_2$ had a capacity of nearly 130 mAh/g due to 4.1 V charging voltage. The early lithium-ion cells used a propylene carbonate-based electrolyte. However, energy density of LIB

Table 2 Patents related lithium-ion batteries

Patents	Patents No. and application date	Name	Company
Transition metal oxides as cathode, $LiCoO_2$	US 4,302,518 (1980/3/31)	J.B. Goodenough	United Kingdom Atomic Energy Authority
Graphite/Li in nonaqueous solvents	Japan 1769661 (1981/6/18)	H. Ikeda, K. Narukawa, H. Nakashima	Sanyo
Graphite/Li in nonaqueous solvents	US 4,423,125 (1982/9/13)	S. Basu	Bell Telephone Laboratories, Inc.
Grapahite/Li in molten salt	US 4,304,825 (1980/11/21)	S. Basu	Bell Telephone Laboratories, Inc.
Graphitized mesophase carbon	Japan 2,943,287 (Sept. 1990)	Kawagoe, Ogino	Bridgistone
Li-Ion battery (battery based on carbonaus material)	Japan 1989293 (1985/5/10)	A. Yoshino, K. Jitsuchika, T. Nakajima	Asahi Chemical Ind.
Cabonous/Li nonaqueous	US 4,959,281 (1989/8/29)	N. Nishi	Sony Co.
Additives for Gr vinylene carbonate	Japan 3059832 (1992/7/27)	M. Fujimoto, M. Takahashi, A. Nishio	Sanyo
Additives for Gr vinylene carbonate	US 5,626,981(May 6, 1997)	A. Simon, J-P. Boeuve	Saft
Additives of propane sulton	US 6,033,809 (1997/8/22)	S. Hamamoto, A. Hidaka, K. Abe	Ube

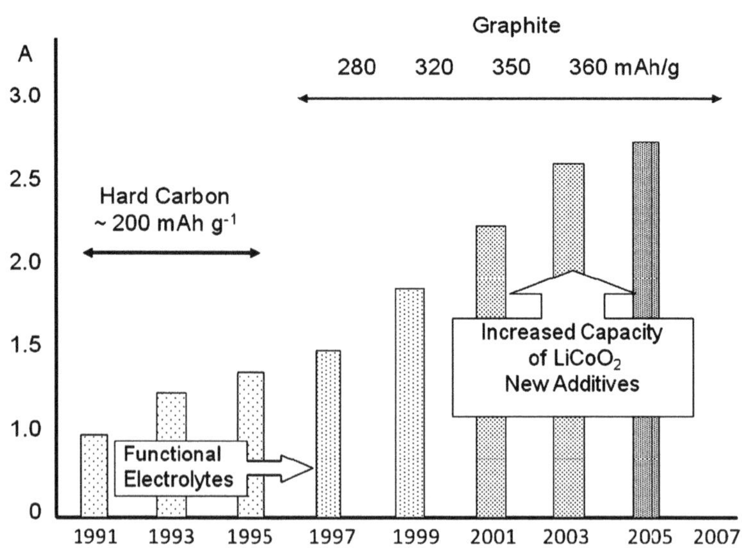

Fig. 1 Capacity increase resulting from changes in materials and charging protocol

improved rapidly and increased on average by 10% per year and has approached 2.6 Ah in 2005.

Responding to pressure from device manufacturers, the cell capacity improved through engineering and the introduction of graphite anodes, improved $LiCoO_2$-based cathode materials, and the introduction of electrolyte additives. The solvent–solute relationships and the control of impurities became very important. Ethylene carbonate replaced propylene carbonate as a principal electrolyte component in the middle 1990s in order to suppress solvent decomposition. Researchers found that increasing the degree of graphitization increased the capacity. But, a high degree of graphitization of carbon increased solvent decomposition. LIB manufacturers could not quickly overcome this obstacle and the early cells employed artificial blends of graphite such as graphitized mesophase carbon of 280 mAh/g (first-generation MCMB 6-28 with a low degree of graphitization) which is still in use. The presence of trace amount water impurities in cell yield HF via hydrolysis of lithium hexafluorophosphate ($LiPF_6$) electrolyte and is particularly harmful for cathode components.

In 1998, Ube Industries Ltd. introduced a high-purity "functional electrolyte" which contained special additives that reacted during the first charge to form a protective solid electrolyte interphase [novel type of solid electrolyte interface (SEI)] surface film.[10,11] This film covers the "active spots" on the graphite materials to prevent electrolyte decomposition. As a result, the degree of graphitization of the anode carbon materials in use first increased up to 320 mAh $g^{-}1$ [graphitized mesophase carbon fiber (MCF) or second generation MCMB-6–28] and now approaches closely the theoretical value of 372 mAh.$g^{-}1$ (for example, the massive artificial battery (MAG), which has a narrow canal in its structure to adsorb electrolyte additives). Another electrolyte additive – cyclohexylbenzene (CHB) – generates hydrogen gas at higher voltages to prevent overcharge in case of electronic control circuitry failure. In such a case gas generation activates the current interrupt device and safety vent to prevent serious safety issues.

In 2003, the capacity of the 18,650 cell reached 2.4 Ah. This corresponds to an energy density of over 200 Wh/kg or 500 Wh/l, respectively. These values were reached in part by increasing the cell-operating voltage higher than 4.2 V due to the availability of improved graphite anode materials, electrolyte additives, and a stabilized $LiCoO_2$. In order to suppress the decomposition of the cell electrolyte at the active sites on the surface of $LiCoO_2$, Ube Co. Ltd. has developed a new generation of additives that eliminated electrolyte decomposition on the active sites of the cathode.[12,13] The concept was based on the formation of a conductive membrane to cover the particles of positive active material just as the SEI formation protects the anode. The commonly used anode additives are 1,3-propanesulton,[14] vinylene carbonate,[15] plus two or three types of additives for anode, cathode additive and the main overcharge protection additive is cyclohexylbenzene (CHB). Although the varieties and amounts of anodic additives are different depending on the graphite type, one of the popular graphite anodes would be MAG, showing a good affinity to additives. The capacity of the 18,650 cell had reached 2.9 Ah in 2006 by using graphite anode, planar Ni-based cathode, and several different types of additives. Other proprietary additives are included in the cell electrolytes in small amounts.

However, charging the cells to higher cutoff voltage highlights safety concerns of any lithium-based power source. Growing an appropriate SEI film, particularly at higher cutoff voltage, becomes essential for both the safety and the battery performance. The process of special SEI film formation by the additives and growth is called the "conditioning process" in the LIB community. This is one of the most important manufacturing operations to suppress the electrolyte decomposition and to maintain safety.

Electrode conditioning, especially containing the well-designed additives, is typically achieved in the following manner. First, several charge-discharge cycles are performed to decompose the additives at rather a low rate (about 1/4 C rate). These cycles initiate the protective film layer formation. After the completion of these initial (conditioning) cycles, the cells are kept in their charged state at room or elevated temperature for several days in order to complete the protective film growing process.

Figure 2 shows the same electrode with SEI on its surface grown in the absence and in the presence of additive (VA). It is clearly seen that SEI film formed on the graphite surface is not a homogeneous formation. A large number of minute spots are observed instead. They form due to the VA addition to the electrolyte. The adequately formed SEI film has a crucial role for improvement of performance and of safety of the LIB in operation.

On the other hand, intercalation of Li^+ into graphite in propylene carbonate (PC)-based electrolytes is possible even in the absence of additives, but only at the expense of a reduced PC ratio, as shown in Fig. 3. However, in this case a voltage plateau at around 0.6–0.8 V appears. The initial coulomb efficiency drops drastically and metallic lithium deposition seems to occur. Such behavior can be explained by the large number of the active sites present on the electrode. These active sites are the net result of the variable electric potential over the anode surface. Such nonhomogeneous voltage distribution facilitates the deposition of metallic lithium at the charged state, because there are many points under potential approaching 0 V vs. Li/Li^+ during the course of Li^+ intercalation.

Figure 4 clearly illustrates that the charge distribution on the graphite surface is not homogeneous. Some parts of the electrode surface resemble exfoliated-expanded graphite. The Li^+ cannot intercalate into the exfoliated graphite. On the other hand it is a good electric conductor on which lithium or another metal ion deposition may take place.

Figure 4 would represent a typical example of the LIB graphite anode in several tens of cycles. Some graphite would be damaged or exfoliated and would not accept Li^+ intercalation, but it would facilitate lithium-metal deposition instead. Conditioning process aims to establish a stable and robust SEI film in order to avoid this scenario.

Graphite has competition reactions especially at low temperature; one is a lithium intercalation into its structure and the other is the lithium deposition on it, which means the rather low intercalation reaction kinetics. Another important and interesting result revealing the importance of the electrolyte additives is shown Fig. 5, representing a graphite anode after the completion of six cycles under the

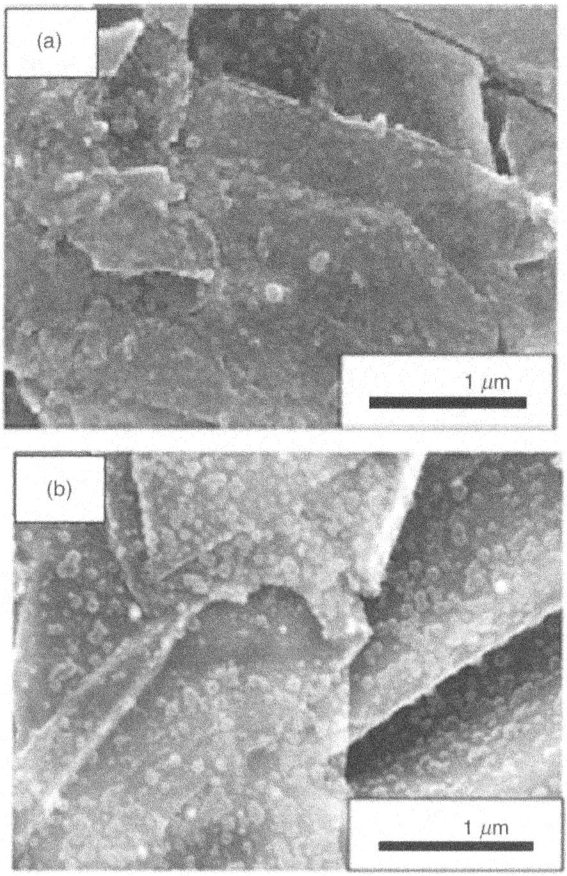

Fig. 2 Scanning electron micrograph (SEM) of natural graphite after charging (a) in the absence of VA, and (b) in the presence of 1% VA

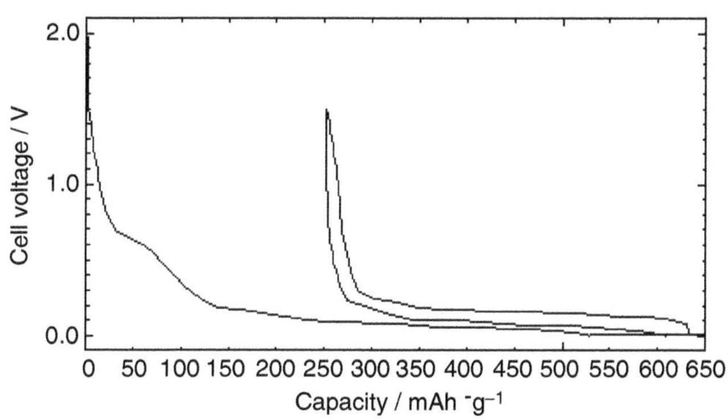

Fig. 3 The voltage profile of graphite vs. Li/Li^+ in 1-M $LiPF_6$ PC:MEC (ethylmethyl carbonate) = 1:4

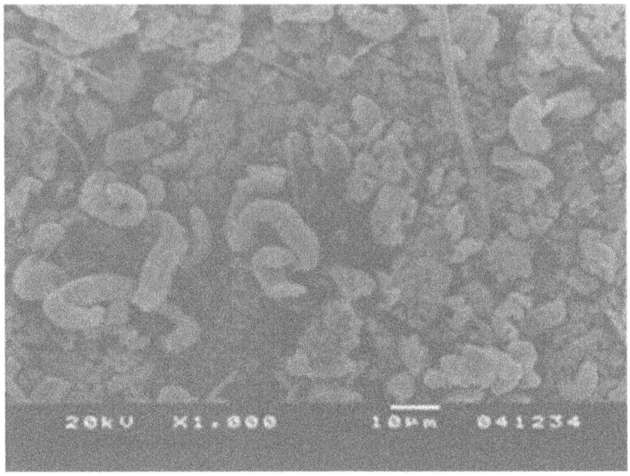

Fig. 4 SEM image of the graphite electrode cycled vs. Li/Li^+ in 1-M $LiPF_6$ PC:MEC = 1:4

Fig. 5 Appearance of the graphite electrode cycled/equilibrated (**a**) in the absence of electrolyte additives and (**b**) in the presence of 0.5 wt% propane sulton, at 0°C vs. Li/Li^+. Cell configuration: Li//1-M $LiPF_6$-EC/DMC(1:2 by vol.)//graphite. Each cycle the cell was equilibrated at 0.005 V for 10 h at 0°C

following conditions: voltage window 0.005–2.500 V, current density 0.4 mA cm^{-2} at 0°C. Cell configuration in this case was Li//1-M LiPF6-EC/DMC (1:2 by vol.)// graphite. Each cycle the cells were equilibrated at 0.005 V vs. Li/Li^+ for 10 h. The color of the electrode shown in Fig. 5 a is not golden, which is the first-stage color of lithiated graphite (LiC_6), because equilibrating the cell at low temperature induces Lilithium-metal deposition, especially in the absence of additives.

When using an appropriate additive, for example, propane sulton (PS), the appearance of the same type of electrode under the same conditions is shown in

Fig. 5b. It has the typical golden color of LiC_6.[16] This means that the additive has decomposed during the equilibration cycles and has deposited at the active sites of graphite, preventing the formation of metallic lithium. We believe that electrolyte additives improve the performance of the graphite electrodes by changing the kinetics of elemental lithium deposition. Because the electrolyte additives make the usage of highly crystalline graphite anodes possible and at the same time suppress the coexistence of elemental Li along with the lithiated graphite, they are considered to be a key material in the LIB industry.

Other recent developments include the incorporation of a fire retardant, which retards the combustion of the solvent, and a new additive to improve the wetting of the separator. It is difficult to use these additives in the gel-type electrolytes employed in lithium-ion polymer cells. This may be one reason for the lower market share experienced by lithium-ion polymer cells.

The capacity of the 18,650 cell appears to have reached its practical limit of 2.9 Ah based on the present graphite and planar nickel-based cathode in 2007. Further improvement in capacity is expected to be realized from the development of a silicon alloy type anode with a capacity of 700 mAh/g or more and the planar lithium-nickel-cobalt-aluminum and nickel-manganese-cobalt cathode materials with capacities approaching 250 mAh/g. New electrolytes and/or additives also are under development.

As society advances, a variety of new technologies, machines, and systems have been developed and more efficient industrial operations have been adopted. The concern for global warming and a clean environment along with the development of advanced electric and hybrid vehicles may be served well by advanced LIB. This situation has resulted in increased demand for high-performance batteries and power generation and storage. At the same time, demand has increased for a cleaner environment and a more efficient energy production coupled with low power consumption systems. In this context, the LIB has made and will continue to make significant contributions and advances.

References

1. A. Yoshino, K. Jitsuchika, T. Nakashima, Japanese Patent 1989293 (issued 1985/5/10)
2. T. Nagura, K. Tozawa, *Progr. Batteries Solar Cells*, 10 (1991) 218
3. P. Dan, E. Mengeritsky, D. Aurbach, I. Weissman, E. Zinigrad, *J. Power Sources*, 68 (1997) 443
4. S. Yata, Y. Hato, H. Kinoshita, N. Ando, A. Anekawa, T. Hashimoto, M. Yamaguchi, K. Tanaka, T. Yamabe, *Synth. Met.*, 73 (1995) 273
5. S. Wang, S. Yata, J. Nagano, Y.Okano, H. Kinoshita, H. Kikuya, T. Yamabe, *J. Electrochem. Soc.*, 147 (2000) 2498
6. H. Ikeda, K. Narukawa, H. Nakashim, Japanese Patent 1769661 (issued 1981/6/18)
7. J.B. Goodenough, U.S. Patent 4,302,518 (issued 1980/3/31)
8. S. Basu, U.S. Patent 4,423,125 (issued 1982/9/13)
9. S. Basu, U.S. Patent 4,304,825 (issued 1980/11/21)
10. H.Yoshitake, Techno-Frontier Symposium, Makuhari, Japan (1999)

11. H. Yoshitake, *Functional Electrolyte in Lithium Ion Batteries* (in Japanese), M.Yoshio, A. Kozawa, Eds., Nikkan Kougyou Shinbunsha, Japan, 2000, pp. 73–82
12. H. Yoshitake, K. Abe, T. Kitakura, J. B. Gong, Y. S. Lee, H. Nakamura, M. Yoshio, *Chem. Lett.*, 32 (2003) 134
13. K. Abe, H. Yoshitake, T. Kitakura, T. Hattori, H. Wang, Masaki Yoshio, *Electrochim. Acta* 49 (2004) 4613
14. S. Hamamoto, A. Hidaka, K. Abe, U.S. Patent 6,033,809 (issued 1997/8/22)
15. M. Fujimoto, M. Takahashi, A. Nishio, Japanese Patent 3059832 (issued 1992/7/27)
16. Gumjae Park, Hiroyoshi Nakamura, Yunsung Lee and Masaki Yoshio *J. Power Sources*, in press (2009)

Chapter 1
Synopsis of the Lithium-Ion Battery Markets

Ralph J. Brodd

1.1 Introduction

Research and development of the lithium-ion (Li-Ion) battery system began in the early 1980s at Asahi Chemicals[1] and was first commercialized in 1990 by Sony Corp. for the Kyocera cellular phone in the 14,500 and 20,500 cell sizes.[2] The following year Sony introduced the 18,650 cell in its camcorder. (The nomenclature for cells size: the first two numbers indicate the cell diameter in millimeters and the last three are the cell length in tenths of millimeters.) Since its introduction, the Li-Ion market has grown to about $4 billion in 2005.

The higher volumetric and gravimetric energy storage capability are key characteristics of the Li-Ion battery system compared to the conventional sealed nickel–cadmium (Ni–Cd), nickel-metal hydride (Ni-MH), and valve-regulated lead acid (VRLA) battery systems (Fig. 1.1). For a given cell size, larger values of Wh/l and Wh/kg translate into smaller and lighter cells. These characteristics became the enabling technology for the proliferation of portable battery-powered electronic devices, especially notebook computers and mobile phone applications.

Safety of the system has been a watchword for Li-Ion batteries. They have the ability to self-destruct if abused. Manufacturers are careful to ensure that the cells are safe in normal operations. In addition, cell designs incorporate features such as devices that shut off current flow when an abuse condition arises. The United Nations[3] as well as the transportation agency in each country have requirements for testing to ensure a safe product for shipping.

Table 1.1 shows the advantages and disadvantages of the Li-Ion and Li-Ion polymer rechargeable batteries.

R.J. Brodd
Broddarp of Nevada, Inc, 2161 Fountain Springs Drive, Henderson, NV 89074, USA
ralph.brodd@earthlink.net

M. Yoshio et al. (eds.), *Lithium-Ion Batteries*,
DOI 10.1007/978-0-387-34445-4_1,

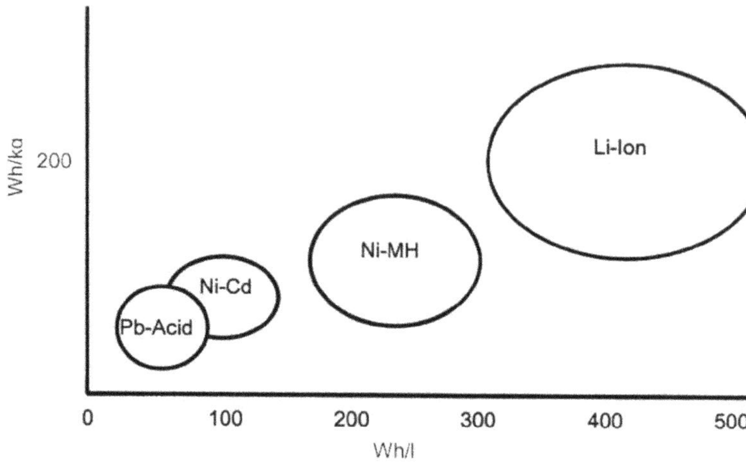

Fig. 1.1 Energy density (Wh/l) and specific energy (Wh/kg) for the major small-sealed rechargeable battery systems

Table 1.1 Advantages and disadvantages of Li-ion and Li-ion polymer rechargeable cells

Advantages	Disadvantages
• Chemistry with the highest energy (Wh/g) and lightest weight (Wh/kg)	• Relatively expensive
• No memory effect	• Lightest weight
• Good cycle life	• Requires protection circuitry for safety and to prevent overcharge and overdischarge
• High energy efficiency	• Nominal 3-h charge
• Good high-rate capability	• Not tolerant of overcharge and overdischarge
	• Thermal runaway concerns
Added advantages and disadvantages of Li-Ion polymer/laminate cells	
• Flexible footprint	• Limited high rate capability
• Plasticized electrolyte	• More expensive
• Internal bonding of anode	• Poor low-temperature performance
• Cathode and separator	

1.2 Present Market for Li-Ion Cells

Lithium ion cells serve the small-sealed rechargeable battery market and compete mainly with the Ni–Cd and Ni-MH cells for the various applications. The Li-Ion cells are available in cylindrical and prismatic format as well as flat plate constructions. The cylindrical and prismatic constructions use a spiral-wrap cell core where the cell case maintains pressure to hold and maintain compression on the anode, separator, and cathode. The lighter-weight polymer constructions utilize the adhesive nature of a polymer/laminate-based electrolyte to bond the anode to the cathode.

As a result, it does not need outside pressure to hold the electrodes in contact with each other. A light-weight polymer-aluminum laminate pouch can substitute for a heavier metal cell enclosure. All three constructions employ the same chemistries.

The sales of Li-Ion cells are shown in Table 1.2.[4] The Li-Ion market is very competitive. The data for the competing Ni–Cd and Ni-MH cells are included for comparison purposes. The market growth for Li-Ion is spectacular and driven by the proliferation of portable electronic devices such as notebook computers and cellular phone applications. In 15 years between 1991 and 2006 the sales and production of Li-Ion batteries experienced double-digit growth. The slower growth period, around 2000, occurred when cell production in China and Korea began to ramp up and may not have been included in the database.

In 1995, an 18,650 cell sold for $8, while in 2006 the same size cell with 2.6 Ah sold for about $4. Over this period the energy density of the cell more than doubled, while the price fell by 50%. The cell producers accomplished the performance improvements through engineering improvements in cell design, new carbon materials for the anode, and automated high-speed production to reduce the cost. The Li-Ion market is expected to continue growing as new technology is introduced and new applications develop.

The major cell manufacturers are listed in Table 1.3. The Japanese manufacturers (Sanyo, Sony, and Matsushita) have a clear lead but the Chinese manufacturers (BYD, Lishen) and Korean manufacturers (Samsung and LG Chemical) are challenging. There are no major Li-Ion manufacturers in the United States (or in Europe), even

Table 1.2 Worldwide Sales (Million of Dollars)[4]

	Year									
Cell type	1991	1992	1994	1996	1998	2000	2002	2004	2005	2006[a]
Ni–Cd	1,535	1,823	2,060	1,695	1,394	1,204	935	1,006	935	939
Ni-MH	39	100	746	863	848	1,245	667	767	726	891
Li-Ion	1	10	152	1,292	1,900	2,805	2,458	4,019	3,899	3,790
Lam Li-Ion	0	0	0	0	2	187	299	487	547	657

[a] Estimated

Table 1.3 2005 Worldwide cell demand (Millions of Cells)[4]

	Cell type				
Application	Ni-Cd	Ni-MH	Cy Li-Ion	Pr Li-Ion	Lam Li-Ion
Cellular		50		898.16	125.85
Notebook		22	422.68	16.34	2.50
Movie	2	4	67.98	11.91	
Digital still camera		56	18.88	48.17	0.94
Power tools	575	53	20.14	0.08	
Audio	80	35	6.99	31.02	45.63
Games				26.82	14.4
Consumer	45	300			
Cordless phones	190	83			
Others	330	178	22.854	28.98	14.42

though they constitute large markets for devices powered by Li-Ion batteries.[5] Activity in the United States is limited to several companies that supply the niche medical and military markets.

1.3 Market Characteristics

The unit cell production for 2005 by product application is given in Table 1.4. Cellular phone applications dominate the unit cell production. The thin, rectangular polymer/laminate cell construction has found favor in the cellular phone market and now accounts for about 13%, with the rest being the prismatic cell sizes. Notebook computers are second followed by cameras.

The period from its introduction in 1991–2002 was a time for establishing the fundamental base for materials and manufacturing processing. During this period, the processing of the materials, cell designs, and production equipment reached a high level of sophistication. The fundamental underpinning of the technology provided a sound basis for future expansion during the next decade. Problems were identified and methods to solve the problems were developed. From a cell engineering viewpoint, the maximum capacity of an 18,650 cell would be 2.5–2.6 Ah with the materials that were available in 2002. Increase in cell capacity and energy storage while maintaining safety would require new materials.

Starting in 2003, a shift in the market applications began to occur, as depicted in Fig. 1.2.[6] In one segment basically the drive to increase capacity and performance for the competitive notebook and cellular phone applications continues. This requires the development and introduction of higher capacity, higher performance anode and cathode materials. Several new high-capacity, safer compositions such as $LiMn_{0.3}Co_{0.3}Ni_{0.5}O_2$ and $LiMn_{0.5}Ni_{0.5}O_2$ cathodes were developed and are in the process of being put into production. In the same line, new anode materials have been developed, based on nanostructured lithium alloy anodes. These materials can drive the 18,650 cell capacity over 2.6 Ah and could approach 3.0 Ah in the future.

Table 1.4 Major Li-ion cell manufacturers, 2005[4]

Manufacturers	Percentage of total
Sanyo	27.50
Sony	13.30
Samsung	10.88
Matsushita	10.07
BYD	7.53
LG Chemical	6.45
Lishen	4.52
NEC	3.60
Maxell	3.26
Others	12.89

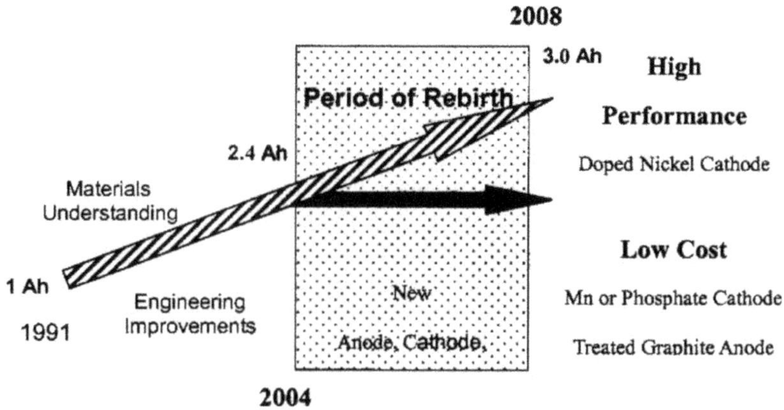

Fig. 1.2 A split develops in the Li-ion market

The other segment consists of applications that do not require a significant increase in energy storage capability but emphasize lower cost and higher power for new applications such as automotive and power tools. These applications also require new cathode and anode materials to meet the market demands for low cost with high-rate performance. An example of a new material is the $LiFePO_4$ cathode materials introduced in the market in 2003 by Valence Technology and followed by A123, which emphasized power tool applications.[7–9]

1.4 Consumer Electronics

Cellular telephones and notebook computer applications drove the market and will continue to dominate cell usage. The Bluetooth and 3G mobile phones should expand the market coupled with expansion as people shift to higher performance devices. The market for portable cellular phone and notebook computers is reaching saturation in the United States and Europe where it is expected to grow in parallel with the gross national product. Large growth areas for cellular phones and notebook computers are in Asia, especially the Chinese and Indian markets.

Movie cameras account for about 25% of cell usage. Often cameras are used intermittently and may sit on the shelf for an extended period before use. The shelf life of the Li-Ion is significantly better than Ni–Cd and Ni-MH cells, the previous battery systems for movie cameras. Digital still cameras are next. They are in the process of transitioning to Li-Ion from alkaline primary and Ni-MH cells. Most primary cells lack the high pulse current required for camera operation. Only the primary $Li–FeS_2$ system gives fully satisfactory performance for digital still camera operation. Notebook computer and cellular phone users have learned to recharge their batteries on a regular basis so that the device gives the expected service. In addition, many mobile phones have a built-in camera and could slow the development of the digital camera market.

1.5 Hand Power Tools

The power tool market is dominated by the Ni–Cd system. In terms of cell volumes it is the third largest. However, it is heavy and has a short run time compared to Li-Ion. Previously, Li-Ion cells could not meet the very high rate capability of the Ni–Cd. Recently, Li-Ion cells with excellent high-rate $LiMn_2O_4$ or $LiFePO_4$ cathode materials were introduced for power tool applications. They are about a third smaller and half the weight of the older Ni–Cd. The phosphate cathode cells have a significantly greater safety characteristic as the cells do not go into thermal runaway until heated to over 600°C. The greater safety, coupled with the superior high-rate capability of the nanostructured phosphate materials, make them ideal for this application. This market segment is price-sensitive. Because the Ni-MH has poor low temperature and poorer very high rate performance, it has not made a significant inroad against Ni–Cd. The introduction of Li-Ion–powered tools by DeWalt and Milwaukee Tool offers a growth opportunity for Li-Ion cells.

1.6 Uninterruptible Power Sources, Stationary Energy Storage

The uninterruptible power source market is about $ 6–10 billion annually and is growing roughly with the gross national product. This market is dominated by lead acid batteries. The technology is slowly shifting from the flooded to the valve-regulated lead acid technology. The valve-regulated lead acid cells are sealed and do not vent hydrogen and acid vapors on charge or stand, but they are more expensive to produce. The main competition to the lead acid is costly pocket plate Ni–Cd, but it has an exceptionally long life. It is not unusual for Ni–Cd to have an operational life of 15 years or more in this application. Increased emphasis on environmental controls has made lead acid and Ni–Cd vulnerable for penetration by the environmentally acceptable, higher-cost Li-Ion batteries.

This market is very price-sensitive and the cost of Li-Ion cells will need to reach $0.30/Wh to penetrate this market. The new lower-cost manganese and phosphate cathode materials could reach this cost goal. It should be noted that several large lead acid battery companies in this market recently have entered into agreements with Li-Ion producers to supply cells and batteries for evaluation purposes. This could be an indication that the more traditional lead acid producers are positioning themselves to supply Li-Ion as an alternative for lead batteries.

1.7 Transportation

There are many emerging market opportunities for Li-Ion cells in transportation. The motive power market is viewed as the largest future growth opportunity for Li-Ion batteries. Once in place, the transportation market will dwarf the present portable device market. The Segway Transporter has shifted from Ni-MH to Li-Ion batteries. The Toyota Prius will shift from Ni-MH to Li-Ion in the next model redesign. These, along with the emerging boating motors for freshwater lakes, have set the stage for penetration into this huge market area. Europe is in the process of banning gasoline motors on all lakes. Several U.S. lakes now bar gasoline-powered boats, as well. Boaters must shift to electric drive and batteries to operate on the lakes. These presently use lead acid batteries. The smaller, lighter, Li-Ion battery is already marketed in Europe for this application.

The introduction of the Tesla electric vehicle in 2007 with a 200 + mile range should set the stage for the transition to electric propulsion. The new cathode and anode systems also offer the potential to replace the present lead acid SLI (starting-lighting-ignition) battery on gasoline-powered vehicles.

References

1. A. Yoshino, *The Chemical Industry*, **146**, 870 (1995)
2. T.Nagaura and K. Tazawa, *Progress in Batteries and Battery Materials*, **10**, 218 (1991)
3. R. J. Brodd, Factors affecting U.S. production decisions: Why are there no volume lithium ion battery manufacturers in the United States? ATP Working Paper 05-01, June, 2005
4. *Recommendations on the Transport of Dangerous Goods*, Manual of Tests and Criteria, Third revised edition, United Nations, New York and Geneva, 2002
5. H. Takeshita, 23rd International Seminar on Primary and Secondary Batteries, Ft. Lauderdale, FL, March 2006
6. R. J. Brodd, Keynote Lecture, IMLB 11, Monterey CA, June, 2002
7. Valence Technology, Inc., Form 10-K, June 30, 2003
8. S.-Y. Chung, J. T.Bloking, and Y.-M. Chiang, *Nature Materials*, **1**, 123 (2002)
9. A. K. Padhi, K. S. Nanjundaswamy, and J. B. Goodenough, *Journal of Electrochemistry Society* **144**, 1188 (1997)

Chapter 2
A Review of Positive Electrode Materials for Lithium-Ion Batteries

Masaki Yoshio and Hideyuki Noguchi

2.1 Recent Cathode Materials

The lithium-ion battery generates a voltage of more than 3.5 V by a combination of a cathode material and carbonaceous anode material, in which the lithium ion reversibly inserts and extracts. Such electrochemical reaction proceeds at a potential of 4 V vs. Li/Li^+ electrode for cathode and ca. 0 V for anode. Since the energy of a battery depends on the product of its voltage and its capacity, a battery with a higher energy density is obtained for a material with a higher voltage and a higher capacity. Therefore, when the same anode material is used, the higher the cathode potential and the larger the capacity of the cathode material, the higher the energy of the battery.

The cathode and anode were packed into a vessel with constant dimensions, so the capacity for unit volume is more important than for weight. The volumetric-specific capacity of $LiCoO_2$ is 808 mAh/cm^3, which is high enough to be used as a cathode material. Nickel-based cathode materials deliver higher capacity of 870–970 mAh/cm^3. Safety problems for this material are overcome by the simultaneous doping of cobalt and aluminum. SAFT Co. has adopted $LiNi_{0.8}Co_{0.15}Al_{0.05}O_2$ supplied by Toda Kogyo Co. (formerly Fuji Chemical Industry Co.) as a cathode material in the lithium-ion battery for an electric vehicle (EV) application. An analogous compound is used in Japan. A press release announced that the capacity of it is 20% higher than that of $LiCoO_2$ and it is safer than $LiCoO_2$ in terms of overcharge problems. Further, nickel-based cathode materials are used for the battery in Toyota's car, without idling.

Manganese spinel cathode materials, although inferior to layered compounds, are cheap and rich in resources. Therefore, it is suitable as a cathode material in large-scale use of lithium-ion batteries. This spinel compound has been used for cellular phones produced by NEC Co. and for EV and hybrid EV produced by Nissan Co. Ltd. However, its share in the market of cathode material is relatively

M. Yoshio (✉) and H. Noguchi
Department of Applied Chemistry, Saga University, Saga 840-8502, Japan
yoshio@cc.saga-u.ac.jp
GBE03502@nifty.com

M. Yoshio et al. (eds.), *Lithium-log Batteries*,
DOI 10/1007/978-0-387-34445-4_2

small. Recently, in 2003, electric mortar bicycles and mortar-assisted bicycles have been commercialized. In such applications, the cathode contains manganese spinel compounds. This trend would lead to the development in large-scale use of batteries using a spinel compound as a cathode material.

The main cathode material, $LiCoO_2$, in the lithium-ion battery has been improved in terms of rate capability and capacity. The rate capability is improved by the control of particle morphology, and high capacity is achieved by increased charge voltage while overcoming safety problem.[1]

2.2 The Structure of Cathode Material

Many of the lithium battery cathode materials have a layered structure, which enables the two-dimensional diffusion of the lithium ion, or a spinel structure, which enables the three-dimensional diffusion. The structures of the cathode materials in the lithium- ion battery are summarized, together with its electrochemical properties and stability of structure, in Table 2.1.

The basic compounds with a layered structure are $LiCoO_2$, $LiNiO_2$, $LiCrO_2$, Li_2MoO_3, and $Li_{0.7}MnO_2$. The initial three compounds have a rhombohedral structure with the symmetry of space group $R\overline{3}_m$. As shown in Fig. 2.1, the rhombohedral unit cell has a geometric feature that has three axes of an equal length, and that each angle between any two axes is the same. Usually, this unit cell is expressed by the hexagonal structure (heavy line in the figure), which has a unit cell volume of three times because the structure of this unit cell is difficult to be imaged. In this structure, Li and transition metal ion (M) occupy alternatively an octahedral site formed by planar O^{2-} ion sheet; as a result, the Li layer and M layer are formed.

The structure of Li_2MnO_3 is slightly deviated from above one, namely, the M ion is replaced by $Li_{1/3}Mn_{2/3}$. Electroactive materials with this structure are Li_2RuO_3, Li_2IrO_3, Li_2PtO_3, and $Li_{1.8}Ru_{0.6}Fe_{0.6}O_3$. These compounds contain an expensive noble metal, so they are unsuitable for practical application. There are many layered electroactive materials prepared by foreign metal ion and lithium doping. These compounds would be classified into two groups. One is a substitution product and the other is the solid solution-type compound. The differences of the two compounds are explained in the $LiNiO_2$–$LiMnO_2$–Li_2MnO_3 quasiternary phase diagram (Fig. 2.2). Three corners in triangles indicate pure phases. The middle point of the $LiNiO_2$–$LiMnO_2$ line has a composition of $LiNi_{1/2}Mn_{1/2}O_2$, where valences of Ni and Mn are estimated to be 2^+ and 4^+, respectively, by the analysis of the XANES spectra of the K edge absorption of Ni and Mn.[2] Therefore, the substitution reaction of Ni^{3+} by $1/2Ni^{2+} + 1/2Mn^{4+}$ proceeds on the line and all the Ni^{3+} ion is consumed at the middle point. A series of single-phase products are easily prepared for the compositions on the $LiNiO_2$–Li_2MnO_3 and $LiNi_{1/2}Mn_{1/2}O_2$–Li_2MnO_3. These products are generally called a "solid solution," in which valences of metal ion are equivalent to those of two end members in all the composition range. Therefore, the shadow areas are solid solutions between Li_2MnO_3 and $LiNi_xMn_{1-x}O_2$ ($x \geq 0.5$).

Table 2.1 Properties of anode and cathode materials for lithium ion battery

Cathode							
Material	Pr. Cap. (mAh/g)	Density (g/cc)	Vol. En. Den. (mAh/cc)	Shape of Dis. curve	Safety	Cost	Comment
$LiCoO_2$	160	5.05	808	Flat	Fair	High	Small-size
$LiNiO_2$	220	4.80	1,056	Sloping	Poor	Fair	Impossible
$LiNi_{0.8}Co_{0.2}0_2$	180	4.85	873	Sloping	Fair	Fair	LIP? Small scale
$LiNi_{0.8}Co_{0.15}Al_{0.05}O_2$	200	4.8	960	Sloping	Fair	Fair	LIP? Small scale
$LiMn_{0.5}Ni_{0.5}O_2$	160	4.70	752	Sloping	Good	Low	?
$Li\ Mn_{1/3}Ni_{1/3}Co_{1/3}O_2$	200	4.7		Sloping	Good	Low	
$Li\ Mn_{0.4}Ni_{0.4}Co_{0.2}O_2$	200	4.7					
$LiMn_2O_4$	110	4.20	462	Flat	Good	Low	HEV, EV
$Li_{1.06}Mg_{0.06}Mn_{1.88}O_4$	100	4.20	420	Flat	Good	Low	HEV, EV
$LiAlMnO_4$							
$LiFePO_4$	160	3.70	592	Flat	Good	Low	Low cond.

Carbon anode	Crystallinity[a] (%)	Practical capacity (mAh/g)	Comment
Spherical graphatized mesocarbon microbeads (MCMB)	82.7	320–330	Easy coating
Graphatized carbon fiber (MCF)	89.7	320–330	Stopped to produce
Pitch base graphite	91.6–94.4	~350	Largest share in the market with electrolyte additives
Carbon-coated natural graphite	100	360 ~ 365	Less decomposition of electrolyte without additives

Alloy anode	Discharge voltage (V vs. Li/Li^+)	Capacity (mAh/g)	Density (g/cm^3)	Capacity/volume (mAh/cm^3)	Comment maximum swelling (MS)
Li	~0	1,840	0.5	1,920	Safety
Sn	0.65	990	7.3	7,230	MS = 360%
Si	0.5	4,200	2.3	9,660	MS = 400%
Al	0.4	990	2.7	2,670	
Sb	1.1	650	6.7	4,360	
$SnB_{0.5}Co_{0.5}O_3$	0.7	600	3.7	2,220	High irreversible capacity
$Li_{2.6}Co_{0.4}N$	0.8	1,200	2.2	2,640	Unstable in air

[a]Crystallinity $(3.44-d_{002})/0.0868$

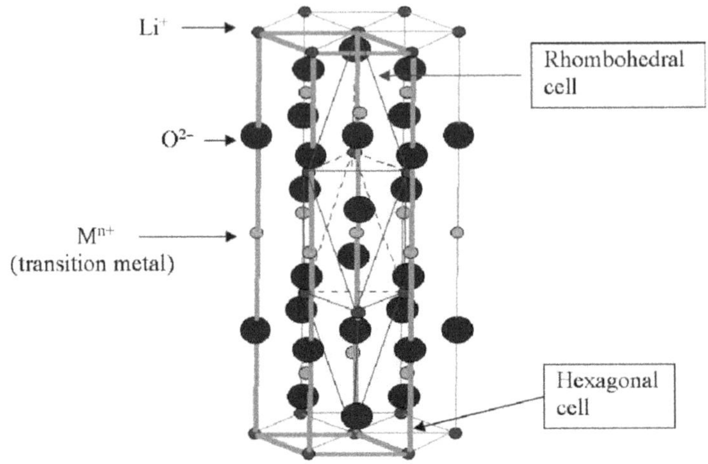

Fig. 2.1 Relation between rhombohedral cell and hexagonal cell

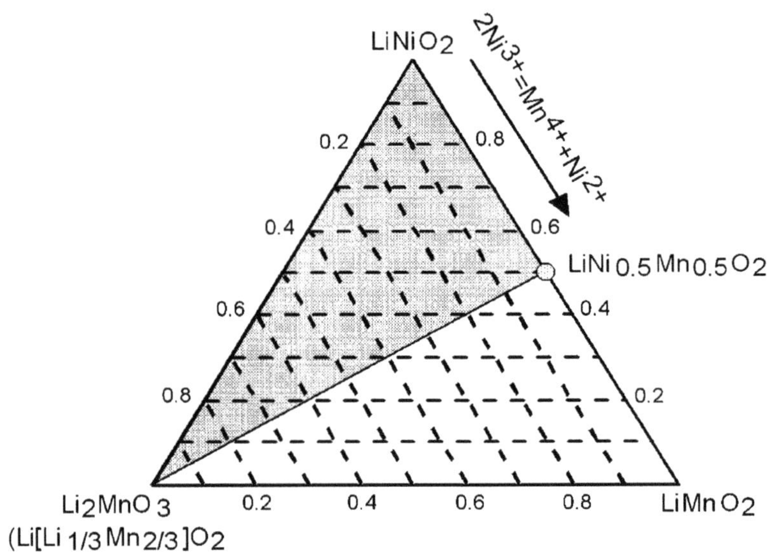

Fig. 2.2 Region of Li_2MnO_3–$LiNi_xMn_{1-x}O_2$ solid solution (painted) in pseudoternary phase diagram

Two types of solid solution are known in the cathode material of the lithium-ion battery. One type is that two end members are electroactive, such as $LiCo_xNi_{1-x}O_2$, which is a solid solution composed of $LiCoO_2$ and $LiNiO_2$. The other type has one electroactive material in two end members, such as $LiNiO_2$–Li_2MnO_3 solid solution. $LiCoO_2$, $LiNi_{0.5}Mn_{0.5}O_2$, $LiCrO_2$, $LiMnO_2$, and $LiFeO_2$ are electroactive end members; on the other hand, Li_2MnO_3, Li_2TiO_3, and $LiAlO_2$ are electrochemically inactive end members. The combination of two different types of end members gives a lot of cathode materials in lithium ion battery. Solid solutions with more complicated combinations composed of end members are $LiNi_{0.8}Co_{0.15}Al_{0.05}O_2$ or $Li_{1+x}Ni_yCo_zMn_{1-y-z}O_{2+\delta}$.

Among manganese-based layered compounds, a layered zigzag-type orthorhombic $LiMnO_2$ and layered $LiMnO_2$ with R_{3m} symmetry are electroactive materials. The latter is prepared by the ion exchange method[3,4] or by the liquid-phase synthesis.[5] Both orthorhombic and layered $LiMnO_2$s are unstable for electrochemical extraction and insertion of the lithium ion, and they change into a spinel structure during the charge and discharge in 3–4 V range. Orthorhombic $LiMnO_2$ synthesized at high temperatures requires many charge/discharge cycles until the transformation to the electrochemically active spinel structure.[6] However, when it is crushed into nanosized particles, it delivers a capacity close to 200 mAh/g in the first cycle. Further, the cyclability at a high temperature also is excellent,[7,8] and it is in the level that the practical application as a cathode material can be allowed.

A series of compounds with lithium-deficient stoichiometry can be synthesized using the ion exchange method. $Li_{0.7}MnO_2$ analogues are synthesized by ion exchange of the $Na_{0.7}MnO_2$ analogue with the Li^+ ion in the molten lithium salt.[9] $Na_{0.7}MnO_2$ has the stacking faults of the oxygen ion sheets. These faults are partly succeeded by $Li_{0.7}MnO_2$ after ion exchange.

The oxygen ions usually make the cubic or hexagonal close-packing (HCP) structure in metal oxides, and metal ions are located in tetrahedral holes or octahedral holes. In the cubic close-packing (CCP) structure, the oxygen ion sheets are divided into three types (A, B, and C) with different phases, and these layers are stacked regularly in the mode of (ABC)*n* (Fig. 2.3). The stacking mode of the HCP structure is described as (AB)*n*. Two oxygen-sheet stacking is typical in metal oxides. However, $Na_{0.7}MnO_2$ has a different stacking sequence and it is described as (ABBA)*n*. Then, two kinds of the hole with different shapes are formed by six oxygen ions: the octahedron and the prism (triangle pole). The manganese ions occupy the octahedral sites C (between A and B in the oxygen sheet) of the same position on a two-dimensional plane with oxygen sheet C, and the sodium ions occupy the prismatic sites between two A sheets or two B sheets. The structures of $Na_{0.7}MnO_2$ and $LiCoO_2$ occasionally were called P2 type and O3 type in the literature, respectively. The relationships between oxygen and the metals in this notation are shown in Table 2.2, where the positions on the two-dimensional planes A, B, and C are equal to a, b, and c, respectively.

$Li_{0.7}MnO_2$ of O2 structure (O2-$Li_{0.7}MnO_2$) has been prepared from P2-$Na_{0.7}MnO_2$ by the ion-exchange reaction, and O3-Li_xMnO_2 is obtained from P3- or O3-$Na_{0.7}MnO_2$.[10,11] The O2-$Li_{0.7}MnO_2$ keeps its structure during the charge/discharge cycles, although O3-Li_xMnO_2 with R_{3m} symmetry transforms to the spinel structure. The charge/discharge capacity of O2-$Li_{0.7}MnO_2$ is considerably as high as 150 mAh/g in the voltage range of 2.5– 4.0 V, and the capacity in the 3 V range is roughly half; therefore, it is slightly inferior to the 4-V class material in energy density. Various analogues, in which manganese was partly substituted by cobalt, nickel, and so forth, were studied and some of them have higher capacity than the original one. However, they may be ranked as future materials because they have a

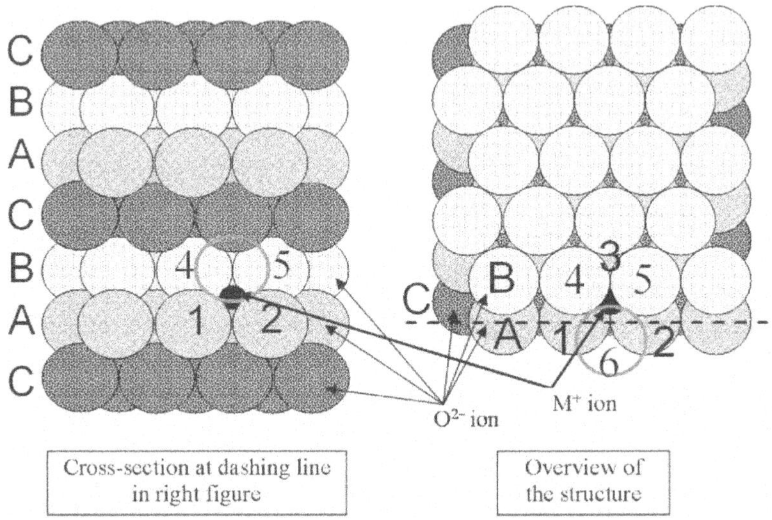

Fig. 2.3 Oxygen stacking in CCP structure

Table 2.2 Relation between oxygen staking and its cation site

Abbreviation	Staking of O^{2-}	Shape of MnO_6	Position of metal[a]	Example
O3	(ABC)*n*	Octahedral	cabcab	$LiCoO_2$
O2	ABCB etc	Octahedral	caac	$Li_{0.7}MnO_2$
O1	(AB)*n*	Octahedral	cc	CoO_2
P3	(AABBCC)*n*	Octahedral and prism	pcp′ap″b	Na_xMnO_2
P2	(AABB)*n*	Octahedral and prism	pcp’c	$Na_{0.7}MnO_2$
P1	(AA)*n*	Prism	pp	–

[a]p, p′, and p″ are different in the position on dimensional plane

poor rate performance; for example, the capacities of $Li_{0.7}Mn_{2/3}M_{1/3}O_2$ (M = Ni, Co) are about 100 mAh/g even at the slow rate of C/20.

It also has been reported that a cobalt-substituted compound (O3-$LiCo_{1/2}Mn_{1/2}O_2$), by the ion exchange method, could be used as the material for the 5-V cathode; however, it has problems such as a large irreversible capacity.

2.3 Electrochemical Characteristics and Structural Changes during Charge/Discharge

2.3.1 Layered Material

The charge/discharge curves of $LiCoO_2$ and $LiNiO_2$ are shown in Fig. 2.4. When the cutoff voltage is selected to be 4.3 V, $LiCoO_2$ has a comparatively smooth curve, while $LiNiO_2$ has a complicated curve with some voltage plateaus. In the following,

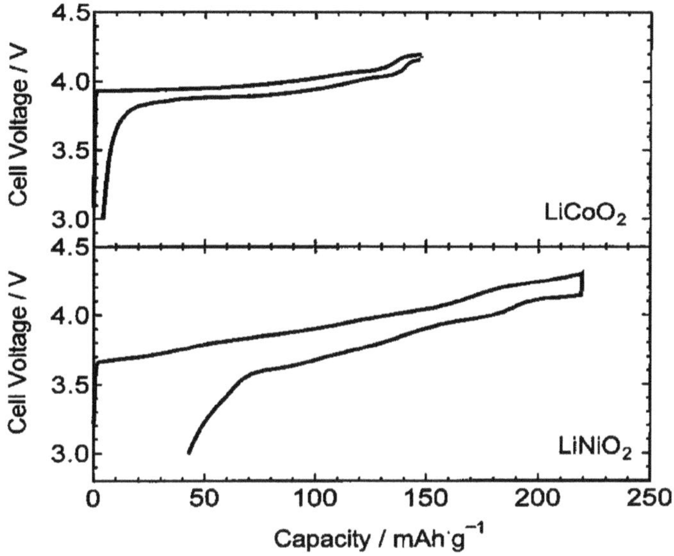

Fig. 2.4 Charge/discharge curves of $LiCoO_2$ and $LiNiO_2$

the composition of the $LiNiO_2$-type compound during the charge/discharge will be expressed as $Li_{1-x}NiO_2$. The origin of this complicated curve is attributed to the structural transformation. $Li_{1-x}NiO_2$ keeps its original structure intially; however, it transforms the monoclinic phase in the range of $0.22 < x < 0.64$. Further delithiation causes the formation of the NiO_2 phase in the range of $x > 0.70$ or more.[12–14] For $LiCoO_2$,[15–17] two crystal phases with $R\bar{3}m$ (original $LiCoO_2$ and $Li_{0.75}CoO_2$) appears in the range of $x < 0.25$. The second rhombohedral phase continues except for a narrow range around $Li_{0.5}CoO_2$, which is a monoclinic phase the same as $Li_{1-x}NiO_2$. However, this phase transforms the rhombohedral phase by increasing the temperature. However, lithium excess $LiCoO_2$ shows different behavior from stoichiometric $LiCoO_2$. It does not have the two-phase range of $x < 0.25$ and the monoclinic phase at around $x = 0.5$, which also is easily judged from the shape of its simple charge/discharge curves. The electrochemical reaction of lithium excess $LiCoO_2$ proceeds in one phase, where the length of the *c*-axis continuously increases and that of the *a*-axis decreases as the degree of the delithiation.[18]

The researchers have reported various values for the formation range of the monoclinic phase and the NiO_2 phase; this would be due to the difference in sample composition because of the easy formation of $Li_{1-x}Ni_{1+x}O_2$ $(x > 0)$. The length of the *a*-axis and the cell volume continuously decrease as the degree of delithiation decreases. These changes can be well explained by the reason that the ionic radius of the transition metal decreases as the increase in the oxidation number increases. However, the length of the *c*-axis (interlayer distance) increases as delithiation increases. There are some differences between $LiCoO_2$ and $LiNiO_2$ in the changes of the length on the c-axis in the range of $x > 0.6$, as shown in Fig. 2.5. As for $LiCoO_2$, the length of the c-axis has a maximum value and then gradually decreases. On the other hand, for $LiNiO_2$, the interlayer distance becomes constant in the range of $x > 0.6$ or more, and the NiO_2 phase appears in the range of 0.7 or more. The CoO_2 phase is formed for $LiCoO_2$ only in the vicinity of $x = 1$. The structure of layered material during the charge/discharge seems to be quite different between $LiCoO_2$ and $LiNiO_2$ as described previously; however, substitution of 20% nickel by manganese in $LiNiO_2$[19] changes the behavior of the c length to that of $LiCoO_2$. Finally, the lithium content and type of transition metal ion have an extreme influence on the structure of charged layered material.

The upper limit of the monoclinic-forming range ($x = 0.75$) in $LiCoO_2$ roughly coincides with lower limit of the NiO_2-forming range in $LiNiO_2$. If the structure of $Li_{1-x}CoO_2$ in this range has stacking faults,[20] the stacking of the oxygen ion sheets in both $LiCoO_2$ and $LiNiO_2$ changes in the range of $x > 0.7$. The structure change within $x < 0.7$ is due to a slight deviation of the position of atoms; on the other hand, that with $x > 0.7$ causes the change in the stacking of the oxygen sheets. This would be the main reason of poor cyclicity for higher-capacity withdrawing in $LiCoO_2$ and $LiNiO_2$.

The shape of the charge/discharge curves also depends on the stacking of oxygen sheets. The O2–$Li_{0.7}MnO_2$ gives complicated charge curve with several voltage plateaus.[21,22] The stacking of the oxygen sheets changes from (ABCB)n to (ABCBCABABCAC)n by the extraction of more than 0.5 Li^+, and this structure change causes the shrinkage of interlayer distance.[22]

In the present study, $LiNi_{0.5}Mn_{0.5}O_2$ and $LiCo_{1/3}Ni_{1/3}Mn_{1/3}O_2$ keep R_{3m} symmetry up to $x = 0.62$ [23] and $x = 0.76$,[24] although a discontinuous shrinkage in the length of the a-axis accompanies the vicinity of 125 mAh/g of charging ($x = 0.45$). If these compounds do not cause the rearrangement in the stacking of oxygen sheets, we

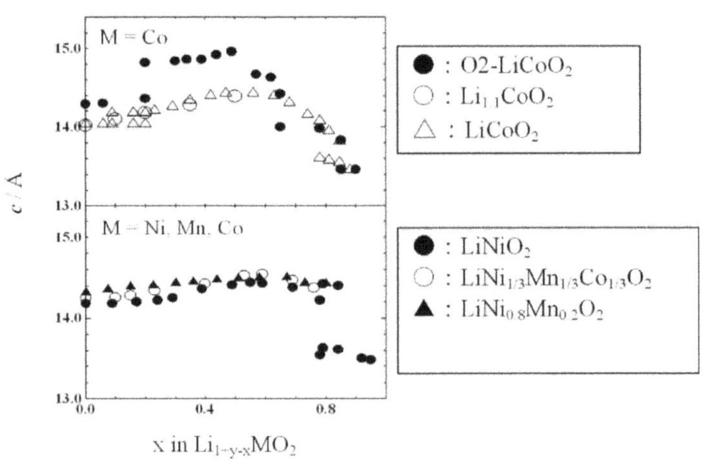

Fig. 2.5 Variation in c-axis of layered cathode materials during the charge

could withdraw higher capacities without loss of cycling performance by increasing charge voltage, although decomposition of electrolyte should be overcome.

The O2–$Li_{0.7}Li_{1/18}Mn_{17/18}O_2$ structure obtained from P2-$Na_{0.7}Li_{1/18}Mn_{17/18}O_2$ by the ion exchange method delivers capacities of about 15 mAh/g in the range of 4.0 to 4.5 V and about 130 mAh/g in the range of 3.0 to 3.5 V, respectively.[24] When a part of the manganese ions is substituted with transition metal ions M (Ni, Co, etc.), the capacity of the 3.0–3.5 V range decreases and a new potential plateau appears at 2.5–3.0 V.[9–11] The oxygen structure is kept in all voltage ranges, and an excellent cycling performance is observed at 30°C; however, its cyclicity deteriorates at 55°C.[9]

2.3.2 *Spinel*

Manganese, whose resource is abundant and inexpensive, is used worldwide as an environmentally friendly and inexpensive dry battery material. Moreover, when a spinel-type manganese-based material is used as the electrode material of a lithium-ion battery, the battery has the advantages of greatly improved safety and an inexpensive battery control circuit. The market trend for the manganese-based cathode material in a lithium-ion battery is roughly divided into two categories. The first category is materials used in portable electronic devices such as the mobile phone. The spinel lithium manganate has been used for the power source of the mobile phone for many years because of its excellent safety and cheaper control circuit,[1] although its market share is low. Moreover, it is said that the capacity fading of the graphite anode during the cycling is prevented by the use of $LiNi_xCo_{1-x}O_2$–$LiMn_2O_4$ mixed cathode because of lower manganese deposition on the anode even at elevated temperatures.

By the way, $LiCoO_2$ has been used mainly as the cathode material in the lithium-ion battery for mobile phones because of high-energy density. However, the rapid rise in price for the lack of the cobalt resources has promoted the hybridization between the cobalt-based material and the manganese-based material with the spinel structure. Further, it has been confirmed that this hybrid material can have the same energy density as that of the conventional one, so the hybrid-type batteries came onto the market since 2004. It is expected that the share of this type of battery will expand further in the future.

The second category is the cathode materials for large-size lithium-ion batteries as power sources for electric vehicles, hybrid vehicles, and so forth. High power, safety, and low cost are strongly required among their performances, so manganese-based cathode materials are suitable for such applications. It overwhelmingly excels in the power density compared to cheaper iron-based cathode material ($LiFePO_4$) and it is used in a large-sized battery. It has been used for 3 years since it appeared in the market as a power source for the hybrid vehicle. Moreover, the spinel-type manganese oxide is used for the main cathode material of the lithium-ion battery for the motor-assisted bicycle and the electric motorcycle, which came onto the market in recent years. In addition, although the iron-based material ($LiFePO_4$), which is expected to succeed the manganese-based cathode material, is being studied all over the world; it has not reached practical use yet because of its poor electric conductivity and its complicated synthesis method.

It is indispensable for the cathode material to contain redox metal ions and Li ions for delithiation at charging. Higher contents of redox ions and Li ions are desirable to withdraw a high capacity from cathode materials. A material with lower formula weight has the advantage getting higher specific energy density per weight; therefore, the oxygen ion with lower mass per charge and free material cost is the most suitable for the charge compensation of cation. In other words, the lithium- manganese oxides are desirable for the cathode materials in lithium-ion batteries. Spinel-type $LiMn_2O_4$, spinel-type $Li_2Mn_2O_4$,[25] orthorhombic $LiMnO_2$,[26] layered $LiMnO_2$,[27] O_2 type $Li_{0.7}MnO_2$,[28] $Li_{0.33}MnO_2$,[29] and so forth are reported to be such manganese-based cathode materials. $Li_{0.33}MnO_2$ is a 3-V type cathode material with high capacity,[29] which was developed by the authors. The Tadiran Co. in Israel commercialized the AAA-type battery composed of this cathode and metallic lithium anode.[30] This battery has a unique safety mechanism, in which the oxolane acts as both solvent for electrolytes, containing amines as inhibitors for polymerization and monomers for the polymerization when the temperature rises to an emergency level.[30] The active material, composite dimensional manganese oxide (CDMO), developed and commercialized by Sanyo Co., also is considered to be $Li_{0.33}MnO_2$.[31] However, the coin-type cell mainly is used for safety in the application of the secondary battery and it is used as a battery for memory backup. Moreover, spinel $Li_2Mn_2O_4$, layered $LiMnO_2$, and $Li_{0.7}MnO_2$ are prepared by chemical reduction or ion-exchange reaction using an expensive reagent; therefore, the manufacturing cost becomes expensive and the advantage of low material cost is lost. Finally, the remaining candidates for the manganese-based cathode material in the lithium-ion battery will be spinel $LiMn_2O_4$, orthorhombic $LiMnO_2$, and a layered manganese-based material, for instance, the $LiMn_xNi_yCo_{1-x-y}O_2$ and $LiMn_xNi_{1-x}O_2$ type materials. All of them can be synthesized by a simple solid-state method. It has been said that orthorhombic $LiMnO_2$ has a poor cyclicity because its crystal structure changes into spinel at the first stage of charge/discharge; afterward, the charge/discharge reaction proceeds, maintaining the spinel structure although it exhibits a capacity of 200 mAh/g or more in the 3 to 4 V range. However, an orthorhombic manganese material, which has an excellent cyclicity even in the 3–4 V range, also has been reported.[32]

Although the layered Ni–Mn-based cathode material has been placed on the market in the form of single or mixed cathode material,[33–35] here we will explain the spinel manganese-based material. This compound also has been placed on the market in the form of single or mixed with a layered nickel- or cobalt-based material.

2.3.2.1 Nonstoichiometry of Manganese Spinel

It was discovered about 20 years ago that the manganese spinel compound can be oxidized electrochemically. The studies on the manganese spinel at an early stage were focused on the properties of 3-V class cathode material, and it was clarified that the spinel compound forms a cation-deficient type ($Li_2Mn_4O_9$) or a lithium-excess type ($Li_4Mn_5O_{12}$). Moreover, the authors have clarified that oxygen-deficient spinel compounds are formed in the high temperature synthesis. As mentioned, the lithium–manganese spinel is a complicated nonstoichiometric compound, in which lithium and manganese are distributed in a cubic close-packed structure of oxygen anions. The authors have classified the spinel compounds into the oxygen stoichiometric spinel and the oxygen-deficient spinel whether the close-packed structure of the oxygen ion was maintained or not, and found that the cyclicity of both spinels are remarkably different. The oxygen stoichiometric compounds with an excellent cyclicity as a cathode in lithium ion batteries are composed of three kinds of oxygen stoichiometric spinel: $LiMn_2O_4$, $Li_4Mn_5O_{12}$ (the molar fraction of $Li_{4/3}Mn_{5/3}O_4$ is expressed by x), and $Li_2Mn_4O_9$ (the molar fraction of $Li_{8/9}Mn_{16/9}O_4$ is expressed by y). Such compounds can be expressed by a general formula as $Li_{1+x/3-y/9}Mn_{2-x/3-2y/9}O_4$. It has the excess oxygen if $y > 0$, because the oxygen stoichiometric spinel is constituted by the oxygen-excess type (cation-deficient) $Li_2Mn_4O_9$. The composition of $Li_{1+x/3-y/9}Mn_{2-x/3-2y/9}O_4$ is plotted inside of triangle ABC in phase diagram of Li–Mn–O spinel (Fig. 2.6), where three kinds of oxygen stoichiometric spinels, $LiMn_2O_4$, $Li_{4/3}Mn_{5/3}O_4$, and $Li_{8/9}Mn_{16/9}O_4$, are located at the corner. The vacancy in cation site (8a and 16d) and anion site (32e) is shown by the same symbol, □. The figure is expressed by using two parameters; one is average oxidation number of manganese (m) and lithium–manganese atomic ratio (n) in spinel. These were directly determined by chemical analysis. Spinels on the parallel line against line AC have the same n value and that against line BC have the same m values. The oxygen stoichiometric spinel without oxygen deficiency presents outside of the

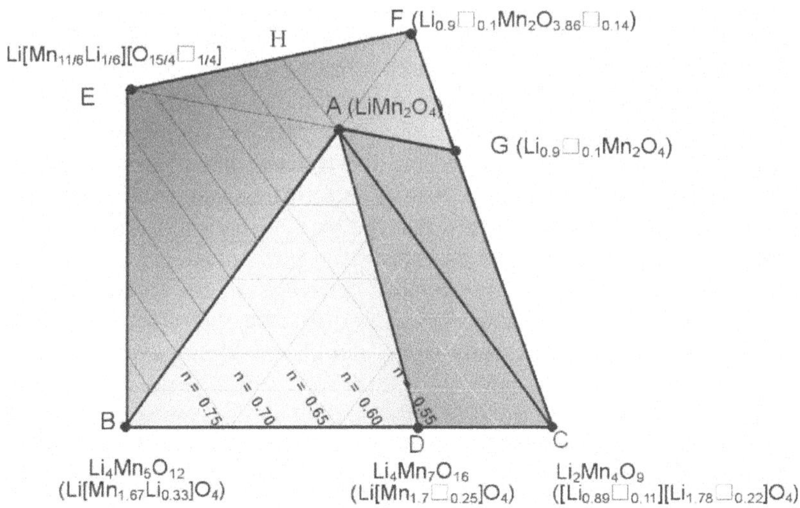

Fig. 2.6 Ternary phase diagram of Li–Mn–O spinels

basic triangle (shown in triangle, AGC), if n is less than 0.5. It is divided into two groups from the structural viewpoint. One is lithium-excess spinel with cation vacancy in only the 8a site (the area of triangle ABD) and the other is the oxygen-excess spinel with vacancy in 8a and/or 16d site (the area of polygon ADCG).

The oxygen-deficient spinels are plotted in the area of polygon ABEFG, where E, F, and G are temporary and hypothetical compounds. Line AG has the relation, $m + n = 4$, and line AE, $3m + n = 11$. Point F is an extrapolation of line AB to $n = 0.45$. The 8a site lithium content is smaller than Mn^{3+} content in the area of triangle ABE and a reverse relation presents in the area of polygon AEFG. The preparation of spinel at higher temperatures causes loss of oxygen; then its composition moves in an upper parallel to AB line with same *n* values by the heating by higher temperatures or prolonged time. Spinel compounds in the area of polygon AGFH have a vacancy in both cation site (8a) and oxygen site (32e); however, those in the area of triangle AEH have only oxygen deficiency. Pure oxygen-deficient spinels with high crystallinity are scattered around corner A.

The first report on the relation of cycle behavior and oxygen deficiency was reported by Yoshio and Xia in 1997.[36] As will be described later, the fact that the 3.2–3.3 V capacity due to the oxygen-deficient spinel is explained by us[37] and supports that the crystal structure of the oxygen-deficient spinel is the $LiMn_2O_{4-\delta}$ type. However, there were several different opinions for the structure of oxygen-deficient spinels such as: it had no substantial oxygen deficiency but had a structure in which the manganese ion shifted to the 8a site; it had a structure in which the manganese ion shifted to the 16c site; and there was Mn_2O_3 as an impurity, although it could not be detected. The structure in which the manganese ion occupies 8a site would be speculated from the structure of tetragonal spinel Mn_3O_4. When the lithium content decreases and the average oxidation number of manganese approaches 3, the crystal distorts from cubic to tetragonal because of the Jahn–Teller effect of Mn^{3+} as it is frequently said, and the transfer of manganese to 8a site occurs at this time.[34] Kanno et al. have clarified that the structure of cubic spinels is the oxygen deficient-type, using the analysis of neutron diffraction data with high reliability for the sake of its high sensitivity to oxygen and lithium.[38] They have also clarified that there is no cation mixing in the 8a site of oxygen-deficient spinel.[37]

The authors systematically carried out the research about the relation of the oxygen content in the spinel with the battery performance and the structural change in detail for the first time, and found that the battery characteristics of the spinel compounds are considerably dependent on the oxygen content. It was found by this study that some of the poor electrochemical behaviors are not common in spinels but peculiar in the oxygen-deficient spinel. For instance, the spinel compound has been considered to give the capacity fading during the cycling even at room temperature as described in most papers, and this capacity fading was attributed to the Jahn–Teller effect. However, we have elucidated for the first time that above spinel compounds are oxygen-deficient spinels. This will be explained in detail in the following. First, the charge/discharge curves of the oxygen stoichiometric spinel and the oxygen-deficient spinel are shown in Fig. 2.7.[37] Samples in Fig. 2.7a, b are an oxygen stoichiometric spinel $LiMn_2O_{4.02}$ and an oxygen- deficient spinel

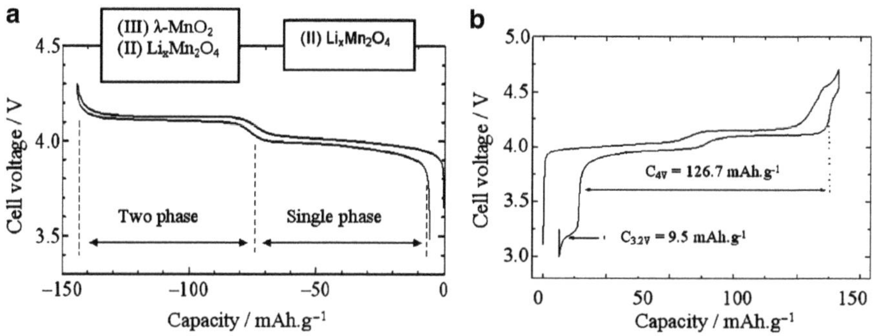

Fig. 2.7 Charge/discharge curves of oxygen stoichiometric spinel (*left*, **a**) and oxygen-deficient spinel (*right*, **b**)

$Li_{1.002}Mn_{1.998}O_{3.981}$, respectively. The 4-V region of $LiMn_2O_{4.02}$ consists of two smooth plateaus: the 4.0-V region (low-voltage plateau) and the 4.15-V region (high-voltage plateau). Here, the charge/discharge product of $LiMn_2O_4$ is expressed as $Li_{1-x}Mn_2O_4$. The low-voltage plateau ($x < 0.5$) is a single-phase region (only cubic II phase exists) where the *a*-axis of spinel $Li_{1-x}Mn_2O_4$ successively shrinks as the increase in x. The high-voltage plateau ($x > 0.5$) is a two-phase region where two cubic phases with different lattice parameter, $Li_{0.5}Mn_2O_4$ (cubic II phase) and λ-MnO_2 (cubic III phase), coexist. On the other hand, the oxygen-deficient spinel, for instance $Li_{1.002}Mn_{1.998}O_{3.981}$, have extra voltage plateaus at around 3.2 V and 4.5 V in addition to the high-voltage plateau and the low-voltage plateau in the discharge curve. The electrochemical reaction is different from that of the oxygen stoichiometric spinel in the low-voltage plateau, it becomes a two-phase mechanism where cubic I and cubic II phases exist.[39] The capacity fading during the cycling is liable to occur because the charge/discharge process accompanies the phase transition in the two-phase region. Figure 2.8 shows the relations between the amount of oxygen deficiency and the cyclicity at room temperature for the spinel compound. Since a linear relation is observed between the amount of oxygen deficiency (δ) and the capacity retention percent, it is clear that the oxygen deficiency dominates the cyclicity at room temperature. Moreover, the extrapolated value of the capacity retention to $\delta = 0$ becomes 100%, and it indicates that the room temperature capacity fading during the cycling does not occur without the oxygen deficiency. Based on the above discussion, it will be understood that the capacity fading of the oxygen-deficient spinel during the cycling even at room temperature is attributed to two sets of the two-phase reaction in both the high- and low-voltage plateaus.

In addition, the capacity fading during the cycling occurs in a simple oxygen stoichiometric spinel because the two-phase reaction proceeds in the high-voltage plateau; however, it is not so serious compared to that in the oxygen-deficient spinel. In other words, the reason for residual capacity fading of the oxygen stoichiometric spinel is the formation of λ-MnO_2 phase (cubic III) at deep delithiation. Finally, if deep delithiation of spinel is inhibited, an electrochemical reaction would proceed under the single-phase mechanism over all of the 4-V range and the cyclicity would

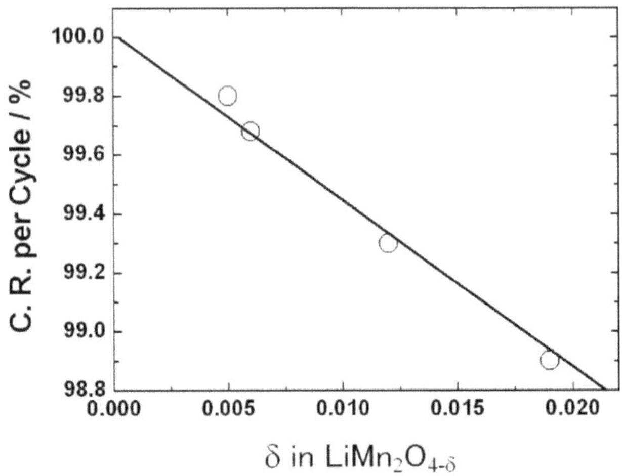

Fig. 2.8 The relation between amount of oxygen deficiency (δ) and capacity retention per cycle (CR per cycle) at room temperature

be improved. The authors have already indicated that it is easily overcome by use of lithium-excess spinel $Li_{1+x}Mn_2O_4$ without oxygen deficiency.[40]

However, oxygen-deficient spinels are formed easily when lithium-excess composition is selected. Such a problem is easily solved by doping foreign metal ions, which accelerates the formation of oxygen-excess-type spinels. In the other words, the oxygen stoichiometric spinels are easily formed using the above doping technique even for lithium-excess composition. The amount of the cation deficiency for foreign metal-doped spinel was estimated from the relations between the amount of the doped metal ion and the capacity (although it depends on the synthesis temperature). These are determined to be 1% for chromium, 1–1.5% for cobalt and aluminum, and 1.5–2% for nickel-doped spinels, and it is confirmed that the foreign metal ion doping promotes the formation of the oxygen stoichiometric spinel. The formation of oxygen deficiency in spinels cannot be neglected for the preparation at ca. 800°C even in the case of the foreign metal doping. Consequently, an excellent synthesis method is desired to fulfill two conditions, i.e., oxygen stoichiometry and prevention of λ-MnO_2 formation at a charged state. The detection of 3.2-V plateau in the discharge curve is convenient to confirm the presence of oxygen deficiency in the spinel compound.

By the way, the battery manufacturers have recommended the high-temperature synthesis to get spinels with lower specific surface area, because they believed the statement in the paper[41] that the capacity fading during the cycling especially at the elevated temperature is dependent on the dissolution of manganese from spinel compounds. As a result, the oxygen-deficient spinel had been offered to the companies, and the physicochemical and electrochemical characteristics of such spinels had been measured. Then, the characteristics of the spinels reported in an earlier study were those of the oxygen-deficient spinel, and such reports gave extreme misunderstandings for the characteristics of the spinel. These misunderstandings are enumerated as follows:

1. The spinel deteriorates during the cycling.
2. The largest deterioration occurs at a depth of discharge of 60–100%,[42] based on the measurement about the effect of storage time on the capacity fading at various discharge depths.
3. The spinel compound changes its structure when it is cooled below the room temperature.[43]

It should be noted that all these descriptions concern the oxygen-deficient spinel and cannot be adopted to the oxygen stoichiometric spinel. That is, it is not overdone how strongly it is emphasized that the oxygen stoichiometric spinel does not have these three features. As mentioned above, it should be understood for the spinel compounds that the data of spinels, in which oxygen content is measured by the chemical analysis, would be reliable.

2.3.2.2 Stabilized Spinel Compounds

As previously described, the synthetic conditions for an excellent spinel compound are considerably restricted. First, it should be an oxygen stoichiometric compound. Second, doping of different kinds ions including Li^+ ion is required in order to proceed the to the charge/discharge process under the single-phase reaction throughout the 4-V region. The chemical composition of desirable spinels is described to be $Li_{1+x}M_yMn_{2-x-y}O_4$. It will be necessary for such spinels to keep roughly $0 < x < 0.06$ and $0.03 < y < 0.15$ in order to withdraw the discharge capacity of around 100 mAh/g. Here, M is one or more kinds of metal ion excluding Li^+ ion. If the counterelectrode is metallic lithium, the cyclicity of the spinel compound is excellent even in the electrolyte of about 60°C. However, it is well known that the insertion and extraction of Li^+ ion for the graphite anode are obstructed by deposited mangnanese from the dissolved manganese ion in the lithium-ion batteries. Since the dissolved manganese ion causes deterioration in the cyclicity of the graphite anode, the performance of the lithium-ion battery also causes deterioration by the influence of the anode.[44] As this deterioration phenomenon of the anode is particularly remarkable in the high temperature electrolyte, the third condition to decrease the concentration of the dissolved manganese ion is required.

The authors have developed a new method (two-step heating; initial firing at 900–1,000°C and second refiring at 600–800°C) for synthesizing the spinel-type cathode material. We have succeeded in the preparation of spinels, which satisfy the above three conditions at the same time.

Well-developed spinel crystallines with lower specific surface area are formed in the initial firing process. Such lower specific surface area is very effective to reduce the amount of manganese dissolution into the electrolyte. It is proven by the chemical analysis that the oxygen-deficient spinel absorbed oxygen and changed to the oxygen stoichiometric spinel in the second refiring process. However, it is

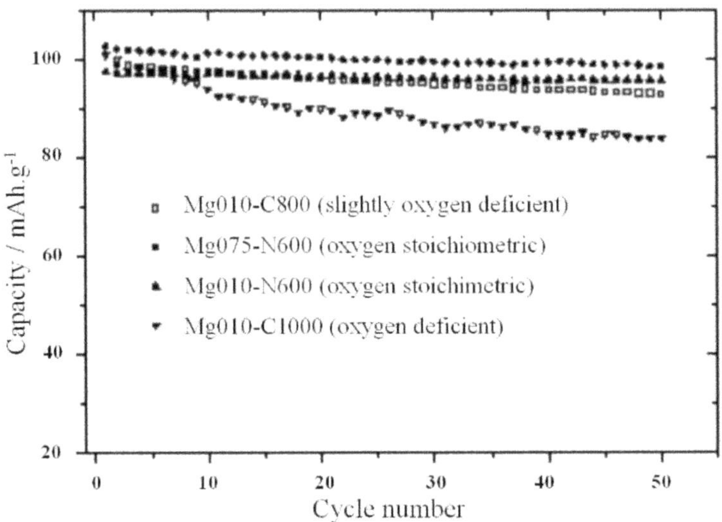

Fig. 2.9 Cycling performance of Mg-doped spinels at 60°C. Mg010-C800: oxygen-deficient $Li_{1.035}Mg_{0.093}Mn_{1.873}O_{3.99}$; Mg075-N600: oxygen stoichiometric $Li_{1.034}Mg_{0.080}Mn_{1.886}O_{4.010}$; Mg010-N600: oxygen stoichiometric $Li_{1.036}Mg_{0.10}Mn_{1.864}O_{4.021}$; Mg010-C1000: oxygen-deficient $Li_{1.033}Mg_{0.090}Mn_{1.877}O_{3.997}$

important in first process to prepare the spinel, in which the amount of oxygen deficiency is as small as possible for the sake of easy conversion to the oxygen stoichiometric spinel in the second process. In that case, it has been found that the magnesium ion, the aluminum ion, the nickel ion, and the cobalt ion are preferable for doping metal ions.

Figure 2.9 shows the cyclicity of the magnesium doping spinel cathode for metallic lithium anode at 60°C.[45] It is understood that the cyclicity of the oxygen stoichiometric spinels is excellent.

In the case of spinel compounds prepared at about 1,000°C with this method, the concentration of manganese ion is roughly 3 ppm even when spinels are stored in the electrolyte at 60°C for 4 weeks. Such value is significantly lower than 100 ppm of $LiMn_2O_4$ prepared at 800°C. Finally, manganese dissolution from spinel can be reduced about 1/30 by the introduction of the initial heating process. As a result, the elevated temperature characteristics of the lithium-ion battery composed of the spinel cathode synthesized by this method and graphite anode are considerably improved, as shown in Fig. 2.10.

2.3.2.3 Structure of Manganese Spinel

In spinel $LiMn_2O_4$, the light element, lithium, occupies the tetrahedral 8a site and the heavy metal element, manganese, occupies the octahedral 16d site; O^{2-} (32e site) forms the cubic close packing. The spinel of such a construction is called a normal spinel. When a foreign metal ion is doped, most of the first transition metal ions substitute for manganese in the16d site, and the crystal becomes normal spinel. However, the zinc ion or the iron ion substitutes for the 8a site according to the synthesis condition, and a random spinel in which the heavy metal element occupies a part of the 8a site is formed.[46] In addition, the gallium ion substitute both the 8a and 16d sites.[47] The interesting point of the zinc-doped spinel is that it forms a cubic structure with the space group P_{4132} in the case of slow cooling and causes the change in XRD pattern as shown in Fig. 2.11, where an additional five peaks in 2θ°=15°–25° are clearly observed. Although manganese-based inverse spinel, in which the heavy metal ion completely occupies 8a site, is not known, $LiNiVO_4$ and $LiCoVO_4$ can be classified into inverse spinel in the cathode material for the lithium battery and the V ion occupies the 8a site.[48,49] The diffusion path of the lithium ion

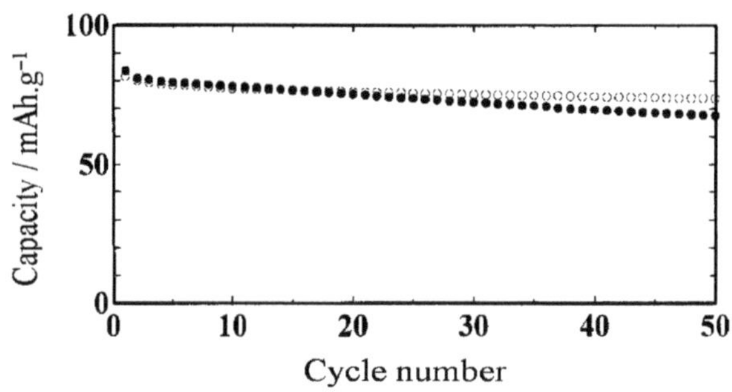

Fig. 2.10 Cycling performance of lithium-ion battery composed of oxygen stoichiometric $Li_{1.06}Al_{0.15}Mn_{1.78}O_4$ and graphite (MCMB6–28) at RT(*open circle*) and 60°C (*filled circle*). Electrolyte contains vinylene carbonate. EC:MEC(3:7), 1-M LiPF6, 4.2–3.3 V

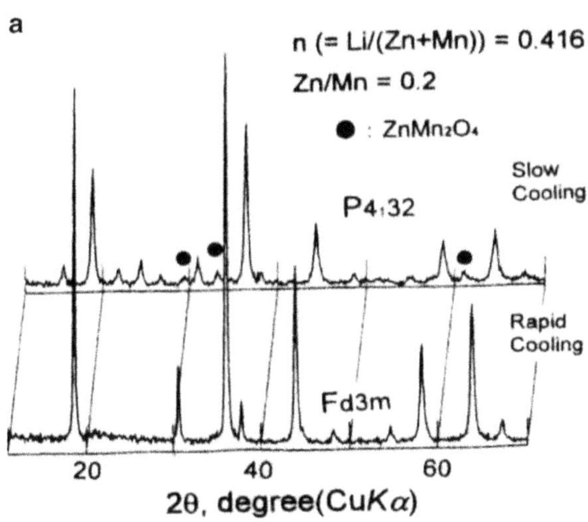

Fig. 2.11 XRD patterns of ordered (P_{4132}) and disordered zinc-doped spinel (F_{d3m})

in this spinel is restricted to the route, such as the octahedral 16c site—the tetrahedral 48f site delithiated 16d. This difficulty in the diffusion of Li^+ ion is considered to restrict the capacity less than 50 mAh/g. In addition, the 8a site where the Li^+ ion exists shares its plane with four 16c sites in the normal spinel, and the Li^+ ion can easily shift from the 8a site to the 16c site.

2.3.2.4 Capacity of Spinel Compound

The authors have confirmed that the measured capacity of the oxygen stoichiometric spinel compound coincides with the capacity calculated from the Mn^{3+} ion content, which was experimentally obtained by chemical analysis within an experimental error. Further, we have established the equations for calculation of capacity from the composition of spinel compounds using the lithium–manganese ratio (n) and the average oxidation number of manganese (m), which are determined experimentally.[36,40,49–52] The classification of spinel compounds, the spinel formula, and the equations for the calculation of theoretical capacity are shown in Table 2.3. The oxygen stoichiometric spinel is classified into the lithium-excess spinel and the oxygen-excess spinel, based on the difference in the site of the cation vacancy. However, the ratio of the Mn^{3+} ion/total Mn above two spinel is commonly described as $(4 - m)/(m + n)$; both equations for the theoretical capacity become equal. On the other hand, the same equation for the capacity is applicable even for the oxygen-deficient spinels with $n<0.5$ and $4 \leq m+n$. Two types of oxygen-deficient spinels with different chemical formulas would form. One type of spinel can be described as the formula of $M_3O_{4-\delta}$ (M is arbitrary cation), which has a deficiency only in the oxygen site. The other would be described as the formula of $Li_{1-z}Mn_2O_{4-\delta}$ ($z>0$), which has additional cation deficiency in the 8a site because manganese occupies only the 16d site in the oxygen-deficient spinel. If the electrochemical oxidation of the spinel compound accompanies the oxidation of Mn^{3+} to Mn^{4+} and the release of the Li^+ ion for the charge compensation, then the capacity

Table 2.3 Classification, spinel formula, and capacities of spinel Li–Mn–O compound[a,b]

Classification		Spinel formula	4 V capacity
Oxygen stoichiometric spinel	Lithium excess spinel	$Li[Mn_{8/(n+m)}Li_{(7n-m)/(n+m)}\square_{(3m-5n-8)/(n+m)}]O_4$ $m/7 < n < (3m-8)/5$, $3.5 < m < 4.0$	$1184(4-m)/(m+n)$
	Oxygen excess spinel	$Li_{8n/(n+m)}\square_{(m-7n)/(n+m)}[Mn_{8/(n+m)}\square_{(2n+2m-8)/(n+m)}]O_4$ $n \leq m/7$, $4.0 < m + n$	
Oxygen-deficient spinel	Mn^{3+} lack type	$Li[Mn_{3/(n+1)}Li_{(2n-1)/(n+1)}]O_{3(m+n)/2(n+1)}\square_{(8+5n-3m)/2(n+1)}$ $n > (3m-8)/5$, $3m + n > 11$	$1184(4-m)/(m+n)$
	Li^+ lack type	$Li[Mn_{3/(n+1)}Li_{(2n-1)/(n+1)}]O_{3(m+n)/2(n+1)}\square_{(8+5n-3m)/2(n+1)}$ $n > (3m-8)/5$, $3m + n < 11$	148
		$Li_{2n}\square_{1-2n}[Mn_2]O_{m+n}\square_{4-m-n}$ $n < 0.5$, $4.0 < m + n$	$296n$

[a]$\square$: Vacancy
[b]Formula weight and Mn content are conventionally assumed as same as those of $LiMn_2O_4$

of spinels would be controlled by the Mn^{3+} content or Li^+ content in the spinel; here, we believe Li^+ in the 8a site is electrochemically active.

We can divide the oxygen-deficient spinel for its capacity. The Mn^{3+} content would control the capacity of one type of oxygen-deficient spinel; the capacity is expressed same equation as the oxygen stoichiometric spinel previously mentioned. If the lithium content in the spinel restricts the capacity, although experimental verification has not been carried out, the capacity of such a spinel would be calculated to be 296*n*.

The oxygen-deficient spinels have two equivalent additional discharge plateaus at 3.2 V and 4.5 V. It already has been experimentally confirmed that the relationship, $C_{3.2\,V}$(mAh/g) = 444δ, exists between the amount of oxygen deficiency, δ, and the capacity at 3.2 V, $C_{3.2\,V}$ (Fig. 2.12). This capacity can be explained by the participation of five O^{2-} ion-coordinated Mn (MnO_5) from the crystallographic viewpoint as follows. Three manganese ions and one lithium ion surround an oxygen anion, and three MnO_5 are formed when an oxygen ion is lost, i.e., formation of one vacancy as shown in Fig. 2.13. Three MnO_6 octahedra bond to three MnO_5, sharing their edges. The redox potential of all 12 Mn would be changed and generate new voltage plateaus at 3.2 V and 4.5 V. Here, we consider the capacity of oxygen-deficient spinels. The capacity for six Mn^{3+} ions per one oxygen deficiency (half of 12 Mn) is reflected by the total capacity of 3.2 V and 4.5 V. The formation of the oxygen deficiency in $LiMn_2O_4$ can be expressed by

$$LiMn_2\,O_4 = LiMn_2\,O_{4-\delta} + \delta/2O_2 \tag{2.1}$$

For a 1-g sample of $LiMn_2O_{4-\delta}$ (formula weight: F_w), the mole of the oxygen deficiency in this sample is δ/F_w. Therefore, the mole of the Mn^{3+} ion that is influenced by the oxygen deficiency becomes $6\delta/F_w$. If δ is small, the value of F_w in $LiMn_2O_{4-\delta}$ can be approximated by the formula weight of $LiMn_2O_4$, and the total capacity of

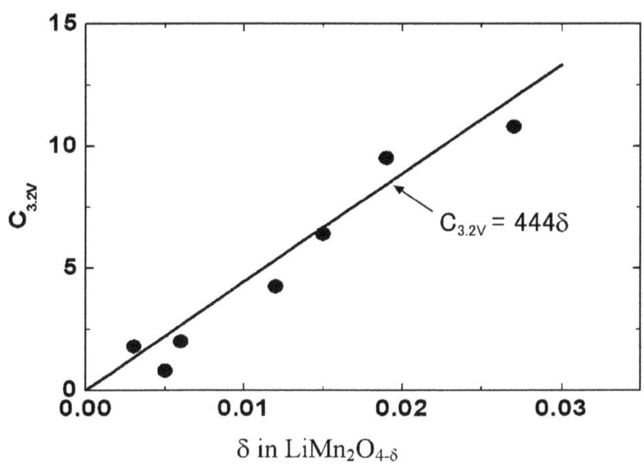

Fig. 2.12 The relation between $C_{3.2\,V}$ and δ in $LiMn_2O_{4-\delta}$

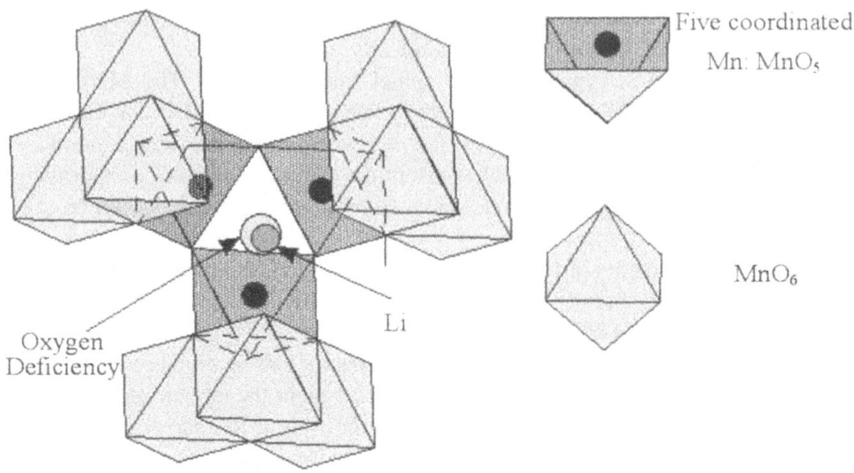

Fig. 2.13 Local structure around oxygen deficiency

3.2-V and 4.5-V region is $6\delta \times 148$ mAh/g = 888δ mAh/g, and one half of it corresponds to the 3.2-V capacity of 444δ mAh/g.

In the spinel material used as a cathode material, the cyclicity is improved by preventing the formation of the λ-MnO_2 phase, and the electrode reaction at a high voltage plateau is changed to the single-phase reaction by introducing the Li^+ ion and the foreign metal ion into the 16d site. However, the control of the composition of the 16d site decreases the Mn^{3+} ion content and causes a decrease in capacity. Here, the capacity of spinels with the ratio of lithium/other metals = 0.5 will be explained. The chemical formula of the oxygen stoichiometric spinel containing foreign metal (M) can be expressed as $Li_xM_yMn_{3-x-y}O_4$ in such cases. Here, the atomic ratio M/(Mn + M) is defined by f [= $y/2$]. When the formula weight of this spinel is approximated by the formula weight of $LiMn_2O_4$, the capacity C (mAh/g) per 1 g of the foreign metal-doped spinel can be calculated using the mole of the Mn^{3+} ion in 1 g of spinel. Then, it can be expressed by (2.2),[52,53] where the charge of foreign metal ion (M) is expressed as v^+:

$$C(\text{mAh/g}) = 148[1-3s-(4v)f] \tag{2.2}$$

That is, the smaller the charge of the foreign metal ion and the larger the f, the capacity decreases. The composition of the spinel can be designed to calculate the biggest capacity, where cation vacancy, s, is equal to zero.

In an ideal 5-V spinel cathode material, the manganese ion is tetravalent[54] and the redox species are foreign metal ions. Nickel, copper, iron, cobalt, and chromium are known as a foreign metal M.[55] The highest 5-V capacity is obtained for the composition of $LiM_{0.5}Mn_{1.5}O_4$ and $LiMMnO_4$, where M is divalent and trivalent, respectively. Although a capacity of 145–147 mAh/g can be expected for the divalent metal (Ni, Cu) under the two-electron transfer mechanism, only $LiNi_{1/2}Mn_{1.5}O_4$

exhibits a discharge capacity of ca. 140 mAh/g in the 5-V region. Although a voltage plateau at the 4-V region based on $Mn^{3+/4+}$ due to the formation of oxygen deficiency appears in this spinel, the introduction of the oxygen absorption process for repairing the oxygen deficiency (annealing under the oxygen atmosphere) is an important technology for increasing the 5-V capacity.[56,57] In the spinel substituted by iron, cobalt, and chromium, the ideal composition with the highest capacity is $LiMMnO_4$ because all three kinds of metals exist as trivalent. Although any samples cannot exhibit a capacity of more than 100 mAh/g in the 5-V region, $LiCrMnO_4$ has no 4-V region because of the absence of oxygen deficiency.

2.3.2.5 Charge/Discharge Mechanism

As it has been already described, it is important for maintaining an excellent cyclability in spinels to keep single-phase mechanism over all 4 V region. Therefore, it is very important to elucidate the charge/discharge mechanism. Two types of reactions are reported for the electrochemical reaction of Li in spinels: The one is the single phase mechanism, where the unit cell simply shrinks or expands during charge or discharge. The other is the two-phase reaction, where two different crystal phases coexist and the ratio of both crystal phases changes. The differences between these two electrochemical reactions are observed in the open circuit voltage (OCV) and the X-ray diffraction profiles. The OCV curve in two-phase mechanism should give flat shape, however, it might give S type curve even in case of the measurement under the non-equilibrium state, and it is likely to misidentify as a single-phase mechanism. Moreover, in the XRD measurement using the CuKα line (two kinds of wave lengths of $K_{\alpha 1}$ and $K_{\alpha 2}$ are irradiated) etc., two diffraction lines of original and produced spinel with closed lattice parameters in the two-phase mechanism are duplicated, and they are regarded as a broad diffraction line. Finally, it might be misunderstood as a single-phase mechanism. Analysis of XRD with high accuracy is possible to use a synchrotron light because monochromatic X-ray with high intensity is obtained. It has been already clarified for the first time by the authors' joint research that the charge/discharge mechanism of the oxygen deficient spinel in the low voltage region is a two-phase reaction.[39]

Figure 2.14 shows the changes in the cubic lattice parameter during the charge for three typical spinel compounds. The reaction of $LiMn_2O_4$ consists of a single-phase reaction in the low-voltage plateau, where the lattice parameter continuously decreases, and a two-phase reaction in the high-voltage plateau, where two cubic phases with different lattice parameters exist. In the oxygen stoichiometric spinel of the metal ion doped-type or the lithium excess type, electrochemical processes at both high-voltage plateaus and low-voltage plateaus proceed in a single-phase mechanism, and the lattice parameter continuously changes. On the other hand, in the oxygen-deficient spinel, two crystal phases with different lattice parameters exist for both low-voltage plateaus and high-voltage plateaus. That is, the cubic phase I of a lattice parameter 8.25 Å and the cubic phase II of 8.17 Å exist in the low-voltage plateau, and it only becomes cubic II at $x = 0.5$. In addition, the cubic

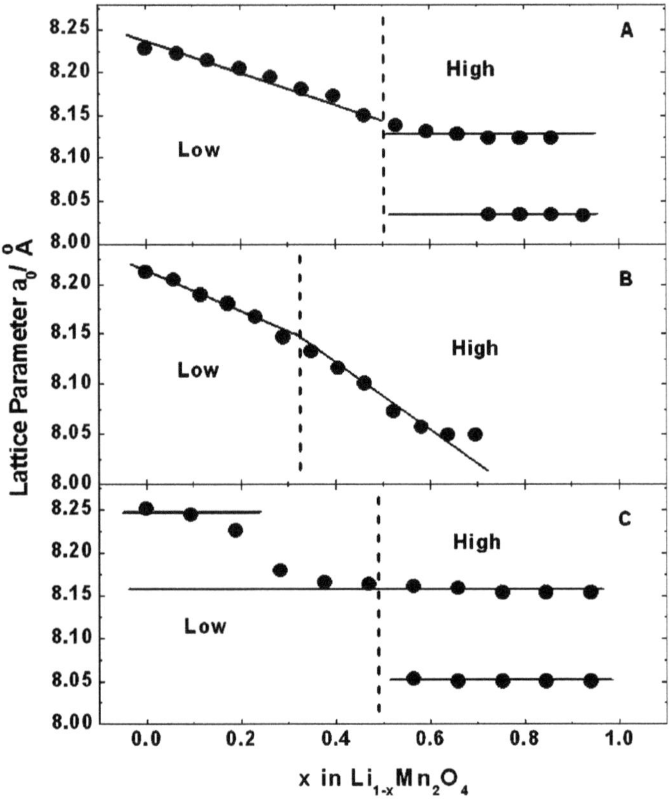

Fig. 2.14 Lattice parameters of delithiated stoichiometric $LiMn_2O_4$ (**a**), oxygen stoichiometric spinel (**b**), and oxygen-deficient spinel (**c**)

phase III of 8.06 Å is formed when the delithiation proceeds, and the crystal phases of cubic phase II and III coexist. The ratio of cubic phase III increases as the delithiation proceeds.

The similar lattice change of such an A type and C type in Fig. 2.14 are reported even for a 5-V cathode material such as $LiNi_{0.5}Mn_{1.5}O_4$.[58] The electrochemical reaction in both low- and high-voltage plateaus are considered to be the two-phase reaction the same as in the oxygen-deficient spinel (C). That indicates the possibility of existence of the oxygen deficiency in the sample used in that report.

2.3.3 Olivine Compound

Orthorhombic $LiFePO_4$ of the olivine structure forms $FePO_4$ during charging/discharging, and two crystal phases exist during charging/discharging; thus it exhibits a flat discharge curve.[59] The accurate crystal structure of orthorhombic

$FePO_4$ formed by charging was obtained using the Rietveld method.[60,61] Substitution of iron by manganese transforms the shape of the charge/discharge curves. The capacities of the lower (3.5 V) and higher voltage region (4.0 V) correspond to the content of iron and manganese. Therefore, a manganese substitution is effective to increase the energy density. However, it causes the decrease in conductivity, which leads to poor rate performance.

2.4 Problems of Cathode Materials (Safety Problem of Layered Material and Spinel $LiMn_2O_4$ Type Materials)

2.4.1 Layered Materials

The cathode materials of lithium batteries have a strong oxidative power in the charged state as expected from their electrode potential. Then, charged cathode materials may be able to cause the oxidation of solvent or self-decomposition with the oxygen evolution. Finally, these properties highly relate to the battery safety.

The self-decomposition reaction of the charged products in $LiCoO_2$ and $LiNiO_2$ is examined in detail using the thermal analysis and the X-ray diffraction method. The thermal decomposition reaction of electrochemically delithiated $Li_{1-y}CoO_2$ ($0.4 < y < 0.6$) is simple, for example, when $y = 0.5$, the reaction obeys the (2.3) and (2.4), and $LiCoO_2$ and Co_3O_4 are observed to be the decomposition products.[62] On the other hand, it is more complicated in $LiNiO_2$. The crystal structure at the charged state becomes rhombohedral or a monoclinic phase, depending on the charged state and the composition of the sample. $Li_xNi_{1-x}O$ with the rock salt structure from $Li_{1-y}NiO_2$ is formed at $y > 0.7$,[63] and a layered product is formed via a spinel phase at $y < 0.7$.[64] The reaction process for $y = 0.5$ is indicated in (2.3) and (2.4). However, it is difficult to determine whether the spinel phase is formed or not because of the similarity in XRD patterns of the spinel and layered structure. Although the formation of the spinel phase is easy in the nickel-excess-type compound,[65] it would be difficult to form $LiNi_2O_4$ in a short time during the thermal analysis by transformation of $Li_{0.5}NiO_2$ to $LiNi_2O_4$, which requires tens of hours in high purity $LiNiO_2$.[66] The difference between both reactions in $Li_{0.5}NiO_2$ is the presence of oxygen evolution; therefore, the oxygen evolution is judged from the weight loss of thermogravimetry:

$$Li_{0.5}CoO_2 = 1/2\ LiCoO_2 + 1/6Co_3O_4 + 1/6O_2 \tag{2.3}$$

$$Li_{0.5}NiO_2 = (1/2\ LiNi_2O_4) = 3/2\ Li_{0.33}Ni_{0.67}O + 1/4O_2 \tag{2.4}$$

The main difference between the above two equations is the amount of evolved oxygen gas. Since $Li_{0.5}NiO_2$ release 1.5 times of O_2 than $Li_{0.5}CoO_2$, $LiNiO_2$ has a greater possibility to generate heat than $LiCoO_2$.

Thermal analysis of the charged product without electrolyte gives information only about self-decomposition; then we can know the thermal property of the decomposition reaction, the oxygen-releasing temperature, and the amount of released oxygen. However, thermal runaway in a charged cell is caused by an endothermic reaction between electrolyte and evolved oxygen; measurements concerning the safety of the cell should be evaluated in the presence of an electrolyte.

Dahn et al.[67] have evaluated the heat generation of various charged cathode materials in the presence of an electrolyte by DSC measurement. These data for various cathode materials are summarized in Figs. 2.15a, b in order to understand the relation of the charging capacity with the heat generation and its rate. The total heat generation increases with an increase in the charging capacity for all cathode materials. Plots of 4-V class cathode materials, $LiCoO_2$, $LiNiO_2$, $LiMn_2O_4$, and $LiNi_{0.8}Co_{0.2}O_2$, roughly lie on a straight line in Fig. 2.15. No extreme deviation is observed among 4-V class cathode materials except for $LiNi_{3/8}Co_{1/4}Mn_{3/8}O_2$. The 3-V class $LiFePO_4$ shows lower heat generation. It would be due to its lower oxidative power or stability of a strong covalent bond in PO_4^{3-}. The behavior of $LiNi_{3/8}Co_{1/4}Mn_{3/8}O_2$ is strange. It shows extraordinary lower total heat generation.

These data would guarantee excellent thermal stability of $LiNi_{0.5}Mn_{0.5}O_2$-based cathode materials, such as $LiNi_{1/3}Co_{1/3}Mn_{1/3}O_2$.

$LiNiO_2$ exhibits uniqueness for the rate of heat generation, i.e., the rate of heat generation suddenly increases at the composition of $Li_{0.3}NiO_2$, which contains a NiO_2 phase. Therefore, if the formation of NiO_2 phase is protected, a sudden increase in the heat generation rate would be protected. In fact, the rate of heat generation greatly decreases cobalt doping by 20%. Many researchers[67,68] have reported excellent effects of cobalt doping into $LiNiO_2$ on the thermal stability. It also is reported that the safety of $LiNiO_2$ is greatly improved by the substitution of 1/4Ni by Al.[69] Since the reaction between charged electroactive material and the

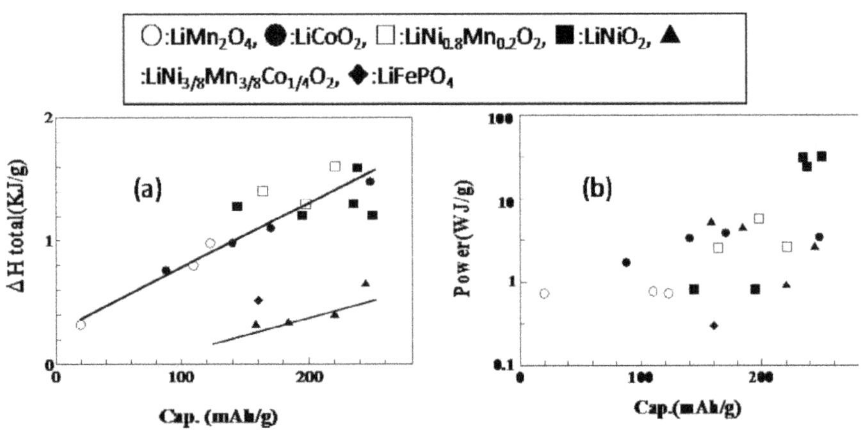

Fig. 2.15 Relation between charge depth and heat generation (**a**) or its power (**b**) for various cathode materials

electrolyte occurs at the interface, the decrease in the particle size of electroactive material causes the increase in the rate of heat generation.[70] Therefore, it is dangerous to use nanosized particle to improve rate capability.

The evaluation of safety for the cathode material also is carried out in the polymer battery system. The $Li_{0.6}CoO_2$, $Li_{0.6}NiO_2$, and $Li_{0.23}Mn_2O_4$ have exothermic peaks of about several W/g in the range of 200 to 300°C[71]; therefore, the reaction between the organic materials and the oxygen generated by the thermal decomposition of cathode material is basically inevitable. In addition, details in the evaluation of safety on lithium-ion batteries are referred to in the paper by Tobishima et al.[72]

2.5 Recent Progress in Practical Cathode Materials for Lithium-Ion Batteries

The battery characteristics, capacities, densities, shapes of the charge/discharge curves, and problems of typical cathode materials, which are used or developed for the lithium-ion battery, are listed in Table 2.1. The $LiCoO_2$ has more than a 90% share in the market of cathode material for the lithium-ion battery, which is applied to cellular phones and portable computers. It has been used since this battery was developed. Manganese-based materials have a share of residual several percent.

The most important factor for a battery is how to increase the active material in the cell with the limited volume. Therefore, the capacity per unit volume is a key factor. The cobalt-based material is still a promising material because an 808 mAh of capacity per unit volume is achieved for the sake of its higher density. Moreover, the shape of the discharge curve also is important because the design of the electric circuit for the charge/discharge control of battery is easier for its sloped curve.

The capacity of lithium-ion batteries has been increased by the improvement of the carbon anode in the initial stage; however, recent improvements in capacity are achieved by the increased charge voltage. Doping foreign ions into $LiCoO_2$ allows the use of a higher charge voltage without capacity fading.[1,73–75] The current 18650-type battery with aluminum- or magnesium-doped $LiCoO_2$ cathode can deliver the capacity of higher than 2.4 Ah.

The practical application of nickel-based materials with large capacity per unit volume has been delayed from the viewpoint of safety; however, the safety of the foreign metal-doped $LiNiO_2$ was confirmed and the practical use of the nickel-based cathode started in fiscal year 2004.

By the way, spinel $LiMn_2O_4$-type materials exhibit excellent safety, so it is considered to be a promising candidate for large-sized batteries such as for hybrid electric vehicles (HEV) and electric vehicles (EV). Battery modules for HEV can be prepared only by connecting ten single cells with the capacity of ca. 10 Ah in series. A lightweight laminated-type battery is suitable for such an application, and the battery price will become more inexpensive. A laminated-type battery containing the manganese-based material appeared in the market as a power source for the motorcycle (metal casing) and the motor-assisted bicycle in 2002. This type battery came into practical use as the power source for HEV in 2003. Capacity fading of a manganese-based cathode at elevated temperatures has been overcome; it then is almost definite that the manganese-based materials will be used for this type of battery. Magnesium- and aluminum-doped spinels, which have the lowest tendency to form oxygen-deficient compounds, are used mainly.

In Sect. 2.5.1 the design policy of the cobalt-based material used for more than 10 years is described in detail. The changes of design policy, which depends on the safety or charge/discharge conditions of battery, are examined and the current states of nickel- and manganese-based materials, which will appear in the market within 1–2 years, are described.

2.5.1 *Recent Cobalt-Based Materials*

The lithium-ion secondary batteries appeared on the market in 1990. The anode material was changed later from hard carbon to graphite; however, the cathode material is consistently $LiCoO_2$. Battery-processing technologies also have been improved and the capacity has been improved by about three times during its history. However, the progress in $LiCoO_2$ should not be neglected. Recently, the properties of $LiCoO_2$ have been considerably improved to fit the pulse discharge with high current, which is demanded for recent cellular phones.

First, let us consider the matters required for $LiCoO_2$. The specification of $LiCoO_2$ supplied by a certain company is shown in Table 2.4 as an example. Naturally, the most important property is the electrode density, which is related to the packing density and the density of the sheet electrode. These data are important for the battery manufacturers in order to stuff the cathode active material, such as $LiCoO_2$, into the battery case with constant volume as much as possible. Currently, it seems that 96 wt% of the cathode mixture is $LiCoO_2$ and the residual 4 wt% is the binder and the conductor, such as carbon. Thus, it is important to stuff electroactive $LiCoO_2$ even 1% more. The electrode density within the battery case is increased by the increase in both cathode sheet density and packing density, which leads to the improvement of cell capacity.

Table 2.4 Specification of a $LiCoO_2$ Sample

Item		Impurities	Content (wt%)
Li content (%)	6.60–7.40	Co_3O_4	≤ 1.0
Co content (%)	59.3–60.7	SO_4	≤ 0.30
Li/Co (in mole)	0.95–1.01	H_2O	≤ 0.20
SSA (Specific Surface Area) (m^2/g)	0.35–0.55	Ni	≤ 0.10
D_{50} (average particle size)	7.0–9.0	Cl	≤ 0.10
Aerated density (g/cm^3)	0.9–1.3	Na	≤ 0.050
Packed density (g/cm^3)	1.9–2.3	Fe	≤ 0.020
pH of 10% slurry	9.5–11.0	K	≤ 0.010

It is considered to be important to control the atomic ratio of Li/Co in the production of $LiCoO_2$. One of the features of this product is that its ratio is generally less than one. The cathode slurry is prepared by dispersing the conductor and electroactive $LiCoO_2$ in the N-methyl-pyrrolidone solution containing polyvinylidene fluoride (PVDF). It is unavoidable that some water is contaminated into the slurry. When the unreacted lithium in $LiCoO_2$ remains as Li_2O, it reacts with water in the slurry, and the cathode slurry becomes basic. The cathode slurry changes to a gel state in such basic media; as a result, the cathode slurry cannot be painted onto the aluminum collector. Therefore, conventional raw $LiCoO_2$ is washed with warm water in order to prevent gel formation of the cathode slurry. In an example of the product supplied by a company, the pH of 10 wt% $LiCoO_2$ aqueous dispersion is adjusted to 9.5–11.0. The recommended pH is less than 10.5. This manufacturer supplies $LiCoO_2$ with Li/Co < 1, so there is a possibility for making the unreacted lithium salt in the product close to 0. It can be considered that the washing process of raw $LiCoO_2$ would be removed for cost reduction.

The specific surface area (SSA) also is one of the guidelines, since the reaction area is large when the surface area is large. It is obvious that cathode materials with a high SSA have improved rate performance for high-current discharge. However, the density of the material inevitably decreases when its surface area increases; then the electrode density becomes small. Finally, there is a limitation in enlarging the SSA. Of course, if nanoparticles are used as an electrode-active material, it causes an extreme decrease in density. Further, the battery safety might be seriously ruined even if the rate performance is improved. Therefore, nanosized cathode material could not be recommended. Such materials are not used practically because they have a high possibility of firing in the nail penetration test (a type of battery short-circuit test) and the hot box test (battery heating test in air bath at 150°C).

The desirable average particle size and its distribution are highly related to the coating process of each battery manufacturer. Then, they are different in each battery manufacturer. Moreover, they are different for the type of batteries: prismatic, cylindrical, or polymer batteries, even if they are produced by the same manufacturer. The mass production of the same kind of material is difficult, because these factors are relative to the battery characteristics. The selection of desirable material is also different corresponding to each use or each machine of the battery manufacturer. This is one of the reasons why the manufacturer of cobalt raw material in the world cannot start the mass production of $LiCoO_2$ with the same properties.

The impurities, which cause a redox reaction in the battery, should be removed from product. Remaining Co_3O_4 in the cathode has been viewed with suspicion in recent years. The Co_3O_4 dissolves into the electrolyte as ions during the repetition of discharge. This ion is reduced on the anode and precipitates in the form of metallic Co(0); then the intercalation of Li is obstructed for the increase in the anode impedance. A considerable amount of preliminary information is obtained from the specification of the cathode material as described.

Next, we will explain the synthetic condition of $LiCoO_2$ and its physical and electrochemical characteristics from the viewpoint of actual industrial products. In general, the easiest method to control the physical property of $LiCoO_2$ is to change its Li/Co ratio. The author already has reported that the structural change from the rhombohedral structure to the monoclinic structure at around 4.1 V disappears by changing the Li/Co atomic ratio to more than 1.[76] In this section, we will explain in detail the physical properties of $LiCoO_2$ with various Li/Co ratio.

SSA decreases with increases in Li/Co, as shown in Fig. 2.16. It is understood that the crystal of $LiCoO_2$ grows easier and sintering of such crystal forms larger particles under the high Li/Co ratio, so it is expected that SSA decreases with increases in the Li/Co ratio. Figure 2.17 shows SEM photographs of $LiCoO_2$ with Li/Co = 1.0 (the left side) and 1.05 (the right side). Although the particle size is about 2–3 µm for the sample with Li/Co = 1.0, the particle size of samples with Li/Co = 1.05 becomes about 10 µm. It is clear that the crystal has grown easily at high Li/Co ratio. The decrease of SSA at high Li/Co ratio also is confirmed as expected.

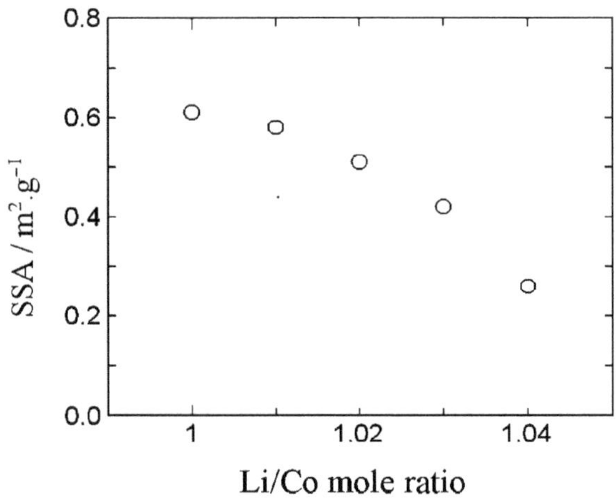

Fig. 2.16 Relation between SSA and Li/Co ratio in $LiCoO_2$

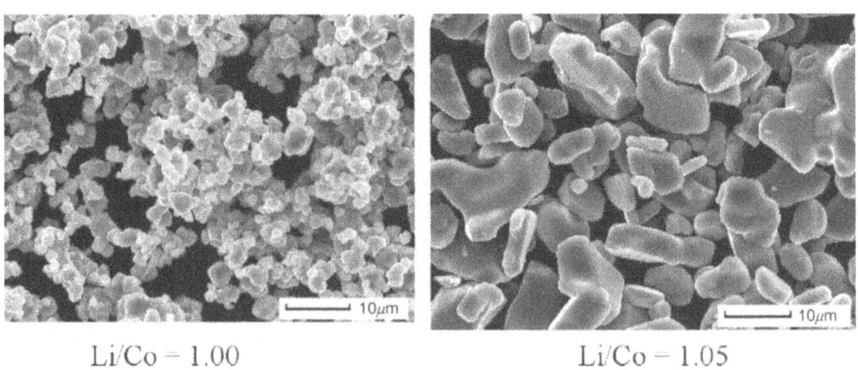

Fig. 2.17 Morphology of $LiCoO_2$ particles with Li/Co mol ratio of 1.00 and 1.05

It would be clear that higher Li/Co ratio is suitable to produce dense material with higher bulk density and higher tap density (Fig. 2.18). Here, it should be noted that the particle size distribution obtained by the wet method is different from that by SEM photograph in some cases. The relations between the Li/Co ratio and the particle size distributions (D_{10}, D_{50}, D_{90}) determined by the microtrack method (wet method) are shown in Fig. 2.19. The particle size of the sample with Li/Co = 1.00 is determined to be as small as 2–3 μm by the SEM photograph; however, the D_{50}, which means average particle size, obtained by the particle size distribution analysis is 10 μm. The differences in particle size by both measurements indicate that the aggregation of the fine particles proceeds under the dispersion in the solution.

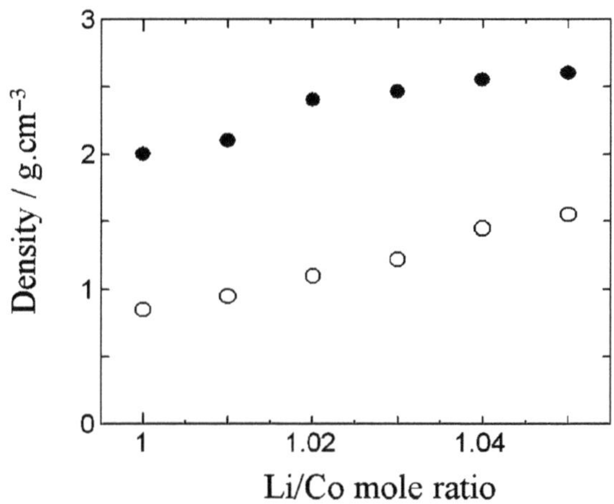

Fig. 2.18 Bulk density (*filled circle*) and tap density (*open circle*) of $LiCoO_2$ with various Li/Co mol ratio

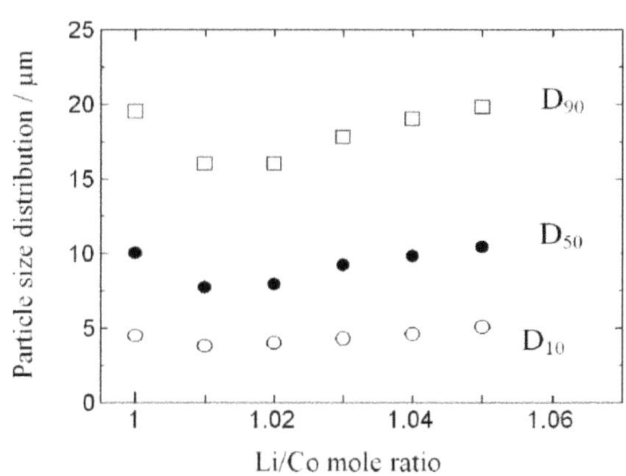

Fig. 2.19 Particle size distribution of $LiCoO_2$ with various Li/Co ratio

In other word, fine particles aggregate in the solution and they are measured as large particles in the case of the wet method. Consequently, measured particle size in the wet method would not reflect the size of primary particles but the size of aggregated secondary particles.

On the other hand, when the particles are growing and become as large as 10 μm in the SEM photograph, the average particle size measured by the microtrack method is D_{50} = 10 μm, as shown in Fig. 2.19. Both methods give the same particle size. The aggregation of particles is protected as the increase in particle size. The particle size distribution reflects the primary particle size in such case. The particle size of the product once decreases and the excess Li remains in product as Li_2O for the synthesis at a higher Li/Co ratio. The amount of remaining Li_2O can be estimated by dispersing $LiCoO_2$ into the aqueous solution and measuring its pH.

The relations between the pH of its 10 wt% suspension and the Li/Co ratio in the raw materials are shown in Fig. 2.20. As the cathode slurry changes to a gel state at the pH > 10.5–10.6, it is necessary to wash the product with warm water for adjusting its pH.

In addition, the effect of the Li/Co ratio on the properties of $LiCoO_2$, such as the crystal size and the pH of the suspension, is described in the paper by the Nippon Chemical Industrial Co., Ltd.[77] Since $LiCoO_2$ with high Li/Co ratio has both a disadvantage and an advantage for the battery characteristics, optimization is required for the synthesis of the product.

Now, let us look back on the history of $LiCoO_2$ synthesis. According to Nishi (Sony Corp.), the synthesis method at the early stage is as follows.[78] The Co_3O_4 and Li_2CO_3 of raw materials are mixed in an aqueous solution containing polyvinyl alcohol to make slurry, and this slurry is baked. In this case, the Li/Co mixing ratio is set at larger than one. The baking temperature is set to be more than 900°C in order to improve the safety of the battery. Moreover, excess Li_2CO_3 remains in the product due to the high Li/Co mixing ratio. However, when this is used as the cathode

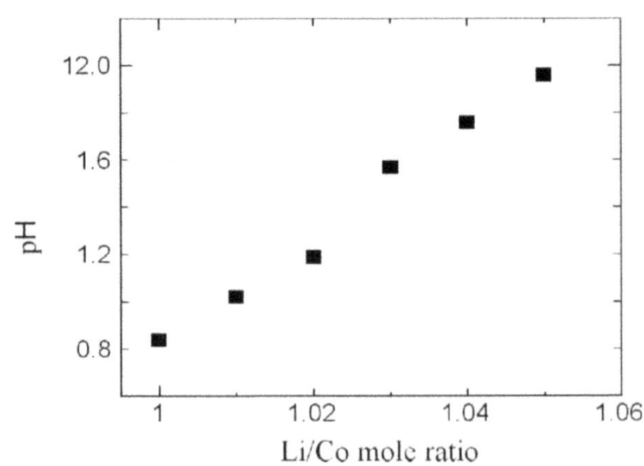

Fig. 2.20 The pH of 10 wt% Li_xCoO_2 slurry

in lithium-ion battery, the remaining Li_2CO_3 generates CO_2 even if the cell is overcharged by any chance. Finally, the battery safety is improved by rupturing the safety valve of the battery by the pressure of generated CO_2 gas. It already has been described that the particle size becomes large and the sintering proceeds more easily at higher Li/Co ratios. It is effective to use the cobalt compound with a large particle size for improving the battery safety, such as the nail penetration test. The $LiCoO_2$ with a large particle size was used in the battery at an early stage. However, the necessity for using such a cathode material with a large particle size has been decreased due to the progress in the battery manufacturing technologies. The recent cathode material has a small particle size as shown in Table 2.4 in order to improve the battery characteristics, especially the high rate performance. Consequently, the highest baking temperature is set lower than 900°C, so that the sintering would not proceed easily. Moreover, it has been found that a part of excess Li_2CO_3 remains in the form of Li_2O after the baking. Recently, the Li/Co ratio is set close to one or slightly lower; thus the remaining Li_2O content after baking has been decreased. In other words, a low-cost process without washing would be used. Such information can be read out from the specification of $LiCoO_2$ supplied by a certain company. The Li/Co ratio decreases very close to 1 and the baking temperature also seems to be decreased. On the other hand, the raw materials, Co_3O_4 and Li_2CO_3, are used now.

However, the production method of $LiCoO_2$ has been changed over the past 2–3 years. Its reason is mainly attributed to the progress in cellular phones. It seems that this new type of $LiCoO_2$ has not been adopted by the battery manufacturer in South Korea and China yet, but it will be used after fiscal year 2003. The pulse discharge mode in the third-generation cellular phone is shown in Fig. 2.21.

In this system, a 2A of maximum current is discharged for 0.6 msec during talking in order to catch the signal for communication. This high current is a considerably excessive requirement for the battery. For instance, much high current, such as a

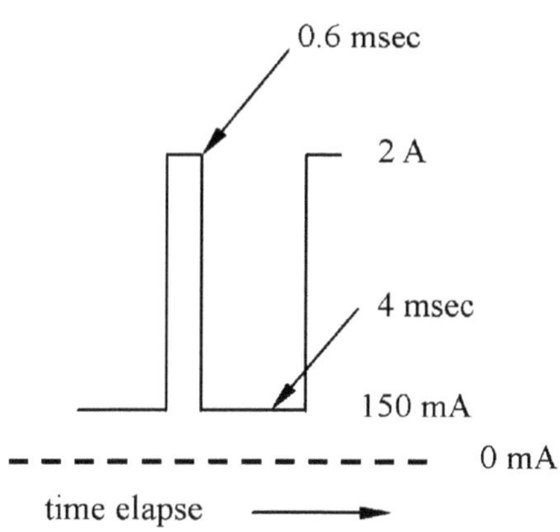

Fig. 2.21 A pulse discharge of GSM-type cellular phone

4-C rate, is discharged for a short time from a battery of 500 mAh/g. Moreover, for the mobile phone, requirements in the cellular phone for high current uses, such as the use of color display, the use of digital camera, and the signal transmission to the satellite capable of an overseas phone call, are steadily increasing. Production methods of $LiCoO_2$ have been improved in order to use the battery at a current of 3-C to 4-C rate. The primary particle size of $LiCoO_2$ is controlled to 1–2 μm for withdrawing such high discharge current. Furthermore, the secondary particle size is controlled to 5–10 μm for keeping a higher density of $LiCoO_2$ and its electrode density. Such increased particle size is also effective in order to avoid the increase in SSA due to the decrease in primary particle size. In order to synthesize this type $LiCoO_2$ with potential of high rate discharge, the synthesis method such as the following hydroxide method has been developed along with the conventional granulation method (solid phase-solid phase method). Currently, the new type of $LiCoO_2$ has been synthesized mainly by the hydroxide method. The raw material would be changed from the conventional cobalt oxide to the cobalt hydroxide. In this method, the precipitate of the hydroxide is synthesized by increasing the pH of aqueous cobalt solution to more than 7. In this case, the precipitation condition, the pH, the aging temperature, and so forth are controlled to get particles with desirable size and shape. Afterward, the precipitate is mixed with the Li_2CO_3 and baked to synthesize $LiCoO_2$. Now, let us introduce an example with a spherical shape in particular. Its SEM image is shown in Fig. 2.22. It can be seen that the primary particles with 1–2 μm aggregate and form the secondary particles with ca. 10 μm. This homogeneous spherical size type of $LiCoO_2$ cannot be used as a cathode of lithium-ion batteries, because the packing density is not enough. A battery using the $LiCoO_2$ with loosely formed secondary particles is shown in Fig. 2.22 (left).

An active material whose physical properties and chemical properties fit the requirements, such as the standard of the targeted battery, the specification of the electrode based on the battery, and the balance with the submaterials except for an electroactive material is selected. The accumulation of data, which can fulfill any requirement, would be important.

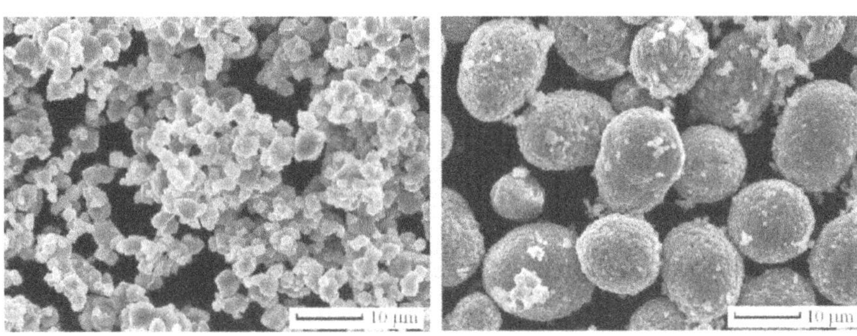

Fig. 2.22 SEM photograph of $LiCoO_2$ with different types of secondary particles

2.5.2 *Recent Layered Nickel-Based Cathode Materials*

In the $LiNiO_2$ type materials, the chemical formula of the prepared sample would be basically described as $Li_{1-x}Ni_{1+x}O_2$ ($x > 0$). A strict preparation condition is required because a part of the nickel ions may occupy the lithium-ion layer. Therefore, the following three synthesis methods, by which the raw materials can be mixed in the atomic level, are usually applied (the solid-state method is basically not suitable to prepare pure material):

- Spray drying method
- Hydroxide method
- Mixed metal carbonate method

All these methods are a kind of co-precipitation method. It is possible to mix several elements at the atomic level by using such a co-precipitation technique. These methods have the advantages in the control of the primary particle size and the secondary particle size against the solid-state method, that is, the fine particle control becomes possible. Furthermore, it has the advantage of enabling the control of crystallinity and the surface morphology. However, in some cases, this method might not be practically used for mass-production because the processing technology is too elaborate.

The sol-gel method is frequently reported as a synthesis method in the laboratory. However, the description about this method is omitted because production cost is too expensive and this method is not suitable for industrial mass production except for an experimental use.

2.5.2.1 Synthesis by Spray-Drying Method

The spray-drying method,[79] developed by Toda Kogyo Co. (formerly Fuji Chemical Industry Co.) is well known. Their sample was shipped to many companies worldwide, and it has been confirmed that it has excellent characteristics. In particular, the safety test has been finished in Saft Co. in France and in Samsung in South Korea in 1999. Those results were disclosed to the public at the international congress. According to what I heard, Saft Co. selects the nickel-based material as the cathode material of the lithium-ion battery for EV, concluding that the capacity fading problem of the manganese-based cathode at elevated temperatures would not be overcome. Moreover, this material has been practically used as the battery material mainly in foreign corporations up to now. According to these patents, metal salt and lithium salt are reacted in an aqueous medium and the obtained slurry is baked at the desirable temperature after spray drying.

In the product of Toda Kogyo Co., aluminum doping is its feature. The features of this product are that the shortcoming of the thermal instability, which is a fault of the nickel-based cathode material, has been overcome by cobalt and aluminum doping. The thermal stability at the charged state has been improved to the $LiCoO_2$ level in particular. The composition of the compound shipped is announced to be

$Li_{0.8}Co_{0.15}Al_{0.05}O_2$. The data about the thermal stability of this compound obtained by Binsan et al. of Saft Co.[80] are shown in Table 2.5.

It can be understood that the temperature of the exothermic peak with oxygen evolution increased from 200°C of $LiNiO_2$ to 310°C of $LiNi_{0.8}Co_{0.15}Al_{0.05}O_2$ due to the decrease in the nickel content and the aluminum doping. The thermal stability of this compound has been considerably improved. The oxygen evolution was not observed below 300°C. On the whole, the thermal stability of this is more excellent than that of $LiCoO_2$ and comparable to that of spinel $LiMn_2O_4$. In addition, the SEM image of this material is shown in Fig. 2.23. It would be possible to control the powder characteristics the same as the co-precipitation method. It is clear from Table 2.5 that this material has a larger capacity than that of the cobalt-based cathode. This cathode delivers more than 180 mAh/g for 4.3 V charge, as shown in Fig. 2.24.

Recently, an example of the laminate-type battery combined this material and the high-capacity graphite was reported from Toshiba Battery Co.[81] The prismatic battery (thickness: 3.8 mm and area: 35 × 62 mm) exhibits a capacity of 920 mAh and its energy density is 200 Wh/g, which is 16% larger than that of the cobalt-based cathode. These data seem to largely contribute to the practical use of the nickel-based battery. The rate performance of this battery is also excellent, as shown in Fig. 2.25. The charge/discharge capacity after 500 cycles of charge/discharge test in 1-C rate keeps 70% of the initial capacity, which is comparable to the cobalt cathode-based batteries.

Table 2.5 Thermal stability of typical cathode materials and $LiNi_{0.8}Co_{0.15}Al_{0.05}O_2$ in charged state

Cathode materials	$LiNiO_2$	$LiCoO_2$	$LiMn_2O_4$	$LiNi_{0.8}Co_{0.15}Al_{0.05}O_2$
Initial charge capacity (mAh/g, 4.2 V)	210	160	130 (4.3 V)	205
Reversible capacity (mAh/g)	165	150	120	160
Temperature of highest heat flow in DSC main peak under the existence of solvent	200	250	300	310
Oxygen evolution temp. from charged cathode material (4.2 V)	200	230	290	300

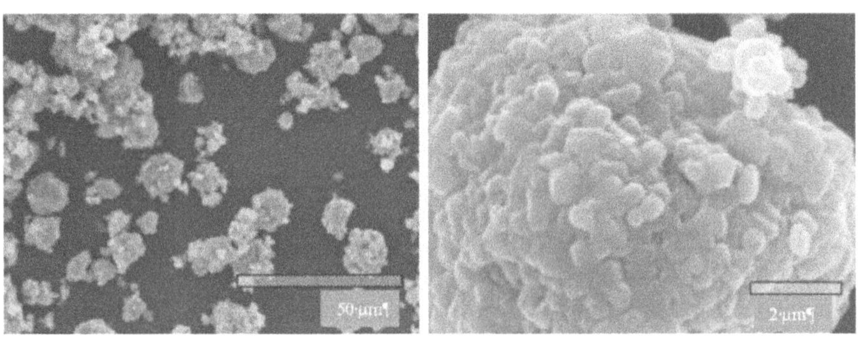

Fig. 2.23 SEM images of morphology controlled $LiNi_{0.8}Co_{0.15}Al_{0.05}O_2$

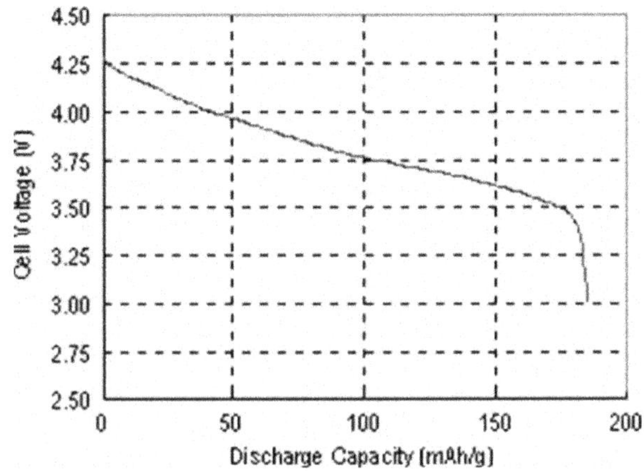

Fig. 2.24 Discharge curve of $LiNi_{0.8}Co_{0.15}Al_{0.05}O_2$.

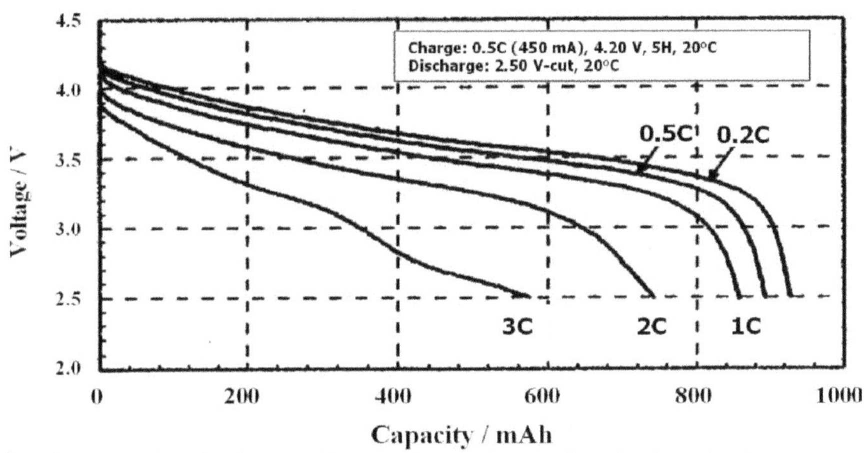

Fig. 2.25 Rate performance of laminate-type lithium ion battery composed of $LiNi_{0.8}Co_{0.15}Al_{0.05}O_2$ cathode and graphite anode

2.5.2.2 Synthesis by the Hydroxide Method

The procedure developed by Tanaka Chemical Co. is in the forefront of these methods. This company occupies almost a monopoly position as a supplier of cathode materials for the nickel–cadmium battery and the nickel–metal hydride battery. Metal hydroxides have been prepared via an amine complex. Spherical material is produced by adjusting the pH of the aqueous solution, the aging temperature, and the introduction rate of the reactant. This technology is applied to the cathode material of lithium-ion batteries and has been patented.[82] Since precipitates with homogeneous distribution of various elements can be obtained using the co-precipitation technique, this method can be considered to be an excellent method. The cobalt hydroxide prepared by the particle shape control technique was developed by Tanaka Chemical Co. Various kinds of particles with different morphorogies, such as the aggregated fine particles, the aggregated plate-like crystals, the aggregated large-size particles, and the isolated large-size particles, can be obtained by controlling the co-precipitation condition. The cathode material for the lithium-ion battery is synthesized by baking after mixing the lithium salt with the raw hydroxide. In this case, it also is important to maintain the particle shapes of raw materials by controlling the heating condition. Fundamentally these types of hydroxides were supplied by Tanaka Chemical Co. and were synthesized to cathode material.

Ohzuku[83] and Dahn in Canada have synthesized $LiNi_{0.5}Mn_{0.5}O_2$ and $LiNi_{1/3}Mn_{1/3}Co_{1/3}O_2$, using the nickel/manganese co-precipitate and the nickel/manganese/cobalt co-precipitate, which are precursors developed in this company. Such cathode materials attract much attention because of the large battery capacity. Here, the data obtained by Ohzuku are shown in Fig. 2.26.[83] Both of them have a capacity of about 200 mAh/g, and they would be promising materials in the future. However, they have a disadvantage in the high-rate performance; improvements in rate capability are being advanced at present. In addition, the nickel/manganese/cobalt co-precipitate with higher tap density, as 2.0–2.3 g/cm^3, is supplied.

2.5.2.3 Mixed Metal Carbonate Method

The hydroxide co-precipitation method is an excellent method as previously described. However, the process for hydroxide co-precipitate containing manganese ion would be difficult in the processing technology. Since hydroxide

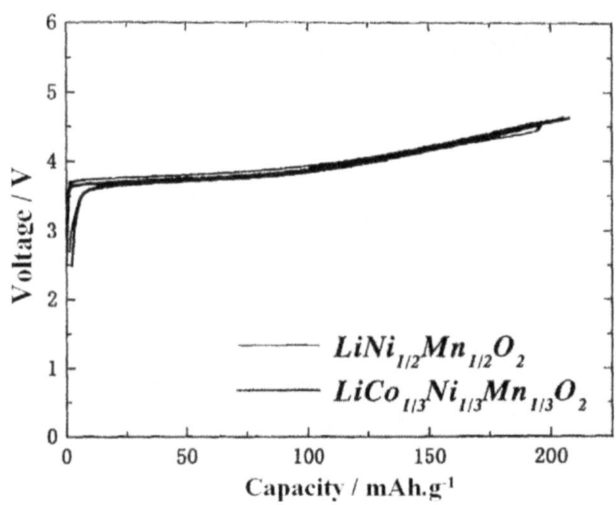

Fig. 2.26 Charge/discharge curves of $LiNi_{1/2}Mn_{1/2}O_2$ and $LiCo_{1/3}Ni_{1/3}Mn_{1/3}O_2$

co-precipitation is carried out in basic conditions, the manganese ion can stay as 2^+, 3^+, or 4^+ ion. As oxygen in air promotes the oxidation of manganese ion, therefore, it seems considerably difficult to synthesize the identical co-precipitate reproducibly. It can be assumed from their patent that at Tanaka Chemical Co. the manganese ion is co-precipitated from the Mn^{2+}-amine complex as $Mn(OH)_2$ under a reducing agent for overcoming this difficulty.

On the other hand, the so-called mixed metal carbonate method developed by Chuo Denki Kogyo Co.[84,85] is a kind of co-precipitation method, where metal salt is co-precipitated as a carbonate or bicarbonate under the presence of bicarbonate in basic solution. This method seems to be an excellent method without the shortcoming of the hydroxide co-precipitation method as mentioned above. As this method also is a kind of co-precipitation method, the control of particle shape is naturally possible and the spherical particles with high density also are obtained. The initial capacity of $LiNi_{0.56}Mn_xCo_{0.44-x}O_2$ synthesized by this method and the peak temperature measured by DSC in coexistence with the electrolyte are shown in Table 2.6. It is obvious that the DSC peak temperature related to the oxygen evolution shifts

Table 2.6 Electrochemical properties and thermal properties of $LiNi_{0.56}Mn_xCo_{0.44-x}O_2$ (x = 0.1, 0.2, and 0.3)

Compounds	Initial capacity (mAh/g)	Average voltage (V)	Peak temp. in DSC (°C)	Heat generation (J/g)
$LiNi_{0.56}Mn_{0.1}Co_{0.34}O_2$	173	3.78	298	728
$LiNi_{0.56}Mn_{0.2}Co_{0.24}O_2$	171	3.81	300	645
$LiNi_{0.56}Mn_{0.3}Co_{0.14}O_2$	167	3.84	316	577
$LiNiO_2$	183	3.84	225	1,398
$LiCoO_2$	156	3.97	251	616

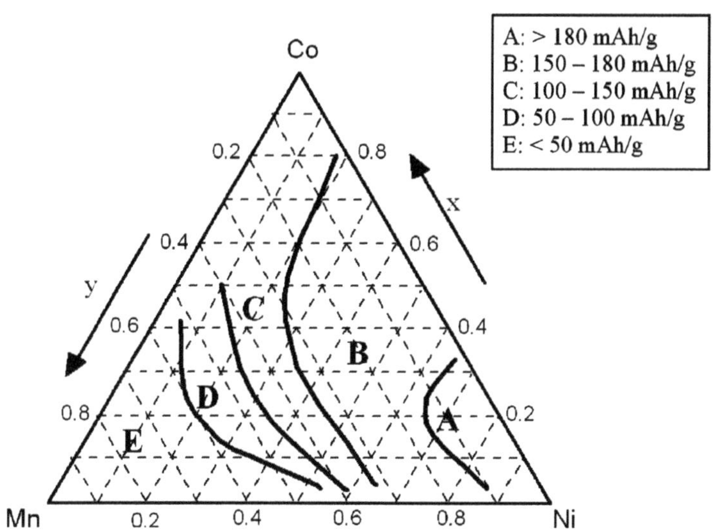

Fig. 2.27 Discharge capacity of $LiNi_{1-x-y}Co_xMn_yO_2$

to the higher temperature as the increase in manganese doping amount and the amount of heat generation decreases extremely. Namely, the safety of the battery is improved by manganese doping. Moreover, its initial capacity is higher than that of $LiCoO_2$. Its energy density (product of capacity and voltage) is higher still because of its higher discharge voltage.

The battery manufacturers using this compound as the cathode-active material prepare large-size batteries with 100 Wh class or 400 Wh class. Their cycle life, thermal behavior, high-rate performance, effect of storage at charged state on the electrochemical property, and so forth have been estimated. The battery shows a capacity retention of 83% after 1,000 cycles. Furthermore, it has been proved that the thermal stability is increased as the manganese content increases, that the manganese content should not exceed 0.35 because of the extreme expansion of the *c*-axis, and that the manganese or cobalt content in it does not have an impact on the charge/discharge curve. Figure 2.27 shows the ternary phase diagram by Terasaki et al., which indicates the relations between capacity and composition of $LiNi_{1-x-y}Co_xMn_yO_2$. They have reported that samples are prepared from mixed metal carbonate. Probably, the mixed metal carbonate precursor would make it possible to prepare $LiNi_{1-x-y}Co_xMn_yO_2$, the wide composition range.

References

1. M. Zou, M. Yoshio, S. Gopukumar, J. Yamaki, *Electrochem. Solid-State Lett.*, **7** (2004) A176
2. W.S. Yoon, Y. Paik, X.Q. Yang, M. Balasubramanian, J. McBreen, C.P. Grey, *Electrochem. Solid-State Lett.*, **5** (2002) A263
3. A.R. Armstrong, P.G. Bruce, *Nature*, **381** (1996) 499
4. G. Vitins, K. West, *J. Electrochem. Soc.*, **144** (1997) 2587
5. M. Tabuchi, K. Ado, H. Kobayashi, H. Kageyama, C. Masquelier, A. Kondo, R. Kanno, *J. Electrochem. Soc.*, **145** (1998) L145
6. Y.I. Jang, B. Huang, H. Wang, D.R. Sadoway, Y.M. Chiang, *J. Electrochem. Soc.*, **146** (1999) 3217
7. Y.S. Lee, M. Yoshio, *Electrochem. Solid-State Lett.*, **4** (2001) A166
8. Y.S. Lee, Y.K. Son, M. Yoshio, *Chem. Lett.*, **30** (2001) 882
9. J.M. Paulsen, C.L. Thomas, J.R. Dahn, *J. Electrochem. Soc.*, **147** (2000) 861
10. J.M. Paulsen, J.R. Dahn, *J. Electrochem. Soc.*, **147** (2000) 2478
11. J.M. Paulsen, D. Larcher, J.R. Dahn, *J. Electrochem. Soc.*, **147** (2000) 2862
12. T. Ohzuku, A. Ueda, M. Nagayama, *J. Electrochem. Soc.*, **140** (1993) 1862
13. W. Li, J.N. Reimers, J.R. Dahn, *Solid State Ionics*, **67** (1993) 123
14. H. Arai, S. Okada, H. Ohtsuka, M. Ichimura, J. Yamaki, *Solid State Ionics*, **80** (1995) 261
15. J.N. Reimers, J.R. Dahn, *J. Electrochem. Soc.*, **139** (1992) 2091
16. T. Ohzuku, A. Ueda, *J. Electrochem. Soc.*, **141** (1994) 2972
17. G.G. Amatucci, J.M. Tarascon, L.C. Klein, *J. Electrochem. Soc.*, **143** (1996) 1114
18. S. Levasseur, M. Menetrier, E. Suard, C. Delmas, *Solid State Ionics*, **128** (2000) 11
19. M. Yoshio, H. Noguchi, K. Yamato, J. Itoh, M. Okada, T. Mouri, *J. Power Sources*, **74** (1998) 46
20. A. Von der Ven, M.K. Aydinol, G. Ceder, *J. Electrochem. Soc.*, **145** (1998) 2149
21. J.M. Paulsen, J.R. Mueller-Neuhaus, J.R. Dahn, *J. Electrochem. Soc.*, **147** (2000) 508
22. D. Carlier, I. Saadoune, M. Menetrier, C. Delmas, *J. Electrochem. Soc.*, **149** (2002) A1310

23. X.Q. Yang, J. McBreen, W.S. Yoon, C.P. Gray, *Electrochemistry Comm.*, **4** (2002) 649
24. D. Li, T. Muta, L. Zhang, M. Yoshio, H. Noguchi, *J. Power Sources*, **132** (1998) 150
25. J.M. Paulsen, C.L. Thomas, J.R. Dahn, *J. Electrochemcal Soc.*, **146** (1999) 3560
26. J.M. Tarascon, D. Guyomard, *J. Electrochem. Soc.*, **138** (1991) 1378
27. R.J. Gummor, D.C. Likes, M.M. Thackeray, *Mat. Res. Bull.*, **28** (1993) 1249
28. G. Vitins, K. West, *J. Electrochem. Soc.*, **137** (1990) 769
29. J.M. Paulsen, J.R. Dhan, *J. Electrochem. Soc.*, **147** (2000) 2478.
30. M. Yoshio, H. Nakamura, Y. Xia, *Electrochim. Acta*, **45** (1999) 273
31. P. Dan, E. Mengeritsky, D. Aurbach, I. Weissman, E. Zinigrad, *J. Power Sources*, **68** (1997) 443
32. E. Levi, E. Zinigrad, H. Teller, M.D. Levi, D. Aurbach, *J. Electrochem. Soc.*, **145** (1998) 3440
33. Y.S. Lee, N. Kumada, M. Yoshio, *J. Power Sources*, **96** (2001) 376
34. N. Imachi, I. Nakane, K. Narukawa, *Sanyo Technical Review*, No.**34** (2004)79
35. S. Yoshimura, *Technical Report (Sanyo),* **2** (2004) 119
36. Y. Xia, M. Yoshio, *J. Power Sources*, **66** (1997) 129
37. T. Nohma, H. Kitao, Manuscript in *Battery Technology Committee* (Japan), 16–3 (2004)
38. S. Komaba, N. Kumagai, Y. Kataoka, *Electrochim. Acta*, **47** (2002) 1229
39. Y. Yagi, Y. Hideshima, M. Sugita, H. Noguchi, M. Yoshio, *Electrochemistry*, **68** (2000) 252
40. X. Yang, X. Sun, M. Balasubramanian, J. McBreen, Y. Xia, T. Sakai, M. Yoshio, Electrochem. *Solid State Lett.*, **4**, (2001) A117
41. Y. Xia, M. Yoshio, *J. Electrochem. Soc.*, **144** (1997) 4186
42. D.J. Jang, Y.J. Shin, S.M. Oh, *J. Electrochem. Soc.*, **143** (1996) 2204
43. H. Yamane, M. Saitoh, M. Sano, M. Fujita, M. Sakata, M. Takada, N. Nishibori, *J. Electrochem. Soc.*, **149** (2002) A1514
44. M. Zou, M. Yoshio, S. Gopukumar, J. Yamaki, *Mat. Res. Bull.*, **40** (2005) 708
45. K. Oikawa, T. Kamiyama, F. Izumi, B.C. Chakoumakos, H. Ikuta, M. Wakihara, J. Li, Y. Matsui, *Solid State Ionics*, **109** (1998) 35
46. B. Deng, H. Nakamura, Q. Zhang, M. Yoshio, Y. Xia, *Electrochim. Acta*, **49** (2004) 1823
47. H. Noguchi, H. Nakamura, M. Yoshio, H. Wang, *Chem. Lett.*, **33** (2004) 546
48. M. Yoshio, H. Noguchi, Y. Todorov and Y. Hideshima, *Denki Kagaku*, **66** (1998) 189
49. G.T.K. Fey, W. Li, J.R. Dahn, *J. Electrochem. Soc.*, **141** (1994) 227
50. G.T.K. Fey, C.S. Wu, *Pure Appl. Chem.*, **69** (1997) 2329
51. Y. Xia, M. Yoshio, *J. Electrochem. Soc.*, **143** (1996) 825
52. Y. Xia, Y. Zhou, M. Yoshio, *J. Electrochem. Soc.*, **144** (1997) 2593
53. M. Yoshio, J. Taira, H. Noguchi, K. Isono, *Electrochemistry*, **66** (1998) 335
54. Y.M. Todorov, Y. Hideshima, H. Noguchi, M. Yoshio, *J. Power Sources*, **77** (1999) 198
55. Y. Terada, K. Yasaka, F. Nishikawa, T. Konishi, M. Yoshio I. Nakai, *J. Solid State Chem.*, **156** (2001) 286
56. H. Kawai, M. Nagata, H. Tsukamoto, A.R. West, *J. Power Sources*, **81–82** (1999) 67
57. T. Ohzuku, K. Ariyoshi, S. Yamamoto, *J. Ceram. Soc. Jpn.*, **110** (2002) 501
58. S.-H. Park, Y.-K. Sun, *Electrochim. Acta*, **50** (2004) 434
59. S. Mukerjee, X.Q. Yang, X. Sun, S.J. Lee, J. McBreen, Y. Ein-Eli, *Electrochim. Acta*, **49** (2004) 3373
60. A.K. Padhi, K.S. Nanjundaswamy, J.B. Goodenough, *J. Electrochem. Soc.*, **144**, (1997) 1188
61. A.S. Andersson, J.O. Thomas, *J. Power Sources*, **97–98** (2001) 498
62. M. Takahashi, S. Tobishima, K. Takei, Y. Sakurai, *J. Power Sources*, **97–98** (2001) 508
63. J.R. Dahn, E.W. Fuller, M. Obrovac, U. von Sachen, *Solid State Ionics*, **69** (1994) 265
64. H. Arai, S. Okada, Y. Sakurai, J. Yamaki, *Solid State Ionics*, **109** (1998) 295
65. K.K. Lee, W.S. Yoon, K.B. Kim, K.Y. Lee, S.T. Hong, *J. Power Sources*, **97–98** (2001) 321
66. M.G.S.R. Thomas, W.I.F. David, J.B. Goodenough, *Mat. Res. Bull.*, **15** (1980) 783
67. R. Kanno, H. Kubo, Y. Kawamoto, T. kamiyama, F. Izumi, Y. Takeda, M. Takano, *J. Solid State Chem.*, **110** (1994) 216
68. D.D. MacNeil, Z. Lu, Z. Chen, J.R. Dahn, *J. Power Sources*, **108** (2002) 8
69. Y. Gao, M.V. Yakovleva, W.B. Ebner, *Electrochem. Solid State Lett.*, **1** (1998) 117

70. J. Cho, H.S. Jung, Y.C. Park, G.B. Kim, H.S. Lim, *J. Electrochem. Soc.*, **147** (2000) 15
71. T. Ohzuku, T. Yanagawa, M. Kouguchi, A. Ueda, *J. Power Sources*, **68** (1997) 131
72. J. Cho, B. Park, *J. Power Sources*, **92** (2001) 35
73. Japan Patent, H11-307094, 2001-128249
74. M. Zou, M. Yoshio, S. Gopukumar, J. Yamaki, *Chem. Mater.*, **15** (2003) 4699
75. M. Zou, M. Yoshio, S. Gopukumar, J. Yamaki, *Chem. Mater.*, **17** (2005) 1284
76. S. Tobishima, K. Takei, Y. Sakurai, J. Yamaki, *J. Power Sources*, **90** (2000) 188
77. K. Ikeda, *Lithium Ion Secondary Battery*, 2nd ed. M. Yoshio and A. Kozawa(editors), p.292, 2000, Nikkan Kogyou Shinbunsha, Tokyo
78. N. Yomasaki, Recent battery technology, *Semin.Electrochem.*, **1**, (1977) 77, Electrochemical Society of Japan
79. M. Nishi, *Topics on Lithium Ion Secondary Battery*, Shokabou, (1977)
80. Patent No. WO/06679, Kokai Tokkyo Kouhou H10-069910
81. Ph. Binsan, *J. Power Sources*, **81–82** (2003) 906
82. N. Takami et al., 11th IMLB, Ab. No. **371** (2002)
83. Japan patent, 2002–201028, H9-270256
84. T. Ohzuku, K. Ariyoshi, S. Yamamoto, Y. Makimura, *Chem. Lett.*, **30** (2001) 1270
85. K. Yamato, K. Kobayashi, S. Ota, K. Hayashi, K. Kitamura, T. Miyashita, 41th Battery Symposium, p.386 (2000)

Chapter 3
Carbon Anode Materials

Zempachi Ogumi and Hongyu Wang

Accompanying the impressive progress of human society, energy storage technologies become evermore urgent. Among the broad categories of energy sources, batteries or cells are the devices that successfully convert chemical energy into electrical energy. Lithium-based batteries stand out in the big family of batteries mainly because of their high-energy density, which comes from the fact that lithium is the most electropositive as well as the lightest metal. However, lithium dendrite growth after repeated charge-discharge cycles easily will lead to short-circuit of the cells and an explosion hazard. Substituting lithium metal for alloys with aluminum, silicon, zinc, and so forth could solve the dendrite growth problem.[1] Nevertheless, the lithium storage capacity of alloys drops down quickly after merely several charge-discharge cycles because the big volume change causes great stress in alloy crystal lattice, and thus gives rise to cracking and crumbling of the alloy particles. Alternatively, Sony Corporation succeeded in discovering the highly reversible, low-voltage anode, carbonaceous material and commercialized the $C/LiCoO_2$ rocking chair cells in the early 1990s.[2] Figure 3.1 schematically shows the charge-discharge process for reversible lithium storage in carbon. By the application of a lithiated carbon in place of a lithium metal electrode, any lithium metal plating process and the conditions for the growth of irregular dendritic lithium could be considerably eliminated, which shows promise for reducing the chances of shorting and overheating of the batteries. This kind of lithium-ion battery, which possessed a working voltage as high as 3.6 V and gravimetric energy densities between 120 and 150 Wh/kg, rapidly found applications in high-performance portable electronic devices. Thus the research on reversible lithium storage in carbonaceous materials became very popular in the battery community worldwide.

In fact, the ability of layer-structured carbon to insert various species was well known by the latter half of the 1800s. The ability of graphite to intercalate anions promoted exploration into the use of a graphite cathode for rechargeable batteries.[3] Juza and Wehle described carbon lithiation studies in the middle of last century.[4]

Z. Ogumi and H. Wang (✉)
Department of Energy and Hydrocarbon Chemistry,
Graduate School of Engineering, Kyoto University, Kyoto 606-8501, Japan
wanghongyu@hotmail.com

M. Yoshio et al. (eds.), *Lithium-Ion Batteries*,
DOI 10.1007/978-0-387-34445-4_3

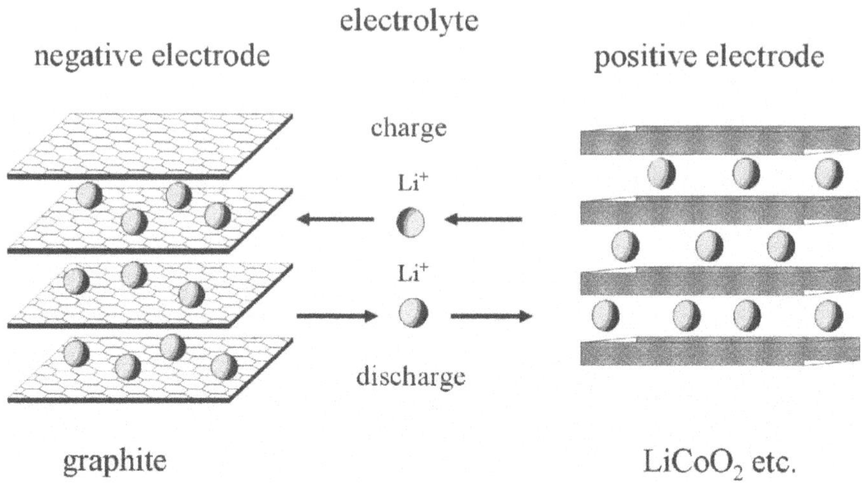

Fig. 3.1 Model for lithium-ion batteries

Guerard and Herold completed pioneering research of lithium intercalation into graphite and other less-ordered carbons such as cokes by a vapor transport method in 1975.[5] In 1976, Besenhard et al.[6,7] tried intercalating Li^+ into graphite electrochemically in the electrolytes of lithium salts dissolved in solvents of DME and DMSO, but obtained Li^+-solvent-graphite ternary intercalation compounds because of the strong affinities between Li^+ and the solvent molecules. In 1980, Basu utilized lithium-graphite intercalation compounds (GIC) in lithium-based secondary batteries for the first time when he used LiCl-KCl melting salts as the electrolytes for high-temperature-type batteries.[8] As for the ambient temperature-type batteries, Ikeda and Basu applied patents on Li-GIC as anode materials in 1981 and 1982, respectively.[9,10] In 1983, Yazami and Ph. Touzain succeeded in synthesizing Li-GIC electrochemically using a solid organic electrolyte.[11] The ease with which lithium can be intercalated and deintercalated from carbon has led to numerous studies on lithiated carbon anodes for battery application.

3.1 Staging Phenomenon of Li-GIC

Graphite intercalation compounds have one significant feature: the staging phenomenon, which is characterized by a periodic sequence of intercalant layers (say, lithium cations) between graphite layers. The nth-stage compound consists of intercalant layers arranged between every n graphite layers. The first-stage lithium graphite intercalation compound has the stoichiometry of LiC_6 with the specific capacity of 372 mAh/g (850 mAh/cm^3), a theoretical saturated value of lithium storage for graphite under normal pressure. The staging phenomenon can be easily monitored and controlled by the electrochemical reactions of carbons in Li^+-containing electrolytes, for instance, galvanostatic (constant current) charge-discharge[12,13] and slow cyclic voltammetry[14–17] (CV) have proven to be particularly useful electrochemical methods. In the case of galvanostatic charge-discharge of graphite electrodes in Li^+-containing electrolytes, the reversible plateaus on the potential curves indicate two-phase regions in the lithium-graphite phase diagram, whereas reversible current peaks demonstrates the two-phase regions in the CV. In conjunction with the electrochemical techniques, some physical methods have been applied to shed light on the stage occurrence and transitions during lithium intercalation into and deintercalation from graphite host. These methods include in situ[12,18] and ex situ[19] XRD, in situ laser raman spectra,[20] STM,[21] and so forth. Different schemes for stage transitions have been proposed. Main discrepancies lay in the ascriptions of higher stages. Actually, more or less turbostratic disorder (randomly stacking of graphene layers) is present in most synthetic graphite samples. Dahn's group found that turbostrtic disorder frustrates the formation of staged phases of Li-GIC, especially for higher stages.[22,23]

3.2 Solid Electrolyte Interface Film Formation

Another important feature for lithium graphite intercalation compounds in Li^+-containing electrolytes is the formation of solid electrolyte interface (SEI) film. During the first-cycle discharge of a lithium/carbon cell, a part of lithium atoms transferred to the carbon electrode electrochemically will react with the nonaqueous solvent, which contributes to the initial irreversible capacity. The reaction products form a Li^+-conducting and electronically insulating layer on the carbon surface. Peled[24] named this film as SEI. Once SEI formed, reversible Li^+ intercalation into carbon, through SEI film, may take place even if the carbon electrode potential is always lower than the electrolyte decomposition potential, whereas further electrolyte decomposition on the carbon electrode will be prevented.

Besenhard et al.[25] proposed the SEI formation mechanism of graphite via intercalation of solvated Li^+ as schematically illustrated in Fig. 3.2. Ogumi's group later verified this assumption in their systematic studies.[26–28] It is generally accepted that SEI film plays a very important role in the electrochemical performance of carbon anodes, especially for graphitic carbons, which are much more sensitive to the electrolyte composition. Ethylene carbonate (EC) and propylene carbonate (PC) are the most widely used high-permittivity solvents for Li^+-ion batteries. PC-based electrolytes demonstrate superior low-temperature performance to EC-based electrolytes mainly because of their different melting points (mp_{PC} −49°C, mp_{EC} 39°C). However, it is well known that EC-based electrolytes are suitable for graphite, whereas PC-based electrolytes are not compatible with graphite anodes, since PC decomposes drastically on graphite surface and exfoliates graphite particles.[29–32] Thus how to successfully apply graphite in PC-based electrolytes becomes a big challenge in the Li^+-ion battery community.

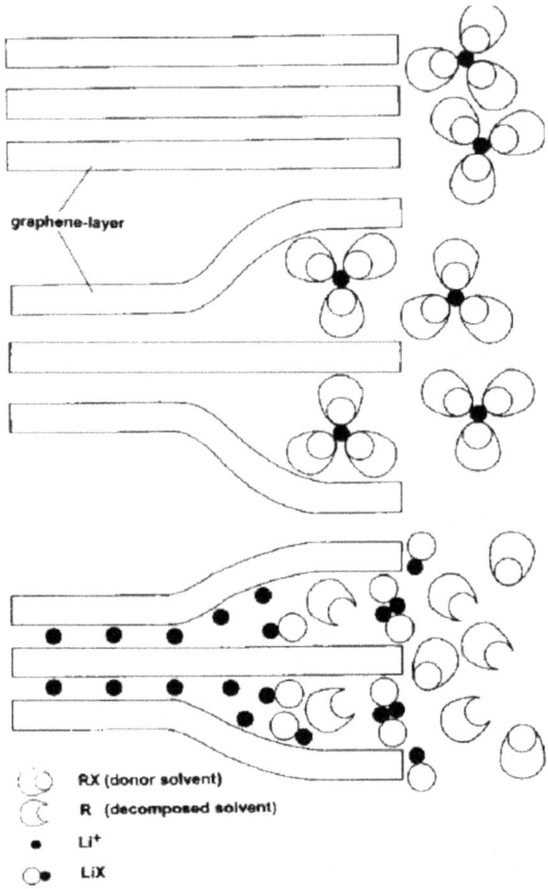

Fig. 3.2 SEI formation mechanism of graphite via intercalation of solvated Li^+. Reprinted from,[25] copyright (1995), with permission from Elsevier Ltd.

In fact, EC is very similar to PC in chemical structure, but only differs by one methyl group. Why the introduction of this single methyl into cyclic carbonate causes graphite exfoliation and electrolyte decomposition aroused the steric effect on solvent cointercalation. Chung et al.[33,34] added in the second methyl group into carbonate structure and got two geometric isomers, cis- and trans- butylene carbonates (BC). In the trans-BC based electrolytes the decomposition of the electrolyte and exfoliation were mild, but very drastic in cis-BC based one. This experiment in turn verified the significance of SEI formation mechanism via Li^+-solvent co-intercalation. Nakamura et al.[35] studied the performance of graphitic carbon in the non-aqueous electrolytes containing the binary solvent mixtures of PC/DEC, PC/DMC and PC/EMC. They found that once the PC concentration decreased to less than $[PC]:[Li^+] \leq 2$, the PC decomposition becomes considerably suppressed. Xu et al.[36] recently showed that the salt of lithium bis(oxalato) (LiBOB) can stabilize graphitic carbon in neat PC and support reversible Li^+ intercalation. Most recently, Jeong et al.[37] succeeded in reversibly intercalate Li^+ into the electrolyte of 2.72 M $LiN(SO_2C_2F_5)_2$ dissolved in 100% PC and suggested that ion-solvent interactions would be a vital factor for SEI formation in PC-based electrolytes. On the other hand, many studies have been devoted to designing effective SEI formation in PC-based electrolytes by some additives, such as vinylene cyclic carbonate,[38] crown ether,[39,40] fluoroethylene carbonat,[41] ethylene sulfite,[42] catechol carbonate,[43,44] vinyl acetate,[45] and so forth. In some cases, the additives will decompose at potentials higher than PC decomposition potential on graphite anode. That means, prior to PC decomposition and graphite exfoliation, the decomposition products of these additives like ethylene sulfite could form a robust SEI film covering graphite anode surface to protect it from direct contact with PC-based electrolytes. In other cases, some solvents like crown ether, DMSO, and tetraglme have much stronger affinity with Li^+ than PC.[46]

Combined with their extensive studies on the passivation films on lithium metal in nonaqueous electrolytes,[47–49] Aurbach's group carried out a series of work on the electrochemical behavior of graphite in Li^+-ion batteries.[50–54] The importance of SEI structure and chemical composition for graphite performance was highlighted. For instance, $ROCO_2Li$ and $(CH_2OCO_2Li)_2$ besides Li_2CO_3 were identified as the key components for the construction of effective SEI passivating graphite electrodes at ambient temperature. These conclusions were verified further by other groups using electron energy loss spectroscopy (EELS),[55] auger electron spectroscopy) (AES), temperature programmed decomposition mass spectroscopy (TPD-MASS),[56] and so forth.

As the performance of Li^+-ion batteries at elevated temperatures (50–70°C) is relevant to their safe utilizations, studies on SEI film properties at elevated temperatures have been pursued recently.[57–62] It was found that metastable species like $ROCO_2Li$ within the SEI layer will decompose into more stable products such as LiC_2O_3 and LiF at elevated temperatures. This leaves more pores in SEI layer and exposes the graphite-lithium surface to electrolytes, causing more irreversible capacities during continuous cycling. Actually, some companies have used the "aging" process to construct stable SEI film on electrodes for Li^+-ion batteries, which is based on the above phenomenon. After fabrication, the Li^+-ion batteries were charged and stored at elevated temperature for certain times before sale in markets. Thus the SEI film is composed mainly of stable species like LiC_2O_3 and LiF and prove to be robust and effective for passivating the carbon electrodes.

3.3 Correlations of Carbon's Structures and Electrochemical Performance

Carbonaceous materials have found wide applications in electrochemical technologies. This fact, in part, may be attributed to their advantages like good thermal and electrical conductivities, low density, adequate corrosion resistance, low thermal expansion, low elasticity, low cost, and high purity; however, to a larger extent, it is due to their flexibility and complexity in functional structures. There are different kinds of structure and each has a profound effect on the electrochemical performance of carbon.

The smallest dimensional structure is the bonding form between C atoms. Carbon atoms bind themselves by sp^3, sp^2, and sp hybrid orbitals. Carbonaceous materials are generally made up from repeating sp^2-bonding C–C atoms, which construct the planar hexagonal networks (honeycombs) of C atoms called graphene layers. There are some cases of doping foreign elements like P, B, N, Si into graphene layer to disturb the sp^2-bonding C–C order and change the lithiation behavior of carbon. For instance, Dahn's group has tried doping B[63] or N[64] into C–C networks and found that N embedded into C lattice shifts the reversible capacity to lower voltage, whereas the substitution of B for C could increase the working voltage. As a result, N decreases the reversible capacity but B enlarges the reversible capacity for Li^+ storage. Besides the bulk of carbon, at the edge of graphene layers or some defects, there are some sp^3 bondings of C–C (dangling carbon) or C–H, C–OH, C–COOH, and so forth; sp^3 bonding C seems more chemically active than sp^2 C–C in the bulk of carbon. Some groups have found the evidence for sewing the adjacent graphene layers' edges into close-edge surface structure for graphitic carbon, very similar to carbon nanotube.[65] The coupling process of adjacent graphene layers' edges is through the joint of the dangling carbon atoms. The close-edge graphitic carbon delivers very small initial irreversible capacity since the chemically stable surface suppresses the SEI formation.

Actually, each graphene layer can be considered to be a superconjugated macromolecule. Van der Waals force stacks these sheets into ordered structures called crystallites. There are two patterns for orderly stacking graphene layers into ideal graphite crystallites, one with the sequence of …ABAB… and the other with the sequence of …ABCABC…., as shown in Fig. 3.3. The former possess a little more thermally stable hexagonal symmetry, while the latter has a rhombohedral symmetry. Graphite usually comprises both crystal structures, but the rhombohedral content seems always less than 30%.[66–68] Several studies have suggested that a high rhombohedral content could suppress the exfoliation effects during the cointercalation of solvated Li^+ into graphene layers, especially for PC-based electrolytes. After a careful investigation on the effects of high-temperature annealing and post-burn-off treatment of synthetic graphite samples, Spahr et al.[69] recently concluded that the presence of rhombohedral stacking pattern in the bulk of graphite crystallite has almost no direct influence on the initial irreversible capacity. The irreversible capacity associated with PC-based electrolytes decomposition and graphite exfoliation appears to be graphite surface-dependent.

The Van der Waals force between graphene layers is so weak that the planes easily slide. Consequently, varying degrees of stacking faults are caused by random rotations and translations of graphene sheets within the carbon matrix. Most carbon atoms deviate from the regular position and the periodic stacking no longer maintains. This type of structure is called turbostratic structure. Its X-ray diffraction (XRD) pattern only shows broad (00l) and asymmetric (hk) diffraction peaks[70]

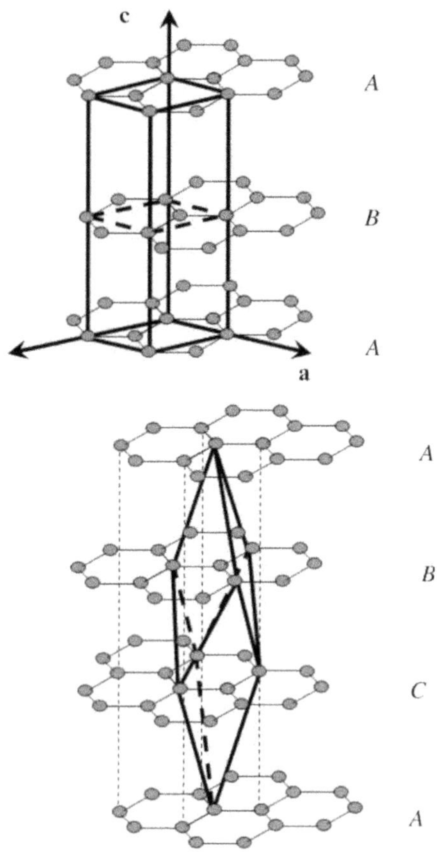

Fig. 3.3 Crystal structures of graphite, hexagonal (*upper*) and rombohedral (*below*)

because the three-dimensional regularity is poor but the roughly parallel and random stacking of graphene layers remain detectable. For turbostratic carbon, interlayer distances are somewhat diffused and on average larger than that of graphite. Disordered carbons fall into two types: soft carbon, whose turbostratic disorder is easily removed by heating to high temperature (near 3,000°C), and hard carbon, for which it is difficult to remove the turbostratic disorder at any temperatures. Franklin has proposed a model for the structures of soft and hard carbons as shown in Fig. 3.4.

There are many studies on the relationship between turbostratic structure and carbon lithiation behavior. In the comprehensive work of Dahn's group,[71,72] an automated structure-refined program has been developed for XRD data collected on disordered carbons. Based on this program to calculate some fundamental parameters, they tried to quantify the Li^+ storage capacity of carbon. This program has been used for more than 40 soft carbons and proved to be valuable. On the other

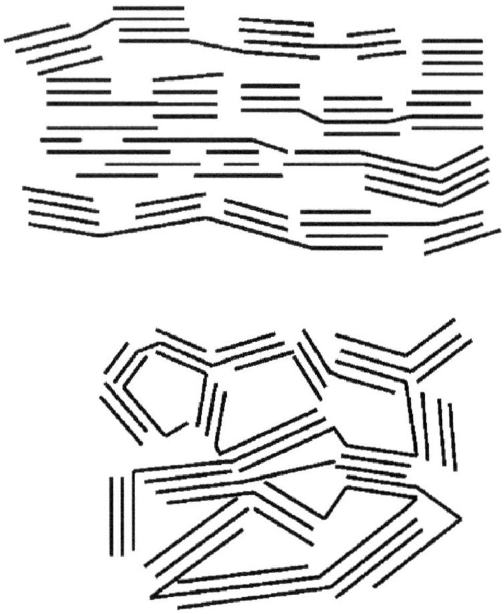

Fig. 3.4 Structure models for soft (*upper*) and hard (*below*) carbons

hand, Osaka Gas research group derived the following equation from the view point of mathematics to predict the attainable capacity of carbon[73]:

$$Q = 372/[1 + d_{002}/L_{c}]\left[1 + 2d_{c-c}(3^{1/2}L_{a} + d_{c-c})/\left(L_{a}^{2} + d_{c-c}^{2}\right)\right]. \qquad (3.1)$$

Toshiba research group discovered the relationship (as shown in Fig. 3.5) between carbon's discharge capacity and the average layer distance (d_{002}) from their experimental data for several series of soft carbons.[74] This figure actually reflects two trends for the choice of carbonaceous materials in Li^+-ion batteries. The d_{002} value of 0.344 nm in fact corresponds to turbostratic disordered carbon. We can take the minimum at 0.344 nm as the starting point and guide our attentions into two directions: one is toward d_{002} value smaller, which means obtaining more graphitic carbon; the other way is toward d_{002} value bigger, which implies the selection of more disordered carbon. Both ways can obtain carbon with high capacity.

Tatsumi et al. found the relationship between P_1 and the capacity in the potential range from 0 to 0.25 V vs. Li/Li^+ for soft carbons.[75] The P_1 stands for the volume ratio of ordered-stacking graphitic crystallites in carbon. In contrast, Dahn et al. used P parameter to indicate the probability of random stacking between the adjacent two graphene layers.[22,71,72,76] It appears that $P_1 = 1 - P$. Moreover, Tstsumi et al. correlated the capacity delivered in the potential range from 0.25 to 1.3 V vs. Li/Li^+ with $1 - P_1$, which is the fraction ratio of turbostratic structure. The relationships between P_1, $1 - P_1$ with the reversible capacities at different potential ranges is shown in Fig. 3.6. Fujimoto et al. calculated the projected probability function and

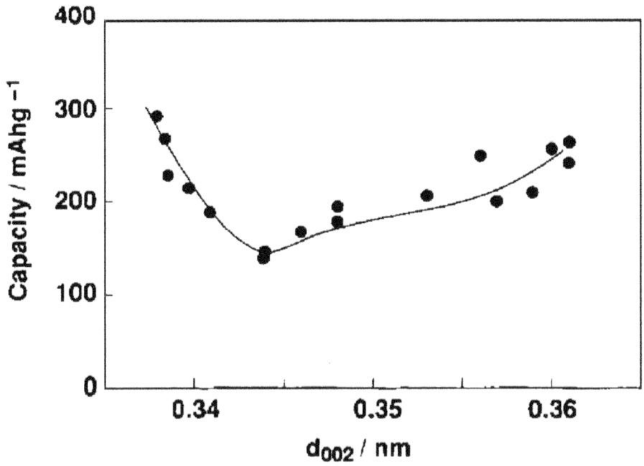

Fig. 3.5 Relationship between carbon's discharge capacity and d_{002}. Reprinted from,[74] copyright (1995) with permission from Elsevier Ltd

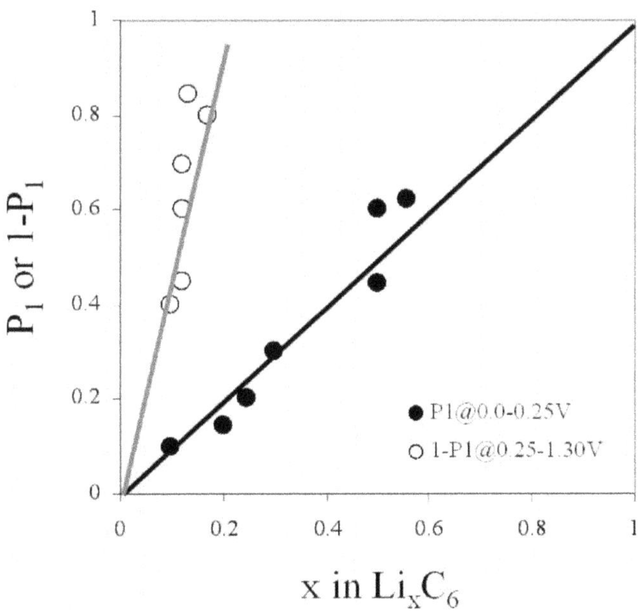

Fig. 3.6 The relationships between P_1 and 1-P_1 with the reversible capacities (x in LixC$_6$) at different potential ranges for soft carbon. Reproduced with permission from,[75] copyright (1995), The Electrochemical Society

measured (hk) XRD peaks of some soft carbons to simulate turbostratic structure in two adjacent parallel graphene layers.[77–80] The calculated pattern of twisted layers demonstrates a "moiré"-like structure with different stacking orders ranging from AB to AA. Li^+ can intercalate into AA stacking "islands," but cannot enter into the AB stacking portion. The maximum Li^+ storage capacity for turbostratic structure can be estimated as $Li_{0.2}C_6$ by extrapolation in Tatsumi's study.

The surface structure of carbon is also important for battery performance. There are two types of surfaces for carbon as shown: one is the basal plane of graphene layers, and the other is the edge plane located at the border of each graphene layers. The studies on Li^+ intercalation into highly oriented pyrolytic graphite (HOPG) showed that the basal plane is inert, whereas the edge plane is active for Li^+ insertion.[81,82] Li^+ intercalated into graphite bulk mainly through the edge planes, and only a small part of Li^+ can pass through the defects of basal planes into graphite bulk. In addition, there are different kinds of functional groups on the edge planes, which can affect the Li^+ intercalation greatly. Some papers also reported the direct relationship between specific surface area and the irreversible capacity.

Another structure for carbons is texture, the ways that the crystallites joined together. Texture is often characterized by the degree of orientation from random to systematic arrangement. If the crystallite size is small enough and there is no specific orientation, the carbon appears to be amorphous. Texture control cannot change the properties of individual crystallites but can alter the properties of the agglomerates of these crystallites like electricity and active surface area. The comparative study[83] of mesophase-pitch-based carbon fibers with different textures showed that the radical texture is more favorable for Li^+ intercalation than the concentric texture, but the radical texture is more easily broken into pieces by solvent cointercalated Li^+.

Aggregations of different types of textures can be considered to be a special structure of carbon in a bigger scale. Because electrochemical properties depend partially on the macroscopic structure of the electrode material, the state of aggregation also plays a vital role in electrochemical performance. In the case of carbon black,[84,85] one parameter to characterize the agglomerate of primary carbon black particles, namely, the adsorbed amount of dibutyl phthalate (DBP) was correlated with electrochemical properties. The neck positions (where the DBP is adsorbed) connecting two primary carbon black particles in the aggregates have parallel graphitic planes for Li+ intercalation, whereas a primary carbon black particle itself has the concentrically spherical lamellar structure that is unfavorable for Li^+ insertion and diffusion. Other examples for the aggregate structure are the composite electrode such as graphite + carbon (acetylene) black,[39,86] carbon fibers bound by carbonized epoxy resin,[87] graphite coated with carbon (core–shell structure),[88–90] or in some carbon fibers, two different textures coexist with one single fiber.[83]

3.4 Li^+ Diffusion in Carbon

Since the slow solid-state diffusion of Li^+ in the bulk of carbon may control the rate-determining step of the intercalation process and consequently affect the power density of Li^+-ion batteries, the chemical diffusion coefficient of Li^+ (D_{Li+}) becomes a very key kinetic parameter. Several electrochemical relaxation techniques such as potentiostatic and galvanostatic intermittent titration technique (PITT and GITT, respectively), current pulse relaxation method (CPR), and electrochemical impedance spectroscopy have been proposed for calculating D.[91–93] The D values might differ by several orders of magnitude, ranging from 10^{-6} to 10^{-13} cm^2/S in the published reports. The accuracy of D values for Li^+ diffusion in carbon depends on many factors,[94] such as carbon structure, potential, surface states of carbon electrodes (edge–plane surface effectively exposed to the electrolytes), SEI properties, the measurement techniques, and so forth, which makes it difficult to compare different D values evaded different research groups for various kinds of carbon. It was found that the activation energy for Li^+ diffusion process in soft carbon decreases with an increase of the graphitization degree given the same x value in Li_xC_6.[95] In contrast, micropores and defects in disordered carbon can retard Li^+ diffusion and enhance the activation energy.[96] The relationship between D values and Li^+ insertion content in carbon has become a focused interest of some research groups.[97–100] Investigations on ultrathin film or single particles of graphitic carbons found that the plots of D vs. the lithiation degree in graphite demonstrate three marked minima located at just the same potentials in which cyclic voltametries show the peaks that correspond to phase transitions of Li^+-graphite intercalation stages, for example, as shown in Fig. 3.7.

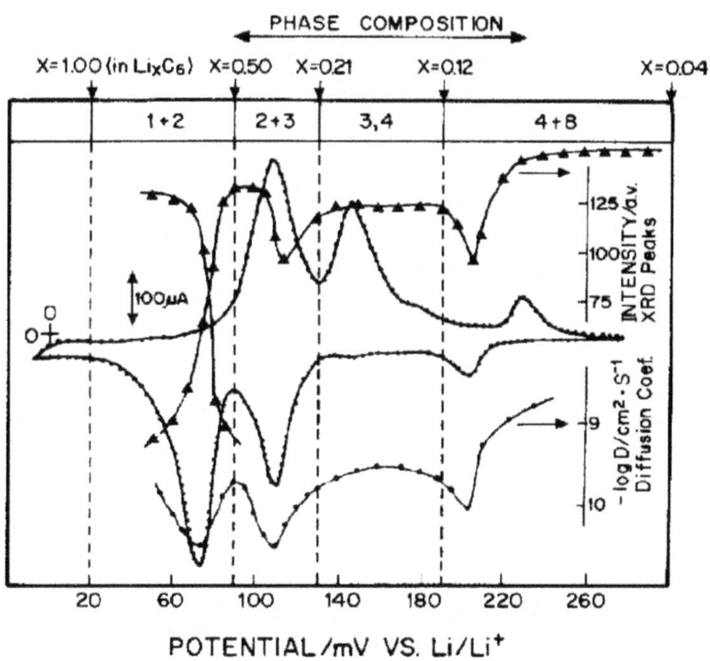

Fig. 3.7 Cyclic voltammogram and the diffusion coeffecient of Li^+ in graphite vs. potential. Reprinted from,[101] copyright (1997) with permission from Elsevier Ltd

3.5 Carbon with Extra-High Capacity

Recently, much enthusiasm and effort have been concentrated on the development of high-capacity carbonaceous materials that are synthesized at relative low temperatures (from 500 to 1,100°C) and deliver reversible capacities over 372 mAh/g. To rationalize the "extra" lithium storage capacity, a variety of models and explanations have been suggested.

Sawai et al. assumed that the high-capacity carbons offer high volume for lithium accommodation, thus only the gravimetric capacity is higher than that of graphite[102]; Yazami et al. proposed the formation of lithium multilayers on the graphene sheets[103]; Peled et al. believed that the extra capacity by mild oxidation of graphite is attributable to accommodation of lithium at edge planes between two adjacent crystallites and in the vicinity of defects and impurities[104]; Sato et al. suggested that lithium occupies the nearest neighbor sites in intercalated carbons[105]; Osaka research group proposed that extra lithium resides into nano-sized cavities[106–108]; Yata et al. discussed the possibility of the formation of LiC_2 in "polyacenic semiconductor" carbons with high interlayer distance (~0.400 nm) [109]; Matsumura et al. assumed that small particle-sized carbons can store considerable amounts of lithium on graphite edges and surfaces in addition to the lithium intercalated between graphene layers[110]; and Xiang et al. ascribed the plateau at about 1 V vs. Li/Li^+ to the lithium doped at the edges of graphene layers.[111]

Dahn and his collaborators have carried out systematic studies on carbon lithiation in detail.[112] They gave a comprehensive set of explanations for high lithium storage capacities in disordered carbons. In the case of both the soft and hard carbons heated below 800°C, large capacities as well as large hysteresis could be obtained.[113,114] The hysteresis capacity is proportional to the hydrogen content in carbon, so that lithium is somehow bound near the hydrogen.[115] Inaba et al. has investigated the thermal behavior of low-temperature-treated MCMB during charge-discharge cycling.[116] The large hysteresis in a voltage profile is accompanied by the large exotherms. This phenomenon was explained in terms of the activation energy barrier, which is in agreement with the above proposal of lithium–hydrogen interactions. As the heating temperature rises, hydrogen is depleted from carbon. The achieved capacities after removal of hydrogen depend mainly on the crystal structure of resulted carbon. Soft carbons heated at temperatures higher than 1,000°C have a lot of turbostratic disorders, so that the capacities are lower than 372 mAh/g. With the rise of heating temperature, graphitization proceeds and the fraction of ordered-stacking graphitic layers increases, so the capacity will increase and approach the value of 372 mAh/g. In contrast, hard carbons obtained near 1,000°C show little hysteresis and deliver capacity exceeding 372 mAh/g at a low potential of a few mV vs. Li/Li^+.[117,118] Dahn suggested lithium is "adsorbed" on both sides of the single-layer sheets that are arranged like a "house of cards."[119]

3.6 Thermal Safety of Lithiated Carbon

Most abusive conditions like short-circuit, crushing, nail-piercing, overcharge/discharge, and so forth will lead to heating of the batteries. Safety problems arise if the batteries exceed a critical temperature, above which thermal runaway occurs. So the thermal stability of the entire Li^+-ion battery and various combinations of battery components are necessary for understanding and improving battery safety. Accelerating rate calorimetry (ARC)[58,59,120–122] and differential scanning calorimetry (DSC)[57,123] are generally used to investigate the causes of thermal run away in Li_xC. The ARC is a sensitive adiabatic calorimeter which tracks the temperature change of reactive samples as they self-heat. In ARC studies, temperatures of the samples and calorimeter are increased to an initial temperature at first, and then the adiabatic self-heating rate of the sample is monitored. ARC studies demonstrated that self-heating of Li_xC_6 depends on at least four factors: initial lithiation degree, the electrolyte, surface area of carbon, and initial heating temperature of the sample. The conversion of the metastable SEI components to stable SEI produce a peak in the self-heating rate profile and the intensity of this peak becomes larger with the increase in carbon surface area. Richard and Dahn have proposed a mathematical model of reactions generating heat in Li_xC_6 samples in the electrolytes.[124] They used this model to calculate the self-heating rate profiles as well as DSC curves (DSC is performed at a fixed heating rate). DSC studies suggested that at first there is an exothermic reaction between 120 and 140°C, which comes from the transformation of the metastable SEI film components into LiF and LiC_2O_3. On further heating, Li_xC_6 reacts with the molten PVDF binder via dehydrofluorination near 200°C. The former reaction depends strongly on the surface area of the carbon and is a linear function of the initial irreversible capacity. The latter reaction depends on PVDF content, lithiation degree, as well as specific surface area of carbon.

To improve the safety of Li^+-ion batteries, some efforts also have been focused on the development of nonflammable electrolytes. One easy approach is the addition of fire retardants into the electrolytes.[125–130]

3.7 Structure Modification of Carbon

In the early stage of the research field of carbons as anode materials for Li^+-ion batteries, most work was devoted to the search for carbon suitable for Li^+-ion batteries' anode materials. This fact is partly due to the diversity of carbon materials and their multifaceted properties. With the development of the studies, deeper, more comprehensive insights have been gained in understanding the key factors that control carbon's electrochemical performance. The enriched knowledge of carbon lithiation helps battery researchers design and modify carbons to meet the practical needs in Li^+-ion batteries. In turn, during the course of "processing" carbons, the understandings of carbon are further tested and rectified. Thus, in the cycles of research, carbon's electrochemical performance is improved step by step to approach the application targets in Li^+-ion batteries.

One series of the carbon modification examples is ascribed to doping foreign elements into carbon bulk matrix (carbon alloy). For instance, Dahn and his co-workers have occluded silicon into carbon by pyrolyzed silicon-containing resins,[131–135] since each silicon can bond with four lithium atoms, the resultant silicon-containing carbon demonstrated very high reversible capacity. Effects of phosphorus, sulfur, and nitrogen doping also have been investigated widely.[136–141] The most prominent doping element for carbon matrix is boron because the boron atom can be considerably dissolved into graphene layers and decreases the structure strain. Boron doping into a carbon matrix can facilitate diffusion of carbon atoms, increase the crystallite size, and thus improve the crystallinity of carbon in the heat treatment. These factors can increase the reversible capacity of lithium storage. Several groups have pursued research on doping boron into different kinds of carbons.[142–148]

Another series in contrast is the modification of the surface. Takamura and his-collaborators have done extensive studies on this, since they realized that initial Li^+ insertion through carbon surface is the prerequisite for the overall intercalation process.[148–154] The simplest way to modify the carbon surface is focused on the effect of functional groups, such as removing some detrimental functional groups such as –OH from the carbon surface[155] or grafting some groups like –COOH,[104] –I,[156] which are facile for the build up of SEI film and Li^+ intercalation. One prominent example is the mild oxidation effect observed by Peled et al.[157] which give rise to nanochannels or micropores as well as a dense layer of oxides on graphite surface. Thus both the reversible capacity and coulombic efficiency can be increased. In addition, mild oxidation upon graphitized MCMB can get rid of a thin carbon layer on the bead's surface and increase the rate capacity.[158,159] The effect of mild oxidation on depressing graphite exfoliation in PC-based electrolytes also has been clarified.[69] Actually, more a popular trend for surface modification is to coat a layer onto carbon surface, which is called the "core-shell" structure. The "core" material is generally graphite because of its sensitivity to the electrolyte, while the "shell" materials may be diversified, including metals or alloys, metal oxides, conducting polymers, and some other kinds of carbons. The "shell" materials may be active or inactive for Li^+ storage. The former type seems more valuable in view of the reversible capacity. Carbon is one of the most promising coating materials partly because it is akin to the "core" material.[160,161]

Another interesting case is the hybrid effect of graphite-cokes and graphite-hard carbon mixtures.[162–164] The most attractive advantage of graphite anode material is its very low and flat working potential, which also can be considered to a disadvantage from another viewpoint since at the end of Li^+ deintercalation process from graphite (corresponding to a discharge process of a Li^+-ion battery), the working voltage of graphite anode will rise abruptly after the three plateaus, which cause the sharp drop down of the whole working voltage of Li^+-ion batteries. This can cause at least one inconvenience for battery users if they are unaware of how much capacity is left inside a Li^+-ion battery. The Sanyo research group has suggested that the rapid change in the charge-discharge potential of a graphite electrode at a lower voltage region in Li^+-ion batteries might cause side reactions such as the electrolyte decompositions, due to the nonuniform potential distribution over the graphite electrode. Thus, disordered carbons with a sloping working voltage are superior to graphitic carbons in this respect. A trade-off was to mix graphite with disordered carbon in proper ratios. Generation of inactive Li^+ species due to electrolyte decompositions during long charge-discharge cycles could be suppressed somehow in big-scale Li^+-ion batteries.

3.8 Practical Carbon Anode Materials

Among so many kinds of carbonaceous materials, the practical anode materials most widely used in commercialized lithium-ion batteries can be roughly classified into three categories: hard carbon, and natural and synthetic graphite. All these types of carbons have different advantages and disadvantages, which will be explained in the following concrete examples.

3.8.1 Natural Graphite

Now, natural graphite is becoming one of the most promising candidates as lithium-ion battery anode materials mainly because of its low cost, low and flat potential profile, high coulombic efficiency in proper electrolytes, and relatively high reversible capacity (330–350 mAh/g). On the other hand, it has two main shortcomings: its low rate capacity and incompatibility with PC-based electrolytes.

The low rate capacity of natural graphite actually comes from its high anisotropy. As schematically shown in Fig. 3.8, graphite flakes typically exhibit a platelike shape with a much shorter dimension along the *c*-axis and much wider dimensions vertical to the *c*-axis. After being spread onto the current substrate (Cu foil) and pressed, these particles have the alignment of the *c*-axis perpendicular to the substrate.[39] The consequence of the above orientation of graphite particles means that Li^+ intercalation occurs in the direction perpendicular to the direction of current flow. In addition, the resistivity of graphite varies with graphite crystal orientation. In the *c*-axis direction the resistivity is $10^{-2}\Omega$ cm, whereas in the *a*-axis direction the resistivity is 4×10^{-5} Ω cm.[165] Thus, the unfavorable orientation of graphite particles would lead to the sluggishness of Li^+ intercalation and inadequate electronic contact between graphite particles and copper substrate. These factors cause the low rate capacity of natural graphite, especially at low temperatures. To solve this problem, mechanical millings have been applied to tear the natural graphite flakes into small pieces.[166] In this way, the preferred orientation of crystallites within each natural graphite flake particle can be distorted to some extent by separated

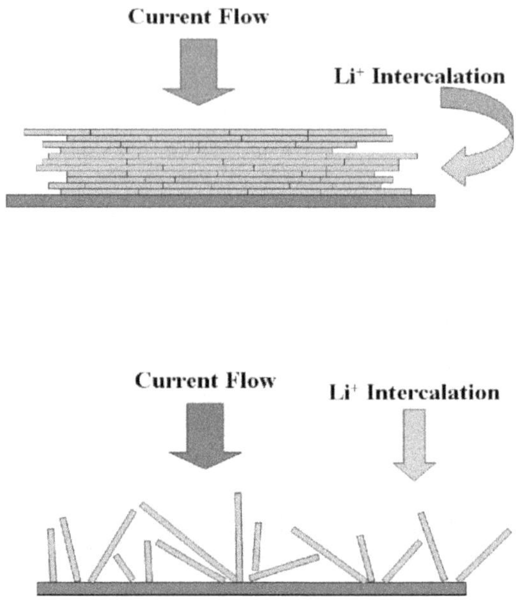

Fig. 3.8 Orientation of graphite flakes and fragments toward the current flow

graphite fragments as shown in Fig. 3.8. Some group also has tried to bind smaller graphite particles with various orientations into bigger graphite particles.[167]

The incompatibility of graphite with PC-based electrolytes has been widely studied. There generally are two ways to solve this problem. One is to modify the electrolytes by additives. The other is to modify the graphite by coating. Saga University and Mitsui Mining Co. have applied the thermal vapor decomposition (TVD) technique to coat carbon onto natural graphite particle surface completely and uniformly.[168–170] High-quality anode materials have been developed and commercialized. Before TVD carbon coating, small natural graphite fragments have been stacked into bigger shuttle-shaped particles as shown in Fig. 3.9, with the edge-plane surface mainly exposed. This kind of morphology is facile for Li^+ intercalation, but becomes very active toward PC decomposition. After TVD carbon coating, the electrolytes decomposition can be considerably suppressed. Moreover, the shuttle shape helps tospread graphite particles with various orientations onto copper foil, which can improve the rate capacity of natural graphite.

Although TVD carbon-coated, shuttle-shaped natural graphite has so many advantages, there is still some margin to improve the electrochemical performance. For instance, the density of TVD carbon coating (1.86 g/cm^3) is smaller than that of natural graphite (2.27 g/cm^3), which decreases the energy density of the anode material. Besides, the TVD process also adds some cost to the preparation of the anode material. So the requisites for carbon-coated natural graphite include not only how to suppress the electrolytes decomposition, but also how to decrease the

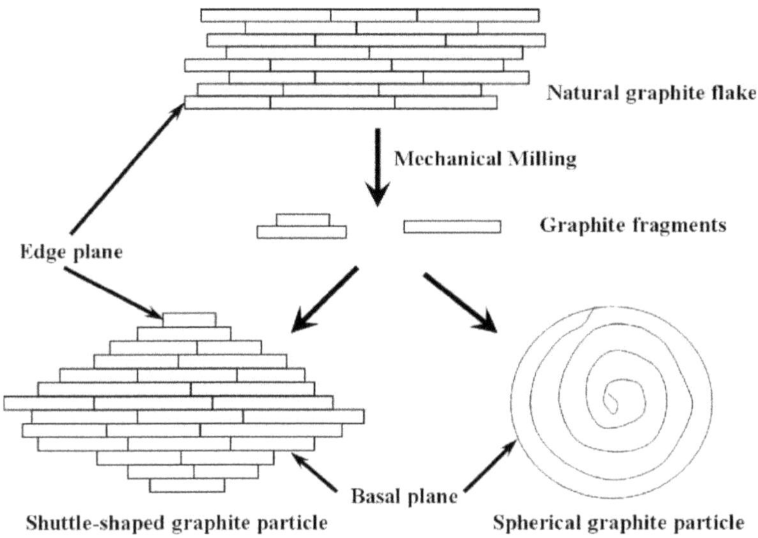

Fig. 3.9 Stacking patterns of graphite fragments. Reprinted from,[171] copyright (2003) with permission from Wiley-VCH Verlag GmbH & Co. KGaA

carbon-coating amount. To cater to the above needs, the "core" part (natural graphite) surface activity has to be lowered. The graphite fragments were rolled into spheres before TVD carbon –coating, as shown in Fig. 3.9. Since most of the edge-plane surfaces are hidden inside the spheres, the inert basal-plane surfaces take over the outer surface before TVD carbon coating. Thus only small amounts of carbon coating are enough to cover the residue-exposed edge-plane surfaces. The TVD carbon-coated spherical natural graphite shows even better electrochemical performance of TVD carbon-coated, shuttle-shaped natural graphite.[172] Furthermore, the high orientation of graphite layers can be highly distorted in each spherical natural graphite particles, the morphology of which is very good at enhancing the rate capacity.

3.8.2 *Synthetic Graphite*

Synthetic graphite has many properties that are the same as those of natural graphite. Besides, it has many unique merits such as high purity, variety of structures suitable for smooth Li^+ intercalation and diffusion, and so forth. Nevertheless, it is more expensive because of the high-temperature treatment (> 2,800°C) on soft carbon precursors, and its reversible capacity is a little smaller than natural graphite. Graphitized MCMB, mesophase-pitch-based carbon fiber (MCF), and vapor grown carbon fiber (VGCF) are the representatives of benchmark synthetic graphite anode materials for Li^+-ion batteries in the market today.

Precursor MCMB generally was separated from the heated pitch containing mesophase spheres before graphitization. There are different kinds of textures in MCMB such as Brooks–Taylor type, Honda type, Kovac–Lewis type, and Huttinger type.[173] In Japan, there are two main companies, Osaka Gas[174–176] and Kawasaki Steel Co. Ltd.,[177–179] who produced MCMB on a large scale. Their MCMB products belong to Brooks–Taylor type, the structure of which is schematically shown in Fig. 3.10. Graphitized MCMB has many advantages:

1. High packing density that guarantees high-energy density.
2. Small surface area that decreases the irreversible capacity corresponding to electrolyte decomposition.
3. Most of the surfaces of MCMB spheres are composed of edge-plane surfaces, thus Li^+ intercalation becomes easier and the rate capacity increases.
4. MCMB can be easily spread onto copper foil.

MCF supplied from Petoca Co. Ltd. was produced from the naphthalene mesophase pitch through the melting-blowing method.[180] Figure 3.11 Shows the SEM image of the typical cross section of graphitized MCF. MCF has a radical-like texture in the "skin" part, together with a lamellar structure in the core.[95,181–183] The radical-like texture in the skin part can make Li^+ intercalation get smooth, and thus enhances rate capacity. On the other hand, the lamellar structure in the core seems to keep the carbon fiber structure stable against volume changes of graphite crystallite during Li^+ intercalation and deintercalation; otherwise carbon fibers are easily deintegrated into small pieces after several cycles of charge and discharge and the cycle ability becomes very poor.

VGCF has been grown by the decomposition of hydrocarbons in the presence of transitional metal particles as a catalyst at temperatures between 1,000 and 1,300°C.[184] These types of carbon fibers are characterized by the concentric alignment of graphene layers along the fiber axis. Figure 3.12 shows the SEM image of VGCF (Grasker ™) produced by Nikkiso. Since the outer surface of the long fibers

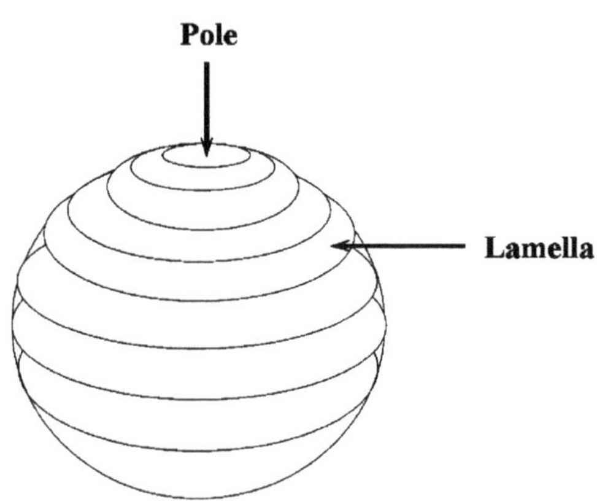

Fig. 3.10 Structure model of Brooks–Taylor type MCMB

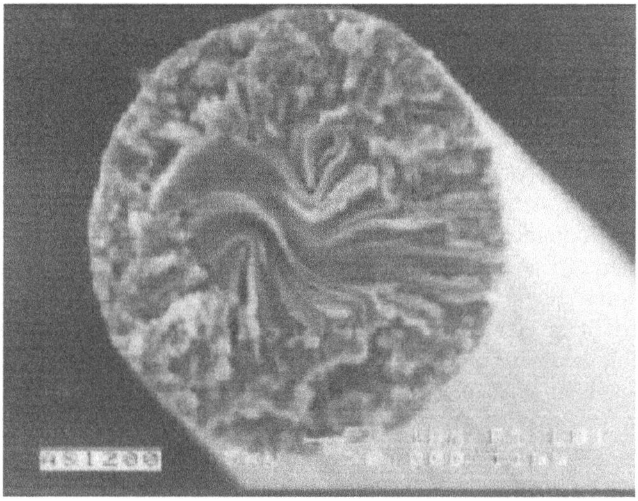

Fig. 3.11 SEM of MCF cross section. Reprinted from,[178] copyright (2000), with permission from Elsevier Ltd

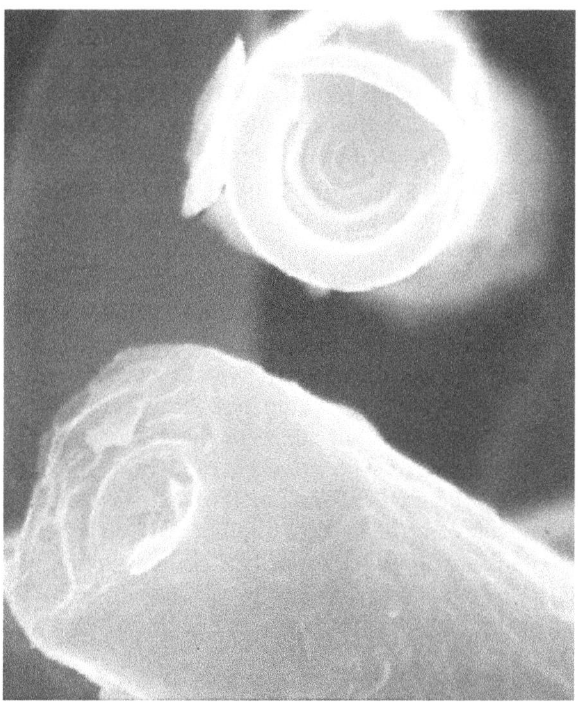

Fig. 3.12 SEM of VGCF cross section. Reprinted from,[186] copyright (1999), with permission from Elsevier Ltd

is generally composed of basal-plane surface, they were chopped to ca. 10 μm in length by the hybridizer to expose more edge-plane surface on the cross-section parts.[15,185–188] Two procedures were applied to prepare the graphitized short VGCFs: the graphitization after chopping and chopping after graphitization. The preparation method and the length and diameter of the fiber are key factors that influence the electrochemical performance.

MAG (Massive Artificial Graphite) from Hitachi Chemical Co. Ltd., Japan is one of the most commonly used synthetic graphite in Li^+-ion batteries since it occupies 70% market share of the cellular phones in Japan to date. MAG particles abound in pores (fine canals) to access the electrolyte solution, thus facilitate the Li^+ movement in the electrode matrix. If the electrolyte solution contains functional addives, stable and safe interface layer is formed elesewhere of inside the MAG particles. Details of MAG can be found in Chapter 18.

3.8.3 Hard Carbon

Hard carbons can deliver high capacity at low potential ranges (<0.2 V vs. Li/Li^+) if they can be adequately charged with lithium. Generally, it is very difficult to fulfill this point in normal time periods. The random alignment of small-dimensional graphene layers in hard carbons provides many voids to accommodate lithium. Nevertheless, the manner in which lithium diffusion occurs inside hard carbons looks like a maze, and thus makes lithium diffusion become very sluggish. So the rate capacity of hard carbon usually is very poor. The voids inside hard carbons also account for certain volume occupations. Although the gravimetric capacity of hard carbon seems very high compared with graphite, the volumetric capacity falls far behind this expectation, indeed. Of course, hard carbon still has some superiority over graphite. For example, its sloping voltage profile vs. capacity at the end of discharge may be very valuable for indicating the residue capacity.

Kureha Chemical has produced hard carbon (Carbotron®P) from phenol resin.[189,190] This hard carbon has the charge capacity and discharge capacity as high as 600 and 500 mAh/g, respectively. The average distance between adjacent graphene layers (d_{002}) in Carbotron®P is as high as 0.38 nm (compared with graphite, which is d_{002} = 0.3354 nm). After full lithiation, the d_{002} value of Carbotron®P only increases by 1% (as compared with graphite, ca. 10% volume expansion). This means that the crystal structure of Carbotron®P is very stable against lithium intercalation and deintercalation. So its cycle ability may be excellent. In addition, Carbotron®P is comparatively stable toward PC decomposition, although difficult to be exfoliated.

References

1. D. Fauteux, R. Koksbang, *J. Appl. Electrochem.*, **23** (1993) 1
2. T. Nagaura, T. Tozawa, *Progr. Batteries Solar Cells*, **9** (1990) 2

3. J. S. Dunning, W. H. Tiedemann, L. Hseuh, D. N. Bennion, *J. Electrochem. Soc.*, **118** (1971) 1886
4. R. Juza, V. Wehle, *Naturwissenschaften*, **52** (1965) 560
5. D. Guerard, A. Herold, *Carbon*, **13** (1975) 337
6. J. O. Besenhard, *Carbon* **14** (1976) 111
7. J. O. Besenhard, H. Molhward, and J. J. Nickl, *Carbon* **18** (1980) 399
8. S. Basu, U. S. Patent 4, **304** (1981) 825
9. H. Ikeda, S. Narukawa, and N. Nakajima, Jpn. Patent 1769661
10. S. Basu, U. S. Patent **4**, 423, 125
11. R. Yazami and Ph. Touzain, *J. Power Sources*, **9** (1983) 365
12. J. R. Dahn, *Phys. Rev., B*, **44** (1991) 9170
13. D. Billaud, F. X. Henry, and P. Willmann, *Mat. Res. Bull.*, **28** (1993) 477
14. A. Mabuchi, H. Fujimoto, K. Tokumitsu, and T. Kasuh, *J. Electrochem. Soc.*, **142** (1995) 3049
15. K. Tatsumi, K. Zaghib, Y. Sawada, H. Abe, and T. Ohsaki, *J. Electrochem. Soc.*, **142** (1995) 1090
16. P. Schoderbock and H. P. Boehm, *Syn. Met.*, **44** (1991) 239
17. D. Aurbach, M. Levi, and E. Levi, *J. Electroanal. Chem.*, **421** (1997) 79
18. D. Aurbach and E. Ein-Eli, *J. Electrochem. Soc.*, **142** (1995) 1746
19. T. Ohzuku, Y. Iwakoshi, and K. Sawai, *J. Electrochem. Soc.*, **140** (1993) 2490
20. M. Inaba, H. Yoshida, Z. Ogumi, T. Abe, Y. Mizutani, and M. Asano, *J. Electrochem. Soc.*, **142** (1995) 20
21. A. Funabiki, M. Inaba, T. Abe, and Z. Ogumi, *Electrochim. Acta*, **45** (1999) 865
22. T. Zheng and J. R. Dahn, *Syn. Met.*, **73** (1995) 1
23. T. Zheng and J. R. Dahn, *Phys. Rev. B*, **53** (1996) 3061
24. E. Peled, *J. Electrochem. Soc.*, **126** (1979) 2047
25. J. O. Besenhard, M. Winter, J. Yang, and W. Biberacher, *J. Power Sources*, **54** (1995) 228
26. M. Inaba, Z. Siroma, Z. Ogumi, T. Abe, Y. Mizutani, and M. Asano, Chem. Lett., (1995) 661
27. M. Inaba, Z. Siroma, Z. Ogumi, T. Abe, Y. Mizutani, and M. Asano, *Langmuir*, **12** (1996) 1535
28. M. Inaba, Z. Siroma, Y. Kawatate, A. Funabiki, and Z. Oguni, *J. Power Sources*, **68** (1997) 221
29. N. Dey and B. P. Sullivan, *J. Electrochem. Soc.*, **117** (1970) 222
30. G. Echinger, *J. Electroanal. Chem.*, **74** (1976) 183
31. M. Arakawa and J. Yamaki, *J. Electroanal. Chem.*, **219** (1987) 273
32. R. Fong, U. von Sacken, and J. R. Dahn, *J. Electrochem. Soc.*, **137** (1990) 2009
33. G. C. Chung, H. J. Kim, S. H. Jun, and M. H. Kim, *Electrochem. Commun.*, **1** (1999) 493
34. G. C. Chung, H. J. Kim, S. I. Yu, S. H. Jun, J. W. Choi, and M. H. Kim, *J. Electrochem. Soc.*, **147** (2000) 4391
35. H. Nakamura, H. Komatsu, and M. Yoshio, *J. Power Sources*, **62** (1996) 219
36. K. Xu, S. Zhang, B. A. Poese, and T. R. Jow, *Electrochem. Solid-State Lett.*, **5** (2002) A259
37. S. K. Jeong, M. Inaba, Y. Iriyama, T. Abe, and Z. Ogumi, *Electrochem. Solid-State Lett.*, **6** (2003) A13
38. S. K. Jeong, M. Inaba, R. Mogi, Y. Iriyama, T. Abe, and Z. Ogumi, *Langmuir* **17** (2001) 8281
39. Z. X. Shu, R. S. McMillan, and J. J. Murray, *J. Electrochem. Soc.*, **140** (1993) 922
40. Z. X. Shu, R. S. McMillan, and J. J. Murray, *J. Electrochem. Soc.*, **140** (1993) L101
41. R. McMillan, H. Slegr, Z. X. Shu, and W. Wang, *J. Power Sources*, **80–81** (1999) 20
42. G. H. Wrodnigg, J. O. Besenhard, and M. Winter, *J. Electrochem. Soc.*, **146** (1999) 470
43. C. Wang, H. Nakamura, H. Komatsu, M. Yoshio, and H. Yoshitake, *J. Power Sources*, **74** (1998) 142
44. C. Wang, H. Nakamura, H. Komatsu, M. Yoshio, and H. Yoshitake, *Denki Kagaku* (Electrochemistry), **66** (1998) 286
45. H. Yoshitake, K. Abe, T. Kitakura, J. B. Gong, Y. S. Lee, H. Nakamura, and M. Yoshio, *Chem. Lett.*, **32** (2003) 134
46. Z. Ogumi, T. Abe, M. Inaba, and S. K. Jeong, *Tanso*, **203** (2002) 136
47. D. Aurbach, A. Zaban, A. Schechter, Y. Ein-Eli, E. Zinigrad, and B. Markovsky, *J. Electrochem. Soc.*, **142** (1995) 2873
48. D. Aurbach and Y. Gofer, *J. Electrochem. Soc.*, **138** (1991) 3529
49. D. Aurbach, Y. Ein-Eli, and A. Zaban, *J. Electrochem. Soc.*, **141** (1994) L1

50. D. Aurbach, Y. Ein-Eli, B. Markovsky, A. Zaban, S. Luski, Y. Carmeli, and H. Yamin, *J. Electrochem. Soc.*, **142** (1995) 2882
51. D. Aurbach, Y. Ein-Eli, O. Chusid, Y. Carmeli, M. Baibai, and H. Yamin, *J. Electrochem. Soc.*, **141** (1994) 603
52. Y. Ein-Eli, B. Markovsky, D. Aurbach, Y. Carmeli, H. Yamin, and S. Luski, *Electrochim. Acta*, **39** (1994) 2559
53. D. Aurbach, M. D. Levi, E. Levi, and A. Schechter, *J. Phys. Chem. B*, **101** (1997) 2195
54. D. Aurbach, B. Markovsky, A. Schechter, Y. Ein-Eli, and H. Cohen, *J. Electrochem. Soc.*, **143** (1996) 3809
55. A. Naji, J. Ghanbaja, B. Humbert, P. Willmann, and D. Billaud, *J. Power Sources*, **63** (1996) 33
56. S. Mori, H. Asahina, H. Suzuki, A. Yonei, and K. Yokoto, *J. Power Sources*, **68** (1997) 59
57. A. N. Pasquier, F. Disma, T. Bowmer, A. S. Gozdz, G. Amatucci, and J. M. Tarascon, *J. Electrochem. Soc.*, **145** (1998) 472
58. M. N. Richard and J. R. Dahn, *J. Electrochem. Soc.*, **146** (1999) 2068
59. D. D. MacNeil, D. Larcher, and J. R. Dahn, *J. Electrochem. Soc.*, **146** (1999) 3596
60. T. Zheng, A. S. Gozdz, and G. C. Amatucci, *J. Electrochem. Soc.*, **146** (1999) 4014
61. A. M. Andersson, K. Edstrom, and J. O. Thomas, *J. Power Sources*, **81–82** (1999) 8
62. A. M. Andersson and K. Edstrom, *J. Electrochem. Soc.*, **148** (2001) A1100
63. B. M. Way and J. R. Dahn, *J. Electrochem. Soc.*, **141** (1994) 907
64. W. J. Weydanz, B. M. Way, T. van Buuren, and J. R. Dahn, *J. Electrochem. Soc.*, **141** (1994) 900
65. K. Moriguchi, S. Munetoh, M. Abe, M. Yonemura, K. Kamei, A. Shintani, Y. Maehara, A. Omaru, and M. Nagamine, *J. Appl. Phys.*, **88** (2000) 6369
66. H. Honbo and Y. Muranaka, The 39th Battery Symposium in Japan, 2D **19** (1998)
67. H. Huang, W. Liu, X. Huang, L. Chen, E. M. Kelder, and J. Schoonman, *Solid State Ionics*, **110** (1998) 173
68. H. Shi, J. Barker, M. Y. Saidi, and R. Koksbang, *J. Electrochem. Soc.*, **143** (1996) 3466
69. M. E. Spahr, H. Wilhelm, F. Joho, J. C. Panitz, J. Wambach, P. Novak, and N. Dupont-Pavlovsky, *J. Electrochem. Soc.*, **149** (2002) A960
70. B. E. Warren, *J. Chem. Phys.*, **2** (1934) 551
71. H. Shi, J. N. Reimers, and J. R. Dahn, *J. Appl. Cryst.*, **26** (1993) 827
72. J. R. Dahn, A. K. Sleigh, H. Shi, J. N. Reimers, Q. Zhong, and B. M. Way, *Electrochim. Acta*, **38** (1993) 1179
73. H. Fujimoto, A. Mabuchi, K. Tokumitsu, T. Kasuh, and N. Akuzawa, *Carbon*, 32 (1994) 193
74. A. Satoh, N. Takami, and T. Ohsaki, *Solid State Ionics*, **80** (1995) 291
75. K. Tatsumi, N. Iwashita, H. Sakaebe, H. Shioyama, S. Higuchi, A. Mabuchi, and H. Fujimoto, *J. Electrochem. Soc.*, **142** (1995) 716
76. T. Zheng, J. N. Reimers, and J. R. Dahn, *Phys. Rev.* **B**, **51** (1995) 734
77. T. Kasuh, A. Mabuchi, K. Tokumitsu, and H. Fujimoto, *J. Power Sources*, **68** (1997) 99
78. H. Fujimoto, *Tanso*, **168** (1995) 179
79. H. Fujimoto, A. Mabuchi, K. Tokumitsu, and T. Kasuh, *Carbon*, **34** (1996) 1115
80. H. Fujimoto, A. Mabuchi, K. Tokumitsu, and T. Kasuh, *Carbon*, **38** (2000) 871
81. T. Tran and K. Kinoshita, *J. Electroanal. Chem.*, **386** (1995) 221
82. A. Funabiki, M. Inaba, and Z. Ogumi, *J. Power Sources*, **68** (1997) 227
83. N. Imanishi, H. Kashiwagi, T. Ichikawa, T. Takeda, O. Yamamoto, and M. Inagaki, *J. Electrochem. Soc.*, **140** (1993) 315
84. K. Takei, N. Terada, K. Kumai, T. Iwahori, T. Uwai, and T. Miura, *J. Power Sources*, **55** (1995) 191
85. T. Uwai, T. Yamada, K. Takei, and T. Iwahori, *Tanso*, **165** (1994) 293
86. T. Takami, Y. Yamazaki, T. Kawamura, T. Takamura, M. Saito, and K. Sekine, The 39th Battery Symposium in Japan, 3D16 (1998)
87. K. Sumiya, M. Saito, K. Sekine, M. Takabatake, and T. Takamura, *Electrochemistry*, **66** (1998) 740
88. I. Kuribayashi, M. Yokoyama, and M. Yamashita, *J. Power Sources*, **54** (1995) 1
89. G. Okuno, K. Kobayakawa, Y. Sato, T. Kawai, and A. Yokoyama, *Electrochemistry*, **65** (1997) 226

90. W. Qiu, G. Zhang, S. Lu, and Q. Liu, *Solid State Ionics*, **121** (1999) 73
91. T. Uchida, T. Itoh, Y. Morikawa, H. Ikuta, and M. Wakihara, *Denki Kagaku* (Electrochemistry), **61** (1993) 1390
92. T. Uchida and M. Wakihara, *Denki Kagaku* (Electrochemistry), **65** (1997) 21
93. T. Uchida, M. Morikawa, H. Ikuta, M. Wakihara, and K. Suzuki, *J. Electrochem. Soc.*, **143** (1996) 2606
94. R. Takagi, M. Yashiro, K. Sekine, and T. Takamura, *Denki Kagaku* (Electrochemistry), **65** (1997) 420
95. N. Takami, A. Satoh, M. Hara, and T. Ohsaki, *J. Electrochem. Soc.*, **142** (1995) 371
96. A. K. Sleigh and U. von Sacken, *Solid State Ionics*, **57** (1992) 99
97. A. Funabiki, M. Inaba, Z. Ogumi, S. Yuasa, J. Otsuji, and A. Tasaka, *J. Electrochem. Soc.*, **145** (1998) 172
98. M. Nishizawa, R. Hashitani, T. Itoh, T. Matsue, and I. Uchida, *Electrochem. Solid State Lett.*, **1** (1998) 10
99. M. D. Levi and D. Aurbach, *J. Phys. Chem*, **101** (1997) 4641
100. M. D. Levi, E. A. Levi, and D. Aurbach, *J. Electroanal. Chem*, **421** (1997) 89
101. D. Aurbach, A. Zaban, Y. Ein-Eli, I. Weissman, O. Chusid, B. Markovsky, M. Levi, E. Levi, A. Schechter, and E. Granot, *J.Power Sources*, **68, 91** (1997)
102. K. Sawai, Y. Iwakoshi, and T. Ohzuku, *Solid State Ionics*, **69** (1994) 273
103. R. Yazami and M. Deschamps, *J. Power Sources*, **54** (1995) 411
104. E. Peled, C. Menachem, D. Bar-Tow, and A. Melman, *J. Electrochem Soc.*, **143** (1996) L4
105. K. Sato, M. Noguchi, A. Demachi, N. Oki, and M. Endo, *Science*, **270** (1994) 590
106. A. Mabuchi, *Tanso*, **165** (1994) 298
107. K. Tokumitsu, H. Fujimoto, A. Mabuchi, and T. Kasuh, *Carbon*, **37** (1999) 1599
108. K. Tokumitsu, H. Fujimoto, A. Mabuchi, and T. Kasuh, *J. Power Sources*, **90** (2000) 206
109. S. Yata, K. Sakurai, T. Ohsaki, Y. Inoue, and K. Yamaguchi, *Synth. Met.*, **33** (1995) 177
110. Y. Matsumura, S. Wang, T. Kasuh, and T. Maeda, *Synth. Met.*, **71** (1995) 1755
111. H. Xiang, S. Fang, and Y. Jiang, *J. Electrochem. Soc.*, **144** (1997) L187
112. J. R. Dahn, T. Zheng, Y. Liu, and J. S. Xue, *Science*, **270** (1995) 590
113. T. Zheng, W. R. McKinnon, and J. R. Dahn, *J. Electrochem. Soc.*, **143** (1996) 2137
114. T. Zheng and J. R. Dahn, *J. Power Sources*, **68** (1997) 201
115. T. Zheng, Y. Liu, E. W. Fuller, S. Tseng, U. von Sacken, and J. R. Dahn, *J. Electrochem. Soc.*, **142** (1995) 2581
116. M. Inaba, M. Fujikawa, T. Abe, and Z. Ogumi, *J. Electrochem. Soc.*, **147** (2000) 4008
117. W. Xing, J. S. Xue, and J. R. Dahn, *J. Electrochem. Soc.*, **143** (1996) 3046
118. J. S. Xue and J. R. Dahn, *J. Electrochem. Soc.*, **142** (1995) 3668
119. W. Xing, J. S. Xue, T. Zheng, A. Gibaud, and J. R. Dahn, *J. Electrochem. Soc.*, **143** (1996) 3482
120. U. von Sacken, E. Nodwell, A. Sundher, and J. R. Dahn, *Solid State Ionics*, **69** (1995) 284
121. U. von Sacken, E. Nodwell, A. Sundher, and J. R. Dahn, *J. Power Sources*, **54** (1995) 240
122. H. Maleki, G. Deng, A. Anani, and J. Howard, *J. Electrochem. Soc.*, **146** (1999) 3224
123. S. Zhang, D. Foucher, and J. R. Rea, *J. Power Sources*, **70** (1998) 16
124. M. N. Richard and J. R. Dahn, *J. Electrochem. Soc.*, **146** (1999) 2078
125. C. W. Lee R. Venkatachalapathy, and J. Prakash, *Electrochem. Solid State Lett.*, **3** (2000) 63
126. X. M. Wang, E. Yasukawa, and S. Kasuya, *J. Electrochem. Soc.*, **148** (2001) A1058
127. X. M. Wang, E. Yasukawa, and S. Kasuya, *J. Electrochem. Soc.*, **148** (2001) A1066
128. K. Xu, M. Ding, S. Zhang, J. Allen, and T. R. Jow, *J. Electrochem. Soc.*, **149** (2002) A622
129. K. Xu, M. Ding, S. Zhang, J. Allen, and T. R. Jow, *J. Electrochem. Soc.*, **150** (2003) A161
130. K. Xu, S. Zhang, J. Allen, and T. R. Jow, *J. Electrochem. Soc.*, **150** (2003) A170
131. A. M. Wilson, J. N. Reimers, E. Fuller, and J. R. Dahn, *Solid State Ionics*, **74** (1994) 249
132. J. S. Xue, K. Myrtle, and J. R. Dahn, *J. Electrochem. Soc.*, **142** (1995) 2927
133. W. Xing, A. M. Wilson, G. Zank, and J. R. Dahn, *Solid State Ionics*, **93** (1997) 239
134. A. M. Wilson, W. Xing, G. Zank, B. Yates, and J. R. Dahn, *Solid State Ionics*, **100** (1997) 259
135. D. Larcher, C. Mudalige, A. E. George, V. Porter, M. Gharghouri, and J. R. Dahn, *Solid State Ionics*, **122** (1999) 71

136. T. D. Tran, J. H. Feikert, X. Song, and K. Kinoshita, *J. Electrochem. Soc.*, **142** (1995) 3297
137. S. Ito, T. Murada, M. Hasegawa, Y. Bito, and Y. Toyoguchi, *J. Power Sources*, **68** (1997) 245
138. Y. Wu, S. Fang, and Y. Jiang, *J. Power Sources*, **75** (1998) 205
139. Y. Wu, S. Fang, and Y. Jiang, *J. Mater. Chem.*, **8** (1998) 2223
140. Y. Wu, S. Fang, and Y. Jiang, *Solid State Ionics*, **120** (1999) 117
141. Y. Wu, S. Fang, Y. Jiang, and R. Holtz, *J. Power Sources*, **108** (2002) 245
142. M. Koh and T. Nakajima, *Electrochim. Acta*, **44** (1999) 1713
143. M. Endo, C. Kim, T. Karaki, Y. Nishimura, M. J. Matthews, S. D. M. Brown, and M. S. Dresselhaus, *Carbon*, **37** (1999) 561
144. C. Kim, T. Fujimoto, T. Hayashi, M. Endo, and M. S. Dresselhaus, *J. Electrochem. Soc.*, **147** (2000) 1265
145. I. Mukhopadhyay, N. Hoshino, S. Kawasaki, F. Okino, W. K. Hsu, and H. Touhara, *J. Electrochem. Soc.*, **149** (2002) A39
146. T. Hamada, H. Shoji, T. Kohno, and T. Sugiura, *J. Electrochem. Soc.*, **149** (2002) A834
147. T. Hamada, K. Suzuki, T. Kohno, and T. Sugiura, *Carbon*, **40** (2002) 2317
148. H. Fujimoto, A. Mabuchi, C. Natarajan, and T. Kasuh, *Carbon*, 40 (2002) 567
149. T. Takamura, *Bull. Chem. Soc. Jpn.*, **75** (2002) 21
150. K. Sumiya, J. Suzuki, R. Takasu, K. Sekine, and T. Takamura, *J. Electroanal. Chem.*, **462** (1999) 150
151. T. Takamura, J. Suzuki, C. Yamada, K. Sumiya, and K. Sekine, *Surf. Eng.*, **15** (1999) 225
152. T. Takamura, H. Awano, R. Takasu, K. Sumiya, and K. Sekine, *J. Electroanal. Chem.*, **455** (1998) 223
153. R. Takasu, K. Sekine, and T. Takamura, *J. Power Sources*, **81–82** (1999) 224
154. T. Takamura, M. Saito, A. Shimokawa, C. Nakahara, K. Sekine, S. Maeno, and N. Kobayashi, *J. Power Sources*, **90** (2000) 45
155. T. Ura, M. Kikuchi, Y. Ikezawa, T. Takamura, 1B11, P. 25, The 35th Battery Symposium in Japan
156. H. Wang and M. Yoshio, *J. Power Sources*, **101** (2001) 35
157. C. Menachem, S. Peled, L. Burstein, and Y. Rosenberg, *J. Power Sources*, **68** (1997) 277
158. X. Cao, J. Kim, and S. Oh, *Carbon*, **40** (2002) 2270
159. M. Hara, A. Satoh, N. Takami, and T. Ohsaki, *Tanso*, **165** (1994) 261
160. H. Y. Lee, J. K. Baek, S. W. Jang, S. M. Lee, S. T. Hong, and K. Y. Lee, *J. Power Sources*, **101** (2001) 206
161. C. Natarajan, H. Fujimoto, K. Tokumitsu, A. Mabuchi, and T. Kasuh, *Carbon*, **39** (2001) 1409
162. K. Yanagida, A. Yanai, Y. Kida, A. Funahashi, T. Nohma, and I. Yonezu, *J. Electrochem. Soc.*, **149** (2002) A804
163. Y. Kida, K. Yanagida, A. Funahashi, T. Nohma, and I. Yonezu, *J. Power Sources*, **94** (2001) 74
164. Y. Kida, K. Yanagida, A. Funahashi, T. Nohma, and I. Yonezu, *Electrochemistry*, **70** (2002) 590
165. K. Kinoshita, *Carbo: Electrochemical and Physico-Chemical Properties*, Wiley, New York (1998), p. 70
166. H. Wang, T. Ikeda, K. Fukuda, and M. Yoshio, *J. Power Sources*, **83** (1999) 141
167. Y. Ishii, A. Fujita, T. Nishida, and K. Yamada, *Hitachikasei Technical Report*, **36** (2001) 27
168. M. Yoshio, H. Wang, K. Fukuda, Y. Hara, and Y. Adachi, *J. Electrochem. Soc.*, **147** (2000) 1245
169. H. Wang and M. Yoshio, *J. Power Sources*, **93** (2001) 35
170. H. Wang, M. Yoshio, T. Abe, and Z. Ogumi, *J. Electrochem. Soc.*, **149** (2002) A499
171. M. Yoshio, H. Wang, and K. Fukuda, *Angew. Chem. Int. Ed.*, **42**, 4203 (2003)
172. M. Yoshio, H. Wang, K. Fukuda, T. Umeno, T. Abe, and Z. Ogumi, *J. Mater Chem.,* **14** (2004) 1754
173. Y. C. Chang, H. J. Sohn, C. H. Ku, Y. G. Wang, Y. Korai, and I. Mochida, *Carbon*, **37** (1999) 1285
174. H. Fujimoto, A. Mabuchi, K. Tokumitsu, and T. Kasuh, *J. Power Sources*, **54** (1995) 440
175. A. Mabuchi, K. Tokumitsu, H. Fujimoto, and T. Kasuh, *J. Electrochem. Soc.*, **142** (1995) 1041
176. H. Fujimoto, *Tanso*, **200** (2001) 243
177. H. Hatano, K. Nakayama, and Y. Tajima, *Kawasaki Steel Giho*, **34** (2002) 140
178. H. Hatano, N. Fukuda, and T. Aburaya, *Kawasaki Steel Giho*, **29** (1997) 233

179. K. Tatsumi, T. Akai, T. Imamura, K. Zaghib, N. Iwashita, S. Higuchi, and Y. Sawada, *J. Electrochem. Soc.*, **143** (1996) 1923
180. F. Watanabe, Y. Korai, I. Mochida, and Y. Nishimura, *Carbon*, **38** (2000) 741
181. N. Takami, A. Satoh, H. Hara, and T. Ohsaki, *J. Electrochem. Soc.*, **142** (1995) 2564
182. T. Ohsaki, M. Kanda, Y. Aoki, H. Shiroki, and S. Suzuki, *J. Power Sources*, **68** (1997) 102
183. K. Yamaguchi, J. Suzuki, M. Saito, K. Sekine, and T. Takamura, *J. Power Sources*, **97–98** (2001) 159
184. M. Endo, Y. A. Kim, T. Hayashi, K. Nishimura, T. Matusita, K. Miyashita, and M. S. Dresselhaus, *Carbon* **39** (2001) 1287
185. K. Tatsumi, K. Zaghib, H. Abe, S. Higuchi, T. Ohsaki, and Y. Sawada, *J. Power Sources*, **54** (1995) 425
186. K. Zaghib, K. Tatsumi, H. Abe, T. Ohsaki, Y. Sawada, and S. Higuchi, *J. Power Sources*, **54** (1995) 435
187. K. Zaghib, K. Tatsumi, H. Abe, T. Ohsaki, Y. Sawada, and S. Higuchi, *J. Electrochem. Soc.*, **145** (1998) 210
188. H. Abe, T. Murai, and K. Zaghib, *J. Power Sources*, **77** (1999) 110
189. N. Sonobe, M. Ishikawa, T. Iwasaki, The 35th Battery Symposium in Japan, 2B09 (1994)
190. M. Ishikawa, N. Sonobe, H. Chuma, T. Iwasaki, The 35th Battery Symposium in Japan, 2B10 (1994)

Chapter 4
Role-Assigned Electrolytes: Additives

Makoto Ue

4.1 Introduction

The role of liquid electrolytes in lithium-ion cells is to act as an ionic conductor to transport lithium ions back and forth between positive and negative electrodes as the cells are charged and discharged. Since the electrodes in lithium-ion cells are the porous composite electrodes, consisting of an active material [carbon in the negative electrode and lithium transition metal (Co, Ni, Mn) oxide in the positive electrode, respectively], a conductive material (carbon black), and a polymer binder as depicted in Fig. 4.1, the liquid electrolyte must seep into the porous electrodes and transfer lithium ions smoothly at the interfaces between the liquid and solid phases. Most lithium-ion cells available in the market utilize nonaqueous electrolyte solutions, where lithium salts are dissolved in aprotic organic solvents. The gelled electrolytes used in lithium-ion polymer cells also are regarded as a liquid electrolyte immobilized with a high molecular weight polymer. Therefore, the same functions are required for the liquid and gelled electrolytes to greater or lesser degrees.

There are many books reviewing the liquid electrolytes for lithium or lithium-ion cells, where the various properties of aprotic solvents, lithium salts, and their mixtures are described.[1–11] This author also reviewed them mainly from the aspect of solution chemistry.[12,13] However, recent researches in the liquid electrolytes are mainly focused on the electrolyte additives, which add extra functions to the liquid electrolytes in addition to a fundamental function as an ionic conductor. This author will attempt to outline the electrolyte additives for the first time by classifying them into several categories,[14,15] although the fragmented information on the additives began to appear in the recent reviews.[8,9,16,17] In addition to the information published in the literature during 1991–2004, some experimental results obtained in our laboratories also are included.

Makoto Ue
Mitsubishi Chemical Group Science and Technology Research Center, Inc.
Battery Materials Laboratory, 8-3-1 Chuo, Ami, Inashiki, Ibaraki, 300-0332, Japan
3707052@cc.m-kagaku.co.jp

M. Yoshio et al. (eds.), *Lithium-Ion Batteries*,
DOI 10.1007/987-0-387-34445-4_4 © Springer Science+Business Media LLC 2009

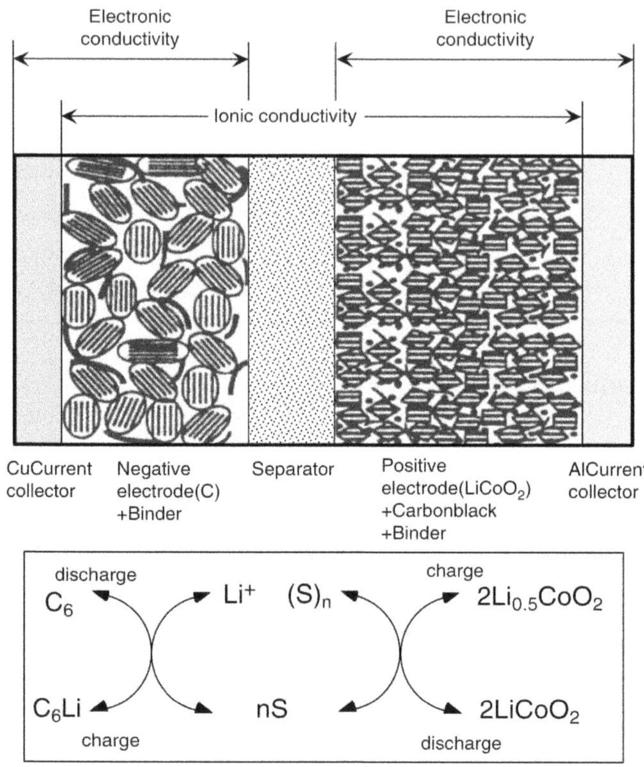

Fig. 4.1 Role of liquid electrolytes in lithium-ion cells

4.2 Role-Assigned Electrolytes

Most of the liquid electrolytes used in the commercial lithium-ion cells are the nonaqueous solutions, in which roughly 1 mol dm^{-3} (≡ M) of lithium hexafluorophosphate ($LiPF_6$) salt is dissolved in the mixture of carbonate solvents selected from cyclic carbonates – ethylene carbonate (EC), and propylene carbonate (PC) –, and linear carbonates – dimethyl carbonate (DMC), ethyl methyl carbonate (EMC), and diethyl carbonate (DEC) –, whose chemical structures are displayed in Fig. 4.2. Recently, another type of liquid electrolyte based on 1.5 M $LiBF_4$/γ-butyrolactone (GBL) + EC came onto the market for the laminated thin lithium-ion cells with an excellent safety performance.[18,19] Many other solvents and lithium salts have limited applications, although much effort has been made to develop new materials. Into the above baseline electrolyte solutions, a small amount of the additives are dissolved, which are so-called "functional electrolytes."[20]

We have been developing various new additives by classifying them into the following categories from their working mechanisms. A specific role is assigned to each additive. A set of specific functions are granted to the liquid electrolytes by

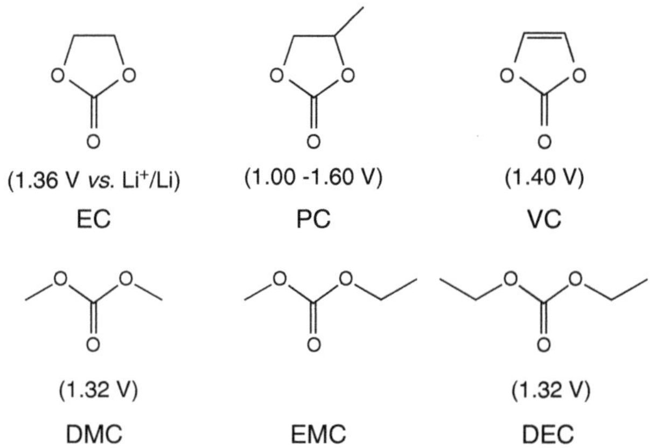

Fig. 4.2 Carbonate solvents and their reduction potentials

formulating each additive at an optimum concentration. These can be called "role-assigned electrolytes":

1. Anode passivation film forming agents
2. Cathode protection agents
3. Overcharge protection agents
4. Wetting agents
5. Flame retardant agents
6. Others

The terms "anode" and "cathode" are traditionally used as negative and positive electrodes in rechargeable cells, respectively, although they are academically incorrect for the charging process. For convenience we call the additive affecting the anode as "A-type" (category 1), the additive affecting the cathode as "C-type" (category 2), and the additive for the bulk solution as "B-type" (categories 3–5). The boundary between the additives and auxiliary solvents (or salts) is sometimes unclear, particularly when the additives are aprotic organic solvents (or lithium salts).

4.3 Anode Passivation Film Forming Agents

It is well known that the solid electrolyte interphase (SEI) formed on the carbon anode determines the important cell performances such as reversible capacity, storage life, cycle life, and safety. This SEI originates from the electrolyte decomposition during the charging process. The electrolyte (particularly EC) begins to decompose around a potential $E = 1.4$ V *vs.* Li^+/Li during the first charge, which results in the creation of an irreversible capacity Q_{irr} (the difference in capacity Q between the charge and discharge) as shown in Fig. 4.3. From the second cycle, this irreversible capacity becomes smaller and the constant reversible capacity Q_{rev} is obtained during charge and discharge cycles.

When PC is used as a solvent, the electrolyte continues to decompose at about 0.9 V *vs*. Li^+/Li, as shown in Fig. 4.4, and graphitic (crystalline) carbons cannot be charged due to their exfoliation caused by the solvent co-intercalation. This is the reason why EC is dominantly used instead of PC, which was adopted for the first commercial lithium-ion cells with a nongraphitic (amorphous) carbon anode. However, the addition of some compounds such as vinylene carbonate (VC) prevents the graphite exfoliation and enables the charge of graphitic carbons, as shown in Fig. 4.4.

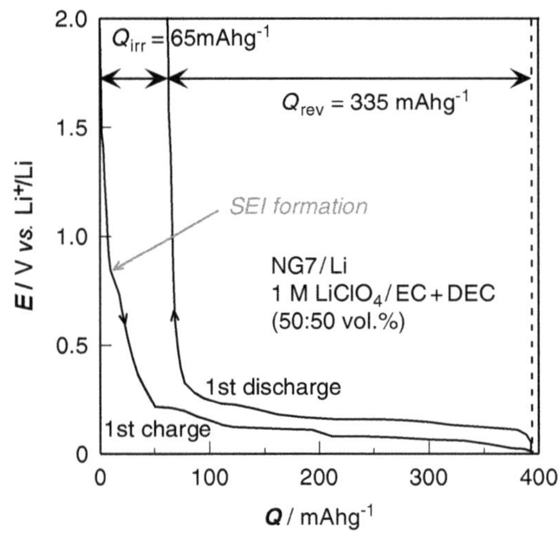

Fig. 4.3 The first charge and discharge curves for a natural graphite

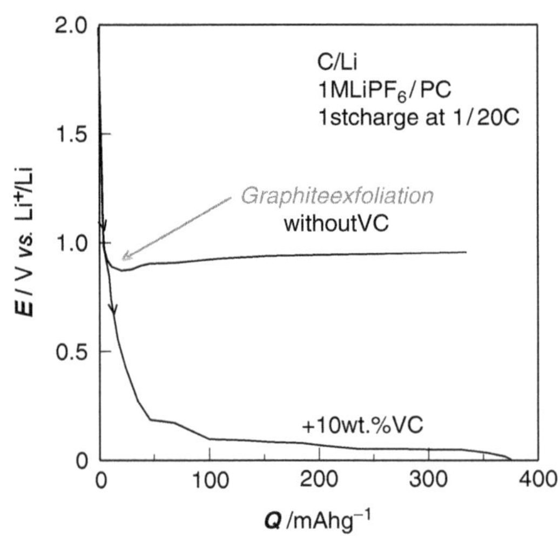

Fig. 4.4 The first charge curves for a graphite in PC-based electrolytes

The SEI produced by the reductive decomposition of EC and linear carbonates is very effective at room temperature but is problematic at elevated temperatures. A large number of compounds have been examined as an additive in order to improve the properties of the passivation films on the graphitic carbon anodes.

4.3.1 Unsaturated Carbon Bond Compounds

The VC was first explored as an electrolyte solvent,[21,22] which affords a good electrolytic conductivity due to its extremely high relative permittivity ($\varepsilon_r = 127$). However, it became a typical compound as an anode passivation film-forming agent,[23] after it was found that the addition of a small amount of VC suppressed gas evolution during the initial charge with the enhanced cycle efficiency,[24,25] and protected the decomposition of reduction-susceptible solvents such as trimethyl phosphate (TMP).[26] The excellent stability of the passivating layer[27] was demonstrated by the fact that the addition of 1 wt% of VC in 1 M $LiPF_6$/EC + DMC + DEC (33:33:33 wt%) improved the cycle life of commercial lithium-ion polymer cells.[28]

Since VC has a smaller lowest unoccupied molecular orbital (LUMO) energy due to the presence of a double bond in its structure, it is considered to be more susceptible to reduction than other carbonates such as EC and DMC.[29] The reduction potential of VC is higher than those of other carbonate solvents, as given in Fig. 4.2, which were measured on a gold electrode in tetrahydrofuran (THF) solvent.[30] It is interpreted that the reductive decomposition of VC precedes the carbonate solvent decomposition, and the resultant good SEI film on the anode protects the further solvent decomposition and the graphite exfoliation by solvent co-intercalation.[29]

We also have examined the composition of evolved gas at various potentials during the initial charge as shown in Fig. 4.5.[31] When VC was absent in 1 M $LiPF_6$/EC + DMC (50:50 vol%), ethylene and carbon monoxide were accumulated as main components, which are the by-products of EC decomposition. On the other hand, carbon dioxide became a main component in the presence of 2 wt% of VC. This is the indication that the reductive decomposition of VC, releasing carbon dioxide, remarkably suppresses the solvent decomposition. Granular products were observed on the surface of a graphitic carbon anode at 1.0 V *vs.* Li^+/Li by a scanning electron microscopy (SEM) and then a gelled organic film covered the surface further as Li^+ intercalation proceeds, as shown in Fig. 4.6.

The characterization of the resultant film by VC was carried out by various methods. The $CH_2 = CHOCO_2Li$ was detected by infrared reflection absorption spectroscopy (IRRAS).[30] It also was speculated that the formation of polymers having $-OCO_2Li$ group affords good adhesiveness and flexibility to the passivation film.[32] The deposition process by VC reduction, which started from 1.3 V *vs.* Li^+/Li, was monitored by *in-situ* atomic force microscopy (AFM) and the film thickness down to 0.8 V was about 10 nm.[30,33] *Ex-situ* AFM also revealed an ultra-thin film (less than 1 nm) on the terrace of the basal plane of a highly oriented pyrolytic graphite (HOPG)

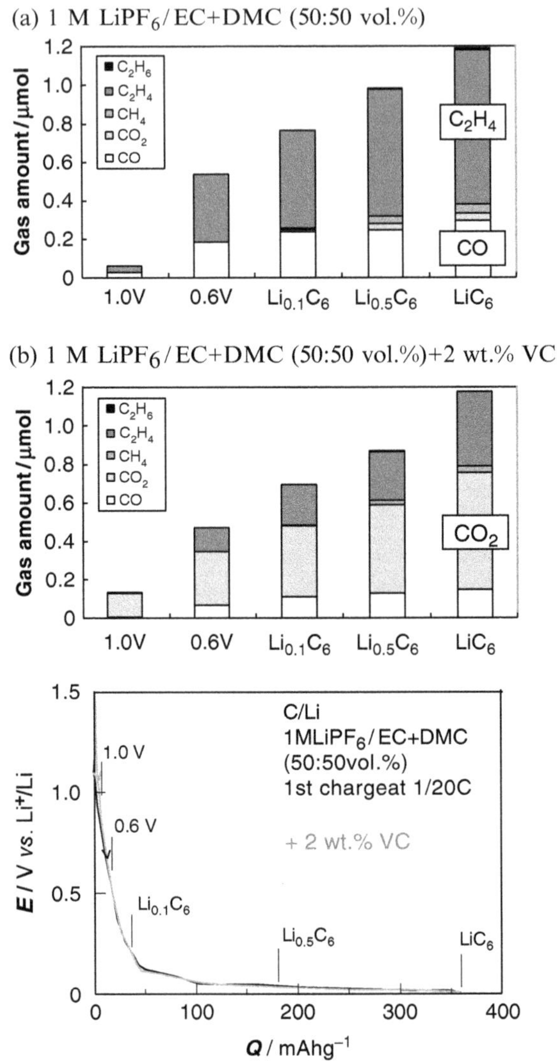

Fig. 4.5 Change of evolved gas composition by the decomposition of electrolytes

anode.[34] The smaller film resistance due to the suppressed formation of LiF on a graphite electrode also was observed by AC impedance analysis.[35–38]

We first have confirmed the presence of polymeric materials in the SEI by Fourier transform infrared spectroscopy (FTIR), X-ray photoelectron spectroscopy (XPS), time of flight-secondary ion mass spectroscopy (TOF-SIMS), and proton nuclear magnetic resonance spectroscopy (^{1}H-NMR), as exemplified in Figs. 4.7 and 4.8.[31] The FTIR and XPS (O1s) spectra indicated the presence of C = C bonds and polymeric materials, respectively. The TOF-SIMS revealed the existence of

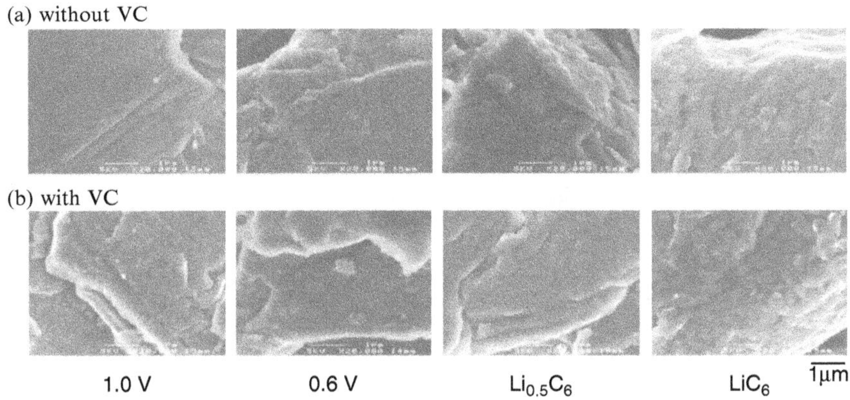

Fig. 4.6 SEM images of the surface of a graphite

polyacethylene in the very top surface of the SEI. To identify the structure of the polymeric materials, the SEI formed in 1 M $LiPF_6$/VC was extracted by dimethyl sulfoxide (DMSO) solvent and characterized by ^{1}H-NMR including two-dimensional analysis. The presence of poly(vinylene carbonate) and other ring-opening compounds was confirmed successfully. The molecular weight of these polymers formed on lithim metal anode, where the addition of VC also enhanced the lithium cycling efficiency,[39] ranged from 1,000 to 3,000 by a gel permeation chromatography (GPC).[40] The thermal stability of the SEI also was examined by temperature-programmed decomposition-mass spectroscopy (TPD-MS).[31] The addition of VC has shifted the decomposition temperature of the SEI to a higher temperature as shown in Fig. 4.9. This enhanced thermal stability can explain the suppression of the reaction between the electrolyte and the anode when the SEI is broken and repaired. From these observations, it is considered that the relatively thinner homogeneous SEI formed by VC is an origin of the better cell performances, although the quality of its SEI is strongly dependent on the kind of graphitic carbons and charging conditions.

The reaction mechanism of VC is not yet determined. From the density functional theory (DFT) calculations of supermolecules, $(EC)_nLi^+$ (VC), the possibility that EC is cleaved through intermolecular electron transfer from VC to EC was proposed.[41,42] Among the cyclic carbonates, VC is most susceptible to a two-electron reduction, possibly one of the characteristics as an additive for SEI film formation, although polymerization of the one-electron reduction product is likely to proceed in parallel for building a foundation for SEI film.[43]

Besides VC, vinyl ethylene carbonate (VEC),[44–49] phenylethylene carbonate (PhEC),[50] phenylvinylene carbonate (PhVC),[50] catechol carbonate (CC),[51,52] ally methyl carbonate (AMC),[53,54] allyl ethyl carbonate (AEC),[55] vinyl acetate (VA), and other vinyl compounds,[53,54,56] acryronitrile (AAN)[57,58] and 2-cyanofuran (CN-F),[59] whose chemical structures are given in Fig. 4.10, showed the similar effect and no graphite exfoliation in PC solvent systems.

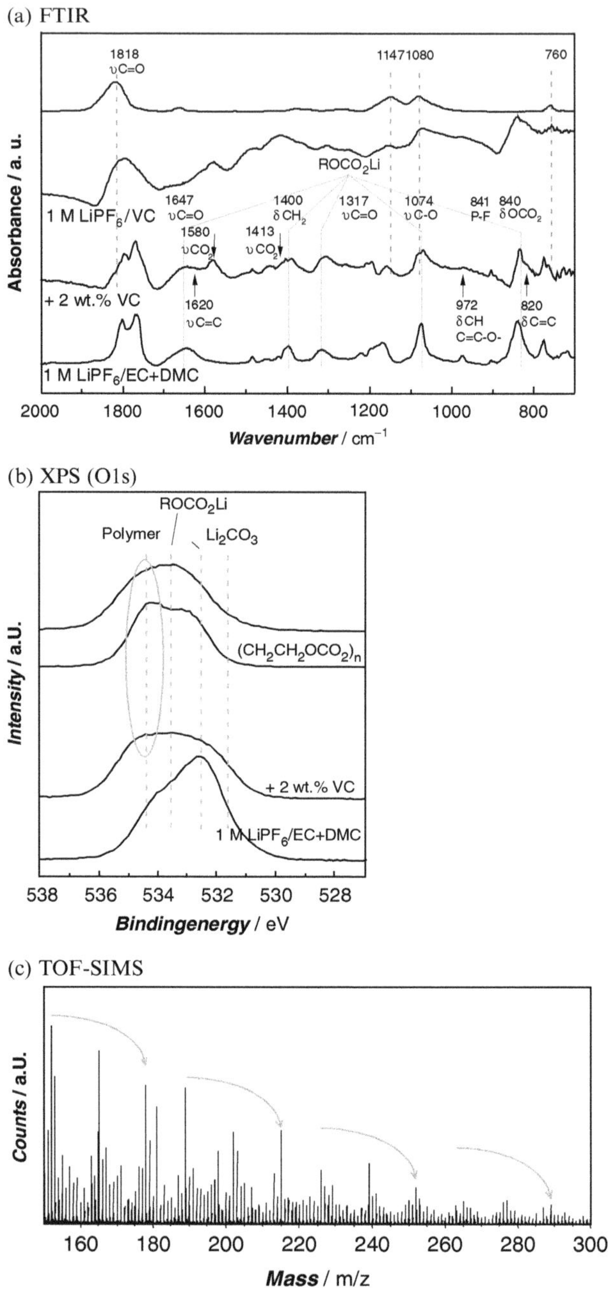

Fig. 4.7 Characterization of SEI formed on a graphite in 1 M $LiPF_6$/EC + DMC (50:50 vol%) + 2 wt% VC

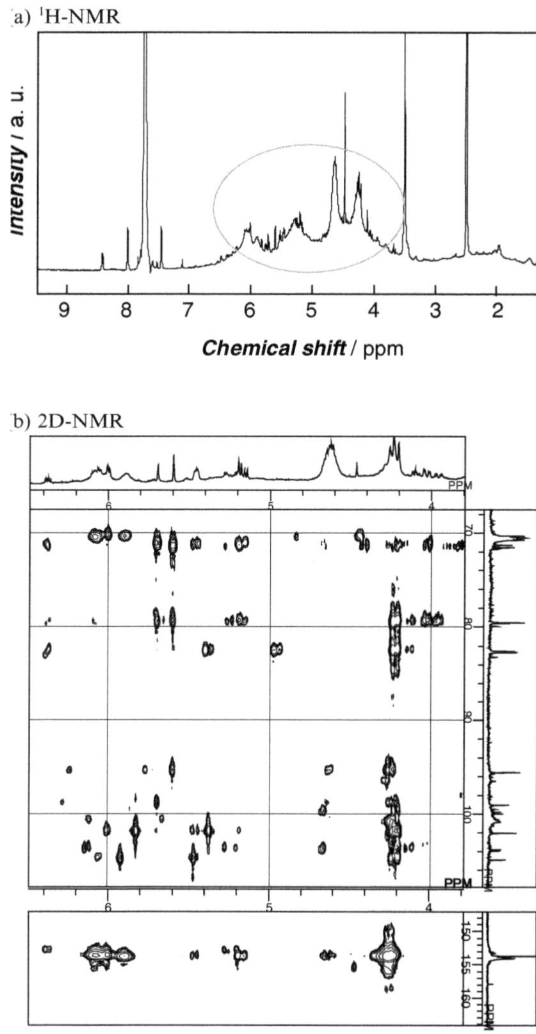

Fig. 4.8 Characterization of SEI formed on graphite in 1 M $LiPF_6$/VC

4.3.2 *Sulfur-Containing Organic Compounds*

Sulfite compounds including ethylene sulfite (ES), propylene sulfite (PS), dimethyl sulfite (DMS), and diethyl sulfite (DES) were examined as passivation film forming agents, and their addition of 5 vol% in a PC electrolyte enabled the charge and discharge of graphite electrodes.[60–62] These compounds were reduced at about 2 V *vs.* Li^+/Li to form the passivation layers, which hindered the PC co-intercalation

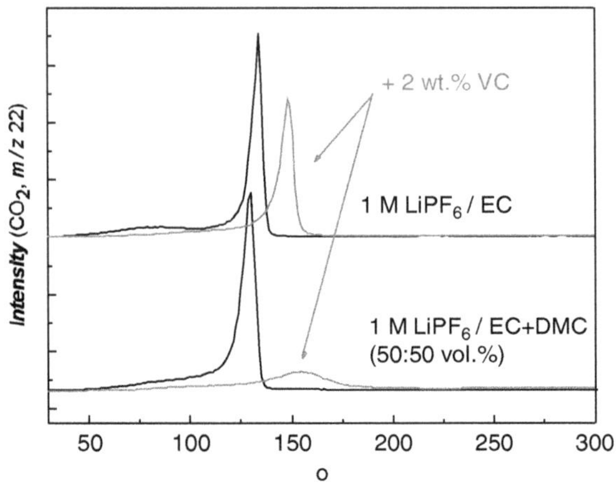

Fig. 4.9 TPD-MS of SEI on a graphite anode

into the graphene layers in the following order: ES > PS > DMS > DES. Besides sulfite compounds, cyclic sulfonate compounds including 1,3-propane sultone (1,3-PS) were reported to give a good SEI film, which resulted in improving cycle and storage performances of cylindrical hard carbon/$LiMn_2O_4$ cells.[63]

The reductive ring opening of ES by one-electron transfer was estimated to be more energetically favorable than carbonates such as EC and VC.[45] *In-situ* AFM observation revealed the swelling of HOPG steps by PC co-intercalation at about 1 V *vs.* Li^+/Li, and the film thickness of the resultant SEI was about 30 nm, which was thicker than that of VC.[33]

We also have characterized the SEI produced in 1 M $LiPF_6$/PC + 5 or 10 wt% ES by various analytical techniques including SEM, XPS, TPD-GC/MS, and TOF-SIMS.[64,65] The ES was reduced at 1.8 V *vs.* Li^+/Li prior to PC, and the composition of the resultant film was significantly dependent on the current density. At high current densities, an inorganic compound ascribed to Li_2SO_3 initially deposited, on which an organic layer comprising from $CH_3CH(OSO_2Li)CH_2OCO_2Li$ was formed. *In-situ* electrochemical impedance spectra showed that the SEI formed by ES showed higher film resistance than that by VC.[38]

4.3.3 Halogen-Containing Organic Compounds

It is easy to know from molecular orbital calculations that the introduction of halogen atoms on cyclic carbonates make them more reducible.[29] Chloroethylene carbonate (ClEC) was reduced at about 1.8 V *vs.* Li^+/Li and produced the passivation film accompanied by CO_2 evolution.[66–71] However, LiCl produced by its reductive cleavage was postulated to become an internal chemical shuttle, which resulted in

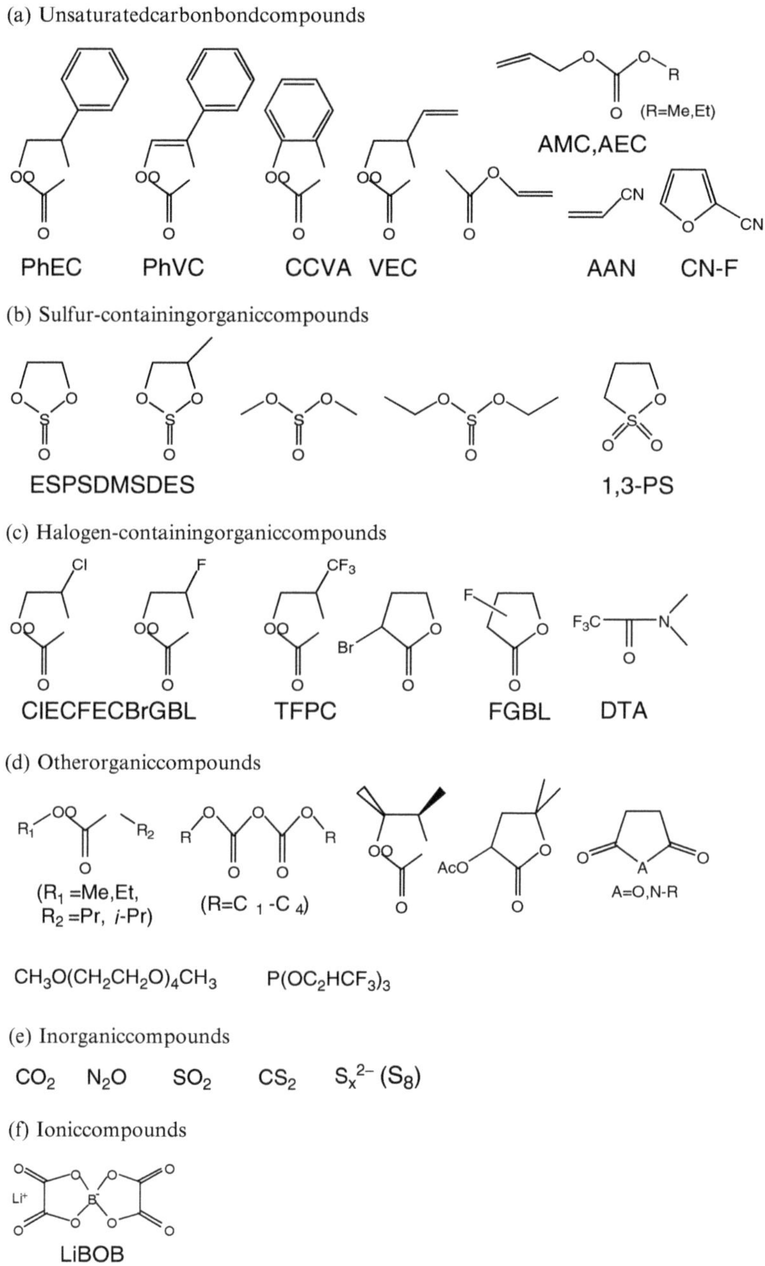

Fig. 4.10 Examples of anode passivation film forming agents

low current efficiency. This disadvantage was solved by using fluoroethylene carbonate (FEC), which produced LiF with much reduced solubility.[72] The film thickness was intermediate between VC and ES by *in-situ* AFM, but the degradation was remarkable during cycling.[33] Trifluoropropylene carbonate (TFPC) also was reduced at about 1.8 V *vs*. Li^+/Li to build up the passivation layer,[73] whose interfacial impedance was larger than that of ClEC.[74,75] Halogenated GBL (BrGBL[76] or FGBL[77]) and *N*,*N*-dimethyltrifluoroacetamide (DTA)[78] also showed the similar effect.

4.3.4 Other Organic Compounds

Some other organic compounds such as asymmetric dialkyl carbonates,[79–81] dialkyl pyrocarbonates,[82–85] and *trans*-2,3-butylene carbonate (*t*-BC)[29,86,87] have been proved to be useful, although these also are regarded as co-solvents. The dialkyl pyrocarbonates contribute the liberation of CO_2 by spontaneous decomposition, which further reacts on the active surface to form Li_2CO_3. The *t*-BC is too large for its solvated Li^+ to intercalate into graphite, thus preventing the exfoliation of graphite. A series of GBL derivatives with various side chains at the 5-position were also examined and some of them were effective in preventing the decomposition of PC due to the decrease of the number of PC molecules coordinated to Li^+.[88] Succinic anhydride[89] and succinimide derivatives[90] also were reported to be effective. Tetra (ethylene glycol) dimethyl ether (TEGME), which has molecular resemblance to organic SEI compounds, formed a stable nonporous passivating layer.[91] An organophosphorous compound, tris(2,2,2-trifluoroethyl)phosphite (TTFP), also was reported to be effective.[92]

4.3.5 Inorganic Compounds

Inorganic compounds including CO_2,[93–97] N_2O,[97] SO_2,[98–100] CS_2,[101] and S_x^{2-} (electrochemical reduction of S_8)[97,102] historically have played an important role in stabilizing the SEI on both lithiumi-graphite and lithium-metal surfaces. Species such as Li_2CO_3, Li_2O, Li_2S, and $Li_2S_2O_4$ formed in the presence of the above additives afford very good passivating layers.

4.3.6 Ionic Compounds

It is well known that the type of lithium salt also affects the composition and quality of the SEI[9,103]; however, the additive effects of lithium salts are not well understood. It was proved that organoboron complexes such as lithium bis(salicylato)borate[36,104] and lithium bis(oxalato)borate (LiBOB) forms stable SEI on graphite anode,[105–111] because their organic moieties work as SEI components. LiI, LiBr, and NH_4I were used to suppress the Mn (II) reduction on the anode in C/$LiMn_2O_4$ cells.[112] There also is a report that the addition of Na^+ reduced an irreversible capacity of a graphite anode at the first charge.[113] The $AgPF_6$ and $Cu(CF_3SO_3)_2$ also were tested in an attempt to form metal-protective layers.[114,115]

However, most of the compounds described in this section have been examined from only one viewpoint whether the addition of these compounds enables the charge and discharge of graphitic carbons in PC-containing electrolytes. Further detailed examinations are mandatory to know whether these compounds work as a practical passivation film forming agent in EC-based electrolyte systems.

4.4 Cathode Protection Agents

There is by far less information available on the cathode interface than the anode interface. However, the reports appeared recently, which insisted that there is a film on the cathode, which may be called a SEI as well as the anode.[116–120] Since the oxidative reactions on the cathode cannot immobilize reaction products like the reductive reactions on the anode, the amount of SEI on the cathode is much smaller than that on the anode, as demonstrated in Fig. 4.11. Due to the analytical difficulties, a very few data in the literature report the effects of electrolyte additives on the cathode. It was reported that the addition of VC reduced the interfacial impedance and improved a bit the rate capability.[32,35] It was speculated that this effect is caused by the polymer formation by VC on the cathode, which suppresses the deposition of lithium fluoride, since this effect disappeared when VC contained polymerization inhibitors such as BHT.[35] This is reasonable because the oxidation potential of VC is lower than those of other carbonate solvents.[121]

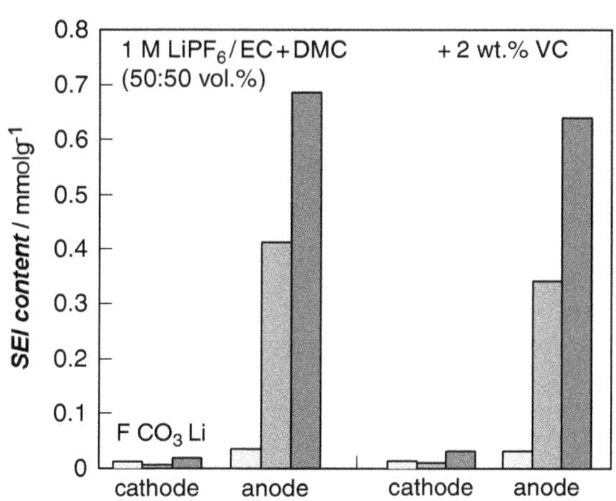

Fig. 4.11 SEI content after three cycles in C/$LiCoO_2$ cells

We have confirmed by an *in-situ* subtractively normalized interfacial FTIR (SNIFTIR) that the polymer by VC begins to form at 4.3 V *vs.* Li^+/Li on Li_xCoO_2 as shown in Fig. 4.12, where the upward peaks at 1,830, 1,805, and 1,750 cm^{-1} below 4.3 V correspond to the consumption of VC, EC, and EMC, respectively.[122] This resultant polymeric film formed at 4.3 V suppressed the further decomposition of EC and EMC above 4.3 V, where the downward peaks at 1,850–1,800 and 1,650–1,550 cm^{-1} can be ascribed to the formation of decomposition products. The chemical structure of the polymer was assumed to be a poly(vinylene carbonate) by XPS. However, the reaction between $Li_{0.5}CoO_2$ and VC at 85°C did not show the existence of polymeric materials.[123] These results indicate that the thermal stability

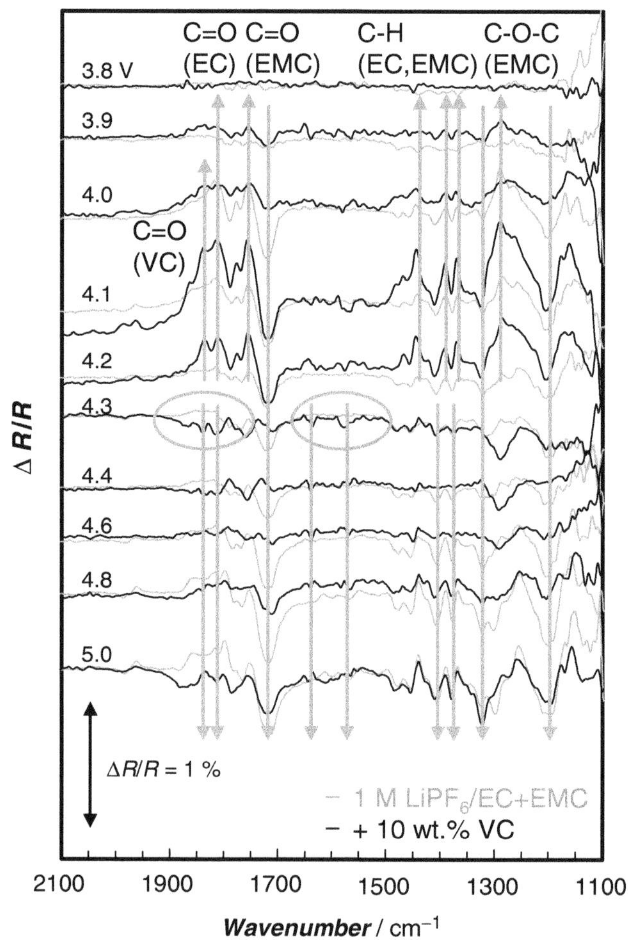

Fig. 4.12 *In-situ* SNIFTIR spectra of 1 M $LiPF_6$/EC + EMC (30:70 vol%) on a $LiCoO_2$ thin film

of passivation film derived by VC on $Li_{0.5}CoO_2$ is low. We have measured the amount of SEI obtained from C/$LiCoO_2$ cells by an ion chromatography (IC); however, the remarkable suppression of lithium fluoride deposition on the cathode by the purified VC was not observed, as shown in Fig. 4.11.[124] *In-situ* electrochemical impedance spectra also showed that the addition of VC increased the film resistance during discharge.[125]

We also have succeeded in characterizing the SEI on the cathode formed in 1 M $LiPF_6$/PC + DMC (50:50 vol%) + 5 wt% ES by an x-ray absorption near edge structure spectroscopy (XANES) for the first time.[65] The compounds deposited on the cathode were different from those on the anode, and Li_2SO_4 was detected as well as other organic sulfur compounds, as shown in Fig. 4.13. It is reasonable that ES is oxidized to SO_4 species on the cathode, while Li_2SO_3 exists on the anode. *In-situ* electrochemical impedance spectra showed that the SEI formed by ES showed higher film resistance than that by VC, as in the case of the anode.[125]

It also was shown that the SEI formed by GBL solvent on the cathode retarded the exothermic reactions between electrolyte and cathode, which is a ground of high safety performance of the cells.[126] By applying these phenomena, it is possible to reduce the reactivity of the cathode surface with the liquid electrolytes by deactivating their active sites.[20] The addition of a very small amount (0.1–0.2%) of aromatic compounds such as biphenyl (BP), *o*-terphenyl (oTP), and *m*-terphenyl (mTP) improved cycleability of the cathode, which originated from the electrolytic polymerization to form a thin and highly Li^+ conductive film covering the cathode surface.[127,128]

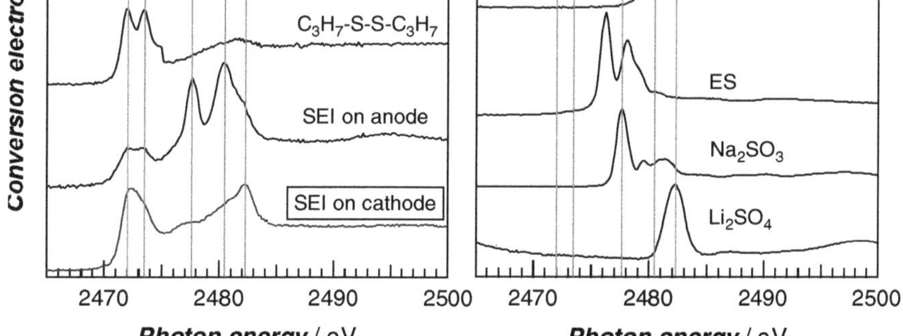

Fig. 4.13 S K-edge XANES spectra of SEI on the electrodes and reference sulfur compounds

We also have found that the addition of some sulfur-containing compounds reduced the reactivity of the cathode without losing its cell performances. Figure 4.14 demonstrates the additive effects of the cathode protection agents. The addition of 1 wt% of sulfonate compounds having higher oxidation potentials (1,3-PS

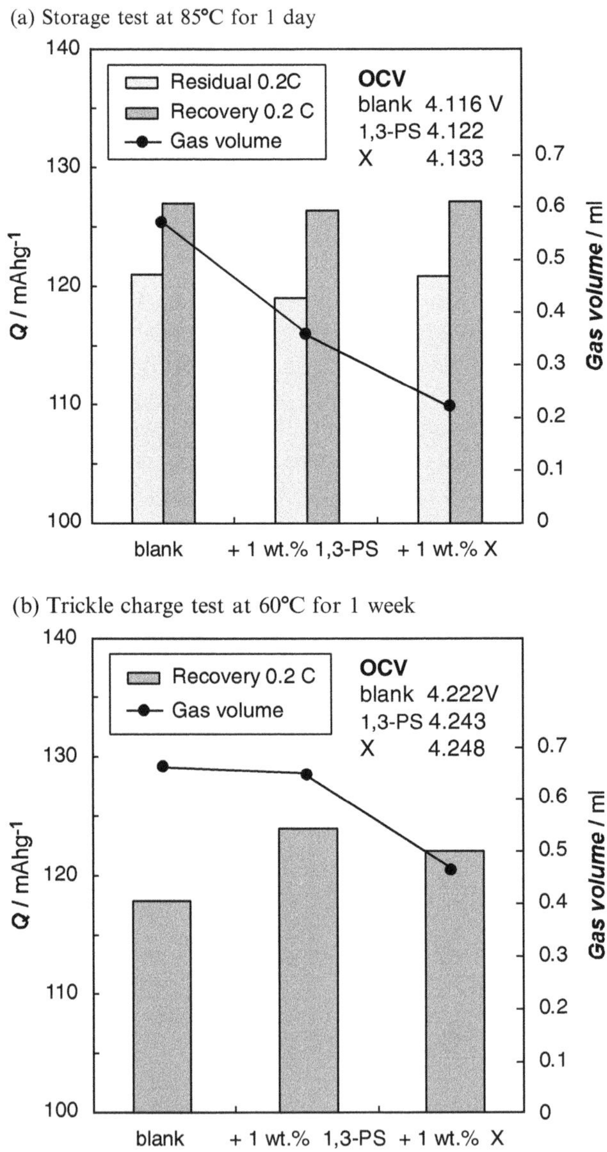

Fig. 4.14 Effect of cathode protection agents in C/$LiCoO_2$ laminated thin cells (70 mAh) using 1 M $LiPF_6$/EC+ DMC+ EMC (30:35:35 vol%) + 2 wt% VC

and a disulfonate X) remarkably reduced the amount of evolved gas and retained the cathode capacity after the storage and trickle charge tests at elevated temperatures.[129] Higher open-circuit voltages (OCV) after these tests indicate that these additives maintain the charged state of the cathodes. The thin film formation attributing to the additives was observed on the cathode, which suppressed the CO_2 gas evolution and the rise in interfacial impedance.

4.5 Overcharge Protection Agents

Nonaqueous electrolyte solutions enable the lithium-ion cells to attain high voltage up to about 4 V, however, no overcharge protection mechanism is included in principle, as the aqueous electrolyte solutions in rechargeable cells such as Pb-acid, Ni–Cd, and Ni–MH have overcharge protection mechanisms, which are the electrolysis of H_2O to $H_2 + O_2$ and their recombination to H_2O. Present lithium-ion cells in the market are sold as a battery pack, where safety devices such as a control integrated circuit and a positive temperature coefficient (PTC) are attached. However, demands for higher safety, higher capacity, and lower cost have been accelerating the movement that safety functions should be included in the battery materials themselves.

The overcharging degrades electrodes, current collectors, electrolytes, and separators, which can result in cell failure due to internal shorting and/or gas evolution. When lithium-ion cells are overcharged, extra lithium ions are subtracted from the cathode and extra lithium metal is deposited on the anode. Then both electrodes become thermally unstable and the input energy is consumed for the Joule heat of the cells, which causes thermal runaway, eventually resulting in fire and explosion, as illustrated in Fig. 4.15. The overcharge protection agent is an additive that prevents the overcharge reactions of the both electrodes by its sacrificial reactions. Therefore, it must be inactive at normal use and work only under overcharged conditions.

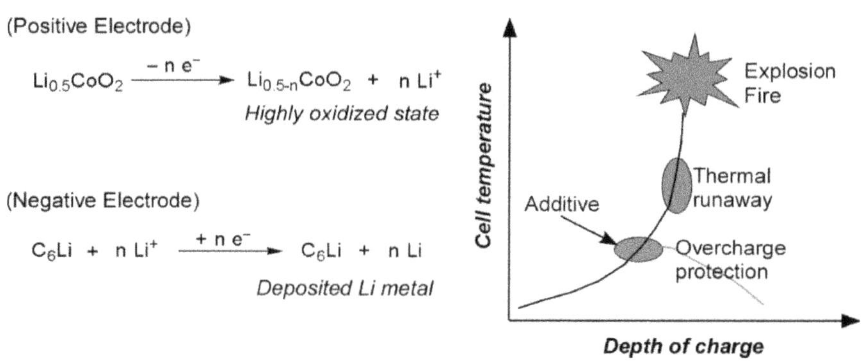

Fig. 4.15 Principle of overcharge protection in lithium-ion cells

4.5.1 Redox Shuttle-Type Compounds

A concept of "redox shuttle" has been proposed before the advent of lithium-ion cells. The principle is depicted in Fig. 4.16. A compound R with a reversible redox potential is added into the electrolyte solution. The R is oxidized to compound O on the cathode, and then O migrates to the anode and is reduced to the original form R. The following properties are required for the compounds[130]:

1. The redox potential should be slightly positive (0.1–0.2 V) to the formal potential of the cathode at the end-of-charge.
2. The redox reactions during overcharge should be kinetically reversible on the negative and positive electrodes (the electrochemical rate constants should be greater than 10^{-5} cm s^{-1}).
3. Redox species should be chemically stable and should not react with other components.
4. The diffusion coefficients and solubilities of redox species should be as high as possible.

A lot of metal complexes[131,132] and aromatic compounds[132–139] have been proposed as an overcharge protection agent; however, their redox potentials were often too low for 4 V class lithium-ion cells. Successful examples are given in Fig. 4.17, where their onset oxidation potentials are given (the values lack consistency due to different measurement conditions).[132,136–139] It is clear from these examples that π-electron conjugated systems are essential.

The limiting current density i_{lim} of one-electron transfer redox shuttle compounds is represented by the following equation:

$$i_{\text{lim}} = FC_0 /(L/2)(1/D_R + 1/D_O), \tag{4.1}$$

where $\boldsymbol{F}$, C_0, L, and D are Faraday constant, initial concentration, separator thickness, and diffusion coefficient, respectively.[136] Since the diffusion coefficient of a

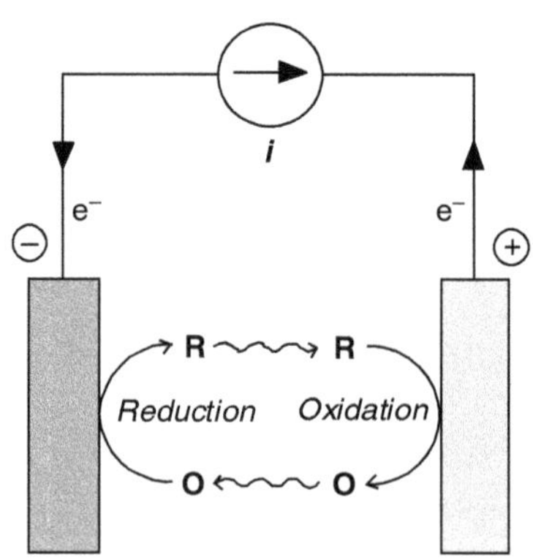

Fig. 4.16 Redox shuttle mechanism during overcharge

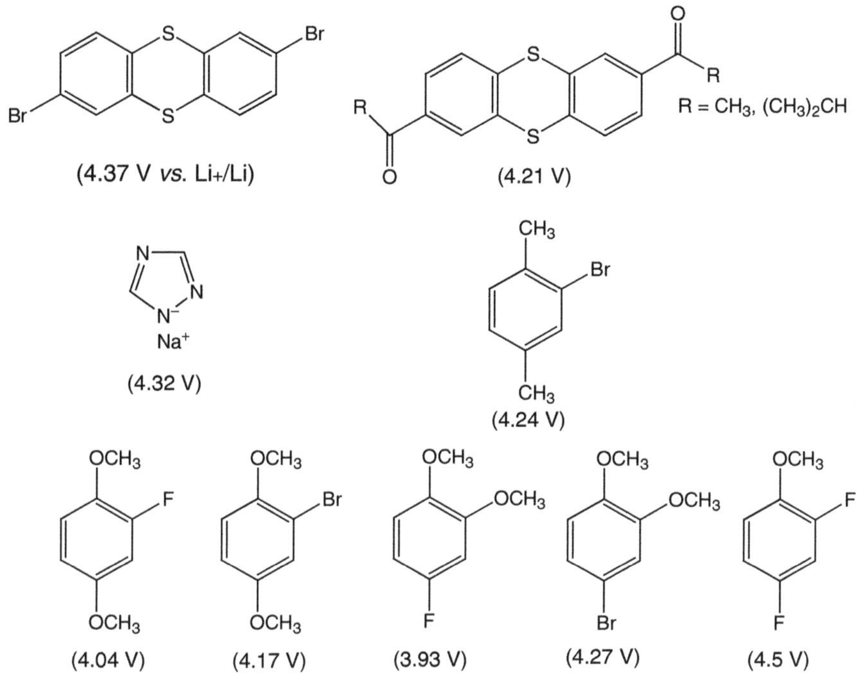

Fig. 4.17 Examples of redox shuttle-type overcharge protection agents and their oxidation potentials

cation radical produced by oxidation is generally smaller than that of an original neutral compound by one order of magnitude, the limiting current density is governed by the diffusion coefficient of the cation radical.[135,136] This is because charged species are solvated and a cation radical tends to interact with a neutral molecule to form a dimer. The diffusion coefficients of the neutral molecules are inversely proportional to the electrolyte viscosity. They are 10^{-5}–10^{-6} cm^2 s^{-1} in 1 M $LiPF_6$/EC + 2DMC,[136] which are similar values to solvated lithium ions.[12] For example, 3 wt% of 2,4-difluoroanisole (DFA)[138] tolerates a relatively large current density i_{lim} = 8 mA cm^{-2}, if we assume C_0 = 0.2 mol dm^{-3}, L = 25 μm, and $D = 10^{-6}$ cm^2 s^{-1}.

It was demonstrated that a redox shuttle compound, 4-bromo-1,2-dimethoxybenzene (0.1 M) dissolved in 1 M $LiPF_6$/PC + DMC, protected the overcharge at 0.02C rate in Li/$LiCoO_2$ 2025 coin cells (25 mAh), as shown in Fig. 4.18.[132] When the additive was absent, the cell voltage V continued to increase and the heat H measured by a C80 calorimeter was generated at about 4.6 V by electrolyte decomposition. On the other hand, when the additive was present, the cell voltage stopped at 4.3 V and the input energy W = 4.3 V × 0.15 mA = 0.65 mW was completely converted to the heat consumed by the redox shuttle reaction.

Another example was reported using a practical prismatic cell (900 mAh), including 2,7-dibromthianthrene (0.05 M) dissolved in 1 M $LiPF_6$/EC + PC + DMC + DEC (35:10:20:35 vol%).[139] The heat flow during overcharge was monitored by an accelerating rate calorimeter (ARC). When the test cell was overcharged at a 1C rate, the cell voltage *V* went up to a preset voltage of 12 V after 1.1 h and then it was maintained until the end of charging, as shown in Fig. 4.19. Although the cell temperature *T* reached to about 110°C, thermal runaway was avoided. However, no

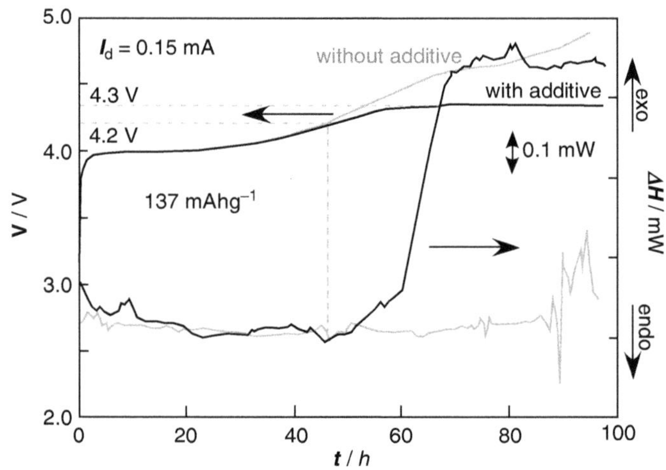

Fig. 4.18 Charge curves and heat flow of Li/$LiCoO_2$ 2025 coin cells (25 mAh) at 30°C

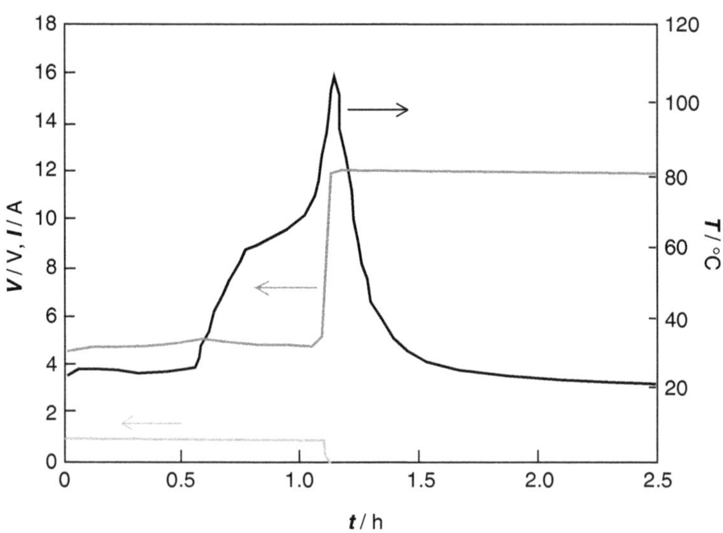

Fig. 4.19 An ARC test of a C/$LiCoO_2$ prismatic cell (900 mAh) at 1C overcharge condition

cells could pass the similar tests over a 2C rate. It was emphasized that the other methods using safety devices also must be employed to prevent overcharge occurring with a large current.

We have found that the alkylbenzene derivatives with no hydrogen atom in benzyl positions can act as a mediator to decompose carbonate solvents, as postulated in Fig. 4.20, because they showed a reversible redox reaction and increased the amount of evolved CO_2 without losing themselves.[140] A *tert*-butylbenzene and 1,3-di-*tert*-butylbenzene showed lower OCVs compared with toluene, ethylbenzene, and cumene after the overcharge test at 60°C, as shown in Fig. 4.21, which is a good indication as an overcharge protection agent.

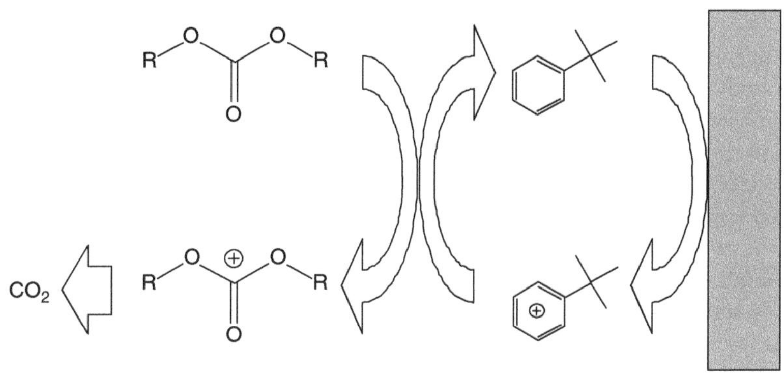

Fig. 4.20 Principle of a redox mediator

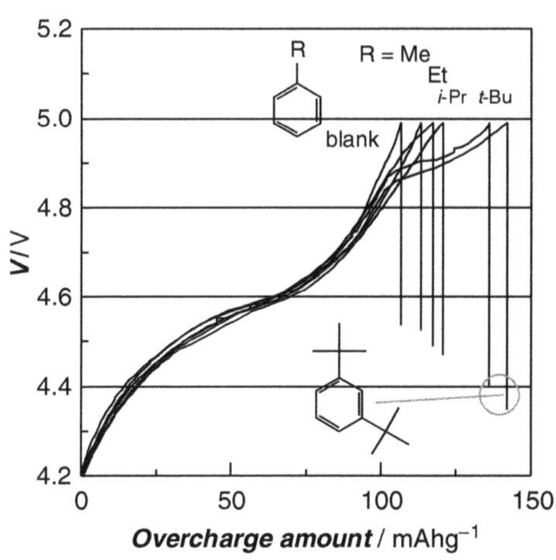

Fig. 4.21 Overcharge curves of C/$LiCoO_2$ 2032 coin cells (3.7 mAh) including 1 M $LiPF_6$/EC + DMC + EMC (30:35:35 vol%) + 2 wt% additives at 60°C, 1C rate

4.5.2 Non-Redox Shuttle-Type Compounds

There is a group of aromatic compounds that shows no reversible electrochemical behavior but gives overcharge protection effects. The first proposed compound of this group is biphenyl.[141] It polymerizes on the cathode during overcharge and the liberated protons migrate to the anode generating hydrogen gas.[138] These phenomena are being used for the overcharge protection, as illustrated in Fig. 4.22. In the case of cylindrical cells with pressure-activated current intermitted devices (CID), the internal pressure rise by H_2 evolution helps to operate them. On the other hand, prismatic cells utilize the current shutdown mechanism by the increase in the internal resistance by the polymerized film and the meltdown of polyethylene separators.[142]

Typical compounds are shown in Fig. 4.23, where oxidation potentials are given.[142,143] Since BP has an disadvantage in that it deteriorates the cell performances during full charge storage at elevated temperatures, partially hydrogenated compounds such as cyclohexylbenzene (CHB), hydrogenated terphenyl (H-TP), hydrogenated dibenzofuran (H-DBF), and tetralin were proposed. Partial hydrogenation shortens π-electron conjugated systems in aromatic compounds and leads to higher oxidation potentials, which are more resistant to the full charge storage at elevated temperatures.

We have found a possibility that partial hydrogenation of mTP gives rise to the compounds having appropriate oxidation potentials by molecular orbital calculations, as shown in Fig. 4.24.[144,145] Depending on the hydrogenated structures, their highest occupied molecular orbital (HOMO) energies vary. Detailed experiments have proved that the oxidation potential of H-mTP (various mixtures containing nine compounds) was continuously controlled by changing the hydrogenation ratio of mTP. The cyclic voltamograms on a platinum electrode of 2 wt% of BP, CHB, and H-mTP (hydrogenation ratio: 42.8%) dissolved in 1 M $LiPF_6$/EC + DMC +

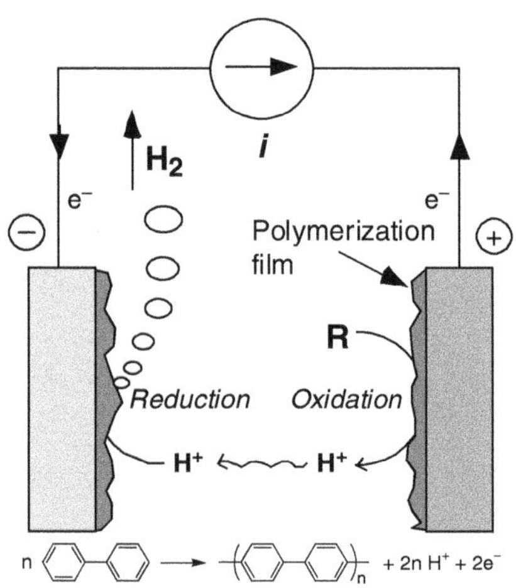

Fig. 4.22 Gas evolution and polymerization mechanisms during overcharge

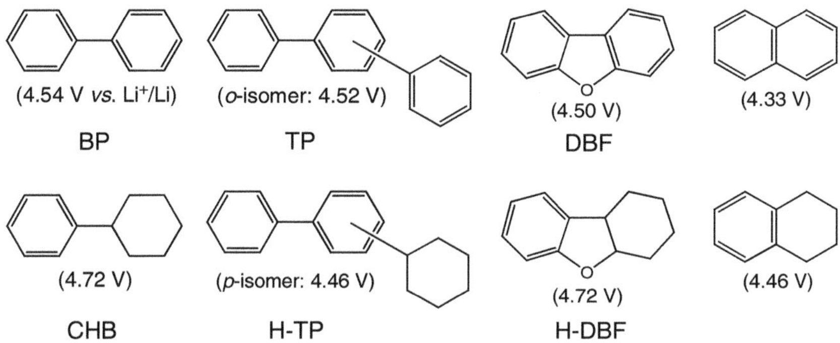

Fig. 4.23 Examples of non-redox shuttle-type overcharge protection agents and their oxidation potentials

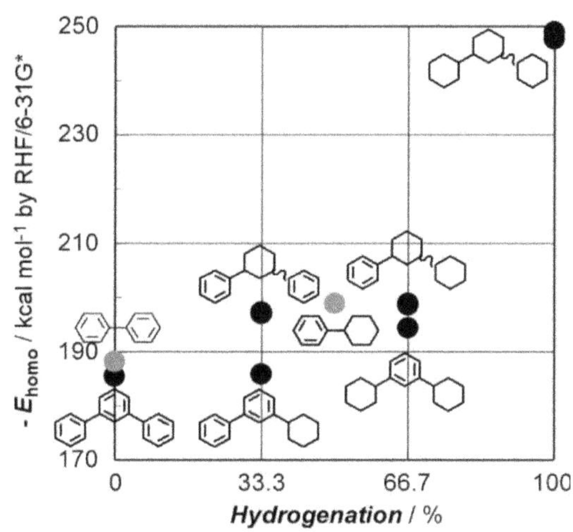

Fig. 4.24 Estimation of oxidation potentials of aromatic compounds by molecular orbital calculation

EMC (30:35:35 vol%) are shown in Fig. 4.25. It was observed that the onset oxidation potentials of second and third cycles were less than that of the first cycle in all compounds, which means that the oxidation products are more oxidizable than the original aromatic compounds.

The overcharge tests at a 1C rate of the six aromatic compounds were carried out by putting 2 wt% of each additive in C/$LiCoO_2$ 2032 coin cells (3.7 mAh) using 1 M $LiPF_6$/EC + DMC + EMC (30:35:35 vol%). The overcharge amount shown in Fig. 4.26 was in the following order: DBF < BP < H-oTP, H-DBF < H-mTP, CHB < blank. The reaction potential of each additive can be estimated from the OCV after the overcharge test, as plotted in Fig. 4.27, and the appropriate voltage region was considered to be from 4.4 to 4.5 V. The OCVs of BP, CHB, and H-mTP fell in

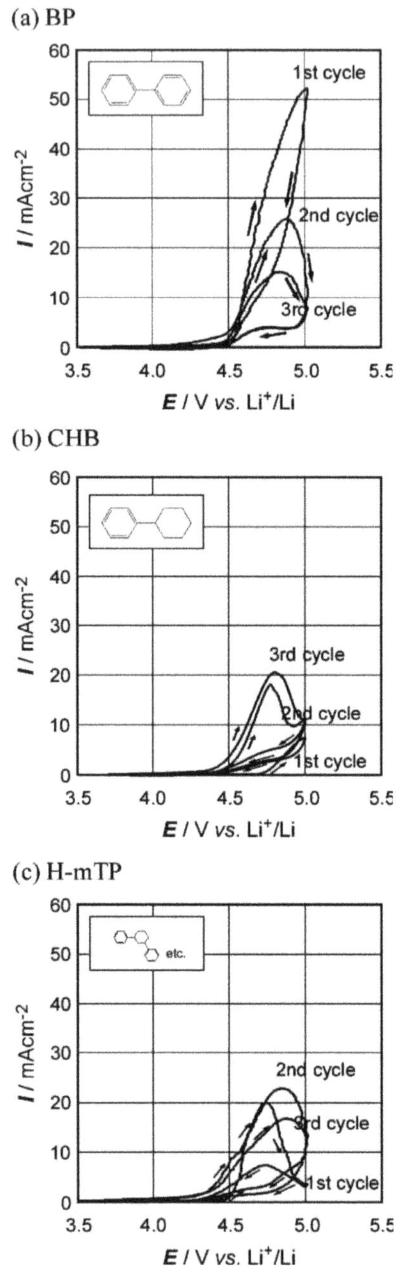

Fig. 4.25 Cyclic voltamograms of 2 wt% aromatic compounds in 1 M $LiPF_6$/EC + DMC + EMC (30:35:35 vol%) on Pt at 5 mVs^{-1}

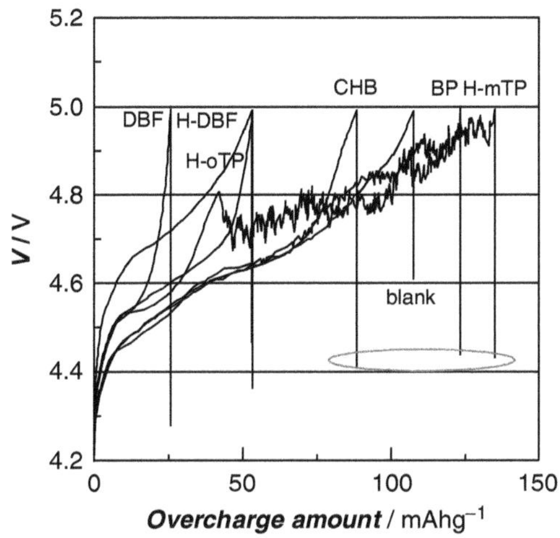

Fig. 4.26 Overcharge curves of C/$LiCoO_2$ 2032 coin cells (3.7 mAh) including 1 M $LiPF_6$/EC+ DMC+ EMC (30:35:35 vol%) + 2 wt% additives at 25°C, 1C rate

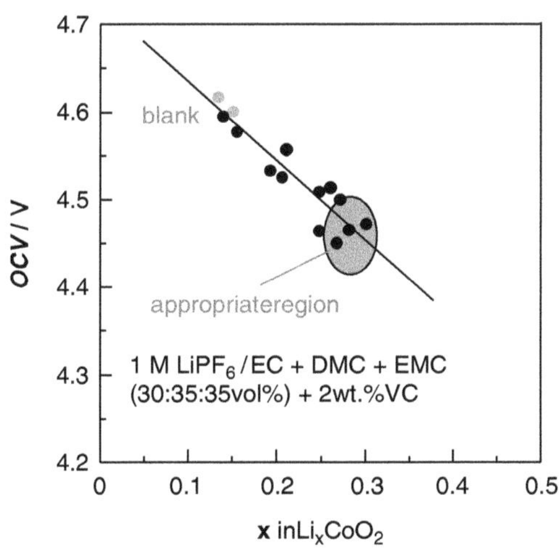

Fig. 4.27 Open circuit voltages after overcharge test

this region. The rise in cell voltage in Fig. 4.26 came from the buildup of internal resistance by film formation, and the voltage vibration came from the internal shorting between the electrodes by lithium dendrites. This soft short circuit was reported to be one mode of overcharge protection mechanisms at a low current rate.[142]

After the overcharge test, black precipitates were observed on the surface of the cathode, whose SEM images are displayed in Fig. 4.28. The morphology of the surface films was dependent on the kind of monomer (BP, CHB, and H-mTP). The

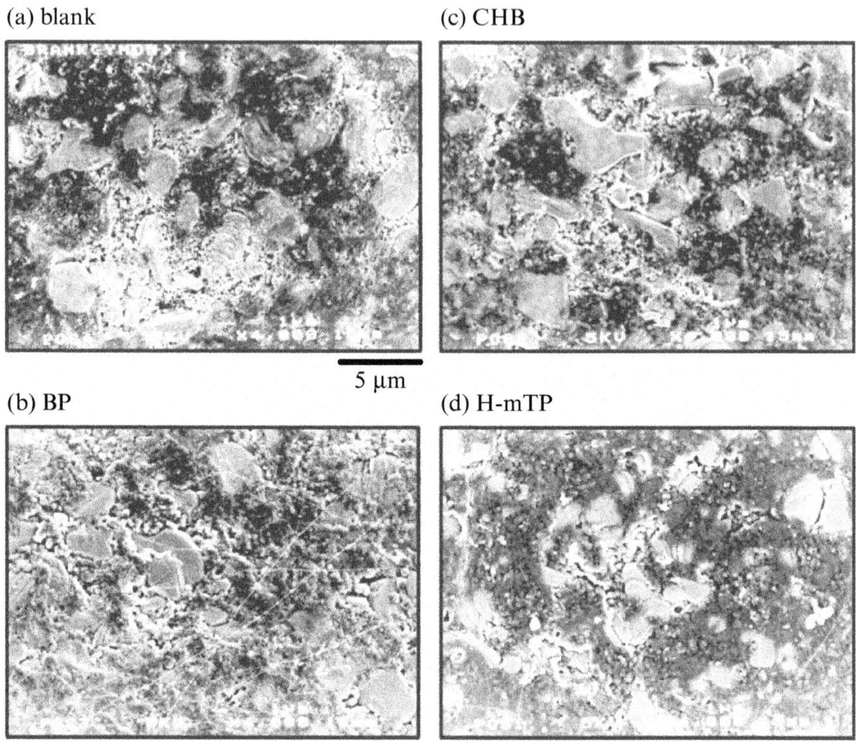

Fig. 4.28 SEM images of the surface of a $LiCoO_2$ cathode after overcharge test

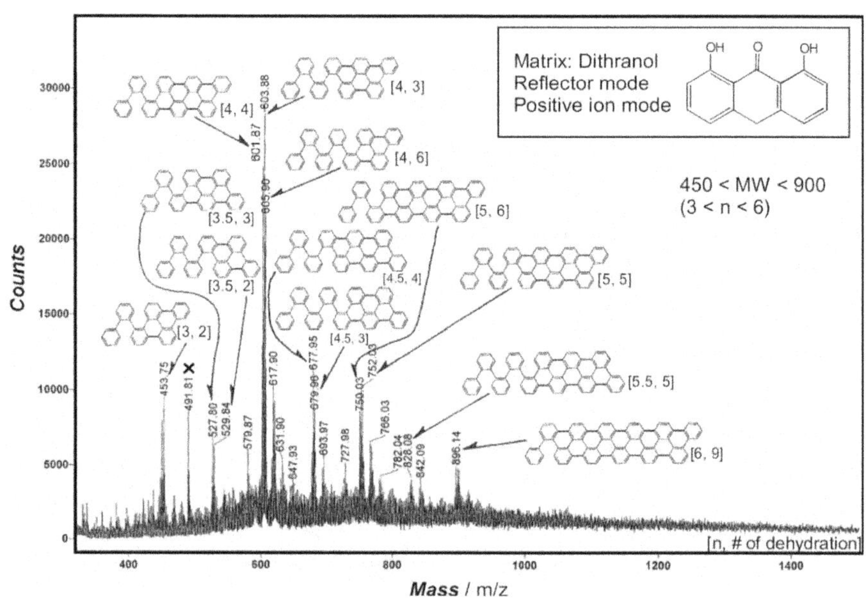

Fig. 4.29 MALDI-TOF-MS of oxidation products of BP on a $LiCoO_2$ cathode

chemical structures of these oxidation products were analyzed by matrix-assisted laser desorption ionization-time of flight-mass spectroscopy (MALDI-TOF-MS) for the first time. They were condensed oligomers having 6–12 benzene rings, as exemplified in Fig. 4.29. It was found that the bond formation between monomers occurs mainly at ortho-positions of the phenyl groups and the molecular distribution is not dependent on the kind of monomer. The cyclohexane structures in CHB and H-mTP were lost by dehydrogenation at 60°C, which was supported by the disappearance of long alkyl chain fragments in TPD-GC/MS.

4.6 Wetting Agents

The topic of wetting is of great importance in battery design. In general, one might expect that the solvents and liquid electrolytes will be attractive to hydrophilic surfaces such as cathode materials, but repulsive to hydrophobic ones such as polyolefin separators and carbonaceous anode materials.[146] For example, the surface treatments of a polypropylene separator by an oxygen glow discharge[147] and a natural graphite by mild oxidation[148] were reported to improve their wetting property. However, there are very few quantitative data of this property in the literature, and only simple qualitative observations were reported on the wettability of microporous separators.[149] Wetting effects are arranged into three classes, as exemplified in Table 4.1: "good" means that the sample can be thoroughly wetted promptly; "poor" implies that the sample cannot be wetted spontaneously unless a reduced pressure is applied; and "fair" stands for the condition lying between. From these data, one can conclude that nonaqueous electrolyte solutions containing no linear carbonate solvent (low viscosity solvent) have a problem in wetting the polypropylene separators. To enhance the wettability of the polypropylene separator, a surface-active component such as tetraethylammonium perfluoro octanesulfonate (TEAFOS) was added to the electrolyte solution.[150]

The wettability of a solid surface is defined in terms of the contact angle θ made by a liquid drop resting on the solid surface. The angle θ can take on values ranging from 0 to 180°. If the solid surface is idealized as smooth, chemically homogeneous,

Table 4.1 Wettability of a PP membrane (Celgard 2400)

Solvent, electrolyte	Wettability
DMC	Good
DEC	Good
EC	Poor
PC	Poor
GBL	Poor
H_2O	Poor
1 M $LiPF_6$/EC + DMC (50:50 vol%)	Fair
1 M $LiPF_6$/EC + DEC (50:50 vol%)	Good
1 M $LiPF_6$/EC + PC (50:50 vol%)	Poor
1 M $LiPF_6$/EC + GBL (50:50 vol%)	Poor

isotropic, and nondeformable and there is no mutual solubility or chemical reaction between the two phases, then there is only one stable, equilibrium contact angle which is given by Young's equation represented in Fig. 4.30, where γ_L, γ_S, and γ_{SL} are the surface tensions of the liquid, solid, and their interface, respectively. The surface tensions concerning the solid on the right cannot be directly measured, and $\gamma_L \cos\theta$ on the left is measurable and used as an index of wettability. The surface tension of the solid, when the contact angle is zero (completely wet), is called a critical surface tension γ_c. If the surface tension of a liquid is smaller than this value, the solid is completely wet with the liquid. Table 4.2 lists the surface tension of the solvents used in liquid electrolytes[151] and the critical surface tension of the polymers used for separators and binders.[149,152].

We have explored the compounds by which the wettability of the electrolyte solutions is improved. The contact angle between PC and graphite was dependent on the kind of graphite, and the addition of methyl phenyl carbonate (MPC) reduced it, as

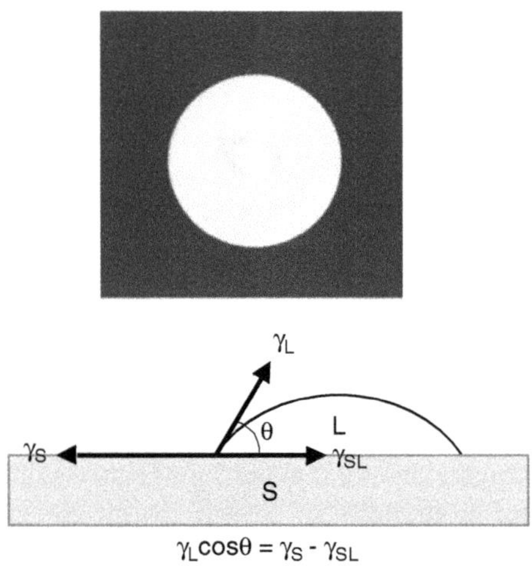

Fig. 4.30 Wettability measurement of a liquid on a solid

Table 4.2 Surface tension of liquids and critical surface tension of solids

Liquid	γ_L/mNm^{-1} (25°C)	Solid	γ_c/mNm^{-1} (20°C)
DMC	28.5	PE	35.6
EMC	26.6	PP	29.8
DEC	25.9	PS	40.6
EC + DMC (30:70 vol%)	33.1	PVDF	40.2
MPC	34.1	PTFE	21.5
DOC	37.7	HFP	14.9
PC	42.3	PVC	44.0
GBL	43.3	PMMA	43.2
EC	51.8 (40°C)	PET	43.8
H_2O	72.8	PAN	44.0

shown in Fig. 4.31, presumably due to the affinity of a phenyl group to the chemical structure of graphite.[153] We also have found that the critical surface tension of the anode and cathode plates is almost determined by the polymer binder included in the electrodes, even though the binder content is less than 10%.[151] Since γ_L of GBL is larger than γ_c of a popular binder, PVDF, the GBL-based electrolytes without a low viscosity solvent have a wetting problem. Only 1 wt% of dioctyl carbonate (DOC) decreased γ_L of the electrolyte to its γ_L, which is lower than γ_c of PVDF, and particularly, a nonionic surfactant containing fluorine atoms (FSO), which has enough electrochemical window, showed the excellent effect as shown in Fig. 4.32.[154]

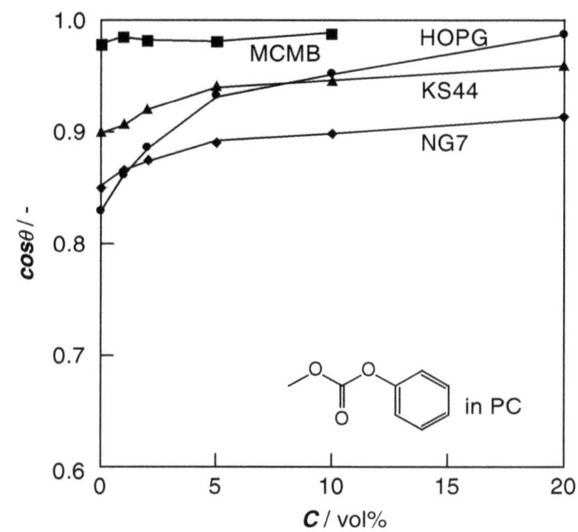

Fig. 4.31 Contact angle between PC containing MPC and various graphites

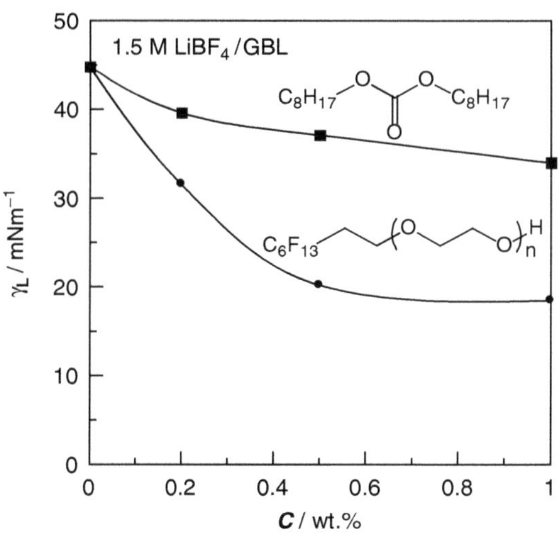

Fig. 4.32 Changes in surface tension of 1.5 M $LiBF_4$/GBL by wetting agents

4.7 Flame Retardant Agents

The author has first proposed the concept of "nonflammable electrolytes" using phosphate solvents, particularly, TMP, which has no flash point.[155] Although it was found later that TMP is not compatible with graphitic carbon anodes, this idea was accomplished by adding anode passivation film forming agents and mixing with another nonflammable fluorinated ether (HFE).[27] The 14500 type cylindrical cells employing 1 M $LiPF_6$/TMP + HFE + VC + DMF (44:40:8:8) passed the nail-test and oven-test. Thereafter, TMP and other nonflammable compounds have been examined as an additive, which affects flame retardation.[156–170] Based on the UL 94 flammability standard, "Test for Flammability of Plastic Materials for Parts in Devices and Appliances," several test procedures have been developed. The linear burn rate of a paper separator[155] or a fiberglass wick[157] soaked with a test solution was measured by horizontal burning. The self-extinguishing time (SET) of a linear shaped glassy filter by vertical burning (Fig. 4.33)[158–160] or SET normalized against the sample weight soaked in a ball-shaped wick[161–166] were recorded. The goal is to find materials that do not support continued combustion when the source of heat, spark, or flame is withdrawn.

We have proposed the flame-retarding mechanism as follows[158]:

1. TMP solvent heated by an external heat source evaporates and reaches the flame in the gaseous form: TMP(l) → TMP(g)
2. The gaseous TMP breaks down in the flame to small radical species containing phosphorous: TMP(g) → [P]·
3. These small radical species containing phosphorous scavenge the H· radicals, which are the main active agent of combustion chain branching reactions: [P]· → [P]H
4. Accordingly, the following combustion chain branching reactions are hindered due to the deficiency of H· radicals: RH → R + H

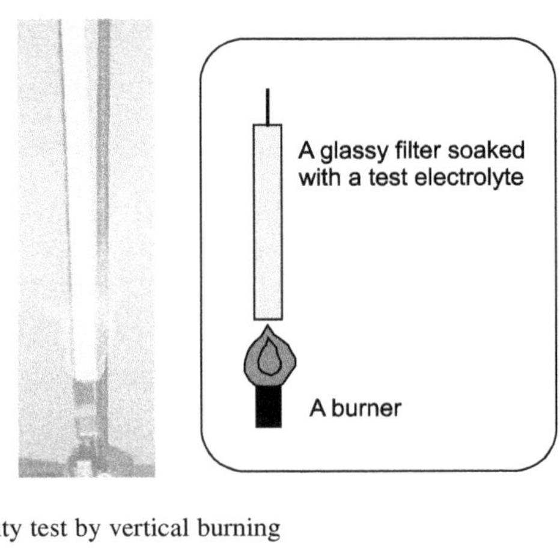

Fig. 4.33 Flammability test by vertical burning

Obviously, two significant factors affect the nonflammability of TMP-based electrolyte systems. One is the TMP content in gas phase. This mainly depends on the boiling points of co-solvents and TMP, and the solvent composition. Another factor is the capacity of co-solvents to produce a hydrogen radical, and the capacity of TMP to produce small radical species containing phosphorous. According to this mechanism, the lowest TMP mol.% in binary electrolytes to maintain nonflammability N_{lim} was represented by the following empirical equation:

$$N_{lim} = 2.6 - 9.3\ (C_P\, T_H / C_H\, T_P), \qquad (4.2)$$

where C_P, T_P, C_H, and T_H are phosphorous content (mol.%) and boiling point (K) of TMP and co-solvents, respectively.

When TMP was used in 1 M $LiPF_6$/EC + PC + TMP and 1 M $LiPF_6$/EC + DEC + TMP ternary systems, more than 20 vol% of TMP and less than 20 vol% of DEC were required at least to keep nonflammability.[160] In these regions no graphitic carbon anode performed well due to the poor reduction stability of TMP[158]; however, utilization of an amorphous carbon in 1 M $LiPF_6$/EC + PC + DEC + TMP (30:30:20:20 vol%)[159] or the addition of 5 wt% ethyl ethylene phosphate (EEP) in 1 M $LiPF_6$/EC + DEC + TMP (60:20:20 vol%)[160] solved the reductive decomposition problem of TMP.

Another family of flame retardant additives is phosphazene derivatives. The addition of hexamethoxycyclotriphosphazene (HMPN) in 1 M $LiPF_6$/EC + DMC (50:50 wt%) depressed exothermic peaks in a differential scanning calorimeter (DSC) and self-heat rate in ARC at elevated temperature due to the passivation layer formed by this additive.[156]

Fair comparison among the organophosphorous compounds, whose chemical structures and properties are given in Fig. 4.34 and Table 4.3, respectively, was carried out by adopting the same baseline electrolyte system and flammability test procedure.[161–164] The results are summarized in Fig. 4.35, where the electrolytes were classified semi-arbitrarily into three categories on the basis of SET. As the content of the additives increases, the flammability drops dramatically. Fluorination made a distinct difference in the flame retarding ability of these phosphates, as indicated by steeper curves. There always is a trade-off between the electrolyte nonflammability and cell performances. More than 20 wt% of the additives was required to obtain nonflammability, however, their high viscosity or inherent electrochemical instability resulted in degradation of cell performances. Among these compounds, tris(2,2,2-trifluoroethyl)phosphate (TFP) showed the best overall performances considering the high flame retarding efficiency, fair ionic conductivity, high reversibility on both anode and cathode materials, and long-term cell stability. Its phosphite form (TTFP) also was effective.[165] The physicochemical properties of $LiPF_6$/EC + PC+ EMC (30:30:40 wt%) containing TFP were thoroughly examined.[166] Polyphosphonate compounds showed preliminary promising results.[168] Since P–N containing flame retardants are generally superior to the corresponding P–O containing ones, hexamethylphosphoramide (HMPA), which is a suspect carcinogen, was examined thoroughly.[169] The flame retarding property of HMPA

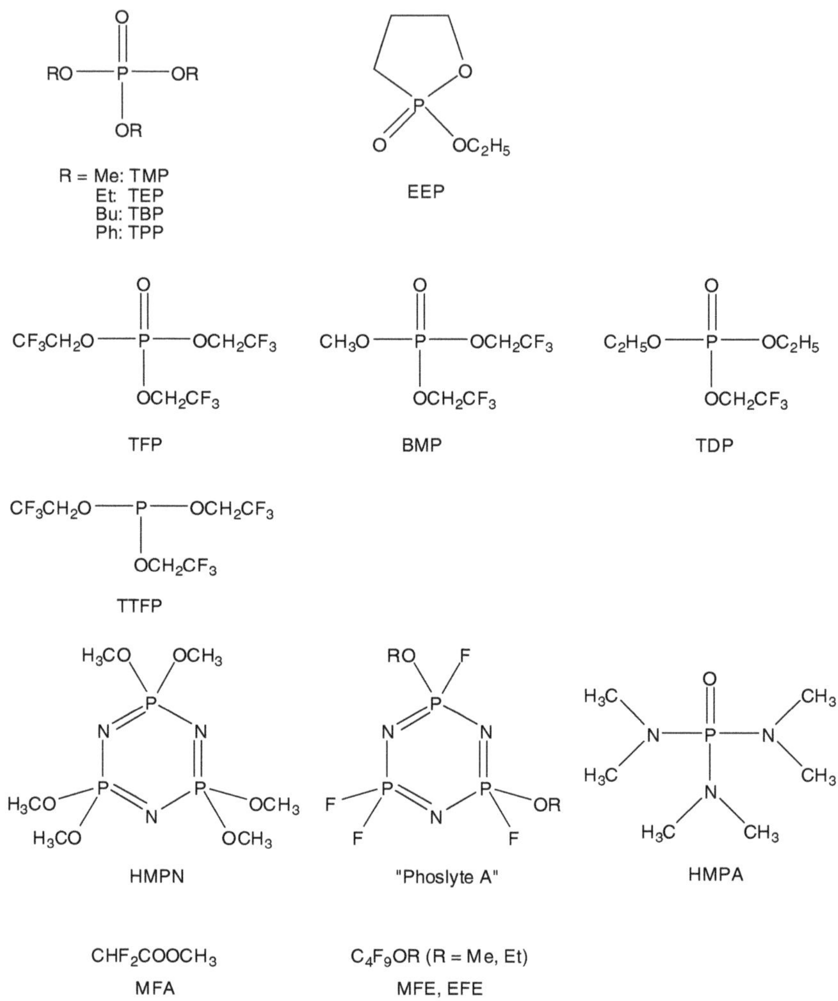

Fig. 4.34 Examples of flame retardant agents

t.3.1 **Table 4.3** Properties of flame retardant additives

	Compound	HMPN	TMP	TEP	TDP	BMP	TFP
t.3.3	B.p. (°C)	> 250	197	215	210	203	178
t.3.4	M.p. (°C)	50	−46	−56		−23	−20
t.3.5	ε_r (20°C)		21	13	15	12	11
t.3.6	C_{SET} (wt%)	> 40	> 40	> 40	40	20	20
t.3.7	η_1 anode (%)[a]	91.5	12.3	82.5	73.4	88.4	91.1
t.3.8	η_1 cathode (%)[a]	80.1	79.9	80.1	76.0	79.0	77.0

t.3.2 (header row)

t.3.9 [a] First cycle coulombic efficiency of 1 M $LiPF_6$/EC + EMC (50:50 wt%) containing 20 wt% of the
t.3.10 additive at graphitic carbon anode and nickel-based cathode

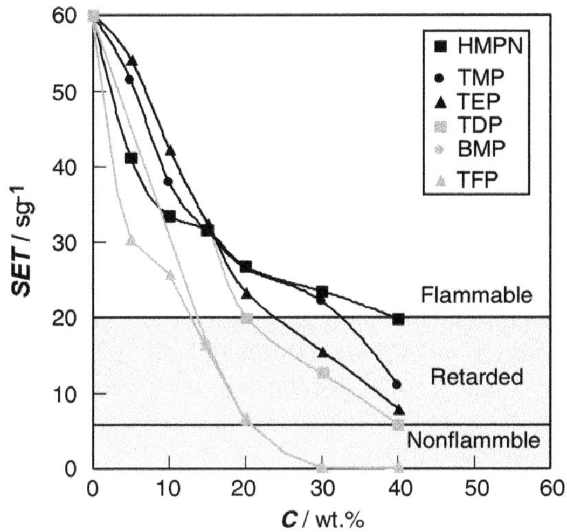

Fig. 4.35 Flammability of 1 M $LiPF_6$/EC + EMC (50:50 wt%) containing organo-phosphorous compounds

was comparable to that of TFP, however, it decreased electrolytic conductivity and electrochemical windows as well as cell performances. Recently, fluorinated cyclic phosphazene derivatives were introduced, and "Phoslyte A" was proposed to improve the performances of non-fluorinated counterparts.[170]

Fluorinated organic solvents have been studied with the intent of improving the low temperature characteristics,[171,172] although halogen atoms are expected to quench radical species in the burning reaction and depress the combustion. It was reported that the addition of partially fluorinated carboxylic acid esters such as methyl difluoroacetate (MFA) suppressed the exothermic reactions of the electrolytes with lithium metal, although the flammability of 1 M $LiPF_6$/EC + DMC + MFA (25:25:50 wt%) and 1 M $LiPF_6$/PC + MFA (50:50 wt%) was not examined.[173–176] Fluorinated ethers such as methyl nonafluorobutyl ether (MFE) also were examined as a main solvent of the electrolyte to reduce the flammability and eliminate the flash point.[177] Because MFE dissolved no lithium salt, the conductivity of the mixed solvent with MFE was always low. Although the cell performances of the 18650 type cylindrical cells employing 1 M $LiN(CF_3SO_2)_2$/EMC + MFE (20:80 vol%) were not good, they did not show thermal runaway when penetrated by a nail, even after the cells were overcharged. Their cell performances were further improved by adding EC and $LiPF_6$.[178,179] Ethyl nonafluorobutyl ether (EFE) also was examined as a cosolvent.[180] Partially fluorinated carbonate and carbamate solvents also were examined with the expectation of the reduced flammability.[181]

Ionic liquids, also known as room temperature molten salts (RTMS), recently attracted much attention as nonflammable electrolytes,[182–184] however, the effect as a flame retardant additive has not yet been reported. Since $LiPF_6$ salt also is important in the flame resistance,[157] fair comparison with ionic liquids in organic solvent systems is necessary.

4.8 Others

It is believed that $LiPF_6$-based electrolytes react with water by the following reactions[185,186] and the produced HF affects the nature of SEI on the electrodes.[187] It was reported that the addition of LiCl and HMPA in $LiPF_6$ electrolytes suppressed the hydrolysis reaction[188] and thermal decomposition,[189] respectively. A Lewis base, which can complex PF_5 to form a stable acid-base adduct, should prevent the generation of the autocatalytic species.

$$LiPF_6 \leftrightarrow LiF + PF_5 \tag{4.3}$$

$$PF_5 + H_2O \rightarrow POF_3 + 2HF \tag{4.4}$$

It is well known that the generated HF enhances the dissolution of Mn^{2+} from $LiMn_2O_4$, which causes capacity fading during cycling at elevated temperatures. Hexamethyldisilazane (HMDS) was used to reduce interfacial resistance.[190] It was considered that HMDS removed water and HF impurities by the following reaction, thus improving cell performance. This kind of additive may be called a "dehydration agent" or "HF scavenger."

$$(CH_3)_3\ SiN\ (H)\ Si\ (CH_3)_3 + H_2O \rightarrow (CH_3)_3\ SiOSi(CH_3)_3 + NH_3 \tag{4.5}$$

$$NH_3 + HF \rightarrow NH_4F \tag{4.6}$$

The addition of a small amount of other silicon compounds such as hexamethylsiloxane and tetraethyl orthosilicate exhibited better cycling behaviors at high temperatures.[191] It was assumed that these compounds reacted with trace HF in solution, and the reaction products were involved in building up the surface films on the cathode, in which Si and Si–F containing species were detected.

Additional improvement relates to solution properties such as ion solvation and conductivity. A typical example is macrocyclic ligands such as crown ethers, whose preferential coordination of Li^+ enhances the Li^+ transference number and cycling efficiency of graphitic anodes.[192,193] However, the studies along these lines have not been continued probably due to the lack of anodic stability.

The "anion receptor" is a similar approach, which strongly coordinates anion, thereby increasing solubility and conductivity. A number of cyclic aza-ether compounds with electron withdrawing CF_3SO_2 groups attached to the aza-nitrogens function with anions in much the same way as crown ethers act with cations.[194–196] The main shortcomings of these compounds are their high molecular weight, low solubility, and instability at elevated temperatures. Another family of anion complexing agents are electron deficient borate, borane, and boronate compounds with various fluorinated aryl or alkyl groups.[197–206] Some of these boron anion receptors can promote the dissolution of normally insoluble salts such as lithium fluoride in several nonaqueous solvents. For example, tris(pentafluorophenyl)borate (TFPBO) increased the solubility of LiF to > 1 M and conductivity more than 3 mS cm^{-1} in EC + DMC (1:2 in volume) at 25°C.[203] These additives can be regarded as a component of the lithium salts, because the following complexation is *in-situ* formation of the new lithium salt:

$$(C_6F_5O)_3B + LiF \rightarrow LiBF\,(OC_6F_5)_3 \tag{4.7}$$

It was shown recently that the addition of tris(pentafluorophenyl)borane (TPFPB) enhanced thermal stability of the graphite anode interface due to a reduced amount of lithium fluoride in the SEI by this complexation.[207] Ethyltriacetoxysilane (ETAS) was used for the same purpose.[208]

4.9 Conclusions

The various functions of the electrolyte additives were explained by classifying them into several categories and by showing their chemical structures. Since small amount of these additives dramatically improve the performances of lithium-ion cells, customers are increasing, who request the liquid electrolytes, including various additives.

Because the degree of the additive effects will change depending not only on the electrode materials but also battery designs, the research and development of the additives become evermore difficult. We have been promoting the research on new electrolyte additives by clarifying their working mechanisms using various analytical techniques, which are specially customized to the analysis of lithium-ion cells, as well as computational micro- (molecular) and macro- (cell) simulations.[209–216] "Role assigned electrolytes" are the electrolytes where additives are formulated by understanding their working mechanisms.

References

1. G. E. Blomgren, in *Lithium Batteries*, J. P. Gabano, ed., Academic Press, New York, NY, Ch. 2 (1983)
2. H. V. Venkatasetty, in *Lithium Battery Technology*, H. V. Venkatasetty, ed., Wiley, New York, NY, Chs. 1 and 2 (1984)
3. L. A. Dominey, in *Lithium Batteries: New Materials, Development and Perspectives*, G. Pistoia, ed., Elsevier Science, Amsterdam, Ch. 4 (1994)
4. J. Barthel, H. J. Gores, in *Chemistry of Nonaqueous Solutions: Current Progress*, G. Mamantov, A. I. Popov, eds., VCH Publishers, New York, NY, Ch. 1 (1994)

5. M. Morita, M. Ishikawa, Y. Matsuda, in *Lithium Ion Batteries: Fundamentals and Performance*, M. Wakihara, O. Yamamoto, eds., Kodansha, Tokyo, Ch. 7 (1998)
6. J. Barthel, H. J. Gores, in *Handbook of Battery Materials*, J. O. Besenhard, ed., Wiely-VCH, New York, NY, Ch. 7 (1999)
7. D. Aurbach, I. Weissman; G. E. Blomgren; D. Aurbach, Y. Gofer; in *Nonaqueous Electrochemistry*, D. Aurbach, ed., Marcel Dekker, New York, NY, Chs. 1, 2, and 4 (1999)
8. G. M. Ehrlich, in *Hand book of Batteries, 3rd ed.*, D. Linden, T. B. Reddy, eds., McGraw-Hill, New York, NY, Ch. 35.2.5 (2001)
9. D. Aurbach, J.-I. Yamaki, M. Saolmon, H.-P. Lin, E. J. Plichta, M. Hendrickson, in *Advances in Lithium-Ion Batteries*, W. A. van Schalkwijk, B. Scrosati, eds., Kluwer Academic/Plenum Publishers, New York, NY, Chs. 1, 5, and 11 (2002)
10. S. Mori, in *Materials Chemistry in Lithium Batteries*, N. Kumagai, S. Komaba, eds., Research Signpot, Kerala, India, p. 49 (2002)
11. M. Nazri; D. Aurbach, A. Schechter; in *Lithium Batteries: Science and Technology*, G.-M. Nazri, G. Pistoia, eds., Kluwer Academic Publishers, Norwell, MA, Chs. 17, and 18 (2003)
12. M. Ue, *Prog. Batteries Battery Mater.*, **16**, 332 (1997)
13. M. Ue, in *Development of Li-ion Rechargeable Battery Materials*, CMC, Tokyo, Ch. 6 (1997); *Material Technologies, Their Evaluation and Application in Advanced Rechargeable Batteries*, Technical Information Institute, Tokyo, Ch. 1.5.3 (1998); *Lithium-Ion Secondary Batteries, 2nd ed.*, M. Yoshio, A. Kozawa, eds., Nikkan Kogyo Shinbunshya, Tokyo, Ch. 6 (2000); *Advanced Technologies for Polymer Batteries II*, K. Kanamura, ed., CMC, Tokyo, Ch. 4.4 (2003) [in Japanese]
14. M. Ue, in *Techno Frontier Symposium 2003*, Makuhari, Japan, April 16–18, Ch. F-5–3 (2003); *35th Semiconference*, Morioka, Japan, December 12–13, Ch. 2 (2003) [in Japanese]
15. M. Ue, in *Extended Abstracts of the Battery and Fuel Cell Materials Symposium*, Graz, Austria, April 18–22, p.53 (2004); *Meeting Abstracts of the 12th International Meeting on Lithium Battery*, Nara, Japan, June 27-July 2, No. 31 (2004); *Meeting Abstracts of the 206th Electrochemical Society Meeting*, Honolulu, HI, October 3–8, No. 308 (2004)
16. G. E. Blomgren, *J. Power Sources*, **119–121**, 326 (2003)
17. K. Xu, *Chem. Rev.*, **104**, 4303 (2004)
18. N. Takami, M. Sekino, T. Ohsaki, M. Kanda, M. Yamamoto, *J. Power Sources*, **97–98**, 677 (2001)
19. N. Takami, T. Ohsaki, H. Hasebe, M. Yamamoto, *J. Electrochem. Soc.*, **149**, A9 (2002)
20. M. Yoshio, H. Yoshitake, K. Abe, in *The Electrochemical Society Extended Abstracts*, Vol. 2003-2, Orlando, FL, October 12–16, No. 280 (2003)
21. H. Kita, A. Kawakami, K. Hasegawa, M. Kimura, JP1993–82138A; JP3095268B [in Japanese]
22. M. Fujimoto, Y. Shouji, T. Nohma, K. Nishio, *Denki Kagaku*, **65**, 949 (1997) [in Japanese]
23. B. Simon, J.-P. Boeuve, US5, 626, 981 (1997)
24. C. Jehoulet, P. Biensan, J. M. Bodet, M. Broussely, C. Moteau, C. Tessier-Lescourret, *Proc. Electrochem. Soc.*, **97–18**, 974 (1997)
25. S. Herreyre, O. Huchet, S. Barusseau, F. Perton, J. M. Bodet, P. Biensan, *J. Power Sources*, **97/98**, 576 (2001)
26. N. Shinoda, J. Ozaki, F. Kita, A. Kawakami, *Proc. Electrochem. Soc.*, **99–25**, 440 (2000)
27. M. Broussely, P. Blanchard, P. Biensan, J. P. Planchat, K. Nechev, R. J. Staniewicz, *J. Power Sources*, **119/121**, 859 (2003)
28. M. Contestabile, M. Morselli, R. Paraventi, R. J. Neat, *J. Power Sources*, **119–121**, 943 (2003)
29. G.-C. Chung, H.-J. Kim, S.-I. Yu, S.-H. Jun, J.-W. Choi, M.-H. Kim, *J. Electrochem. Soc.*, **147**, 4391 (2000)
30. X. Zhang, R. Kostecki, T. J. Richardson, J. K. Pugh, P. N. Ross Jr., *J. Electrochem. Soc.*, **148**, A1341 (2001)
31. H. Ota, Y. Sakata, A. Inoue, S. Yamaguchi, *J. Electrochem. Soc.*, **151**, A1659 (2004)

32. D. Aurbach, K. Gamolsky, B. Markovsky, Y. Gofer, M. Schmidt, U. Heider, *Electrochim. Acta*, **47**, 1423 (2002)
33. S.-K. Jeong, M. Inaba, R. Mogi, Y. Iriyama, T. Abe, Z. Ogumi, *Langmuir*, **17**, 8281 (2001)
34. O. Matsuoka, A. Hiwara, T. Omi, M. Toriida, T. Hayashi, C. Tanaka, Y. Saito, T. Ishida, H. Tan, S. S. Ono, S. Yamamoto, *J. Power Sources*, **108**, 128 (2002)
35. R. Oesten, U. Heider, M. Schmidt, *Solid State Ionics*, **148**, 391 (2002)
36. B. Markovsky, A. Rodkin, Y. S. Cohen, O. Palchik, E. Levi, D. Aurbach, H.-J. Kim, M. Schmidt, *J. Power Sources*, **119–121**, 504 (2003)
37. M. Itagaki, N. Kobari, S. Yotsuda, K. Watanabe, S. Kinoshita, M. Ue, *J. Power Sources*, **135**, 255 (2004)
38. M. Itagaki, S. Yotsuda, N. Kobari, K. Watanabe, S. Kinoshita, M. Ue, *Electrochim. Acta*, **51**, 1629 (2006)
39. H. Ota, K. Shima, M. Ue, J. Yamaki, *Electrochim. Acta*, **49**, 565 (2004)
40. H. Ota, Y. Sakata, Y. Otake, K. Shima, M. Ue, J. Yamaki, *J. Electrochem. Soc.*, **151**, A1778 (2004)
41. Y. Wang, S. Nakamura, K. Tasaki, P. B. Balbuena, *J. Am. Chem. Soc.*, **124**, 4408 (2002)
42. Y. Wang, P. B. Balbuena, *J. Phy. Chem. B*, **106**, 4486 (2002)
43. K. Tasaki, *J. Phys. Chem. B.*, **109**, 2920 (2005)
44. M. Kotato, T. Fujii, N. Shima, H. Suzuki, WO00/79632
45. Y.-K. Han, S. U. Lee, J.-H. Ok, J.-J. Cho, H.-J. Kim, *Chem. Phys. Lett.*, **360**, 359 (2002)
46. J. M. Vollmer, L. A. Curtiss, D. R. Vissers, K. Amine, *J. Electrochem. Soc.*, **151**, A178 (2004)
47. Y. Hu, W. Kong, H. Li, X. Huang, L. Chen, *Electrochem. Commun.*, **6**, 126 (2004)
48. M. Kotato, Y. Sakata, S. Kinoshita, M. Ue, in *Meeting Abstracts of the 206th Electrochemical Society Meeting*, Honolulu, HI, October 3–8, No. 335 (2004)
49. K. Tasaki, K. Kanda, T. Kobayashi, S. Nakamura, M. Ue, *J. Electrochem. Soc.*, **153**, A2192 (2006)
50. H. Suzuki, T. Sato, M. Kotato, H. Ota, H. Sato, US6, 664, 008 (2003)
51. C. Wang, H. Nakamura, H. Komatsu, H. Noguchi, M. Yoshio, H. Yoshitake, *Denki Kagaku*, **66**, 287 (1998) [in Japanese]
52. C. Wang, H. Nakamura, H. Komatsu, M. Yoshio, H. Yoshitake, *J. Power Sources*, **74**, 142 (1998)
53. H. Yoshitake, K. Abe, T. Kitakura, J. B. Gong, Y. S. Lee, H. Nakamura, M. Yoshio, *Chem. Lett.*, **32**, 134 (2003)
54. K. Abe, H. Yoshitake, T. Kitakura, T. Hattori, H. Wang, M. Yoshio, *Electrochim. Acta*, **49**, 4613 (2004)
55. J.-T. Lee, Y.-W. Lin, Y.-S. Jan, *J. Power Sources*, **132**, 244 (2004)
56. P. Ghimire, H. Nakamura, M. Yoshio, H. Yoshitake, K. Abe, *Electrochemistry*, **71**, 1084 (2003)
57. H. J. Santner, K.-C. Möller, J. Ivančo, M. G. Ramsey, F. P. Netzer, S. Yamaguchi, J. O. Besenhard, M. Winter, *J. Power Sources*, **119–121**, 368 (2003)
58. K.-C. Möller, H. J. Santner, W. Kern, S. Yamaguchi, J. O. Besenhard, M. Winter, *J. Power Sources*, **119–121**, 561 (2003)
59. C. Korepp, H. J. Santner, T. Fujii, M. Ue, J. O. Besenhard, K.-C. Möller, M. Winter, *J. Power Sources*, **158**, 578 (2006)
60. G. H. Wrodnigg, J. O. Besenhard, M. Winter, *J. Electrochem. Soc.*, **146**, 470 (1999)
61. G. H. Wrodnigg, T. M. Wrodnigg, J. O. Besenhard, M. Winter, *Electrochem. Commun.*, **1**, 148 (1999)
62. G. H. Wrodnigg, J. O. Besenhard, M. Winter, *J. Power Sources*, **97–98**, 592 (2001)
63. Y. Kusachi and K. Utsugi, in *Extended Abstracts of the 44th Battery Symposium in Japan*, Sakai, Japan, November 4–6, p. 526 (2003) [in Japanese]
64. H. Ota, T. Sato, H. Suzuki, T. Usami, *J. Power Sources*, **97–98**, 107 (2001)
65. H. Ota, T. Akai, H. Namita, S. Yamaguchi, M. Nomura, *J. Power Sources*, **119–121**, 567 (2003)

66. Z. X. Shu, R. S. McMillan, J. J. Murray, I. J. Davidson, *J. Electrochem. Soc.*, **142**, L161 (1995)
67. Z. X. Shu, R. S. McMillan, J. J. Murray, I. J. Davidson, *J. Electrochem. Soc.*, **143**, 2230 (1996)
68. M. Winter, P. Novák, *J. Electrochem. Soc.*, **145**, L27 (1998)
69. M. Winter, R. Imhof, F. Joho, P. Novák, *J. Power Sources*, **81–82**, 818 (1999)
70. A. Naji, P. Willmann, D. Billaud, *Carbon*, **36**, 1347 (1998)
71. A. Naji, P. Willmann, D. Billaud, *Mol. Cryst. Liq. Cryst.*, **310**, 371 (1998)
72. R. McMillan, H. Slegr, Z. X. Shu, W. Wang, *J. Power Sources*, **81–82**, 20 (1999)
73. M. Inaba, Y. Kawatate, A. Funabiki, S.-K. Jeong, T. Abe, Z. Ogumi, *Electrochim. Acta*, **45**, 99 (1999)
74. H. Katayama, J. Arai, H. Akahoshi, *J. Power Sources*, **81–82**, 705 (1999)
75. J. Arai, H. Katayama, H. Akahoshi, *J. Electrochem. Soc.*, **149**, A217 (2002)
76. A. Naji, J. Ghanbaja, P. Willmann, D. Billaud, *Electrochim. Acta*, **45**, 1893 (2000)
77. M. Takehara, R. Ebara, N. Nanbu, M. Ue, Y. Sasaki, *Electrochemistry*, **71**, 1172 (2003)
78. K.-C. Möller, T. Hodal, W. K. Appel, M. Winter, J. O. Besenhard, *J. Power Sources*, **97–98**, 595 (2001)
79. Y. Ein-Eli, S. F. McDevitt, D. Aurbach, B. Markovsky, A. Schechter, *J. Electrochem. Soc.*, **144**, L180 (1997)
80. Y. Ein-Eli, S. McDevitt, R. Laura, *J. Electrochem. Soc.*, **145**, L1 (1998)
81. J. Vetter, P. Novák, *J. Power Sources*, **119–121**, 338 (2003)
82. F. Coowar, A. M. Christie, P. G. Bruce, C. A. Vincent, *J. Power Sources*, **75**, 144 (1998)
83. M. C. Smart, B. V. Ratnakumar, S. Surampudi, *Proc. Electrochem. Soc.*, **99–25**, 423 (2000)
84. W. Xing, C. Schlaikjer, in *Proceedings of the 39th Power Sources Conference*, p. 294 (2000)
85. M. D. Levi, E. Markevich, C. Wang, M. Koltypin, D. Aurbach, *J. Electrochem. Soc.*, **151**, A848 (2004)
86. G.-C. Chung, H.-J. Kim, S.-H. Jun, M.-H. Kim, *Electrochem. Commun.*, **1**, 493 (1999)
87. G. C. Chung, *J. Power Sources*, **104**, 7 (2002)
88. Y. Matsuo, K. Fumita, T. Fukutsuka, Y. Sugie, H. Koyama, K. Inoue, *J. Power Sources*, **119–121**, 373 (2003)
89. H. Suzuki, M. Kotato, JP2000–268859A; JP3658517B [in Japanese]
90. K. Abe, H. Yoshitake, T. Tsumura, H. Nakamura, M. Yoshio, *Electrochemistry*, **72**, 487 (2004) [in Japanese]
91. D. L. Foster, B. K. Behl, J. Wolfenstine, *J. Power Sources*, **85**, 299 (2000)
92. S. S. Zhang, K. Xu, T. R. Jow, *Electrochem. Solid-State Lett.*, **5**, A206 (2002)
93. O. Chusid, Y. Ein Ely, D. Aurbach, M. Babai, Y. Carmeli, *J. Power Sources*, **43–44**, 47 (1993)
94. D. Aurbach, Y. Ein-Eli, O. Chusid, Y. Carmeli, M. Babai, H. Yamin, *J. Electrochem. Soc.*, **141**, 603 (1994)
95. Y. Ein-Eli, B. Markovsky, D. Aurbach, Y. Carmeli, H. Yamin, S. Luski, *Electrochim. Acta*, **39**, 2559 (1994)
96. B. Simon, J. P. Boeuve, M. Broussely, *J. Power Sources*, **43–44**, 65 (1993)
97. J. O. Besenhard, M. W. Wagner, M. Winter, A. D. Jannakoudakis, P. D. Jannakoudakis, E. Thodoridou, *J. Power Sources*, **43–44**, 413 (1993)
98. J. O. Besenhard, M. Winter, J. Yang, W. Biberacher, *J. Power Sources*, **54**, 228 (1995)
99. Y. Ein-Eli, S. R. Thomas, V. R. Koch, *J. Electrochem. Soc.*, **143**, L195 (1996)
100. Y. Ein-Eli, S. R. Thomas, V. R. Koch, *J. Electrochem. Soc.*, **144**, 1159 (1997)
101. Y. Ein-Eli, *J. Electroanal. Chem.*, **531**, 95 (2002)
102. M. W. Wanger, C. Liebenow, J. O. Besenhard, *J. Power Sources*, **68**, 328 (1997)
103. S. Mori, H. Asahina, H. Suzuki, A. Yonei, K. Yokoto, *J. Power Sources*, **68**, 59 (1997)
104. D. Aurbach, J. S. Gnanaraj, W. Geissler, M. Schmidt, *J. Electrochem. Soc.*, **151**, A23 (2004)
105. K. Xu, S. Zhang, B. A. Poese, T. R. Jow, *Electrochem. Solid-State Lett.*, **5**, A259 (2002)
106. K. Xu, S. Zhang, T. R. Jow, *Electrochem. Solid-State Lett.*, **6**, A117 (2003)

107. K. Xu, S. Zhang, T. R. Jow, *Electrochem. Solid-State Lett.*, **6**, A144 (2003)
108. J. Jiang, J. R. Dahn, *Electrochem. Solid-State Lett.*, **6**, A180 (2003)
109. S. S. Zhang, K. Xu, T. R. Jow, *J. Power Sources*, **129**, 275 (2004)
110. G. V. Zhuang, K. Xu, T. R. Jow, P. N. Ross Jr., *Electrochem. Solid-State Lett.*, **7**, A224 (2004)
111. K. Xu, U. Lee, S. Zhang, J. L. Allen, T. R. Jow, *Electrochem. Solid-State Lett.*, **7**, A273 (2004)
112. S. Komaba, B. Kaplan, T. Ohtsuka, Y. Kataoka, N. Kumagai, H. Groult, *J. Power Sources*, **119–121**, 378 (2003)
113. S. Komaba, T. Itabashi, B. Kaplan, H. Groult, N. Kumagai, *Electrochem. Commun.*, **5**, 962 (2003)
114. M.-S. Wu, J.-C. Lin, P. J. Chiang, *Electrochem. Solid-State Lett.*, **7**, A206 (2004)
115. M.-S. Wu, P. J. Chiang, J.-C. Lin, J.-T. Lee, *Electrochim. Acta*, **49**, 4379 (2004)
116. D. Aurbach, M. D. Levi, E. Levi, H. Teller, B. Markovsky, G. Salitra, *J. Electrochem. Soc.*, **145**, 3024 (1998)
117. T. Eriksson, T. Gustafsson, J. O. Thomas, *Proc. Electrochem. Soc.*, **98–16**, 315 (1999)
118. D. Ostrovskii, F. Ronci, B. Scrosati, P. Jacobson, *J. Power Sources*, **94**, 183 (2001)
119. Y. Matsuo, R. Kostecki, F. McLarnon, *J. Electrochem. Soc.*, **148**, A687 (2001)
120. Y. Wang, X. Gao, S. Greenbaum, J. Liu, K. Amine, *Electrochem. Solid-State Lett.*, **4**, A68 (2001)
121. X. Zhang, J. K. Pugh, P. N. Ross, *J. Electrochem. Soc.*, **148**, E183 (2001)
122. A. Inoue, S. Kinoshita, N. Maekita, H. Ota, T. Nakamura, T. Akai, K. Kanamura, in *The Electrochemical Society Extended Abstracts*, Vol. 2003-2, Orland, FL, October 12–16, No. 307 (2003)
123. M. Onuki, S. Kinoshita, M. Ue, in *Meeting Abstracts of the 205th Electrochemical Society Meeting*, San Antonio, TX, May 9–13, No. 60 (2004)
124. H. Ota, M. Ue, unpublished results
125. M. Itagaki, N. Kobari, S. Yotsuda, K. Watanabe, S. Kinoshita, M. Ue, *J. Power Sources*, **148**, 78 (2005)
126. M. Deguchi, T. Matsui, in *Extended Abstracts of the 44th Battery Symposium in Japan*, Sakai, Japan, November 4–6, p. 518 (2003) [in Japanese]
127. K. Abe, T. Takaya, H. Yoshitake, Y. Ushigoe, M. Yoshio, H. Wang, *Electrochem. Solid-State Lett.*, **7**, A462 (2004)
128. K. Abe, H. Yoshitake, T. Takaya, H. Nakamura, M. Yoshio, T. Hirai, *Electrochemistry*, **73**, 199 (2005) [in Japanese]
129. M. Onuki, JP2003-331920A [in Japanese]
130. S. R. Narayanan, S. Surampudi, A. I. Attia, C. P. Bankston, *J. Electrochem. Soc.*, **138**, 2224 (1991)
131. C. C. Cha, X. P. Ai, H. X. Yang, *J. Power Sources*, **54**, 255 (1995)
132. M. Adachi, K. Tanaka, K. Sekai, *J. Electrochem. Soc.*, **146**, 1256 (1999)
133. T. J. Richardson, P. N. Ross Jr., *J. Electrochem. Soc.*, **143**, 3992 (1996)
134. F. Tran-Van, M. Provencher, Y. Choquette, D. Delabouglise, *Electrochim. Acta*, **44**, 2789 (1999)
135. T. J. Richardson, P. N. Ross Jr., *J. Power Sources*, **84**, 1 (1999)
136. T. J. Richardson, P. N. Ross Jr., *Proc. Electrochem. Soc.*, **99–25**, 687 (2000)
137. Y. Yan, Z. Zhou, in *Solid State Ionics: Materials and Devices*, B. V. R. Chowdari, W. Wang, eds., World Scientific, Singapore, p. 467 (2000)
138. H.-J. Kim, S.-I. Yoo, J.-J. Cho, in *Extended Abstracts of the 43rd Battery Symposium in Japan*, Fukuoka, Japan, October 12–14, p.78 (2002)
139. D.-Y. Lee, H.-S. Lee, H.-S. Kim, H.-Y. Sun, D.-Y. Seung, *Korean J. Chem. Eng.*, **19**, 645 (2002)
140. K. Shima, M. Ue, J. Yamaki, *Electrochemistry*, **71**, 1231 (2003)
141. L. Xiao, X. Ai, Y. Cao, H. Yang, *Electrochim. Acta*, **49**, 4189 (2004)
142. S. Tobishima, Y. Ogino, Y. Watanabe, *Electrochemistry*, **70**, 875 (2002) [in Japanese]
143. S. Tobishima, Y. Ogino, Y. Watanabe, *J. Appl. Electrochem.*, **33**, 143 (2003)
144. K. Shima, K. Shizuka, M. Ue, WO03/012912

145. K. Shima, K. Shizuka, M. Ue, H. Ota, T. Hatozaki, J. Yamaki, *J. Power Sources*, **161**, 1264 (2006)
146. G. E. Blomgren, *J. Power Sources*, **81–82**, 112 (1999)
147. J. J. Auborn, H. Schonhorn, *Proc. Electrochem. Soc.*, **81–4**, 372 (1981)
148. C. Menachem, E. Peled, L. Burstein, Y. Rosenberg, *J. Power Sources*, **68**, 277 (1997)
149. J. Y. Song, Y. Y. Wang, C. C. Wan, *J. Electrochem. Soc*, **147**, 3219 (2000)
150. A. Chagnes, B. Carré, P. Willmann, R. Dedryvère, D. Gonbeau, D. Lemordant, *J. Electrochem. Soc*, **150**, A1255 (2003)
151. M. Iwade, D. Noda, H. Suzuki, M. Ue, unpublished results
152. A. Kitazaki, T. Hata, *J. Adhesion Soc. Jpn.*, **8**, 131 (1972) [in Japanese]
153. H. Suzuki, N. Shima, K. Hasegawa, Y. Yoshida, JP1996–306387A; JP3893627B [in Japanese]
154. D. Noda, M. Kotato, T. Fuji, H. Suzuki, JP2002–319433A [in Japanese]
155. M. Ue, JP1992–184870A; JP3274102B [in Japanese]
156. C. W. Lee, R. Venkatachalapathy, J. Prakash, *Electrochem. Solid-State Lett.*, **3**, 63 (2000)
157. D. Peramunage, J. M. Ziegelbauer, G. L. Holleck, *Proc. Electrochem. Soc.*, **2000-21**, 306 (2001)
158. X. Wang, E. Yasukawa, S. Kauya, *J. Electrochem. Soc.*, **148**, A1058 (2001)
159. X. Wang, E. Yasukawa, S. Kauya, *J. Electrochem. Soc.*, **148**, A1066 (2001)
160. H. Ota, A. Kominato, W.-J. Chun, E. Yasukawa, S. Kasuya, *J. Power Sources*, **119-121**, 393 (2003)
161. K. Xu, M. S. Ding, S. Zhang, J. L. Allen, T. R. Jow, *J. Electrochem. Soc.*, **149**, A622 (2002)
162. K. Xu, S. Zhang, J. L. Allen, T. R. Jow, *J. Electrochem. Soc.*, **149**, A1079 (2002)
163. K. Xu, M. S. Ding, S. Zhang, J. L. Allen, T. R. Jow, *J. Electrochem. Soc.*, **150**, A161 (2003)
164. K. Xu, M. S. Ding, S. Zhang, J. L. Allen, T. R. Jow, *J. Electrochem. Soc.*, **150**, A170 (2003)
165. S. Zhang, K. Xu, T. R. Jow, *J. Power Sources*, **113**, 166 (2003)
166. M. S. Ding, K. Xu, T. R. Jow, *J. Electrochem. Soc.*, **149**, A1489 (2002)
167. Y. E. Hyung, D. R. Vissers, K. Amine, *J. Power Sources*, **119–121**, 388 (2003)
168. B. G. Dixon, R. S. Morris, S. Dallek, *J. Power Sources*, **138**, 274 (2004)
169. S. I. Gozales, W. Li, B. L. Lucht, *J. Power Sources*, **135**, 291 (2004)
170. T. Ogino, M. Otsuki, S. Endo, in *46th New Battery Committee Seminar*, Tokyo, April 23, p. 17 (2003) [in Japanese]
171. T. Nakajima, K. Dan, M. Koh, *J. Fluorine Chem.*, **87**, 221 (1998)
172. T. Nakajima, K. Dan, M. Koh, T. Ino, T. Shimizu, *J. Fluorine Chem.*, **111**, 167 (2001)
173. J. Yamaki, I. Yamazaki, M. Egashira, S. Okada, *J. Power Sources*, **102**, 288 (2001)
174. K. Sato, I. Yamazaki, S. Okada, J. Yamaki, *Solid State Ionics*, **148**, 463 (2002)
175. M. Ihara, B. T. Hang, K. Sato, M. Egashira, S. Okada, J. Yamaki, *J. Electrochem. Soc.*, **150**, A1476 (2003)
176. J. Yamaki, T. Tanaka, M. Ihara, K. Sato, I. Watanabe, M. Egashira, S. Okada, *Electrochemistry*, **71**, 1154 (2003)
177. J. Arai, *J. Appl. Electrochem.*, **32**, 1071 (2002)
178. J. Arai, *J. Electrochem. Soc.*, **150**, A219 (2003)
179. J. Arai, *J. Power Sources*, **119–121**, 388 (2003)
180. M. Morita, T. Kawasaki, N. Yoshimoto, M. Ishikawa, *Electrochemistry*, **71**, 1067 (2003)
181. M. C. Smart, B. V. Ratnakumar, V. S. Ryan-Mowrey, S. Surampudi, G. K. S. Prakash, J. Hu, I. Cheung, *J. Power Sources*, **119–121**, 359 (2003)
182. A. Weber, G. E. Blomgren, in *Advances in Lithium-Ion Batteries*, W. A. van Schalkwijk, B. Scrosati, eds., Kluwer Academic/Plenum Publishers, New York, NY, Ch. 6 (2002)
183. M. Ue, in *Ionic Liquids*, H. Ohno, ed., CMC, Tokyo. Ch. 7.3 (2003) [in Japanese]
184. M. Ue, in *Electrochemical Aspects of Ionic Liquids*, H. Ohno, ed., Wiley, New York, NY, Ch. 17 (2005)
185. U. Heider, R. Oesten, M. Jungnitz, *J. Power Sources*, **81–82**, 119 (1999)
186. K. Tasaki, K. Kanda, S. Nakamura, M. Ue, *J. Electrchem. Soc.*, **150**, A1628 (2003)
187. T. Sato, M. Deschamps, H. Suzuki, H. Ota, H. Asahina, S. Mori, *Mat. Res. Soc. Sym. Proc*, **496**, 457 (1998)
188. T. Kawamura, T. Sonoda, S. Okada, J. Yamaki, *Electrochemistry*, **71**, 1139 (2003)

189. C. L. Campion, W. Li, W. B. Euler, B. L. Lucht, B. Ravdel, J. F. DiCarlo, R. Gitzendanner, K. M. Abraham, *Electrochem. Solid-State Lett.*, **7**, A194 (2004)
190. H. Yamane, T. Inoue, M. Fujita, M. Sano, *J. Power Sources*, **99**, 60 (2001)
191. B. Markovsky, A. Nimberger, Y. Talyosef, A. Rodkin, A. M. Belostotskii, G. Salitra, D. Aurbach, H.-J. Kim, *J. Power Sources*, **136**, 296 (2004)
192. Z. X. Shu, R. S. McMillan, J. J. Murray, *J. Electrochem. Soc.*, **140**, L101 (1993)
193. M. Inaba, Y. Kawatate, A. Funabiki, S.-K. Jeong, T. Abe, Z. Ogumi, *Electrochemistry*, **67**, 1153 (1999)
194. H. S. Lee, X. Q. Yang, J. McBreen, L. S. Choi, Y. Okamoto, *J. Electrochem. Soc.*, **143**, 3825 (1996)
195. H. S. Lee, X. Sun, X. Q. Yang, J. McBreen, J. H. Callahan, L. S. Choi, *J. Electrochem. Soc.*, **147**, 9 (2000)
196. K. Tasaki, S. Nakamura, *J. Electrochem. Soc.*, **1487**, A984 (2001)
197. X. Sun, H. S. Lee, S. Lee, X. Q. Yang, J. McBreen, *Electrochem. Solid-State Lett.*, **1**, 239 (1998)
198. H. S. Lee, X. Q. Yang, C. L. Xiang, J. McBreen, L. S. Choi, *J. Electrochem. Soc.*, **145**, 2813 (1998)
199. X. Sun, H. S. Lee, X. Q. Yang, J. McBreen, *J. Electrochem. Soc.*, **146**, 3655 (1999)
200. H. S. Lee, X. Q. Yang, J. McBreen, *J. Power Sources*, **97–98**, 566 (2001)
201. X. Sun, H. S. Lee, X. Q. Yang, J. McBreen, *Electrochem. Solid-State Lett.*, **4**, A184 (2001)
202. X. Sun, H. S. Lee, X. Q. Yang, J. McBreen, *J. Electrochem. Soc.*, **149**, A355 (2002)
203. H. S. Lee, X. Sun, X. Q. Yang, J. McBreen, *J. Electrochem. Soc.*, **149**, A1460 (2002)
204. X. Sun, H. S. Lee, X. Q. Yang, J. McBreen, *Electrochem. Solid-State Lett.*, **5**, A248 (2002)
205. X. Sun, H. S. Lee, X. Q. Yang, J. McBreen, *Electrochem. Solid-State Lett.*, **6**, A43 (2003)
206. H. S. Lee, Z. F. Ma, X. Q. Yang, X. Sun, J. McBreen, *J. Electrochem. Soc.*, **151**, A1429 (2004)
207. M. Herstedt, M. Stjerndahl, T. Gustafsson, K. Edström, *Electrochem. Commun.*, **5**, 467 (2003)
208. M. Herstedt, H. Rensmo, H. Siegbahn, K. Edström, *Electrochim. Acta*, **49**, 2351 (2004)
209. S. Zhang, A. Tsuboi, H. Nakata, T. Ishikawa, *J. Power Sources*, **97–98**, 584 (2001)
210. P. Kolar, H. Nakata, J.-W. Shen, A. Tsuboi, H. Suzuki, M. Ue, *Fluid Phase Equilibria*, **228–229**, 59 (2005)
211. T. Kawai, *Chemical Engineering*, **69**, 26 (2005) [in Japanese]
212. Y. Wang, S. Nakamura, M. Ue, P. B. Balbuena, *J. Am. Chem. Soc.*, **123**, 11708 (2001)
213. T. Tasaki, *J. Electrochem. Soc.*, **149**, A418 (2002)
214. M. Ue, A. Murakami, S. Nakamura, *J. Electrochem. Soc.*, **149**, A1385 (2002)
215. M. Ue, A. Murakami, S. Nakamura, *J. Electrochem. Soc.*, **149**, A1572 (2002)
216. M. Ue, in *High Energy and Power Technologies for Electric Double-layer Capacitors and Lithium-ion Rechargeable Batteries*, Technical Information Institute, Tokyo, Ch. 12.2 (2005) [in Japanese]

Chapter 5
Carbon-Conductive Additives for Lithium-Ion Batteries

Michael E. Spahr

5.1 Introduction

5.1.1 Basic Relationships

Carbon materials like carbon black and graphite powders are widely used in positive and negative electrodes to decrease the inner electrical resistance of several electrochemical systems.[1] The attractive features of carbon materials in electrochemical systems are the high electrical conductivity and, in the case of graphite, the high thermal conductivity. Besides, carbon materials are nontoxic and environmentally benign; they are available in high purity and large quantity and can be handled easily. A sufficiently high corrosion resistance to acidic and basic media, the low weight, and the relatively low production costs are advantages of carbon to conductive metal powders.

Carbon materials are added to decrease the electrical resistivity of the electrode mass but generally are not involved in the electrochemical redox process which delivers the energy of the electrochemical cell. To optimize the specific charge of the electrode as well as the energy density of the electrochemical cell, the amount of carbon in the total electrode volume needs to be minimized. The carbon quantity typically applied is below 10 wt% of the total electrode mass. Thus, in relation to the electrochemically active electrode material, the conductive carbon component can be considered to be electrode additive.

Methods to determine the electronic conductivity of powdered battery materials and their mixtures have been studied intensively.[2–4] To mathematically describe the electronic conductivity of the active electrode material mixed with different amounts of conductive carbon, logarithmic equations,[5] the percolation theory (PT),[6,7] and the effective medium theory (EMT)[7–9] may be considered to be rules.

M.E. Spahr
TIMCAL Ltd, Strada Industriale, CH-6743, Bodio, Switzerland
m.spahr@ch.timcal.com

M. Yoshio et al. (eds.), *Lithium-Ion Batteries*,
DOI 10.1007/978-0-387-34445-4_5

Logarithmic equations are specific for cases in which the conductivity differences between the individual components of the mixture are small. The latter two theories better comply with binary mixtures in which the electronic conductivity of the two components significantly differ. Both theories define the binary mixture of particles with different conductivities as spatial resistance lattice. The EMT uses a type of perturbation calculation to determine the conductivity of a network that describes the particle arrangement. The conductivity relationship of the mixture is mathematically described for proportions between 0 and 100% of the better-conducting component. In contrast, the PT simplifies the conductivities of the components in the binary mixture as largely different. The PT examines a minimum amount of the component with the higher conductivity. The percolation threshold is indicated by a sharp rise in specific conductivity when a certain volume percentage of the better conductor is reached. At this critical volume, the amount of particles of the good conductor, being close enough to assure electrical contact to each other either by electron tunneling or direct contact, is sufficient to form a conductive path that spans through the entire electrode. From the percolation threshold up to 100 vol%, the approximated PT formula describes the conductivity in dependence of proportions of the particles with the good conductivity. The electronic resistivity dependence of a binary powder mixture on the volume fraction of the higher conductive component is shown schematically in Fig. 5.1.

In a real electrode, the required amount of the conductive carbon depends on both the nature of the carbon and the active electrode material as well as on their particle size distributions. For an optimum electrical resistivity, the conductive carbon might be used slightly above the percolation threshold. Usually, the practical concentrations of conductive additive are higher than the percolation thresholds due to other electrochemical parameters and electrode-processing factors.

To ensure a high retention of the cell capacity at high current drain, both the electronic and ionic conductivity of the electrodes must be optimal.[7,10,11] A high ionic conductivity is guaranteed, besides a high ionic conductivity of the active electrode material, by a good contact of all electrode particles to the electrolyte and by a high ion transport rate through the electrode pores. To avoid concentration

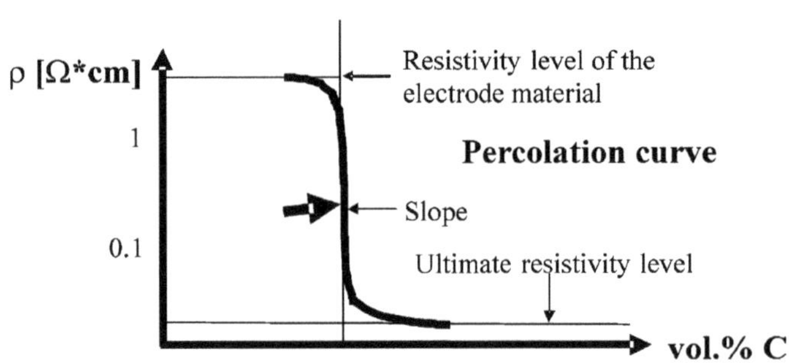

Fig. 5.1 Schematic presentation of a percolation curve ideally describing the effect of a conductive additive at various carbon concentrations on the electronic resistivity of an electrode

polarization effects, an optimal pore structure of the electrode is required. The conductive carbon influences the size and shape of the voids between the active particles in the electrodes. The ability of conductive carbon powders to absorb electrolyte additionally influences the penetration of electrolyte into the electrode. An increased electrode density could improve the electronic resistance of the electrode by an improved electrical contact between the particles of the electrode mass. However, by increasing the electrode density, the ion transport properties within the electrode pores may deteriorate due to an unfavorable electrode tortuosity. For an optimum electrochemical performance, a good balance between the electrode density and the electrolyte penetration into the electrodes must be found.

In addition, the carbon additive may influence the mechanical aspects of the electrode, which is important for both the electrochemical cell performance as well as the electrode manufacturing process. Carbon properties such as compressibility and polymer binder absorption affect the mechanical stability of the electrode, and thus the electrode manufacturing process and production yield.

5.1.2 Specific Requirements for the Conductive Additive in the Lithium Ion Technology

Carbon conductive additives are applied in both the positive and the negative electrode of commercial lithium ion batteries. The electrode design and manufacturing process deduces specific electrical and mechanical requirements for the carbon conductive additive. Lithium-ion battery electrodes are film electrodes of about 50–100 μm thickness that are attached on both sides of a copper foil (negative electrode) or an aluminum foil (positive electrode) current collector. These film electrodes are manufactured by coating a liquid dispersion of the electrode mass on the metal foil using a blade coating or printing process. The mechanical stability of the electrodes is achieved by adding several percentages of binder materials such as polyvinylene difluoride (PVDF) or other nonfluorinated polymers to the electrode formulation as well as by roll pressing. The electrodes are stacked on each other, separated by a thin, porous polymer foil soaked with liquid electrolyte or, in the case of a lithium polymer battery, by a gelled electrolyte. Cells are assembled by stacking, folding, or rolling the sandwich structurelike arrangement.

Due to the particular thin film design of lithium battery electrodes, it might be misleading to select the optimal type and amount of conductive carbon exclusively based on its percolation threshold. It is important to note that the PT only applies for electrical resistivity relationships of the bulk volume. Thus for an optimal electrical electrode performance, specific thin film parameters such as the electrode thickness must be taken into consideration. Nevertheless, electrical resistivity measurements of blends of conductive carbons and the active electrode material provide useful comparative information about the electronic properties of different carbons. In general, conductive carbons provide the conductive matrix in which the active component is ideally dispersed. They optimize the electrical contact between the redox-active particles and may improve the contact of the electrode to the current collector foil. Due to the particular cell design, the electrical resistivity of the electrode perpendicular to the current collector foil is particularly important for the electrochemical cell properties.

In addition to the electronic conductivity, a high ionic conductivity of both the positive and negative electrode is required for the cell performance, specifically at high current rates. A good ionic contact of all electrode particles as well as a sufficiently high lithium ion transport rate in the electrode pores are controlled by the amount of electrolyte that is retained in the electrodes as well as by the size and shape of the electrode pores (tortuosity).[12–15] Due to the limited cell volume, besides the separator, the electrodes also function as electrolyte reservoirs. From this point of view, the porosity should be as high as possible, which contradicts the demand of charge density and a sufficient mechanical strength of the electrodes. The carbon-conductive additive influences the structure as well as the electrolyte retention in the electrodes, and thus is a key factor for the cell performance in the high-current drain regime.

Since the introduction of the first lithium-ion battery by Sony in 1990 up to now, the energy density of newly introduced generations of portable lithium-ion batteries has been increased continuously. This has been achieved by improved electrode materials, especially in the negative electrode. At the same time, the cell designs have been optimized to be able to use higher amounts of active electrode material in the cell. Higher electrode densities as well as thinner current collector and separator foils allow the possibility of packing more active material in the cell volume. Electrode porosities below 30% have been achieved in practical cells. However, these low electrode porosities as well as extremely thin separator foils limit the amount of electrolyte in the cell. To achieve a good cycling stability and power density, a trade-off between the charge density and the electrolyte retention in the electrodes needs to be found. In this context, the selection of the conductive carbon has been gaining increasing importance. Highly compressible carbon materials help to increase the press density of the electrodes. At the same time, carbon additives are selected whose electrolyte absorption is balanced to maintain a sufficiently high wetting of the electrode by the electrolyte with a minimum volume of electrolyte. To avoid excessive electrolyte penetration and thus a large swelling of the electrode, the amount and type of binder material is another important factor.

Trace element impurities in the electrode materials deteriorate the energy density, cycling stability, and calendar life of the cell by provoking undesired electrochemical side reactions that lead to charge losses in the cell. In addition, metal particles as well as electrode material particles larger than the electrode thickness ("oversized particles") can generate local short circuits between the electrodes, leading to self-discharge of the cell. A minimum amount of trace element impurities and no contamination of the conductive carbon material with large particles and metal particles are prerequisites for good battery storage properties as well as long cycle and battery life.

Besides the electrochemical parameters, the carbon additive influences the electrode-manufacturing process. Even though used in small quantities, the carbon additive influences the rheology of the liquid dispersion, which is prepared to coat the electrode mass on the current collector foil. Also, the mechanical stability and flexibility of the film electrodes are influenced by the carbon additive. For sufficiently high film cohesion and adhesion on the current collector foil, the polymer binder should work as an adhesive but flexible link between the electrode particles as well as between the particles and the current collector foil. A minimum amount of polymer binder is applied in both electrodes because, as insulating and electrochemical inactive material, it increases the electrode polarization and decreases the energy density of the cell. The proportion of polymer that is absorbed by the carbon does not contribute to the mechanical strength of the electrode. Thus, for a good film adhesion and cohesion of an electrode with minimal binder content, the absorption of polymer within the carbon material should be low, but the adsorption of polymer at the carbon surface should be high. Usually, carbon materials with low polymer absorption also show low electrolyte absorption capability. A good balance between polymer absorption and electrolyte absorption allows good mechanical stabilities and sufficiently high electrolyte penetration in the electrode, and thus a good electrochemical performance.

In modern commercial batteries, a variety of graphite, carbon black, and special fibrous materials are applied in both the positive and the negative electrodes. So far, no standardization has occurred for the types of the active electrode materials used in secondary lithium battery systems. Every individual active electrode material imposes special requirements on the conductive additive for an optimum battery performance. In addition, existing lithium battery manufacturing processes show differences that require special adjustments of the electrode formulations and material properties. The numerous types of lithium-ion batteries with specific characteristics, which today are produced and sold to the market, require specific electrode formulations. For example, a typical lithium polymer battery containing a polymer (gel-type) electrolyte system contains a different conductive carbon matrix to a lithium ion battery containing a liquid electrolyte system.[16] In the following, the characteristic material and battery-related properties of graphite, carbon black, and other specific carbon-conductive additives are described.

5.2 Graphite Powders

5.2.1 Preparation

The types of graphite carbon powders, which primarily are applied as conductive additives in the positive and negative electrodes of lithium batteries, belong to the family of highly crystalline graphite materials. These graphite materials show real densities of 2.24–2.27 g cm^{-3} (values based on the xylene density according to DIN

12 797 and DIN 51 901-X) and average interlayer distances of $c/2 = 0.3354–0.3360$ nm. These materials are either based on natural graphite or primary synthetic graphite sources.[17] Primary synthetic graphite materials are synthesized specifically for the desired application. In contrast, secondary synthetic graphite (so-called "scrap" graphite) is manufactured from defect graphite electrodes or graphite electrode parts that were produced for the aluminum or steel production process.[18,19] Secondary synthetic graphite usually is not suitable for electrochemical applications because of its poor crystallinity and high impurity level.

Primary synthetic graphite powders are derived from mixtures of selected carbon precursors such as petroleum cokes and coal tar-based cokes (Fig. 5.2). These coke materials are graphitized by a heat treatment at temperatures above 2,500°C under exclusion of oxygen. During this heat treatment the amorphous coke material is transformed to crystalline carbon. Various continuous and batch graphitization processes are possible for this heat treatment. A traditional batch process is the Acheson furnace technology.[18] In this process the carbon raw material is positioned between two electrodes and covered by refractory material to protect it from the air atmosphere. An electric current passes through the carbon bulk, which represents an electric resistance between the electrodes. Using this Joule effect, temperatures above 3,000°C can be generated. In alternative batch or continuous graphitization processes the carbon precursor is heat-treated in resistance, induction, or plasma furnaces in an inert helium gas atmosphere. Some graphitization processes use graphitization catalysts to facilitate the transformation from the amorphous to the crystalline carbon and to improve the crystallinity. Due to the high amount of electrical energy needed to graphitize carbon, electrical power is one of the main cost factors for the synthetic graphite production. If starting coke materials with high

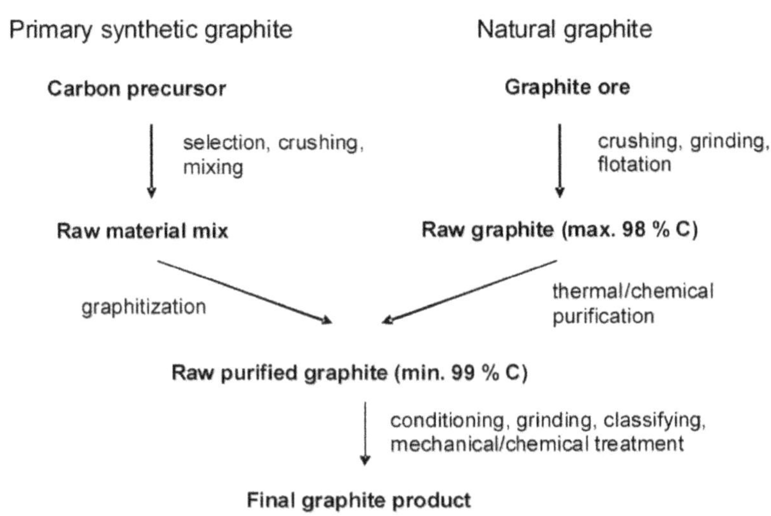

Fig. 5.2 Main steps of the synthetic and natural graphite manufacturing process

purity are used in the graphitization process, purity levels above 99.9% can be achieved. To reach the final particle size distribution, the graphite raw material is conditioned by applying mechanical treatments such as specific grinding, sieving, or classification processes.

Whereas in the case of the synthetic graphite production the final product purity is controlled by the starting material, the natural graphite process primarily is a purification of the graphite ore taken from a mine. Geologic graphite deposits in nature are the result of the transformation of organic matter under high metamorphic pressure combined with high temperatures. The graphite contents of graphite ores available from exploited graphite mines can vary between 3% and more than 90%, and mainly depend on the location and quality of the mine. The graphite is retrieved from the ore in various subsequent crushing, grinding, and flotation steps.[17] Modern flotation processes achieve purities of more than 98%. The remaining impurities give a fingerprint of the deposit from which the graphite originates. To achieve the purities above 99.9%, which are required for lithium batteries, the graphite material is further purified either by thermal or chemical purification. Thermal purification processes are performed at high temperatures, mostly above 1,500°C. Depending on the nature of the impurities, reactive gases such as chlorine may be added. Chemical treatment is performed by either acid or alkaline leaching, using either mineral acids or aqueous NaOH. The efforts to purify natural graphite increase with the purity level required. Nowadays, the costs for natural graphite powders with purities above 99.9% are in a similar range as for synthetic graphite with the same purity.

5.2.2 *Graphite Material Properties*

For the selection of a graphite-conductive additive, the question of its artificial or natural origin plays a minor role. The graphite material properties are of higher importance. A variety of natural graphite materials with different material properties exist. Synthetic graphite processes simulate the natural graphite formation and allow manufacturing graphite materials with similar properties as those of natural graphite materials. The main categories of graphite properties that distinguish graphite materials are crystallinity, texture and porosity, particle size and shape, surface properties, as well as purity.

5.2.2.1 Crystallinity

The chemical crystal structure of graphite consists of condensed six-membered rings of carbon atoms forming planar layers. As shown in Fig. 5.3, these layers are stacked in a way that every third layer has an identical position to the first layer, resulting in a stacking sequence A, B; A, B; A, B; etc.[20,21] This structure type represents the thermodynamically stable hexagonal crystal structure. Besides, a

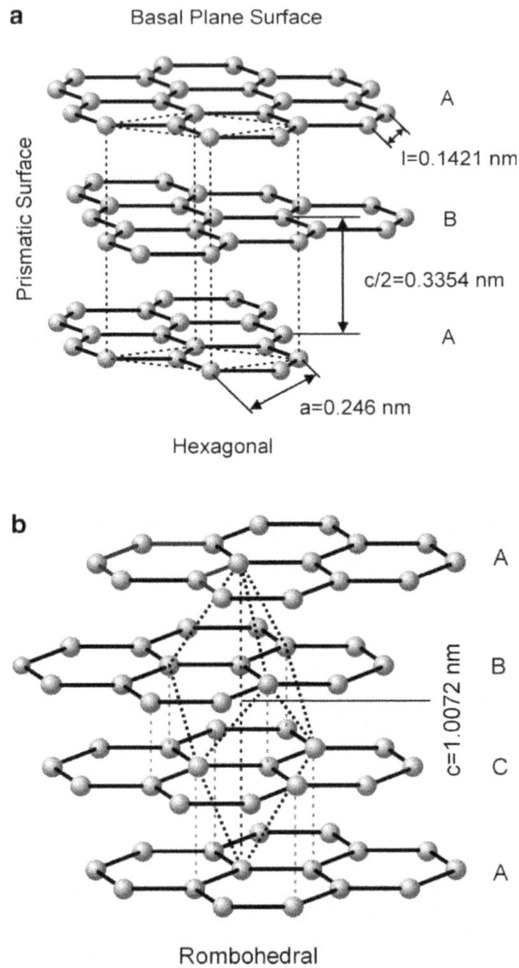

Fig. 5.3 The hexagonal and rhombohedral graphite crystal structure

rhombohedral structure exists in which only every fourth graphite layer has an identical position to the first layer.[22] This rhombohedral structure appears as statistical stacking defects and can be formed by mechanical deformation of hexagonal crystals. Heat treatment completely transforms the rhombohedral structure to the hexagonal form. The carbon atoms are bound by covalent bonds with a bond length of 0.1421 nm. Besides the three localized σ-bonds to their nearest neighbors, every carbon atom participates with one valence electron to a delocalized π molecular orbital along the graphite layer, which is responsible for the metallic behavior parallel to graphite layers. The graphite layer stacks are bound by relatively weak van der Waals forces, which are the reason for a relatively large theoretical inter-layer distance of 0.3353 nm. The specific electrical conductivity parallel to the layer is found to be $2.6.10^4\ \Omega^{-1}\ cm^{-1}$ at room temperature and decreases with increasing temperature.[21,23] Perpendicular to the layer, the electrical conductivity is smaller by a factor of 10^{-4} and increases with increasing temperature, which is typical for a semiconductor.

Crystallinity of graphite powders is characterized by X-ray diffraction. The interlayer distance $c/2$ gives information about the degree of crystallinity of the single crystals or the degree of graphitization. High crystallinity is a prerequisite for a graphite powder to be used as a conductive additive. For this reason, conductive graphite powders usually show $c/2$ values between 0.3354 and 0.3356 nm. The size of the crystals parallel to the graphite layers and perpendicular to them is important for the distinction of different graphite materials. The L_a describes the average crystal size parallel to the graphite layers, and the L_c describes the average crystal size in c direction and gives information about the average number of graphite layers stacked on each other in a single crystal. It is important to note that both L_c and L_a characterize average crystal sizes but generally do not give information about the degree of graphitization and the graphite crystallinity. Examples of highly crystalline graphites with very small L_c and L_a values are the naturally occurring amorphous and vein graphites.

5.2.2.2 Texture, Porosity, and Particle Shape of Graphite Powders

Graphite powders consist of polycrystalline particles, which have the shape of platelets.[24] These platelets are agglomerates of intergrown single crystals. The graphite texture describes the orientation of the single crystal in the particle (mosaicicity). Two extreme cases of graphite textures are schematically illustrated in Fig. 5.4. The first case contains few, relatively large, single crystals, which are aligned along the platelet plane and give rise to relatively strong anisotropic material properties. Such types of textures typically can be found in graphites with flaky or anisometric particle shapes. The second case contains many relatively small single crystals that are randomly oriented in the particle, giving rise to more isotropic properties. Such types of textures typically can be found in graphite materials with isometric particle shapes. Figure 5.5 shows a high-resolution transition

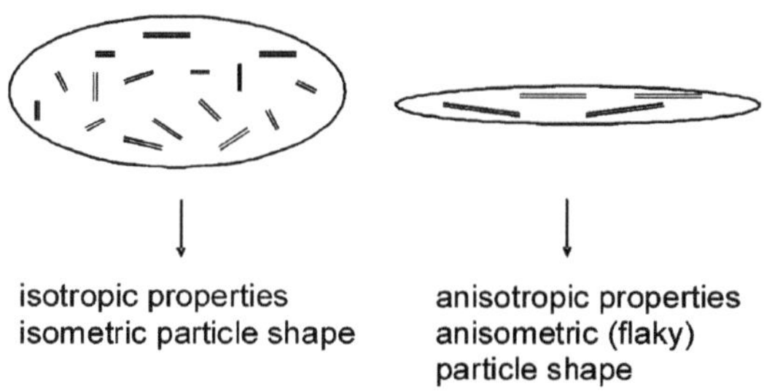

Fig. 5.4 Schematic of two graphite particle models with an extremely isotropic and an anisotropic graphite texture

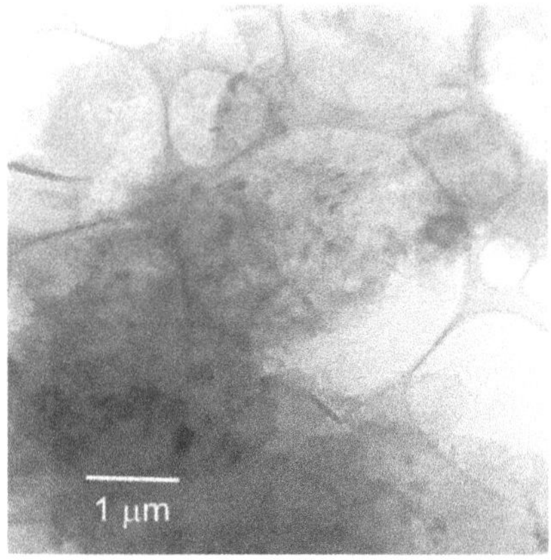

Fig. 5.5 High-resolution transition electron microscope (HRTEM) picture of a synthetic TIMREX® KC44 graphite particle describing the graphite texture as agglomeration of intergrown single crystal domains

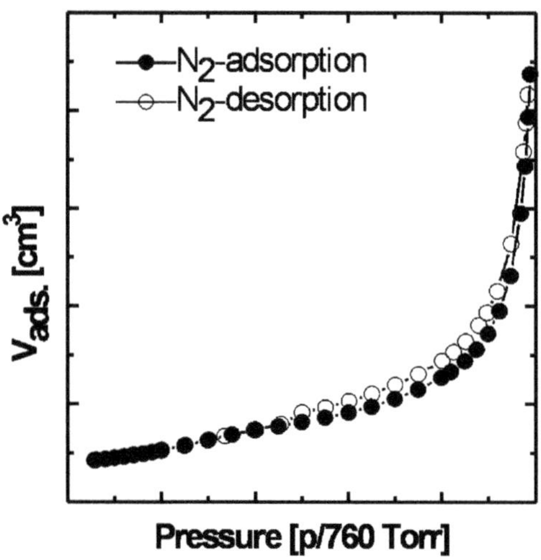

Fig. 5.6 Nitrogen adsorption isotherm of TIMREX® SFG44, which indicates the mesoporous character of highly crystalline graphite materials

electronmicroscope (HRTEM) picture of a TIMREX® KC44 synthetic graphite particle, which illustrates a practical example of a graphite texture. Most of the graphite single crystals in this particle are oriented parallel to the *xy*-plane of the platelet. However, crystals oriented perpendicular to the platelet plane also can be found. The space between the single crystals in the particles form slit pores with random pore size distribution.

Figure 5.6 shows an isotherm obtained from nitrogen absorption measurements of TIMREX® SFG44, which is a typical highly crystalline graphite. The shape of the graphite adsorption isotherm according to the Brunauer, Emmet, and Teller theory as well as the HRTEM picture of the TIMREX® KC44 graphite particle describe the graphite texture as agglomeration of intergrown single crystal domains. Theory as well as the hysteresis of the adsorption and desorption curve at higher pressure values according to the Brunauer, Joyner, and Halenda theory is characteristic for a mesoporous material.[25]

The specific (BET) surface area, which is measured by nitrogen adsorption, consists of surface area fractions contributed by the geometrical surface area, by the surface roughness and surface defects as well as the mesopores. The geometric surface area is related to the particle dimensions. It increases with decreasing particle-sized distribution. For fine graphite powders, the increase of the geometric surface area is the main reason for the increase of the specific BET surface area.

5.2.2.3 Graphite Surface Properties

The total graphite surface distinguishes three basic surface types, which can be derived from the graphite crystal structure. The outer graphite layers of the graphite single crystal form the basal plane surfaces. The edges of the exposed graphite layers form the higher energetic prismatic surfaces (polar edges). The sp^2-carbons terminating the graphite layers at the edges have free valences, which are saturated by surface groups. Besides hydrogen, typically surface oxygen groups such as carboxyl, carboxylic anhydride, lactone, lactol, phenolic hydroxyl, carbonyl, *o*-quinone, or ether-type groups can be found at the prismatic edges.[26–28] The third type of surface are low-energy defects such as dislocation lines created by imperfections of the graphite structure. The fraction of prismatic, basal plane surfaces, and low-energy defects, which can be measured especially by krypton gas adsorption measurements, is a key material parameter that explains performance differences of graphite materials in the various applications. In addition, superficially disordered sp^3 carbon is considered to be another defect type that can influence the graphite performance. The degree of crystallinity at the surface or the surface-near regions that corresponds to the intensity ratio of the D- and G- band of RAMAN spectroscopy measurements is another parameter to distinguish surface properties of graphite powders.[29,30]

The concept of the active surface area (ASA) is a useful parameter to describe graphite surfaces. It is based on the fact that during chemisorption of oxygen at outgassed carbon surfaces at 300°C and an oxygen partial pressure of 50–100 Pa, surface oxygen complexes are formed on a specific part of the graphite surface called the "active surface area."[31,32] The ASA is composed of active sites that exist on the carbon surface where the carbon atom valency is not saturated. On a "clean" graphite surface, these active sites would be located on the edges of the exposed graphene layer planes (prismatic surfaces) as well as at points of imperfection in the graphite structure including vacancies, dislocations, and steps in the outer basal plane surfaces.[33] They can be attributed to structural features, heteroatoms (O, S, N), and mineral matter. The ASA and d_{ASA} correspond to the amount of active sites and the density of active sites, respectively. The d_{ASA} is defined as the fraction of ASA related to the total graphite surface area, and therefore is calculated by normalizing the ASA by the total surface area (TSA) of the graphite material, determined by the BET method. The ASA and d_{ASA} are index values that are used to characterize the reactivity of graphite surfaces toward gaseous and liquid media.

5.2.2.4 Purity

The amount and nature of trace element impurities are decisive parameters that decide whether a graphite material can be used in a secondary lithium battery. Especially, metal impurities can affect the electrochemical processes and create side reactions that lead to charge losses in the cell, a deterioration of the cycling stability, and reduced battery life.

5.2.3 *Graphite as Conductive Additive*

5.2.3.1 Graphite in the Positive Electrode

From the theoretical standpoint of the PT and EMT, the required amount of conductive carbon in the positive electrode for a sufficiently low resistivity should be lower, the lower the apparent (bulk) density of the carbon is. The low apparent density, besides the requirements given by the dimensions of the electrode film, is the reason why fine graphite powders are used mainly as conductive additives. An average particle size below 10 μm is typical for graphite conductive additives in the positive electrode. Graphite materials with such particle sizes only show small differences in apparent density (see Scott density values in Table 5.1). Therefore, additional material properties must be considered to understand the different performance of graphite-conductive additives. Figure 5.7 shows the resistivity of pressed pellets containing a $LiCoO_2$ positive electrode material mixed with different types of fine TIMREX® graphite powders at different graphite concentrations. At high graphite concentrations, all graphite materials provide similar electrical pellet resistivities. The KS15, SFG15, and MX15 are different graphite types with identical particle size distribution (average particle size distribution of about 8 μm) and show differences in pellet resistivity at low graphite concentration. For the best performing graphite, MX15, the required resistivity level of the electrode is achieved at lower graphite concentration than for SFG15 and KS15. No significant difference of the pellet resistivity values could be observed for different particle size distributions of the same graphite types in the case of maximum particle sizes below ca. 20 μm Although KS6 ($d_{90\%}$ = 6.5 μm) is significantly finer than KS15 ($d_{90\%}$ = 17 μm), and thus has a smaller apparent density, a similar performance can be observed for the two graphite materials regarding the electrical resistivity at different graphite concentrations.

Table 5.1 Typical material parameters of fine TIMREX® graphite powders used as conductive additives[a]

Material parameter	Particle size $d_{90\%}$ [μm]	Particle shape	BET SSA $m^2\ g^{-1}$	DBPA gDBP _100 g graphite (ASTM 281)	OAN mL DBP _100 g graphite (ASTM D2414)	Scott density g cm^{-1} (ASTM B329)
KS4	4.8	Isometric irregular spheroids	26	200	116	0.05
KS6	6.5	Isometric irregular spheroids	20	170	114	0.07
KS15	17.0	Isometric irregular spheroids	12	140	104	0.10
SFG6	6.5	Anisometric flakes	16	180	117	0.06
SFG15	18.0	Anisometric flakes	7	150	110	0.09
SFG44	48.0	Anisometric flakes	5	120	90	0.19
MX15	18.0	Strongly anisometric flakes	9	190	120	0.06

[a] The Scott density is a standardized method to measure the apparent density of a powder

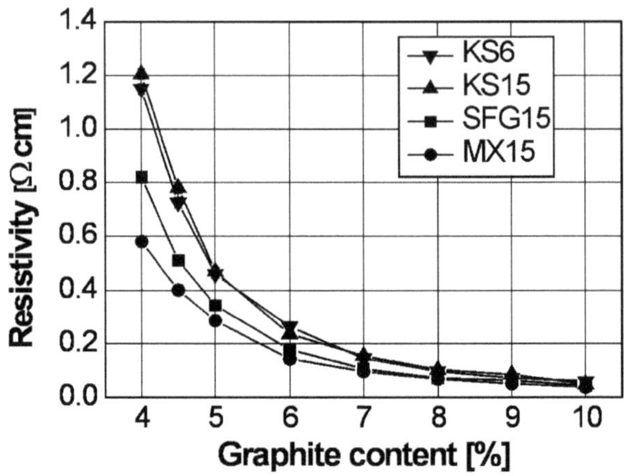

Fig. 5.7 Electrical pellet resistivity of a mixture of a $LiCoO_2$-positive electrode material with different graphite materials at different graphite concentrations (pellets were pressed at 3 tons cm^{-2} and the resistivity was measured by a 4-point method)

The additional material parameters that need to be considered to explain the different electrical performances are the aspect ratio of the particles and the graphite texture. The MX15 has an extremely anisometric (flaky) particle shape, whereas the particle shape of KS15 is isometric. As can be seen from the TEM pictures in Fig. 5.8, the particles of MX15 consist of a few large, single crystals aligned parallel to the *xy*-plane of the graphite platelet. The shape of the platelets is extremely anisometric. Electron diffraction at single MX15 graphite platelets showed separated diffraction peaks, indicating the small number of crystals grown together in the particle. A similar behavior is observed for the SFG15 material. In contrast, the KS platelet contains a large number of small crystals that are oriented more randomly. The particle shape is more isometric. Electron diffraction shows no separated diffraction peaks but Debye rings, indicating the distinct polycrystalline morphology of the KS particle.

For ultrafine graphite powders with maximum particle sizes below 10 μm, the effect of graphite texture and particle shape on the electrode resistivity vanishes. For instance, different types of ultrafine graphite materials show almost no difference in the electrical performance. The similar electrical performance of KS6 and KS15 in the electrode mass can be explained by the higher apparent density of KS6, which compensates the stronger influence of the aspect ratio of the coarser KS15 particles on the pellet resistivity.

The particle aspect ratio not only has an influence on the electrode performance, but also on the DBP absorption (DBPA), which affects the electrode manufacturing process. The DBPA gives the amount of dibutyl phthalate (DBP) oil in gram which is absorbed by 100-g carbon (ASTM 281). In contrast, the oil absorption number (OAN) is based on a modified method giving the volume of DBP, which is absorbed by 100-g carbon (ASTM D2414). The OAN, which is commonly used to compare

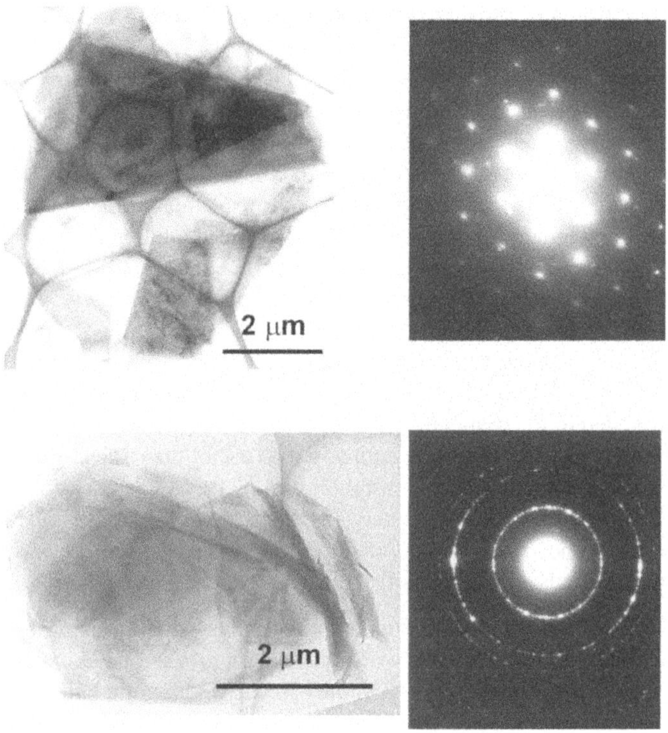

Fig. 5.8 TEM pictures as well as electron diffraction patterns of MX15 and KS15 graphite particles

different carbon blacks, is not always feasible for graphite materials. Despite its limits for graphite materials, Table 5.1 shows OAN values for the sake of direct comparison with the conductive carbon blacks mentioned later in this text. The DBPA value correlates with the amount of polymer that a graphite is able to absorb. Graphite powders with high DBPA also require a higher amount of solvent to be dispersed in liquid media. The DBPA increases with decreasing particle size distribution and increasing aspect ratio. The more anisometric the particles of a graphite powder are, the higher is its DBPA. The specific BET surface area does not have a direct influence on the DBPA value. This could be explained by the fact that the DBP oil, like other liquids or dissolved polymer binder molecules, cannot enter all graphite pores in which the nitrogen gas used for the BET measurement penetrates. The higher the DBPA is, the more polymer binder is necessary to achieve a sufficient cohesion and adhesion of the electrode film on the metal current collector. In addition, graphite materials with higher DBPA cause higher viscosities when dispersed in *N*-methyl pyrolidone, dimethyl amide, or acetone. This is an important factor for the slurry preparation of the electrode coating process. The advantage in the electrode manufacturing process might be the reason why graphite materials with low DBPA, despite slightly lower electrical performance, preferentially are used as conductive additive.

5.2.3.2 Graphite as Conductive Additive in the Negative Electrode

Electrical and mechanical aspects. At first glance, carbon-conductive additives do not seem to be necessary for negative electrodes, which contain a carbon material as active electrode material. However, typical commercial carbon electrode materials such as graphitized mesocarbons (e.g., mesocarbon microbeads, MCMB, or other similar mesocarbons) as well as hard carbon-coated graphite materials (e.g., MPG, MAG) show complementary properties compared to graphite powders optimized as conductive additive.[34–39] The particular texture of the graphitized mesocarbon as well as the hard carbon coating of graphite materials with core-shell structure cause particular surface morphologies and a low BET surface area ranging from 1 to 2 $m^2\ g^{-1}$, thus fulfilling an important requirement of negative electrode materials in portable lithium batteries. Such electrode materials typically show relatively low intrinsic conductivities due to the relatively low crystallinity in the case of graphitized mesocarbon and due to the hard carbon coating in the case of the surface-treated graphite. In addition, the particular surface morphologies of these active electrode materials could cause high contact resistances between the particles leading to impedance problems of the electrode. Thus, carbon materials are applied in the negative electrodes for the same functions as in the positive electrode.

The positive effect of a graphite-conductive additive on the electrode impedance, cycling stability, and rate capability of a mesocarbon-based negative electrode has been reported.[40,41] As shown in Fig. 5.9, the capacity fading of a half-cell containing an MCMB-negative electrode could be significantly improved by adding 10% of SFG6. The SFG6 improves the electrode impedance, thus improving the capacity retention of the cell, especially at high currents.

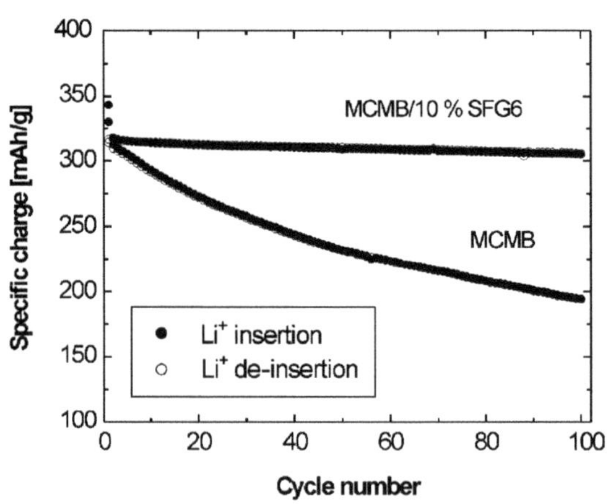

Fig. 5.9 Capacity fading of a lithium-ion cell containing an MCMB electrode with and without the addition of 10% SFG6[42]

In addition to the electrical performance, the graphite-conductive additive could improve the electrode density. Typically, the graphitized mesocarbon or surface-treated graphite electrode show a relatively large spring back after pressure release, which is the reason for the relatively low compressibility. In addition, high compaction pressure during the electrode manufacturing process may lead to particle breakage.[43] In contrast, a conductive graphite powder with high crystallinity show high compressibility. Adding conductive graphite to the mesocarbon electrode mass could improve the electrode density, as shown in Fig. 5.10.

Electrochemical aspects. The use of highly crystalline graphite as a conductive additive in the negative electrode is a particular case because of its double function: the graphite material works both as a conductivity enhancer and an active electrode component. Due to the high degree of crystallinity, graphite materials are able to electrochemically insert lithium up to a chemical composition of LiC_6. Structurally, LiC_6 corresponds to the stage-1 lithium graphite intercalation compound in which all graphite interlayer spaces are occupied by lithium.[44–49] Thus, at an appropriate current density and with a suitable electrode design, an equivalent specific charge of 372-mAh g^{-1} graphite is achievable using highly crystalline graphite in the negative electrode. Figure 5.11 shows as an example a galvanostatic potential curve describing the first electrochemical lithium insertion and deinsertion in a highly crystalline graphite. A high reversible capacity of about 365 mAh g^{-1} (at *C*/5) is obtained in the measurement. The steplike part of the potential curve below 0.25 V vs. Li/Li^+ indicate the reversible formation of distinct lithium graphite intercalation compounds during the electrochemical lithium insertion process. Thus, the steplike shape potential curve can be obtained only in the case of graphite materials with a high crystallinity degree. Carbon with lower crystallinity shows a less structured potential curve.

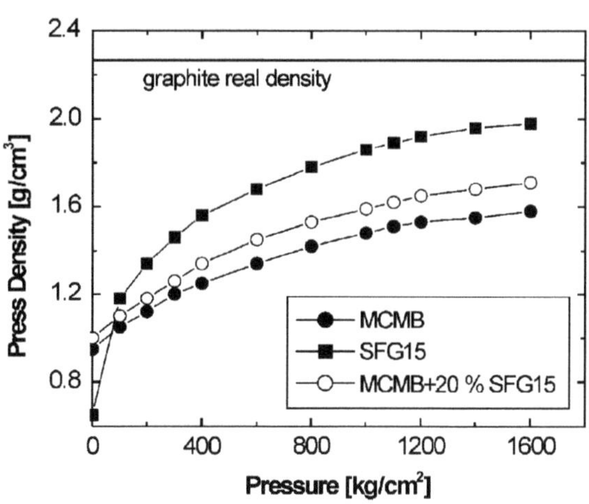

Fig. 5.10 Electrode density as a function of the compaction pressure of a pure MCMB electrode, a pure TIMREX® SFG15 electrode, as well as an electrode containing MCMB blended with 20% SFG15

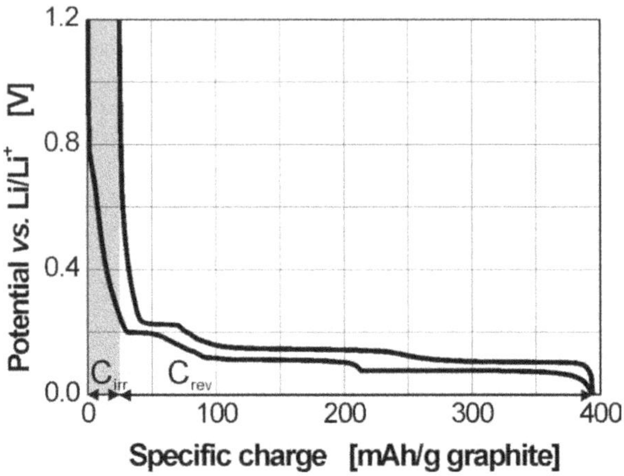

Fig. 5.11 First electrochemical lithium insertion in graphite with high degree of graphitization (TIMREX® SFG44)

The irreversible part of the potential curve in Fig. 5.11 is the charge loss, which corresponds to the formation of a passivation layer at the graphite surface. This so-called solid electrolyte interphase (SEI) layer is formed by electrolyte decomposition products. The SEI layer is essential for the reversible electrochemical process because it passivates the graphite surface against further electrolyte decomposition.[50–52] However, the charge loss related to the SEI layer formation needs to be minimized, since it decreases the energy density of the cell.

Numerous publications have studied the structure and formation mechanism of the SEI layer as well as its impact on the cell performance.[52–57] It is a general agreement that the charge loss of a lithium cell is increasing with the specific BET surface area of the graphite-negative electrode material. Figure 5.12 shows the linear relation of the specific charge loss of different synthetic and natural graphite materials with a high degree of crystallinity with their specific BET surface area. For this graphite family, a linear relation of the irreversible capacity with the double-layer capacitance also exists. The double-layer capacitance measured by impedance spectroscopy in the low-frequency range corresponds to the active electrode surface area wetted by the electrolyte. The resulting linear relation of the irreversible capacity with the specific BET surface area is typical for the family of highly crystalline graphites. It indicates that the liquid battery electrolyte penetrates into the graphite mesopores, which participate at the electrochemical processes occurring in the negative electrode.[58]

An optimal SEI layer formation not only minimizes the irreversible capacity, and thus the energy density of the cell, it also affects the cycling stability, rate capability, and safety aspects of the cell.[59] In addition, an effective SEI layer suppresses the electrochemical exfoliation of graphite caused by the co-intercalation of solvated lithium between the graphite layers.[46,47,49,60–62] The unstable solvated lithium intercalation compound formed as a short-living intermediate is decomposed

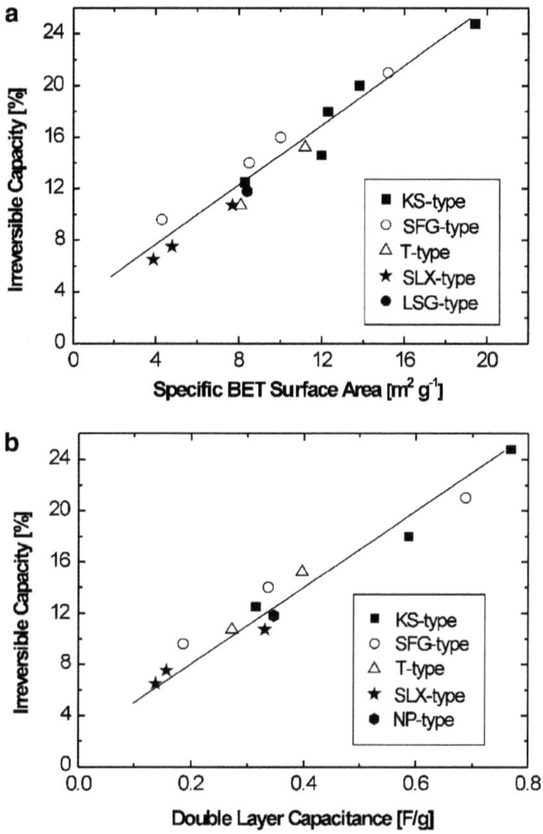

Fig. 5.12 Irreversible capacity of various highly crystalline graphite materials as a function of the specific BET surface area and the double-layer capacitance

electrochemically. The decomposition reaction generates gases between the graphite layers and in microcracks of the particles that exfoliate the graphite structure and destroy the particles. Electrochemical graphite exfoliation leads to a decomposition of the graphite structure and finally battery failure.[53,54] Exfoliation easily happens for graphite materials with high crystallinity especially in the case that the battery electrolyte contains propylene carbonate as solvent.[63]

An effective SEI formation being controlled by a suitable electrolyte system suppresses graphite exfoliation. Usually, electrochemical exfoliation of highly crystalline graphite can be avoided in an ethylene carbonate-based electrolyte systems.[64,65] Also, electrolyte additives can support the formation of an optimal SEI layer. A typical example is vinylene carbonate, which forms protective polymeric passivation films at the graphite surface.[66–68] Graphite exfoliation can be avoided during the first electrochemical reduction in a suitable electrolyte system, if the formation of the SEI layer is accomplished at more positive potentials than the potential at which the graphite exfoliation takes place.

Besides the electrolyte system, the graphite texture and surface morphology influence the passivation and exfoliation tendency.[69–73] It could be shown that the active surface area (ASA) and the density of active surface sites (d_{ASA}) give information about the reactivity of the graphite surface toward the electrolyte system.[74] Thus, the ASA could be used to estimate and compare the passivation properties of different graphite materials: Graphite materials with large ASA or d_{ASA} show a higher reactivity toward the electrolyte system than graphite materials with smaller ASA or d_{ASA} values. For graphite materials with the higher d_{ASA} value, passivation can be expected at more positive potentials. Graphite materials with a low d_{ASA} value show electrochemical exfoliation even in ethylene carbonate electrolyte systems due to the hindered graphite passivation caused by the low surface reactivity toward the electrolyte system.[74–76] The ASA and d_{ASA} values can explain the stabilities of different graphite-negative electrode materials toward exfoliation in mixed ethylene carbonate/propylene carbonate systems. SFG6 is a graphite-conductive additive with a high stability in electrolyte systems containing high fractions of propylene carbonate. Even at 50 wt.% propylene carbonate content in the electrolyte, SFG6 shows no exfoliation effect, as indicated in Fig. 5.13. In contrast, at this propylene carbonate level, SFG44 and SFG15 show exfoliation which is indicated in the potential curve by the irreversible plateau at about 0.8 V vs. Li/Li^+.

Differential electrochemical mass spectrometry (DEMS) measurements indicate significant differences of the passivation mechanism of SFG6 compared to SFG15 and SFG44, which could be correlated with the different surface reactivities (Fig. 5.14). Since ethylene and propylene are gaseous decomposition products formed during the first reduction of the graphite-negative electrode from the ethylene carbonate and propylene carbonate electrolyte components, they can be used as tracer gases to monitor the graphite passivation process in the DEMS measurement.

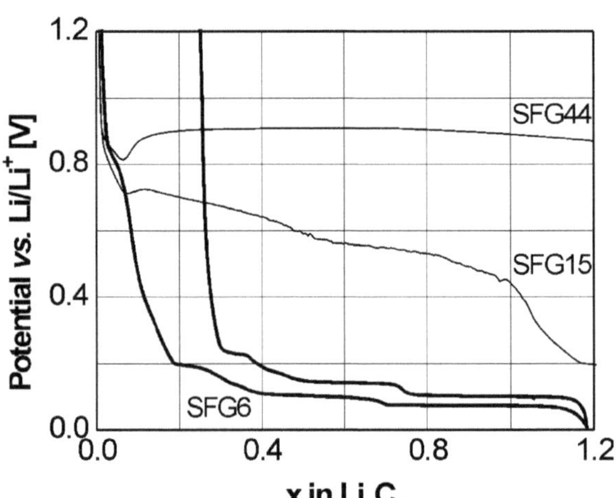

Fig. 5.13 First electrochemical lithium insertion into different SFG-type conductive additives using 1-M $LiPF_6$ in ethylene carbonate/propylene carbonate 1:1 (w/w) as electrolyte system

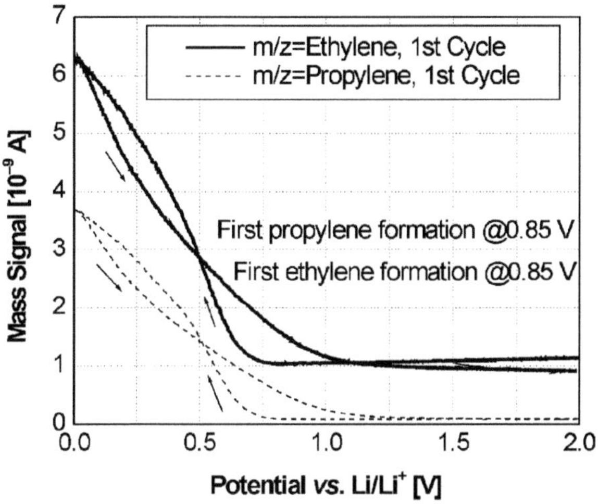

Fig. 5.14 Differential electrochemical mass spectrometry measurements in half-cells containing TIMREX® SFG6 and SFG44 as electrodes and 1-M $LiPF_6$ in ethylene carbonate/propylene carbonate 1:1 (w/w) as electrolyte. The mass signal $m/z = 27$ and $m/z = 44$ corresponding to the ethylene and propylene formation, respectively, was monitored as a function of the potential applied to the graphite electrode at a scan rate of 0.4 mV s^{-1}.[42]

Table 5.2 Active surface area and density of active sites, d_{ASA}, of TIMREX® SFG44, SFG15, and SFG6

Graphite material	Active surface area (ASA)	Density of active sites (d_{ASA})
TIMREX® SFG44	0.5	0.08
TIMREX® SFG15	0.9	0.10
TIMREX® SFG6	1.8	0.12

The ASA and d_{ASA} values shown in Table 5.2 indicate a higher surface reactivity for SFG6 than for SFG15 and SFG44. This higher surface reactivity could explain why the potential at which the formation of ethylene and propylene gas starts is more positive for SFG6. During the further electrochemical reduction, the amount of gas formed at the SFG6 electrode rate proceeds through a maximum and stops before the electrochemical oxidation half-cycle. In the case of the SFG15 and SFG44 electrodes, the gas formation rate increases with decreasing potential. In contrast to SFG15 and SFG44, the passivation of the SFG6 graphite surface is accomplished prior to the potential at which the graphite exfoliation occurs.[77]

Although typical graphite-conductive additives show promising electrochemical properties like a high reversible capacity, an excellent cycling stability, and a high electronic conductivity, only limited fractions can be applied in the electrode mass of lithium batteries. One reason is the relatively high specific BET surface area for these graphite materials, leading to elevated specific charge losses. In case of lithium batteries containing $LiCoO_2$ as positive electrode material, specific charge losses at the negative electrode determine the total charge losses of the cell, and thus need to be minimized for an optimal energy density of cell. Another reason for the limited amount of graphite-conductive additive in the negative electrode is the relatively high polymer absorption leading to a lack of cohesion and adhesion of the electrode at the current collector. The use of styrene butadiene rubber-based latex systems, which have replaced PVDF as a binder system in some of the new battery generations, allow the use of a larger amount of graphite-conductive additive.[78–80] The good compatibility of SBR with highly crystalline graphite materials leads to improved mechanical stabilities of the electrode, lower specific charge losses, and improved cycling stability of the cell.

5.3 Carbon Black Conductive Additives

5.3.1 *Preparation of Carbon Black*

Carbon black is the generic name for a family of small particle size carbon and their fused aggregates which are formed in the gas phase by the thermal decomposition of hydrocarbons. The numerous existing processes principally differ in the operating conditions. They can be classified into three categories: partial combustion, thermal decomposition, and the more recent plasma process.[81,82]

5.3.1.1 Thermal Black Process and Acetylene Process

Thermal blacks are manufactured by the decomposition of natural gas or oil in the absence of oxygen at temperatures above 1,000°C:

$$\text{Cracking}: C_xH_{2x+z} + \text{Energy} \rightarrow C_xH_y + \left(x + \frac{z}{2} - \frac{y}{4}\right)H_2 (z > y) \quad (5.1)$$

$$\text{Synthesis}: C_xH_y \rightarrow xC + \frac{y}{2}H_2 \quad (5.2)$$

Generally, the thermal process consists of two subsequent production steps. The energy for the pyrolysis and carbon black synthesis is generated by the combustion of the oil or gas feedstock and remaining carbon black in a separate cycle after the synthesized carbon black was removed from the furnace chamber. This process is in fact a partial combustion process where combustion and cracking are separated.[82,83]

The acetylene process is a particular form of the thermal process because acetylene thermally decomposes at about 800°C in an exothermic reaction. Once the reaction is started, the acetylene decomposition reaction autogenously provides the energy required for the cracking of acetylene to carbon followed by the synthesis of the carbon black:

$$nC_2H_2 \rightarrow 2nC + nH_2 + \text{Energy} \tag{5.3}$$

The reaction produces temperatures exceeding 2,500°C at the carbon black surface. The carbon black formation takes place in the temperature region below 2,000°C; above 2,000°C a partial graphitization occurs.[84] The Shawinigan process is a typical example of an acetylene process.[82,83]

5.3.1.2 Partial Combustion Process

A partial combustion of the hydrocarbon feedstock by co-injecting air in the reaction compartment provides the energy required for the thermal decomposition of the hydrocarbon. The generated energy cracks the hydrocarbon at temperatures above 1,000°C, followed by the synthesis of the carbon black:

$$\text{Combustion}: C_xH_{2x+z} + \left(\frac{3x}{2} + \frac{z}{4}\right)O_2 \rightarrow xCO_2 + \left(x + \frac{z}{2}\right)H_2O + \text{Energy} \tag{5.4}$$

$$\text{Cracking}: C_xH_{2x+z} + \text{Energy} \rightarrow C_xH_y + \left(x + \frac{z}{2} - \frac{y}{4}\right)H_2 (z > y) \tag{5.5}$$

$$\text{Synthesis}: C_xH_y \rightarrow xC + \frac{y}{2}H_2 \tag{5.6}$$

Usually after the carbon black synthesis a cooling step is applied to stop the growth of the particles. Various types of oils are used as raw materials in the different industrial processes. Some processes apply natural gas as fuel to produce the required energy for the cracking reaction. The carbon black types, which are produced based on the particle combustion technology, are lamp black, furnace black, the Super™/ENSACO™ products, and the carbon black produced as by-products of the shell gasification process.[81–84]

5.3.1.3 Plasma Process

In the plasma process the energy required for the thermal decomposition of the hydrocarbon is delivered by a plasma generated by electrical energy in a plasma generator[85]:

$$\text{Cracking}: C_xH_{2x+z} + \text{electrical energy} \rightarrow C_xH_y + \left(x + \frac{z}{2} - \frac{y}{4}\right)H_2 (z > y) \tag{5.7}$$

$$\text{Synthesis}: C_xH_y \rightarrow xC + \frac{y}{2}H_2. \tag{5.8}$$

Gaseous or liquid hydrocarbons are used as starting materials to produce various types of carbon blacks. The advantages of the plasma process is the high flexibility on raw materials, the 100% yield on carbon, and the very low level of atmospheric emissions.

5.3.1.4 Carbon Black Formation and Carbon Black Texture

Although not yet fully elucidated, the carbon black formation seems to follow a basic formation mechanism consisting of different steps for the carbon black growth, which essentially applies for all types of manufacturing processes[86–91]:

1. Vaporization of the feedstock (in case of an oil feedstock) and pyrolysis down to C_1 or C_2 units.
2. Formation of nuclei or growth centers for the primary carbon particles.
3. Growth and fusion of the nuclei to concentric primary particles.
4. Aggregation of the primary particles to primary aggregates.
5. A step of secondary growth in some cases, which relates to formation of a pyrolytic deposit on the surface of the aggregate.
6. Agglomeration of the aggregates by van der Waals forces.
7. Eventual aggregation of the aggregates followed by subsequent coating of carbon sometimes observed in the plasma process.[85]

This mechanism explains the texture of carbon black and deduces the fundamental properties that distinguish between different carbon black types. The distinct differences in morphology and surface chemistry of the carbon black types are caused by the manufacturing process and raw materials used in this process. Nowadays, a large number of carbon black types for different applications are produced. The worldwide carbon quantities produced annually exceed 9 million tons. Among these carbon blacks only a small number can be considered to be conductive carbons, and only a small fraction of these conductive carbons are suitable for lithium battery application because of the strict requirements of the lithium battery technology for purity, electrical conductivity, and inertness. The following sections mainly will focus on conductive carbon blacks.

5.3.2 Morphological Carbon Black Properties

5.3.2.1 Size and Microstructure of the Primary Particles

Carbon blacks are distinguished from most other carbons by their extremely small particle sizes and by the very broad range of particle sizes that can be produced. Primary particles are defined as the roughly spherical regions of rotational crystalline domains within amorphous carbon.[92] The primary particle diameter (fineness) is used to describe the primary particle dimension. Prior to any graphitization treatment, the microstructure of the primary particle is turbostratic. However, significant differences exist between carbon black types that are related to the sizes and quality of the graphene layers (extent of the "aromatic" structural unit) in the primary particles. Bourrat et al. classified different sizes of layers and their mode of stacking into four different categories[93]:

1. Small isometric coherent domains with $L_a \approx L_c$ (isometric turbostratic piles).
2. Columnar turbostratic stacking with $L_a < L_c$ (columnar turbostratic pile).
3. Wavy turbostratic stacking (large distorted layers).
4. Planar graphite layers (large straight layers).

In Fig. 5.15 high-resolution transition electron microscope (HRTEM) pictures compare the primary particle microstructure of an acetylene black (A) showing the typical planar graphite layers with the microstructure of Super P™ (B) showing the large distorted graphene layers.

5.3.2.2 Structure

The carbon black structure describes the pattern of how the primary particles are fused together into aggregates. Carbon black aggregates can vary in shape from individual spheroidal particles to more clustered and fibrous aggregate types. Figure 5.16 shows the highly structured aggregate of ENSACO 250.

Packing of the aggregates creates voids. The resulting void volume depends on the size and shape of the aggregate, the aggregate agglomeration, and the porosity of the primary particles. Therefore, the carbon black structure can be considered to be the sum of a number of accessible voids by unit weight:

1. The interaggregate space.
2. The interstices within the aggregates.
3. The porosity of elementary particles.

The higher the structure level of the aggregate, the higher is the volume of the voids. DBP absorption (DBPA) today is replaced by the oil absorption number

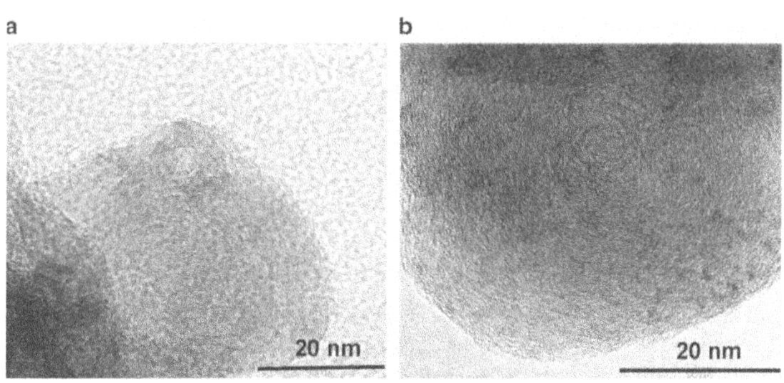

Fig. 5.15 HRTEM of a typical primary particle of an acetylene black (**a**) and of Super P™ Li (**b**)[94]

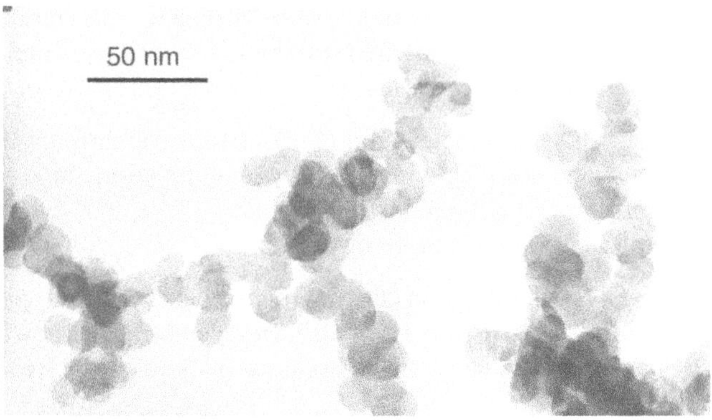

Fig. 5.16 TEM picture of ENSACO™ 250 showing a highly structured aggregate of fused primary particles[94]

Table 5.3 Typical DBP absorption, specific BET surface area, relative conductivity, dispersibility and purity of various conductive carbon blacks from different manufacturing processes

Carbon black type	BET SSA ($m^2\ g^{-1}$)		OAN (mL DBP/100 g)		Conductivity	Dispersibility	Purity
Conductive furnace black (P-type)	120	Medium	102	Low	+	+	+
ENSACO™150G	50	Low	165	Medium	++	+++	++
SUPER™ S	45	Low	280	High	++++	++++	++
ENSACO™210G	55	Low	155	Medium	++	+++	++
Acetylene black	80	Low	250	High	++	+++	++++
Supraconductive furnace black N-472	250	High	180	High	+++	++	+
ENSACO™250G	65	Low	190	High	+++	+++	++++
SUPER P™ Li	60	Low	290	High	++++	+++++	++++
ENSACO™260G	70	Low	190	High	++++	+++	++++
ENSACO™350G/ Ketjenblack EC300	800	Very high	320	Very high	+++++	+	+++
Ketjenblack EC-600 JD	1,270	Super high	495	Very high	++++++	+	+++

(OAN) (ASTM D2414), which is employed to measure the void volume, and thus the average structure level. The more complex the structure of the aggregates, the higher the OAN. Table 5.3 gives an overview about the OAN of conductive carbon black types obtained from different manufacturing processes. Carbon blacks with high structure levels show an OAN of more than 170-mL/100 g carbon.

A recent approach suggests that one differentiate between structures of conductive carbon blacks by using the resistivity vs. density or volume fraction of the carbon black.[95] High shear forces that are generated by intensive mixer or extruders during the dispersion of the carbon black in a liquid or in elastomers do not change the primary particles, but may break down the primary agglomerates of the carbon black. This breakdown of the aggregates is accompanied by a decrease of the DBP absorption. Recent investigations experimentally proved that the mechanical properties of carbon blacks mainly are governed by their electrical state.[95] The collapse of the carbon black structure provoked by shear energy could be interpreted as an agglomeration of electrically charged aggregates. According to this approach, the comparison of the densification behavior and the resistivity vs. press density of Super™ P and acetylene black, as shown in Fig. 5.17, could explain the higher stability of Super™ P toward shear energy generally observed in the electrode manufacturing process.

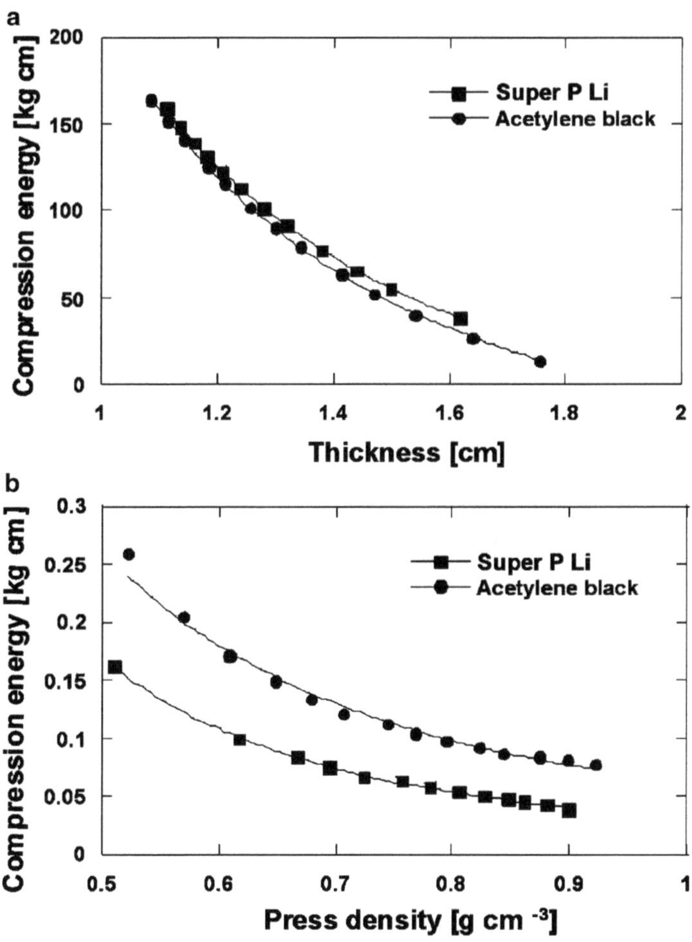

Fig. 5.17 Densification behavior and the resistivity vs. press density of Super P™ Li and acetylene black

5.3.2.3 Porosity

Porosity is created in carbon black by pores in the primary particles and ranges from mild surface pitting to the actual hollowing of the particles. Mild oxidation with air or steam shift reactions are typical processes to create oxygen groups on the surface in a first step and remove carbon atoms by evacuation of CO_2 in a second step, resulting in the creation of porosity in carbon black.[83] These reactions preferentially occur at the most active sites, which are the amorphous parts of the particle or, in the case of a particle that consists of concentric graphene layers, the higher strained inner graphene layers. The typical carbon black porosity is described by nanopores that cannot be detected necessarily by nitrogen adsorption measurements. The TEM picture of Fig. 5.18 indicates as an example the pores in the primary particles of the ENSACO™ 350 G carbon black.

5.3.2.4 Surface Properties

The surface activity is the tendency of carbon black to interact with its surroundings and is linked to the surface microstructure. Surface heterogeneities given by graphitic planes at the surface, amorphous carbon, crystallite edges, and slit-shaped cavities representing adsorption sites of different energies describe the surface microstructure. Also, the nature and amount of surface functional groups attached to superficial carbon control the interaction with the surrounding media. Oxygen functional groups at the surface are subject to acidic reactions in solvents or aqueous media. Such surface oxides can react as Lewis bases, which may act as anchor atoms for Lewis acids. It also can happen that some higher-molecular-weight organic molecules are adsorbed on the surface.

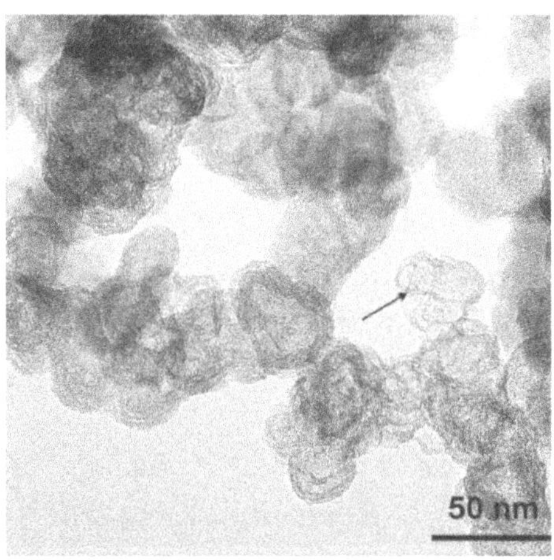

Fig. 5.18 Transition electron microscope picture of ENSACO™350G[94]

5.3.2.5 Purity

Metal impurities and anion impurities such as halogenides, sulfates, nitrates, and phosphates are the most important impurities that occur in carbon black and could have an influence on the electrochemical system. Besides, sulfur, inorganic residue of refractory material, coke particles, and organic molecules formed during the carbon black synthesis are possible contaminants in carbon blacks.

5.3.3 *Carbon Black as Conductive Additive*

The most popular carbon blacks used as conductive additives in the positive and negative electrode are typically highly conductive carbon blacks such as the Super™ and ENSACO™ products,[96–98] acetylene blacks (e.g., Denka™ and Shawinigan carbon black),[99] as well as carbon blacks taken from the Shell gasification process (e.g., Ketjenblack™).[100] Super™, ENSACO™, and acetylene black show a low surface area, a high structure, and, related to other carbon black types, a high crystallinity. Ensaco™ 350 and Ketjenblack™ show a very high specific surface area and structure. Due to the highly graphitic surface, these carbon blacks exhibit high chemical and electrochemical inertness and contain a low level of oxygen volatiles. The low impurity levels required by lithium battery technology typically are achieved for acetylene blacks, since acetylene gas used as a raw material easily can be purified. In case of the other suitable conductive carbon blacks, the high purity level mainly is due to the absence of quenching gas, the highly pure feedstock, and the adequate design of process and equipment.

For an optimal performance of the electrochemical cell, the nature and amount of carbon black in the positive and negative electrodes play an important role. However, the selection of the optimum carbon black grade and amount requires a compromise between several cell parameters, which depend on the cell specifications. Figure 5.19 gives the electrical resistivities of laminated films of $LiCoO_2$ and $LiMn_2O_4$ mixed with several carbon blacks at different concentrations. It can be shown that the critical carbon black concentration, which is necessary for an optimal resistivity, is lower the higher the structure level of the carbon black is. Therefore, an optimized cell impedance at a minimum amount of carbon black, which is required for a maximum charge density of the positive and negative electrode, is obtained if a carbon black with a high OAN is applied.[101,102] Also, in the presence of a polymer binder such as PVDF in the electrode mixture, a carbon black with a high OAN gives the optimum electronic performance.[102,103]

However, the disadvantage of carbon blacks with a high OAN is their high absorption of polymer binder and of liquid or polymer electrolyte. For a sufficient mechanical stability of the electrode, carbon blacks with a high OAN require a higher amount of binder than carbon blacks with a low OAN. However, the higher binder amount decreases the energy density of the cell. In addition, the higher level of electrolyte absorption might improve the electrolyte penetration into the

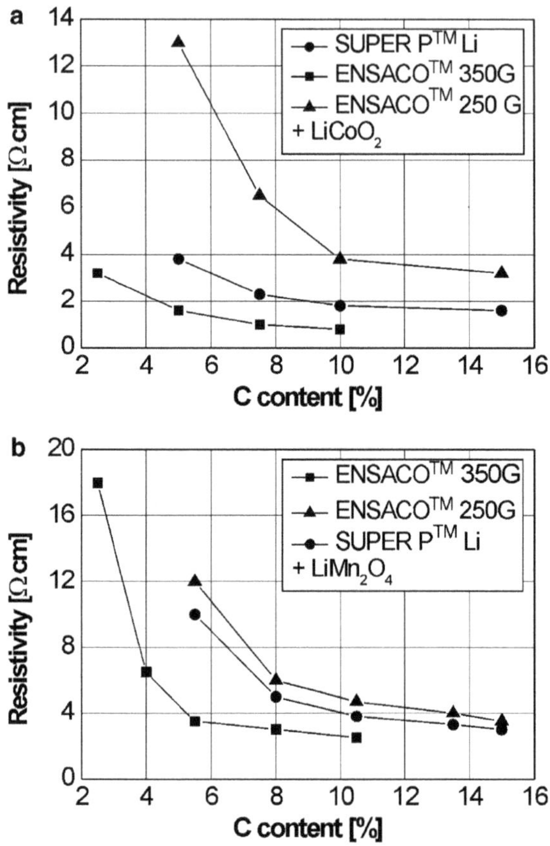

Fig. 5.19 Electrical resistivity of positive $LiCoO_2$ and $LiMn_2O_4$ electrodes containing different conductive carbon blacks as a function of the carbon black concentration

electrode, but at the same time might increase the swelling of the electrode, leading to a further decrease of the energy density of the cell. Besides, a high-structure level generally is combined with a high specific BET surface area of the carbon black such as in the case of Ensaco™ 350 and Ketjenblack™. This could lead to an increased rate of chemical and electrochemical side reactions, which may occur in both the positive and negative electrodes, leading to charge losses, an increased capacity fading during cycling of the cell, and a deterioration of the calendar life. In the negative electrode, charge losses increase with increasing surface area of the negative electrode mass due to the SEI formation.[58] In the positive electrode, a higher surface area carbon black increases the electrolyte oxidation rate, particularly occurring in the charged state of the electrode.[104,105] It was found for a $LiMn_2O_4$ spinel electrode that a high-surface area carbon black accelerates the dissolution of manganese from the charged $Li_xMn_2O_4$ electrode in the electrolyte. This manganese dissolution is accompanied by the corrosion of the carbon material and leads to a severe fading of the cell capacity during cycling of the cell.[106] These negative effects of carbon black with a high OAN on the cell performance might be negligible due to the low quantity, which is required for an optimal electrode impedance.

In addition to the electrochemical performance, parameters related to the dispersion preparation and the electrode coating process should be considered for the carbon black selection. Carbon blacks with stable agglomerates require high shear energy in order to disperse them into the slurry. However, high shear energy might cause a collapse of the carbon black structure, leading to a deterioration of their electrical performance. Thus, easily dispersible carbon blacks with highly stable structures should be preferred. Besides, the absorption properties of the carbon black additive influence the rheology of the slurry. Carbon blacks with high OAN generally cause higher viscous liquid dispersions at a given solid content. As described above, the polymer absorption properties of the carbon black additive also influence the mechanical stability of the electrode.

5.4 Graphite or Carbon Black?

The question of whether graphite or carbon black should be preferred as conductive carbon is badly posed. Graphite and carbon blacks show rather complementary properties. The decision of which carbon type should be selected basically depends on the cell requirements and the type of active electrode materials used in the electrodes. Figure 5.20 compares the morphology of a typical carbon black and graphite conductive additive and illustrates the dimensional differences of the primary particles of a factor of about 10. The smaller particle sizes of the carbon black

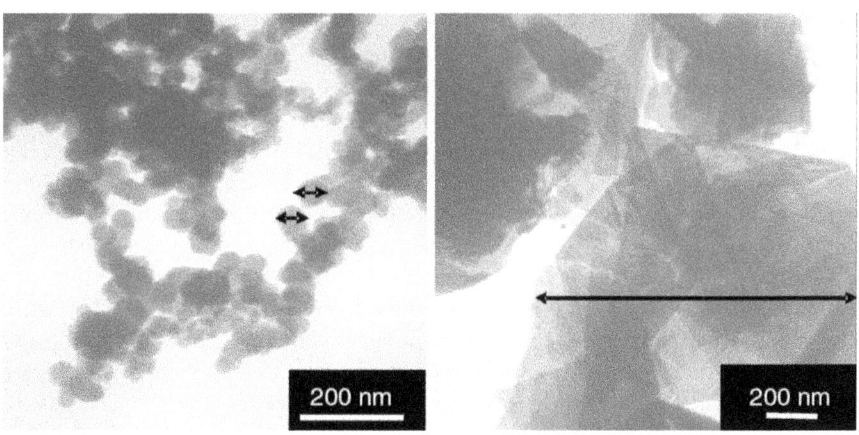

Fig. 5.20 TEM pictures of Super P™ Li (*left*) and TIMREX® KS4 (*right*)[94]

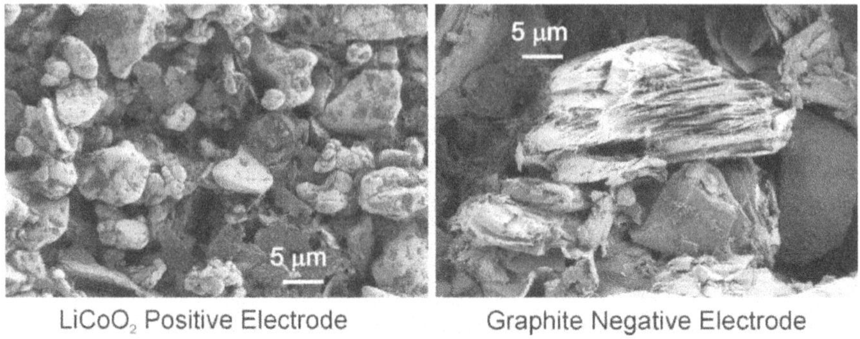

Fig. 5.21 SEM pictures of a $LiCoO_2$-positive electrode and of a surface-treated graphite negative electrode both containing TIMREX® KS6 graphite and Super P™ Li conductive additives[42]

combined with the complex aggregate structure are the reason for the lower volumetric density of conductive carbon blacks.

The highly structured carbon network allows the use of a lower amount of carbon in the electrode compared to graphite conductive powders. On the other side, graphite may improve the electrode density by its higher compressibility. Due to its generally lower DBP absorption, graphite may minimize the amount of binder material necessary for a suitable mechanical electrode stability and at the same time reduces the electrolyte retention in the electrode. The higher thermal conductivity of graphite might cause advantageous heat dissipation in high power cells. Figure 5.21 shows the SEM picture of a positive electrode containing $LiCoO_2$ as well as the SEM picture of a negative electrode based on a surface-treated graphite. In both electrodes a mixture of graphite and carbon black are used for the conductive matrix. The carbon black preferentially is attached at the surface of the active electrode materials, whereas the fine graphite particles fill the voids between the coarser active electrode particles. It may be concluded that carbon black and graphite fulfill complementary electrical functions in the electrode: Carbon black improves the contact between the particles of the active electrode material, whereas graphite creates the conductive path through the electrode as has been reported.[107] The beneficial synergy of both carbon types for an optimal conductive matrix, which lead to an improved cell impedance causing an improved cycling stability at high discharge currents, is shown in the case of a positive and negative electrode in Fig. 5.22. In addition, Cheon et al. recently reported optimized cell performance with regard to the energy and power density by using a binary mixtures TIMREX® KS6 graphite and Super™ P (1:1) as a conductive matrix in a positive $LiCoO_2$ electrode.[108] High-rate performance was obtained in the case of a $LiMn_2O_4$ spinel electrode containing TIMREX® SFG6 graphite and ENSACO™ 250 P.[109] The positive effect of a mixed graphite/carbon black conductive additive also has been claimed in several patents.[110,111]

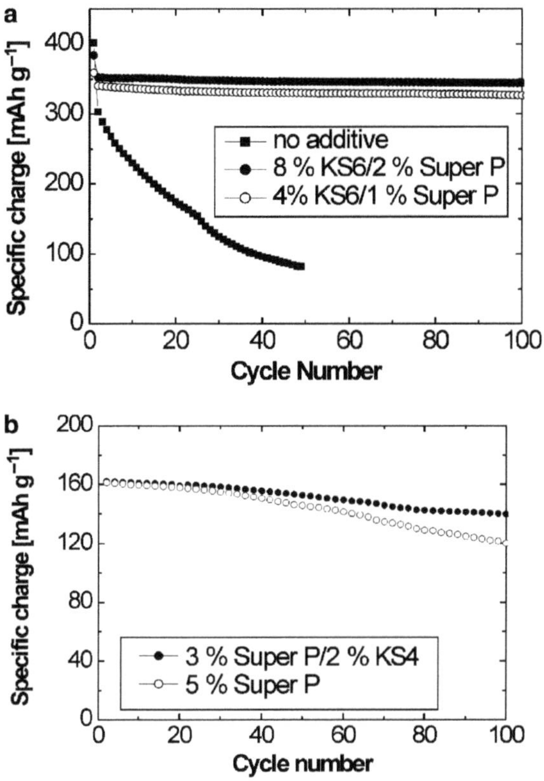

Fig. 5.22 Capacity retention of half-cells containing a surface-treated graphite-negative electrode (*top*) and a positive $LiCoO_2$, respectively, with different fractions of TIMREX® KS graphite and Super P™ Li as conductive additive (*bottom*) (electrode porosity: ca. 35%, electrolyte: 1-M $LiPF_6$ in ethylene carbonate/ethyl methyl carbonate 1:3 (v:v))[42]

5.5 Other Carbon Conductive Additives with Fibrous Morphology

Vapor-grown carbon fibers such as VGCF® (Showa Denko KK.) or similar carbon filaments are materials with low bulk densities below 0.05 g cm^{-3} that show high intrinsic electronic (typically ca. 50 Ω^{-1} cm^{-1} at 0.8 g cm^{-3} compaction density) and thermal conductivity. These properties explain the excellent performance of the vapor-grown carbon fibers as a conductive additive in the positive electrode of lithium-ion cells.[112] Besides a strong conductivity enhancement, VGCF®-containing electrodes show an improved dissipation of the heat locally evolved at high current rates. Typical practical fiber concentrations range below 1 wt% of the total electrode mass. The texture of the VGCF® carbon fibers consisting of concentrically wound graphite layers is illustrated by the tunneling electron microscope picture in Fig. 5.23. This texture explains the high electronic and thermal conductivity along the carbon fiber axis.

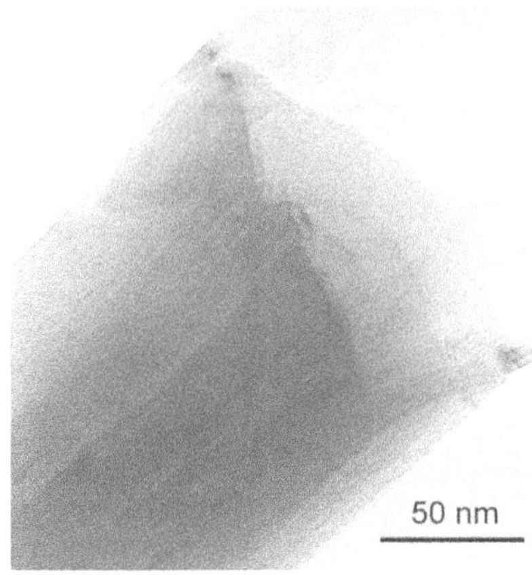

Fig. 5.23 Tunneling microscope picture of a VGCF® showing the concentrically ordered graphite sheets resulting in the fibrous morphology.[113]

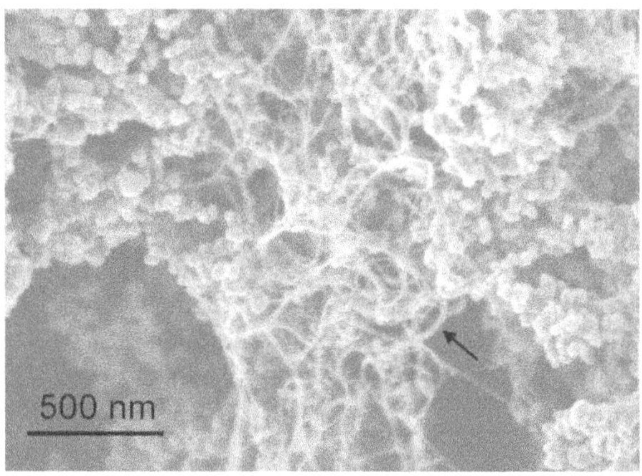

Fig. 5.24 TEM picture showing the fibrous morphology of single-walled nanotubes[94]

In addition to the electronic and thermal properties, the fibrous morphology results in a high flexural modulus and a low coefficient of thermal expansion, which explains the increased flexibility and mechanical stability of the VGCF®-containing electrode film.[114] Vapor-grown fibers are produced by exposing a metallic catalyst to a mixture of hydrocarbon gases and hydrogen at temperatures above 1,000°C. Decomposition of the hydrocarbon gas provides carbon from which filaments grow.[114–116] The complex manufacturing process is the reason for the high material costs that have been a considerable drawback for vapor-grown fibers as a conductive additive in lithium batteries and have allowed only the use of this carbon material either in traces or in particular niche applications.

Recently, nanostructured carbon materials have been receiving increasing attention as conductive additives. A positive effect of single-walled and multiwalled carbon nanotubes as a conductivity enhancer has been reported in the case of positive and negative electrodes.[117–119] Single- and multiwalled carbon nanotubes are obtained by treating carbon precursors in a plasma or in the electric arc in an inert helium atmosphere and a subsequent separation from the carbon soot. Figure 5.24 illustrates the typical morphology of single-walled nanotubes. However, an application of such materials as conductive additives in commercial lithium-ion batteries requires significantly decreased material costs by an optimized or more selective production process.

Acknowledgments The support of TIMCAL Ltd. is gratefully acknowledged. Thanks are due to Drs. Nicolas Probst, Henri Wilhelm, Alain Monnier, and Jean-Yves Huot, for fruitful discussions and the proofreading of the manuscript. Showa Denko Co. and Osaka Gas Chemical Co. are thanked for the samples of VGCF® and MCMB, respectively.

References

1. K. Kinoshita, Carbon-Electrochemical & Physiochemical Properties, Ch. 7.3, Wiley, New York (1988) pp. 403–430
2. K.J. Euler, *J. Power Sources*, **3** (1978) 117
3. A. Espinola, P. Mourente Miguel, M. Roedel Salles, A. Ribeiro Pinto, *Carbon*, **24** (1986) 337
4. L.R. Brodd, A. Kosawa, in: *Techniques of Electrochemistry*, E. Yeager, A.J. Salkind (Eds.), Vol. 3, Wiley, New York (1978) p. 222
5. K.-J. Euler, *Elektrotech. Z. Ausg. B*, **28** (1076) 45
6. S.R. Broadbent, J.M. Hammersley, *Proc. Cambridge Philos. Soc., Mat. Phys. Sci*, **53** (1957) 629
7. K.-J. Euler, R. Kirchhof, H. Metzendorf, *J. Power Sources*, **5** (1980) 255
8. D.A.G. Bruggemann, *Ann. Phys. (Leipz.)*, **24** (1935) 636
9. D. Adler, L.P. Flora, S.D. Senturla, *Solid State Commun.*, **12** (1973) 9
10. L.G. Austin, H. Lerner, *Electrochim. Acta*, **9** (1964) 1469
11. I. Roušar, K. Micka, A. Kimla, *Electrochemical Engineering*, Vol. 2, Academia, Prague (1986) p. 107–222
12. T. Katan, H.F. Baumann, *J. Electrochem. Soc.*, **122** (1975) 77
13. F. Joho, B. Rykart, A. Blome, P. Novák, H. Wilhelm, M.E. Spahr, *J. Power Sources*, 97–98 (2001) 78
14. S.-I. Lee, Y.-S. Kim, H.-S. Chun, *Electrochim. Acta*, **47** (2002) 1055
15. K. Sawai, T. Ohzuku, *J. Electrochem. Soc.*, **150** (2003) A674
16. B. Scrosati, in: *Lithium Ion Batteries—Fundamentals and Performance*, Ch. 10 M. Wakihara, O. Yamamoto (Eds.), Kodansha, Tokyo, Wiley, New York (1998), pp. 218–247
17. *Ullmanns Enzyklopàdie der technisches Chemie*, W. Foerst (Ed.), 3rd Edition, Vol. 9, Urban & Schwarzenberg, München, Berlin (1957) pp. 778–800
18. M.B. Redmount, E.A. Heintz, in: *Introduction to Carbon Technologies*, H. Marsh, E.A. Heintz, F. Rodriguez-Reinoso(Eds.), ch. 11Alicante: Universidad (1997)
19. H. Marsh, K. Fiorino, in: *Introduction to Carbon Technologies*, H. Marsch, E.A. Heintz, F. Rodriguez-Reinoso (Eds.), ch. 12Alicante: Universidad (1997)
20. A.W. Hull, *Phys. Rev*, **10** (1917) 661
21. A.F. Hollemann, E. Wiberg, *Lehrbuch der Anorganischen Chemie*, Walter de Gruyter, New York (1985) p. 701
22. W. Primak, L.H. Fuchs, *Phys. Rev.*, **95** (1954) 1
23. H. Lipson, A.R. Stokes, *Proc. Roy. Soc.*, **181** (1942) 101
24. M. Bastick, P. Chiche, J. Rappeneau, in: *Les Carbons*, Tome II, Masson et Cie (Eds.), Ch. 15, Collections des Chimie Physique (1965) pp. 163–178

25. M. Bastick, P. Chiche, J. Rappeneau, in: *Les Carbons*, Tome II, Masson et Cie (Eds.), Ch. 14, Collections des Chimie Physique (1965), pp. 24–160
26. P.E. Fanning, M.A. Vannice, *Carbon*, **31** (1993) 721
27. H.P. Boehm, *Carbon*, **32** (1994) 759
28. H.P. Boehm, in: *Graphite and Precursors*, P. Delhës (Ed.), Gordon and Breach, Amsterdam (2001), pp. 141–178
29. F. Tuinstra, J.L. Koenig, *J. Chem. Phys.*, **53** (1970) 1126
30. H. Wilhelm, M. Lelaurain, E. McRae, *J. Appl. Phys.*, **84** (1998) 6552
31. N.R. Laine, F.J. Vastola, P.L. WalkerJr., *J. Phys. Chem.*, **67** (1963) 2030
32. P.J. Harat, F.J. Vastola, P.L. WalkerJr., *Carbon*, **5** (1967) 363
33. W.P. Hoffman, F.J. Vastola, P.L. Walker, *Carbon*, **22** (1984) 585
34. S. Flandrois, B. Simon, *Carbon*, **37** (1999) 165
35. I. Kuribayashi, M. Yokoyama, M. Yamashita, *J. Power Sources*, **54** (1995) 1
36. M. Yoshio, H. Wang, K. Fukuda, Y. Hara, Y. Adachi, *J. Electrochem. Soc.*, **147** (2000) 1245
37. M. Yoshio, H. Wang, K. Fukuda, T. Umeno, T. Abe, Z. Ogumi, *J. Mater. Chem.*, **14** (2004) 1754
38. M. Yoshio, H. Wang, K. Fukuda, *Angew. Chem. Int. Ed.*, **42** (2003) 4203
39. H. Wang, M. Yoshio, *J. Power Sources*, **93** (2003) 123
40. T. Takamura, M. Saito, A. Shimokawa, C. Nakahara, K. Sekine, S. Maeno, N. Kibayashi, *J. Power Sources*, **90** (2000) 45
41. M. Nishizawa, H. Koshika, T. Itoh, M. Mohamedi, T. Abe, I. Uchida, *Electrochem. Comm*, 1 (1999) 375
42. Unpublished results of a cooperation between TIMCAL Ltd. and The Paul Scherrer Institut, Switzerland, 2003
43. C.-W. Wang, Y.-B. Yi, A.M. Sastry, J. Shim, K.A. Striebel, *J. Electrochem. Soc*, **151** (2004) A1489
44. J.R. Dahn, *Phys. Rev. B*, **44** (1991) 9170
45. T. Ohzuku, Y. Iwakoshi, K. Sawai, *J. Electrochem. Soc*, **140** (1993) 240
46. M. Winter, J.O. Besenhard, in: *Handbook of Battery Materials*, J.O. Besenhard (Ed.),Ch.5 Wiley, New York (1999), pp. 383–418
47. M. Winter, J.O. Besenhard, in: *Lithium Ion Batteries – Fundamentals and Performance*, M. Wakihara, O. Yamamoto (Eds.), Ch. 6, Kodansha, Tokyo, Wiley, New York (1998), pp. 127–155
48. N. Imanishi, Y. Takeda, O. Yamamoto, in: *Lithium Ion Batteries – Fundamentals and Performance*, M. Wakihara, O. Yamamoto (Eds.), Ch. 6, Kodansha, Tokyo, Wiley, New York (1998), pp. 98–126
49. M. Winter, J.O. Besenhard, M.E. Spahr, P. Novák, *Adv. Mater*, **10** (1998) 1
50. E. Peled, *J. Electrochem. Soc*, **126** (1979) 2047
51. E. Peled, in: *Lithium Batteries*, G.P. Gabano (Ed.), Academic, New York, 1983, pp. 43–72
52. E. Peled, D. Golodnitsky, J. Penciner, in: *Handbook of Battery Materials*, J.O. Besenhard (Ed.), Wiley, New York, 1999, pp. 419–456
53. D. Aurbach, M.D. Levi, E. Levi, A. Schechter, *J. Phys. Chem. B*, **101** (1997) 2195
54. D. Aurbach, B. Markovsky, I. Weissman, E. Levi, Y. Ein-Eli, *Electrochim. Acta*, **45** (1999) 47
55. D. Aurbach, A. Zaban, Y. Ein-Eli, I. Weissman, O. Chusid, B. Markovsky, M. Levi, E. Levi, A. Schechter, E. Granot, *J. Power Sources*, **68** (1997) 91
56. Y. Ein-Eli, *Electrochem. Solid-State Lett.*, **2** (1999) 212
57. D. Aurbach, E. Zinigrad, L. Zhongua, A. Schechter, M. Moshkovich, in: *Lithium Batteries*, S. Sarampudi, R. Marsh (Eds.), The Electrochemical Society Proceeding Series PV98-16, Pennington, NJ (1999), p. 95
58. F. Joho, B. Rykart, A. Blome, P. Novák, H. Wilhelm, M.E. Spahr, *J. Power Sources*, 97–98 (2001) 78–82
59. F. Joho, P. Novák, M.E. Spahr, *J. Electrochem. Soc.*, **149** (2002) A1020–A1024
60. J.O. Besenhard, H.P. Fritz, *J. Electroanal. Chem.*, **53** (1974) 329

61. G.-C. Chung, H.-J. Kim, S.-I. Yu, S.-H. Jun, J.-W. Choi, M.-H. Kim, *J. Electrochem. Soc.*, **147** (2000) 4391
62. J.O. Besenhard, M. Winter, J. Yang, W. Biberacher, *J. Power Sources*, **54** (1995) 228
63. A.N. Dey, B.P. Sullivan, *J. Electrochem. Soc.*, 137 (1990) 222–224
64. R. Fong, U.V. Sacken, J.R. Dahn, *J. Electrochem. Soc.*, **137** (1990) 2009
65. J.R. Dahn, A.K. Sleigh, H. Shi, B.M. Way, W.J. Weydanz, J.N. Reimers, Q. Zhong, U. von Sacken, in: *Lithium Batteries, New Materials, Developments and Perspectives*, G. Pistoia (Ed.), Ch. 1, Elsevier, Amsterdam (1994)
66. M. Jinno, Y. Shoji, N. Nishida, K. Nishio, T. Saito, Jpn. Kokai Tokkyo Koho JP 09147910 A2 19970606 (1997)
67. M. Broussely, S. Herreyre, F. Bonhomme, P. Biensan, P. Blanchard, K. Nechev, G. Chagnon, *Proc. Electrochem. Soc.*, 2001–21 (Batteries and Supercapacitors), (2003) 75
68. R. Mogi, M. Inaba, S.-K. Jeong, Y. Iriyama, T. Abe, Z. Ogumi, *J. Electrochem. Soc.*, **149** (2002) A1578
69. K. Guerin, A. Fevrier-Bouvier, S. Flandrois, M. Couzi, B. Simon, P. Biensan, *J. Electrochem. Soc.*, **146** (1999) 3660
70. B. Simon, S. Flandrois, A. Fevrier-Bouvier, P. Biensan, *Mol. Cryst. Liq. Cryst.*, **310** (1998) 333
71. B. Simon, S. Flandrois, K. Guerin, A. Fevrier-Bouvier, I. Teulat, P. Biensan, *J. Power Sources*, **81–82** (1999) 312
72. H. Huang, W. Liu, X. Huang, L. Chen, E.M. Kelder, J. Schoonman, *Solid State Ionics*, **110** (1998) 173
73. H. Buqa, A. Würsig, D. Goers, L.J. Hardwick, M. Holzapfel, P. Novák, F. Krumeich, M.E. Spahr, *J. Power Sources*, **146** (2005) 134–141
74. M.E. Spahr, H. Buqa, A. Würsig, D. Goers, L. Hardwick, P. Novák, F. Krumeich, J. Dentzer, C. Vix-Guterl, *J. Power Sources*, **153**, (2006) 300–311
75. M.E. Spahr, T. Palladino, H. Wilhelm, A. Würsig, D. Goers, H. Buqa, M. Holzapfel, P. Novák, *J. Electrochem. Soc.*, **151** (2004) A1383
76. M.E. Spahr, H. Wilhelm, F. Joho, J.-C. Panitz, J. Wambach, P. Novák, N. Dupont-Pavlovsky, *J. Electrochem. Soc.*, **149** (2002) A960–A966
77. H. Buqa, A. Würsig, J. Vetter, M.E. Spahr, F. Krumeich, P. Novák, *J. Power Sources*, **153** (2006) 385–390
78. Y. Ishihara, Y. Mochizuki, H. Nose, Jpn. Kokai Tokkyo Koho JP 04370661 A2 19921224 Heisei (1992)
79. K. Inoe, H. Koshina, Y. Ozaki, Jpn. Kokai Tokkyo Koho JP 08306391 A2 19961122 Heisei (1996)
80. Y. Shoji, M. Uehara, M. Yamazaki, T. Noma, K. Nishio, Jpn. Kokai Tokkyo Koho JP 09283119 A2 19971031 Heisei (1997)
81. *Ullmann's Encyclopedia of Industrial Chemistry*, Vol. 5, Wiley, Weinheim (1986), p.140
82. J.B. Donnet, R.P. Bansal, M.J. Wang, in: *Carbon Black Science and Technology*, 2nd Ed., Marcel Dekker, New York (1993)
83. R. Taylor, in: *Introduction to Carbon Technologies*, H. Marsch, E.A. Heintz, F. Rodriguez-Reinoso (Eds.),Ch. 4 Alicante, Universidad (1997)
84. V. Schwob, in: *Chemistry and Physics of Carbon*, Vol. 15, P.L. Walker, P.A. Thrower (Eds.), Marcel Dekker, New York (1982), p. 109
85. L. Fulcheri, N. Probst, G. Flamant, F. Fabry, E. Grivei, X. Bourrat, *Carbon*, **40** (2002) 169
86. H.C. Ries, *SRI Report 90*, Stanford Research Institute, Menlo Park, California (1974)
87. J. Lahaye, *Carbon*, 30 (1992) 309
88. J. Lahaye, G. Prado, *Chemistry and Physics of Carbon*, Vol. 14, P.L. Walker, P.A. Thrower (Eds.), Marcel Dekker, New York (1978), p. 168
89. E.H. Homann, *Combustion Flame*, **11** (1967) 265
90. H.F. Calcote, *Combustion Flame*, **42** (1981) 215
91. X. Bourrat, in: *Science of Carbon Materials*, H. Marsh, F. Rodriguez-Reinoso (Eds.), Ch. 1 Alicante, Universidad (2000)
92. K.A. Burgess, C.E. Scott, W.M. Hess, *Rubber Chem. Technol.*, **36** (1063) 1175

93. X. Bourrat, *Carbon*, **31** (1993) 287
94. Measurements obtained at the Département de Physico-Chimie et Physique des Matériaux, University of Louvain, Belgium, by request of Timcal Ltd, 2004
95. E. Grivei, N. Probst, Belg. KGK, *Kautschuk Gummi Kunststoffe*, **56** (2003) 460
96. J.N. Reimers, J.R. Dahn, *J. Electrochem. Soc.*, **139** (1992) 2091
97. J.N. Reimers, E.W. Fuller, E. Rossen, J.R. Dahn, *J. Electrochem. Soc.*, **140** (1993) 3396
98. I. Koetschau, M.N. Richard, J.R. Dahn, J.B. Soupart, J.C. Rousche, *J. Electrochem.l Soc.*, **142** (1995) 2906
99. J. Shigetomi, Jpn. Kokai Tokkyo Koho JP 06333558 A2 19941202 Heisei (1994)
100. S. Kuroda, N. Tobori, M. Sakuraba, Y. Sato, *J. Power Sources*, 119–121 (2003) 924
101. E. Grivei, J.B. Soupart, H. Lahdily, *ITE Battery Lett.*, **2** (2001), 1, B17
102. N. Probst, E. Grivei, H. Smet, J.B. Soupart, *ITE Battery Lett.*, **1** (1999), 1, 64
103. F. Carmona, *Ann. Chim. Fr.*, 13 (1988) 395
104. P. Novák, J.-C. Panitz, F. Joho, M. Lanz, R. Imhof, M. Coluccia, *J. Power Sources*, **90** (2000) 52
105. R. Imhof, P. Novák. *J. Electrochem. Soc.*, **146** (1999) 1702
106. D.H. Jang, S.M. Oh, *Electrochim. Acta*, **43** (1998) 1023
107. M. Nishizawa, H. Koshika, T. Itoh, M. Mohamedi, T. Abe, I. Uchida, *Electrochem. Commun.s*, **1** (1999) 375
108. S.E. Cheon, C.W. Kwon, D.B. Kim, S.J. Hong, H.T. Kim, S.W. Kim, *Electrochimica Acta*, **46** (2000) 599
109. M. Lanz, C. Kormann, H. Steininger, G. Heil, O. Haas, P. Novák, *Journal of the Electrochemical Society*, **147** (2000) 3997
110. E.S. Takeuchi, W.C. Thiebolt, Eur. Pat. Appl. EP 978889 A1 20000209 (2000)
111. K. Uchitomi, A. Ueda, S. Aoyama, Jpn. Kokai Tokkyo Koho JP 2002279998 A2 20020927 (2002)
112. C.A. Frysz, X. Shui, D.D.L. Chung, *J. Power Sources*, **58** (1996) 41
113. Unpublished results of a cooperation between TIMCAL Ltd. and The ICSI-CNRS, Mulhouse, France, 2004
114. N. Murphie, in: *Introduction to Carbon Technologies*, H. Marsch, E.A. Heintz, F. Rodriguez-Reinoso (Eds.), ch. 14 Alicante: Universidad (1997) p. 603
115. D.D. Eddie, R.J. Diefendorf, in: *Carbon–Carbon Materials and Composites*, NASA vol. 1254, ch. 2 (1992) p. 19
116. J. Baptiste Donnet, R.C. Bansal, *Carbon Fibers*, Marcel Dekker, New York (1984)
117. L.J. Rendek, R.E. Doe, M.J. Wagner, Abstract No. 85, 205th Meeting of the Electrochemical Society Inc., 2004
118. S.-C. Han, M.-S. Song, H. Lee, H.-S. Kim, H.-J. Ahn, J.-Y. Lee, *J. Electrochem. Soc.*, **150** (2003) A889
119. R. Ochoa, I. Kerzhner-Haller, H. Maleki, US 2003099883 A1 20030529 Pat. Appl. Publ. (2003)

Chapter 6
Applications of Polyvinylidene Fluoride-Related Materials for Lithium-Ion Batteries

Aisaku Nagai

6.1 Introduction

The ionic conductivity of the nonaqueous electrolyte used in lithium-ion batteries (LIB) is much lower than that of the aqueous electrolyte used in nickel–cadmium (Ni–Cd) and nickel-metal hydride (Ni–MH) batteries. Thus, to obtain a high-current output, LIB require much thinner and much wider electrodes than do Ni–Cd and Ni–MH batteries. This is one of the reasons why binders for LIB are much more important than those for other batteries.

Table 6.1 shows the list of required characteristics for effective binders in an LIB. The items in the first section are related to the battery's performance. The design rule for secondary batteries for portable devices is very simple: the greater the amounts of active materials filling up the restricted volume, the more power can be obtained. Thus, any components other than active materials, such as binders, electrodes, separators, and conductive additives, should be reduced as much as possible. However, even if the amount of binder is reduced, it should still adhere well to the active materials and to the metallic electrodes. Furthermore, it must be inactive over a wide potential range and should not be oxidized at the cathode or be reduced at the anode. It is desirable that the binder have a high melting point and that the composite structure of the active materials and binders remain stable in an electrolyte, even at higher temperatures. If the binder should swell in the electrolyte beyond a certain extent, the electrical contact between the active materials and the electrode will be lost; at that point, a capacity fading will be evident. A potential drawback of a binder is that it may coat the surface of active materials. So, it is very important that lithium ions can pass through the binder. The amorphous region in polyvinylidene fluoride (PVdF) is a good matrix for polar molecules, and lithium ions can pass through a thin layer of swollen PVdF.[1] Finally, if the binder could conduct electricity well, the battery performance would be further improved.

A. Nagai
Research and Technology Division, Kureha Corporation, 3-3-2, Nihonbashi-Hamacho, Chuo-ku, Tokyo 103-8552, Japan
a-nagai@kureha.co.jp

M. Yoshio et al. (eds.), *Lithium-Ion Batteries*,
DOI 10.1007/978-0-387-34445-4_6

Table 6.1 Required characteristics for LIB binders

For battery performance
Good retention of active materials
Excellent adhesion to metal electrodes
Electrochemical stability over a wide potential range
High melting points
Low swelling rate in nonaqueous electrolytes
Good lithium-ion conductivity
Good electron conductivity
For productivity
Slurry viscosity remains constant over long periods
Soluble in highly concentrated form in solvents that have a low heat of vaporization
Easily shaped in a roll press without spring back
Resistant to chipping during electrode slitting

The items listed in the second section pertain to various aspects of the production process. Nowadays, battery manufacturers raise the coating speed higher and higher to increase productivity, without investing in additional drying chambers. So, to reduce the heat of vaporization, the binder should dissolve easily at high concentrations in solvents. After the drying process, the electrodes are pressed by rolling at a certain temperature to increase their density. But, if the binder is at all rubbery, the pressed electrode will expand again. This phenomenon is called "spring back." Needless to say, it is desirable that the binder should not change its shape after the pressing process. Finally, the pressed electrodes material will be slit to a certain width. Should so much as one particle of the active material be removed during the slitting operation, a phenomenon called "chipping," that particle could attach to the surface of the electrode, penetrate the separator film, and cause a short circuit. Chipping, therefore, will fatally decrease the yield of the winding process.

Unfortunately, there is no one binder that meets all the criteria listed in Table 6.1. At present, PVdF and styrene-butadiene rubber (SBR) are used for anode binders, while for cathode binders, PVdF and poly(tetrafluoroethylene) (PTFE) are employed. The drawbacks of each of these binders are overcome by various methods.

6.2 Electrochemical Stability of Polymers

During the 1970s, propylene carbonate (PC) was found to be a suitable solvent for lithium batteries, but this does not mean that PC is stable on lithium metal. PC is decomposed by a reduction process, after which a passivation layer [so-called solid electrolyte interface (SEI)] is formed on the surface of the lithium metal. In fact, most organic solvents are not stable at the potential of lithium metal. In addition, during the 1980s, no one believed that any organic solvent could be stable at more than 4 V. Thus, the electrochemical environment in LIB produces severe demands on the properties of organic materials used as binders. Binders should adhere to the active materials at all times and should be stable in anodic and cathodic environments. However, it is not easy to measure the redox windows of polymers. Therefore, we developed a prediction method and presented it at the battery symposium held in 1998.[2] In this method, we calculated the Highest Occupied Molecular Orbital (HOMO) and Lowest Unoccupied Molecular Orbital (LUMO) of various organic solvents and found several relationships between the HOMO and the oxidation potential and between the LUMO and the reduction potential. Next, we calculated the HOMOs and LUMOs of various polymers having different conformations and molecular weights.

Figure 6.1 provides one of the set of results obtained for various polymers and ethylene carbonate (EC). EC is chosen as a typical solvent molecule. Each bar shows a calculated redox window. The bottom part of each bar represents the energy level of the HOMO of a certain molecule and its top represents the energy level of the LUMO of the same molecule. The HOMOs of PTFE, PVdF, and polyacrylonitrile (PAN) are lower than those of other polymers. This result means that it is very difficult to remove an electron from PTFE, PVdF, and PAN; that is, these polymers can be stable in a cathodic environment.

On the other hand, the highest LUMO is that of polyethylene oxide (PEO). PEO and polypropylene oxide (PPO) are very familiar as polymer electrolytes. According to this calculated result, PEO and PPO can be stable in an anodic environment but may not be stable in a cathodic environment, because the HOMO of both polymers is very high. In addition, PEO and PPO are soluble in organic solvents and cannot be used as a binder. The LUMO of polyethylene (PE) is high and we expect it to be stable in an anodic environment. But it is difficult to dissolve PE in organic solvents. The LUMOs of PVdF and SBR are almost the same as that of PE, and thus these compounds may be used as a binder. So, the prediction based on the theoretical calculations is consistent with the actual choice of binder for both electrodes.

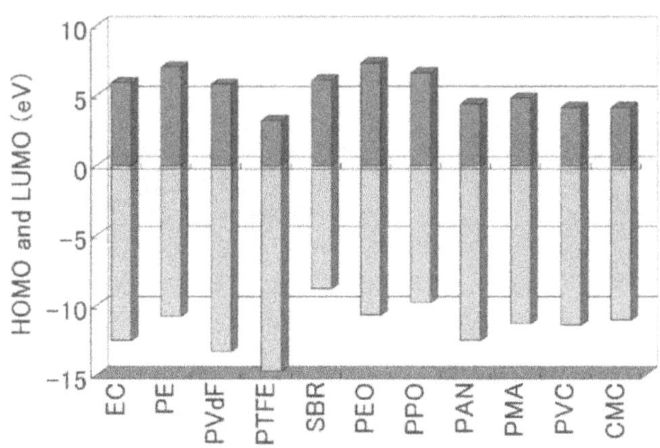

Fig. 6.1 Calculated redox windows of various polymers

6.3 Physical Properties of PVdF

A monomer of PVdF has two hydrogen atoms and two fluorine atoms (–CH_2–CF_2–). Hydrogen tends to donate its electron to other atoms, and thus is called a donor. In contrast, a fluorine atom tends to accept electrons from other atoms, and thus is called an acceptor. So, each monomer of PVdF has a dipole moment. With polymerization, each monomer lines up by dipole-dipole interaction as in –CH_2–CF_2–CH_2–CF_2–. This bond sequence is called a regular bond. However, thermal perturbation may change the bond sequence to be –CH_2–CF_2–CF_2–CH_2–. This latter sequence is called a head-to-head bond or simply an irregular bond. Such bond sequences can have much effect on the crystallinity and other physical properties of the material through strong intra- and intermolecular dipole-dipole interactions.

An important characteristic of PVdF is its crystallinity. PVdF has several crystalline forms, the most stable of which is the α-form. X-ray diffraction data show that about 50% of PVdF has the α-form structure and the rest is amorphous. Figure 6.2 shows the melting points of various PVdF samples polymerized at different temperatures. When the polymerization temperature is low, the melting point of the PVdF is increased. Figure 6.2 also demonstrates the relationship between the molar ratio of regular bonds and the polymerization temperature. When the polymerization temperature is low, the fraction of irregular bonds is lower. Thus, in order to get good crystallinity, the PVdF should be polymerized at a lower temperature.

Kureha KF polymer is a PVdF product developed by Kureha Corporation, Tokyo Japan. KF polymer has a high chemical resistance and desirable mechanical properties. One of its most remarkable characteristics is that the irregular bonding of its molecular chain is lower than that of any of the other PVdFs, an attribute that leads to a perfectly crystallized polymer. These valuable properties allow lithium-ion secondary batteries to combine long-term, stable performance with a minimum amount of swelling by the organic electrolyte.

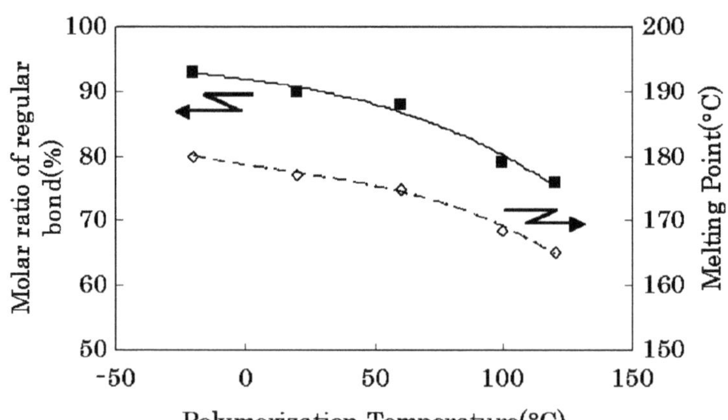

Fig. 6.2 Molar ratio of regularly arranged bonds and the melting points of various PVdF samples polymerized at different temperatures

Another unique characteristic of PVdF is its dielectricity. Because these dipole moments in the amorphous region will move easily along the direction of an applied electric field, the dielectric constant of PVdF is the highest among polymers.

6.4 Product Range of KF Polymer

Since lithium ion secondary batteries have come on the market, KF polymer is the most popular binder used throughout the world. KF polymer is available as the "W series" as powder, and in the "L series" in which the "W series" powder is dissolved in NMP (*N*-methyl-2-pyrollidone). Table 6.2 shows the different grades available for the two series.

In general, the higher the molecular weight of the KF polymer, the less the required volume. In other words, it is possible to increase the amount of active

Table 6.2 Grades for the powder and solution KF polymer

W series (powder type)	
W#1100	Standard grade
W#1300	Standard grade
W#1700	High molecular weight grade
W#7200	Ultra high molecular weight
W#7300	Ultra high molecular weight
W#9100	High adhesion grade for anode
W#9200	High adhesion and high molecular weight grade for anode
W#9300	High adhesion and ultra high molecular weight grade for anode
L series (solution type)	
L#1120	Standard grade (W#1100)
L#1320	Standard grade (W#1300)
L#1710	High molecular weight grade (W#1700)
L#7208	Ultra high molecular weight grade (W#7200)
L#7305	Ultra high molecular weight grade (W#7300)
L#9130	High adhesion grade for anode (W#9100)
L#9210	High adhesion and high molecular weight grade for anode (W#9200)
L#9305	High adhesion and ultra high molecular weight grade for anode (W#9300)

Parenthesis indicates the powder type

Table 6.3 Standard choice of KF polymers for anode and cathode

	Molecular weight				
	About 280,000	About 350,000	About 500,000	About 630,000	About 1 million
For cathode	W#1100	W#1300	W#1700	W#7200	W#7300
	L#1120(12%)	L#1320(12%)	L#1710(10%)	L#7208(8%)	L#7305(5%)
For anode	W#9100		W#9200		W#9300
	L#9130(13%)		L#9210(10%)		L#9305(5%)

Parenthesis of L series shows the powder content

Table 6.4 Properties of powder and solution type KF polymer

Property	W series (powder type)							
Property	W#1100	W#1300	W#1700	W#7200	W#7300	W#9100	W#9200	W#9300
Molecular weight, ×1,000	280	350	500	630	1,000	280	500	1,000
Inherent viscosity, dL/g	1.10	1.30	1.69	2.17	3.10	1.12	1.74	3.10
Melting temperature, C	175	175	174	173	172	173	170	163
Crystallization temperature, C	138	138	136	136	136	136	137	134
Moisture content, %	<0.1	<0.1	<0.1	<0.1	<0.1	<0.1	<0.1	<0.1
	L series (solution type)							
Property	L#1120	L#1320	L#1710	L#7208	L#7305	L#9130	L#9210	L#9305
Grade of raw resin	W#1100	W#1300	W#1700	W#7200	W#7300	W#9100	W#9200	W#9300
Resin content, %	12.0	12.0	10.0	8.0	5.0	13.0	10.0	5.0
Solution viscosity, MPa s	550	1,150	1,750	2,350	1,800	950	2,500	2,100
Moisture content, %	<0.1	<0.1	<0.1	<0.1	<0.1	<0.1	<0.1	<0.1

materials if you use a KF polymer with a high-molecular weight. This choice will lead to a reduction of production costs and an improvement of the battery capacity, both of which are important requirements in the current IT (information technology) market.

The W#1100 (L#1120), and W#1300(L#1320) standard types of KF polymer and the higher molecular weight materials W#1700 (L#1710), W#7200(L#7208), and W#7300(L#7305) are recommended for the cathode (Table 6.3).

On the other hand, anode materials have difficulties in bonding to the PVdF in the anode. Kureha Corporation developed the adhesion-improved grades, W#9100 (L#9130), W#9200 (L#9210), and W#9300 (L#9305), featuring a functional group introduced into the PVdF molecule. Thanks to the functional group, these grades show a higher adhesive strength compared to other grades of the same molecular weight. The adhesion-improved grades also can be used for the cathode. However, users of these materials should be on the alert for the gelation problems, which might occur, depending on the type of cathode materials.

The above adhesion-improved grades will contribute to a cost reduction, an increase in electrical capacity, and an improvement of high discharge capacity at high current drain rate. On the other hand, because of the high molecular weight, the slurry viscosity might increase, a change possibly affecting its handling, mixing, and coating abilities in the production process. Kureha Corporation has the know-how to produce good cell performance using the KF Polymer discussed above and can provide one with a wide range of technical solutions, including the best choice of KF polymer for your application and troubleshooting, if needed. For reference, the typical properties of each grade are listed in Table 6.4.

References

1. K. Tsunemi, H. Ohno, E. Tsuchida, *Electrochimica Acta*, **28**(6), 591(1983).
2. A. Kurihara, A. Nagai, *The 39th Battery Symposium in Japan*, 3C09, 309 (1998).

Chapter 7
SBR Binder (for Negative Electrode) and ACM Binder (for Positive Electrode)

Haruhisa Yamamoto and Hidekazu Mori

7.1 Introduction

An electrode binder for lithium-ion (Li-Ion) secondary batteries is used either to bind active material layers or between active material layers and collectors. The binder in general is preferably electrochemically inert. Although the binder is not essential for the battery, it has an important role in the facilitation of producing batteries and in their properties. Therefore, the binder is said to be one of the major elements of the battery.

Table 7.1 shows a number of excellent binders developed to date. Binders properties include – besides the actual binding – flexibility when used for electrodes, insolubility in the electrolyte, compactness, chemical and electrochemical stability, and easy application to electrode paints. The binder should be able to satisfy all of these properties simultaneously. This is a difficult task, and only two binders that comply with all these requisites have been found: polyvinylidene fluoride (PVDF) and styrene-butadiene copolymer (SBR).

Initially PVDF was the main binder employed for negative electrodes[1] but now the use of SBR has become more popular.[2] SBR is now used in almost 70% of all batteries. Compared to PVDF, SBR provides better battery properties. For example: more flexible electrode; higher binding ability with a small amount; larger battery capacity; and higher cyclability. SBR also is suitable for graphite with a larger relative surface area and it is operation-environment friendly because SBR uses aqueous solvent. Due to these advantages, Korea and China recently have begun switching from PVDF to SBR.

A binder for positive electrodes demand more rigorous qualities than those for negative electrodes. For example, the binder has to be oxidation-resistant in the unstable oxidation atmosphere that occurs in charging batteries. SBR is suitable for negative electrodes used in a reduction atmosphere; however, it is not suitable for positive electrodes with double bonds, which are prone to be oxidized. These requirements

H. Yamamoto (✉) and H. Mori
ZEON Corporation, Yako 1-2-1, Kawasaki, 210-9507, Japan
h.mori@zeon.co.jp

M. Yoshio et al. (eds.), *Lithium-Ion Batteries*,
DOI 10.1007/978-0-387-34445-4_7

Table 7.1 Binder examples.[7] Reproduced by permission of CMC Publishing Co. Ltd

Binder	General formula and characteristics	Papers
PVDF	$-(CH_2-CF_2)_n-$ • Binding strength to collector or active material is improved	AH 06-093025
PTFE	$-(CF_2-CF_2)_n-$ • To obtain electrode paint with a good current collection and binding strength • Discharge performance and storage life are improved	AH 08-106897
FKM	$-(CF2-CH_2-CF(CF_3)-CF_2)_n-$ • Improvement of cycle performance	AH 10-027601
SBR	$-(CH_2-CH=CH-CH_2)_l-(CH_2-CH(C_6H_5))_m-(X)_n-$ • Binding ability is improved • Charge/discharge performances are improved	AH 07-037619
NBR	$-(CH_2-CH=CH-CH_2)_m-(CH_2-CH(CN))_n-$ • Good binding strength at a small amount of binder	AH 11-121014
BR	$-(CH_2-CH=CH-CH_2)_n-$ • Large initial discharge capacity • Good cycle performance	AH 04-255670
PAN	$-(CH_2-CH(CN))_n-$ • Conductivity of electrode is improved	AS 63-121264
EVOH	$-(CH_2-CH_2)_m-(CH_2-CH(OH))_n-$ • Paint property is improved • Good cycle performance	AH 11-250915
EPDM	$-(CH_2-CH_2)_l-(CH_2-CH(CH_3))_m-(M_1)_n-$ • Good cycle performance	AH 05-062668
Polyurethane	$-(R_1-O-C(=O)-N(H)-R_2-N(H)-C(=O)-O)_n-$ • Electrode productivity is improved • Battery capacity and durability are improved	AH 11-001676
Polyacrylic acid	$-(CH_2-CH(COOH))_n-$ • Used together with PVDF • Good cycle performance • Battery capacity is enhanced	AH 11-045720
Polyamide	$-(CH_2-CH(COOR_5))_m-(X)_n-$ • Battery capacity is improved • Good cycle performance	AH 10-134816
Polyacrylate	$-(CH_2-CH(OR_6))_n-$ • Good binding strength • Good cycle performance • Battery capacity is improved	AH 08-287915

Table 7.1 (continued)

Binder	General formula and characteristics	Papers
Polyvinyl ether	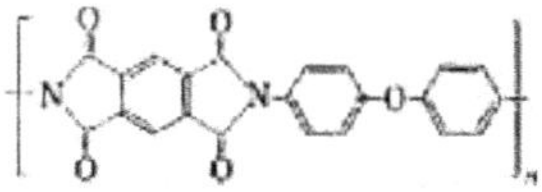	AH 10-069911
	• Reaction between electrolyte and lithium is prevented	
Polyimide	• Electrode with strength and flexibility is prepared	AH 10-302771

Note: These structures are the basic formula of each polymer. Most polymers are actually denatured

have caused delays in its development. Nowadays, PVDF is still used for positive electrodes, although more recently, the highly flexible acrylate-type copolymer (ACM) has started to be used in prismatic batteries.[3]

This report introduces BM-400B and BM-500B, which Zeon Corporation has developed as SBR binder for negative electrodes and ACM binder for positive electrodes, respectively. SBR and ACM binders are very different from PVDF binder in their usage. Once accustomed to the use of PVDF, it becomes difficult to make electrode paint when switching from PVDF to SBR or ACM. Therefore, this report also describes in detail how to use SBR and ACM binders.

7.2 Properties of SBR and ACM Binders

SBR is the general term for a copolymer that mainly consists of styrene and butadiene (Table 7.1). The properties of SBR vary greatly according to the ratio of its components. Many SBRs are elastomers.

BM-400B is a dispersion system of SBR fine particles in water. These particles are random copolymer molecules, i.e., styrene and butadiene, containing some other minor elements such as acrylic ester and organic acids. The copolymer is an elastomer with a glass transition temperature of -5°C. Its chemical formula is,

Table 7.2 shows the properties of BM-400B, the state of electrode paint, and an example of electrode properties. BM-400B is designed to be differentiated from other SBR binders by optimizing its type and ratio of components and optimizing manufacturing conditions and developed to exert its excellent properties as a binder for negative electrodes.

ACM, on the other hand, is the general term for a copolymer that mainly consists of acrylic ester and also contains some other minor elements. The properties of ACM vary greatly according to the type and component ratio of acrylic ester and type and component ratio of other minor elements; however, many ACMs are elastomers.

BM-500B is a dispersion system which uniformly disperses ACM fine particles in NMP (N-methyl-2-pyrrolidone). These particles are random copolymer which mainly consists of 2-ethyl-hexylacrylate and acrylonitrile, and also contains some other minor elements. This copolymer is elastomer with a glass transition temperature of –40°C and chemical formula:

Table 7.2 Binders developed by Zeon Corporation.[7] Reproduced by permission of CMC Publishing Co. Ltd

Item	BM-400B (for negative electrode)	BM-500B (for positive electrode)
Polymer properties		
Type of polymer	SBR	ACM
Glass transition temperature (T_g, °C)[a]	-5	–40
Particle size (nm when dried)	130	170
Thermal decomposition starting temperature(°C)[b]	248 (in air) 342 (in nitrogen)	308 (in air) 363 (in nitrogen)
Binder properties		
Dispersion medium	Water	NMP
Solid content (wt%)	40	8
Viscosity (mPa s)	12	150
pH	6	–
Binder properties		
Resistance to electrolyte[c]		
Swelling [weight (times)]	1.6	1.6
Chemical reactivity	No change in color	No change in color
Electrochemical stability	Excellent reduction resistance	Excellent oxidation resistance
Example of electrode paint mixture ratio (wt%)		
	BM-400B (solid matter) 1.5	BM-500B (solid matter) 0.53
	CMC 1.0	Thickener – A (solid matter)[d] 0.27
	Graphite (MCMB) (SSA 0.9 m²/g) 100	Acetylene black 2
	Water 51.25	$LiCoO_2$ 100 NMP 25.7
Example of electrode paint properties		
Solid content (wt%)	66.7	80
Viscosity (60 rpm, mPa s)	3,000	2,000
Storage stability	No settlement (7 days)	No settlement (1 day)
Example of electrode properties		
Binding ability to current collector (g/cm, not pressed)	3	2
Flexibility[e]		
Stiffness (g, H = 10 mm)	2	4
Cracking point (mm)	1	2
Surface roughness of electrode (Ra, μm, not pressed)	3	0.8

[a]Measured with the DSC method
[b]Measured with TGA (10°C/min)
[c]Electrolyte used (EC/DEC = 1/2, 1-M/L $LiPF_6$); binder film is immersed at 60°C for 72 h
[d]Thickener-A: Nitrile type polymer in NMP solvent
[e]Refer to Fig. 7.11

Table 7.2 shows the properties of BM-500B, the state of electrode paint and an example of the electrode properties. BM-500B is the first nonfluorine polymer to be practically applied to batteries as a binder for the positive electrode. BM-500B is designed to exert its excellent properties as a binder for positive electrodes by optimizing its components, their ratio, and manufacturing conditions.

Compared to PVDF binder, the main advantages of this binder is that it works in small amounts, provides flexibility to electrodes, and is stable at high temperatures. These advantages provide a large improvement in battery properties.

The binder is required not only to provide binding ability in the active material layers of the electrode and between the active material layer and collector but also to have electrochemical stability and binding durability under battery-operating conditions (Table 7.2). Also, the binder has to satisfy various conditions required in the electrode and battery manufacturing processes.

In general, electrode paint is prepared by mixing water or NMP with the active material, the binder, and other additives (such as thickener or conductive carbon). The paint is then applied to the collector using the doctor-blade method and so on, and then dried, pressed, and cut to prepare electrodes. The properties of the electrode paint and electrodes themselves change according to the type of active material and additives used. Therefore, the composition must be optimized to the conditions under which the paint is applied to the collector.

7.3 Binder for Negative Electrode: BM-400B

BM-400B, a binder for negative electrodes, is a water dispersion system that contains fine SBR gel particles (~130 nm diameter) and a small amount of gel-particle polymer (>130 nm diameter) (Fig. 7.1). These two polymers are elastomer-containing butadiene.

Fig. 7.1 BM-400B TEM image.[7] Reproduced by permission of CMC Publishing Co. Ltd

The binder has two roles: to bind active material particles to each other and to bind the metal-foil collector to active material particles. We think, however, that when PVDF – which is used for the negative electrode and NMP-soluble type – is used, the binding structure of active material particles (graphite) is different from that when BM-400B is used. We assume that graphite is held in the PVDF network because PVDF is poorly adsorbed on graphite surfaces, as shown in Fig. 7.2. On the other hand, SBR in BM-400B is elastomer-containing butadiene, providing good absorbance to graphite. We suspect, therefore, that SBR fine particles have point-contact to the graphite surface. It also can be thought that if a large amount of BM-400B is used, SBR fine particle – an electrical insulating material – may cover up the active material to prevent the battery from performing sufficiently. If SBR content in the active material of the electrode is 3 wt% or less to graphite, the battery works without problems. If SBR fine particles can be spread uniformly on a graphite surface, a good binding ability for the electrode can be maintained even if SBR content is about 1 wt%.

The binder is inherently unnecessary. The less the binder is used, the higher the density of the active material, providing a high-capacity battery. If the quantity of the binder is reduced, however, the performance of the paint for the negative electrode is deteriorated, resulting in nonuniform application to the collector. Making full use of BM-400B is crucial to manufacture good negative electrode paint, even with a small amount of binder.

The production of a good electrode regardless of its sign depends greatly on its paint. Good electrode paint satisfies the following three conditions:

1. Active material does not settle. The paint has an appropriate viscosity, giving a uniform application of paint.
2. The binder is well dispersed so as not to cause a second aggregation of the active material.

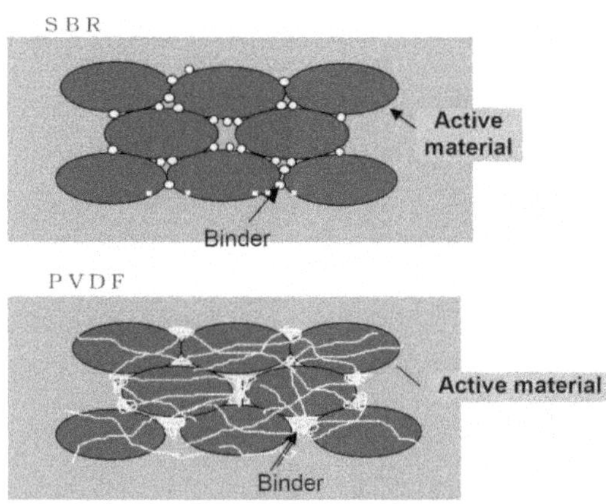

Fig. 7.2 Conceptual illustration of binding between binder and active material.[7] Reproduced by permission of CMC Publishing Co. Ltd

3. The binder is uniformly dispersed over the whole of the active material surface.

We do not know how battery manufacturers actually produce electrode paints. The utilization method of the BM-400B binder developed by Zeon Corporation is described next.

7.3.1 BM-400B Utilization Method

Since BM-400B is a particle-dispersion type binder, the negative electrode paint with a good flowability cannot be obtained only by mixing BM-400B, graphite, and the necessary water. Water-soluble polymer has to be added as a thickener. Carboxymethyl cellulose (CMC) is the best water-soluble polymer to use. The general chemical formula of CMC is:

CMC can be used exclusively because it is a crystalline polymer with adhesive properties. However, when used in this way, electrodes become hard and rigid and break often when rolled. To avoid this problem, CMC is used together with BM-400B, giving the electrodes an adequate elasticity.

7.3.2 Manufacturing of Negative Electrode Paint

The following points are important in manufacturing negative electrode paint:

1. Select CMC appropriate to the active material because many types of CMC are available. Since CMC and binder are not essential in the manufacturing of negative electrode paint, their amount should be limited to the minimum required amount.
2. Mix the active material uniformly with CMC water solution under suitable conditions, and then add in BM-400B.

The first step in the manufacturing of negative electrode paint consists in dividing the prescribed amount of CMC water solution into several portions (Fig. 7.3). The solution is added to graphite and kneaded into a thick paste. Water is added to distill the paste if necessary until a moderate flowability is obtained (no water is added in Fig. 7.3). At this point BM-400B is thoroughly mixed into the paste. Finally, more water is added if necessary (no water is added in Fig. 7.3) to adjust the viscosity and flowability of the electrode paint (Fig. 7.4). CMC is selected to match the type of graphite.

It is recommended to review beforehand the conditions needed to disperse the active material and to obtain a good flowability. The following is an easy method developed by Zeon Corporation.

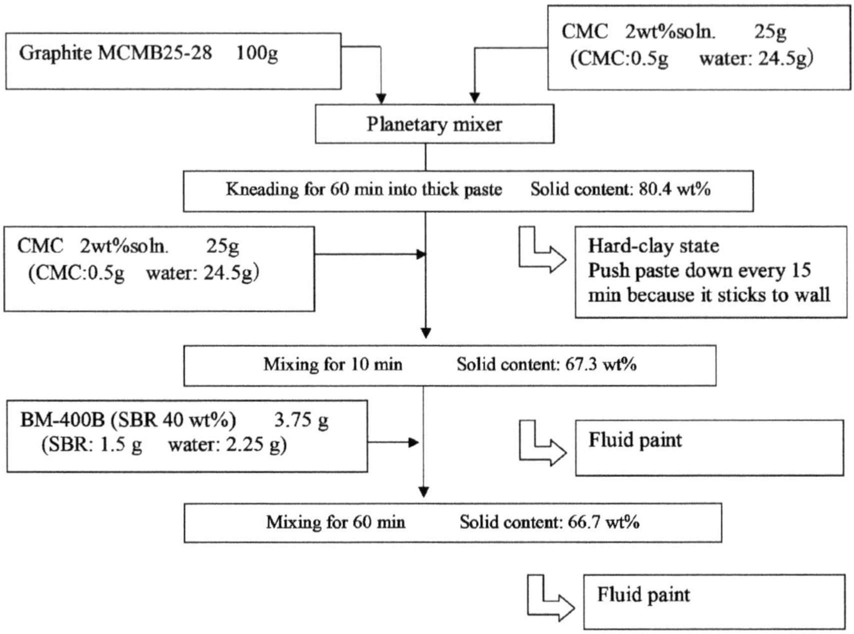

Fig. 7.3 Flow of manufacturing negative electrode paint.[7] Reproduced by permission of CMC Publishing Co. Ltd

Fig. 7.4 Paint for negative electrodes.[7] Reproduced by permission of CMC Publishing Co. Ltd

It starts by weighing 10 g of graphite (active material) and mixing it little by little with 1 wt% low viscous CMC water solution (Cellogen 7A by Dai-ichi Kogyo Seiyaku Co., Ltd.). The flowability level (Table 7.3) is obtained through visual inspection and plotted versus the solid content (Fig. 7.5). The flowability levels 2 and 8 are of importance since level 2 is the guideline point of the solid content to obtain a thick paste with uniformly dispersed graphite, and level 8 is the guideline of the solid content to obtain the negative electrode paint with the targeted good flowability.

Figure 7.6 shows a liquid-absorption behavior map plotting level 8 vs. level 2 using various kinds of graphite, the active material for the negative electrode. The unnamed points in Fig. 7.6 are results obtained from using graphite used in batteries offered by battery manufacturers, and the named points are results obtained from using the major graphite of each graphite manufacturer.

Table 7.3 Flowability level (visual observation).[7] Reproduced by permission of CMC Publishing Co. Ltd

Flowability level	Paint state (visual observation)
Level 1	Particles, not coagulating
Level 2	*Hard cake (optimum thick paste)*
Level 3	Soft cake
Level 4	Some liquid is observed on the surface
Level 5	Surface tends to flow to flat
Level 6	Slipping down from spatula in lump
Level 7	Paint spreads when slipping down from spatula
Level 8	*Sufficient flowable state (optimum flow state)*

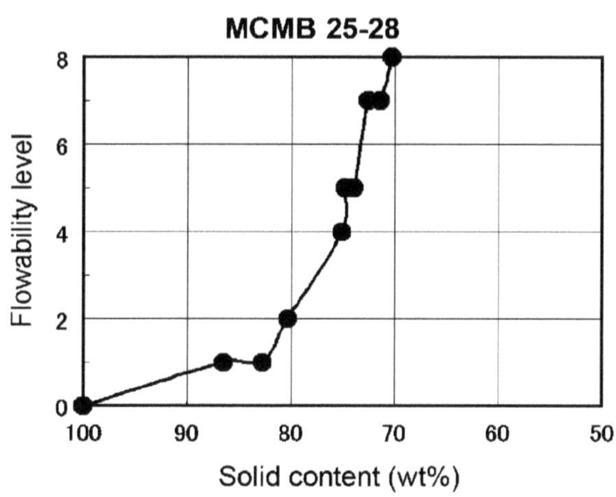

Fig. 7.5 Liquid-absorption curve (MCMB).[7] Reproduced by permission of CMC Publishing Co. Ltd

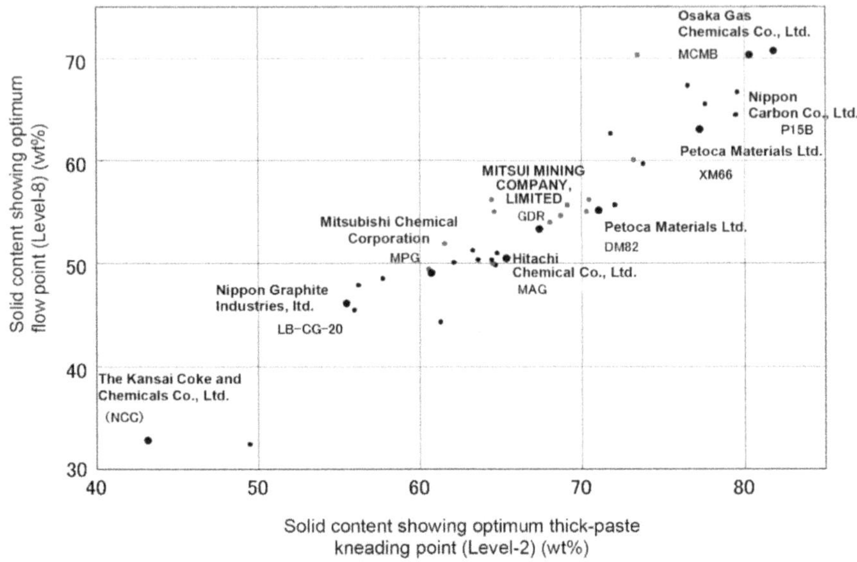

Fig. 7.6 Liquid-absorption behavior map for various kinds of graphite.[7] Reproduced by permission of CMC Publishing Co. Ltd

The solid content in the negative electrode paint for level 8 varies widely from 30 to 70 wt% depending on the graphite type. When graphite requires a solid content ≥55 wt% for level 8, PVDF can be used. However, when the solid content needed is ≤55 wt%, PVDF is inappropriate and BM-400B should be used.

When selecting CMC for level 8, a low-molecular-weight CMC (viscosity 100–500 mPa s at 1 wt%) is recommended for graphite requiring a solid content of ≥60 wt%; on the other hand, high-molecular-weight CMC (viscosity 1,000–2,000 mPa·s at 1 wt%) is advisable for a solid content ≤60 wt%.

For level 2, a thick paste at solid content is most suitable to disperse the active material. The electrode surface roughness can be used as an index to show the graphite dispersion state in electrode paint. Figure 7.7 shows the relationship between the solid content of thick paste and the surface roughness of the electrode. We think that smaller surface roughness and higher concentration of thick paste lead to a better dispersion of the mix.

During the preparation of the negative electrode, the negative-electrode paint is applied to the copper-foil collector using the doctor-blade method; then the paint is left to thoroughly dry and expel any reminisce of water, and finally pressed to the density designed for the active material. Allowing the paint to dry in two stages provides the electrode with a good binding ability. Namely, the first stage should be performed below 100°C (preferably 50–60°C) to remove free water, while during the second stage the temperature is raised above 100°C to thoroughly remove water. It should be noted that drying the paint at 100°C or above during the first stage may

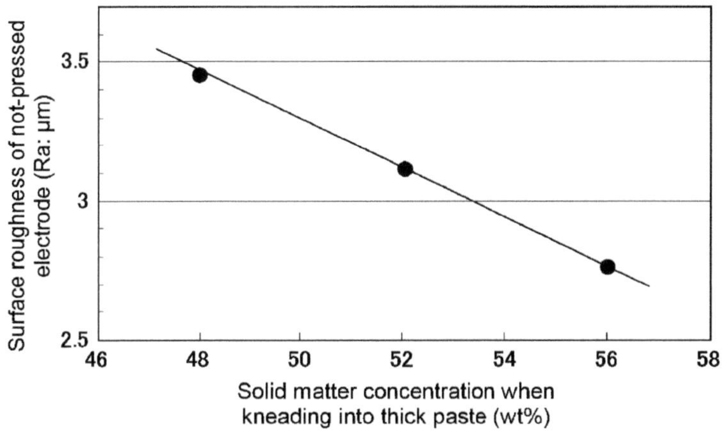

Fig. 7.7 Solid content and electrode surface roughness when kneading into thick paste.[7] Reproduced by permission of CMC Publishing Co. Ltd

cause bumping and migration of the SBR particles to the electrode surface, lowering the binding ability and increasing the electroresistance of the electrode surface.

SBR of BM-400B has a glass transition temperature (Tg) of -5°C, providing the electrode with flexibility. Furthermore, its high thermal decomposition temperature allows the use of high temperatures for the drying process at the second stage.

The application of the binder is important, since it determines the quality of the finished electrode, e.g., its surface uniformity (presence or absence of cracks, chipping, fractures, craters, lines, and rings) before pressing, its flexibility, its surface roughness, and the ability to bind the collector with the active material. Battery performance is the final index to check the binder usage.

7.4 Binder for Positive Electrodes: BM-500B

The binder for positive electrodes must possess even better properties than that for negative electrodes. The binder used for batteries, for example, is required to be electrochemically stable over a long period under an oxidation environment when a battery is charged fully at 60°C. Furthermore, the binder also has to pass various safety tests such as the battery-nailing test. This poses a problem since there are not many polymers available that pass these tests and can work as a binder. As with binders for negative electrodes, a high binding ability is necessary even with small amounts. In addition, to cope with the recent trend toward the use of thin prismatic batteries, the electrode must be very flexible so that no cracks are induced even if the electrode is bent at an acute angle.

Ten years have passed since lithium-ion batteries were put on the market and many patents for binders for positive electrodes have been presented (Table 7.1). However, no practical binders that satisfy these tough conditions have been found yet with the exception of PVDF, which is still mainly employed.

Recently Zeon Corporation has developed BM-500B as a binder for positive electrodes in which ACM fine particles are dispersed in NMP. This binder satisfies the tough conditions mentioned above.

BM-500B is an elastomer consisting of gel ACM with a diameter of about 170 nm in a dry state. This is a particle-dispersion type elastomer in which swelled ACM is dispersed in NMP, and has a polymer structure that concurrently possesses resistance against electrolyte and dispersive ability in NMP solution.

Since BM-500B is a particle dispersion type, it does not increase the viscosity of the solution, and when used exclusively provides insufficient properties for paint. Therefore, it is important to employ a thickener such as the CMC used with BM-400B for negative electrodes. Zeon Corporation has been developing various thickeners for positive electrodes depending on the types of the active material. The details are not described here due to lack of space.

The electrochemical stability of the binder for positive electrodes is determined in the following way. The polymer used for the binder is mixed with carbon to prepare a mixed electrode. A voltage of 3–5 V is applied and the oxidation current is measured using the cyclic voltammetry method. Polymers with small oxidation current at applied voltages of up to 4.6 V are considered to be suitable. It has been reported that the oxidation resistance of polymers can be predicted using molecular-orbital energy calculations.[4] This (HOMO) energy (maximum occupancy orbital energy) is calculated by the semiempirical molecular orbital calculation method where the AM1 method is used as Hamiltonian. Zeon Corporation thinks that polymers with a HOMO energy less than -10 eV are appropriate to be used as binders for positive electrodes. BM-500B is a polymer whose main carbon-carbon saturated bond has a small HOMO energy of –11.4 eV. Therefore, we think this polymer has an oxidation-resistant structure.

As for the negative electrodes, a particle dispersion type binder can be used to uniformly disperse the active material and the conductive carbon. An example of that type of binder is the BM-500B.

An important point in the preparation of the positive electrode is the appropriate dispersion of its components – i.e., active material, conductive carbon, binder, and thickener – since this affects directly the performance of the electrode. Another important condition in the preparation of the electrode is the need to prevent the settling of the active material. This is crucial for the positive electrode since its active material has a large density. The affinity and aggregation of BM-500B to the active material and conductive carbon are strong; however, their strength differs individually.

The simultaneous addition and kneading of the components – i.e., active material, binder, thickener, and conductive carbon – as in the case of a PVDF solution for the binder, tends to intensify the dispersion failure of the active material and conductive carbon. In order to avoid this phenomenon Zeon Corporation proposes the paint manufacturing method shown in Fig. 7.8. In this method the dispersion of the conductive carbon and active material are performed independently. The pastes of the carbon and active material are adjusted separately before mixing them to use as binder for the positive electrode. This method facilitates the preparation of uniformly and stable mixed paint. It also allows the uniform dispersion of the conductive carbon which provides adequate conductivity with a minimum amount.

The high-dispersion technology for conductive carbon has been studied.[5] In our method, the appropriate dispersion is achieved by kneading the conductive carbon and thickener to produce a thick paste. For example, using a planetary mixer, carbon and thickener are first kneaded into a clay state with a solid matter of 35 wt%; the process continues until obtaining a carbon paste of 60–70 wt% with carbon particles of < 1 μm of diameter (refer to Fig. 7.9). Use of a two-roll-type kneader can provide a higher dispersion carbon paste of 80–90 wt% with carbon particles of < 1 μm of diameter. The diameter distribution of the carbon particles can be controlled by adjusting the solid content as well as the kneading time.

BM-500B and the active material can be fully kneaded into a mixture of high solid content (about 85 wt%, hard clay state) using a planetary mixer. Combining

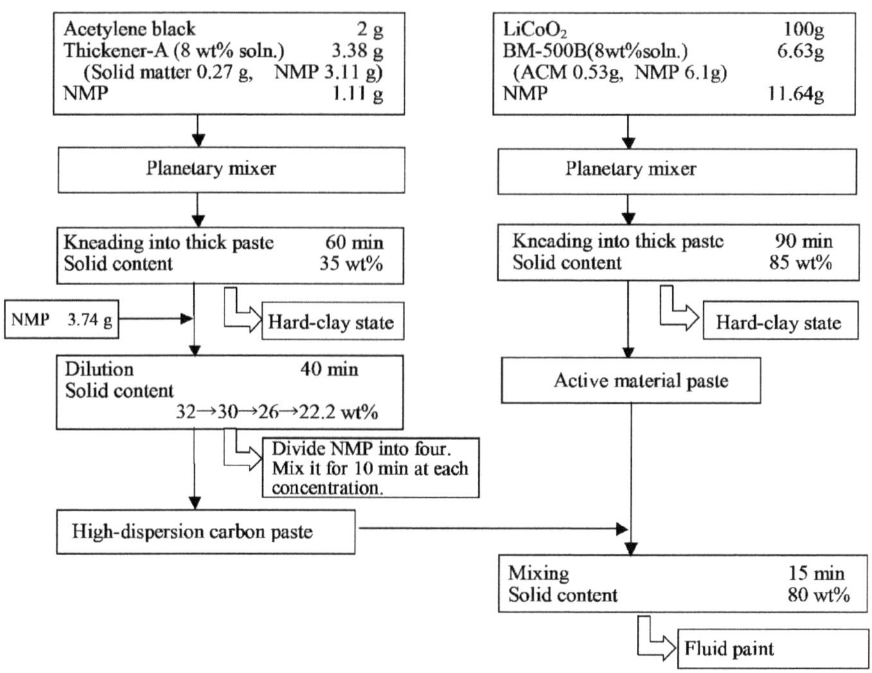

Fig. 7.8 Flow of manufacturing positive electrode paint.[7] Reproduced by permission of CMC Publishing Co. Ltd

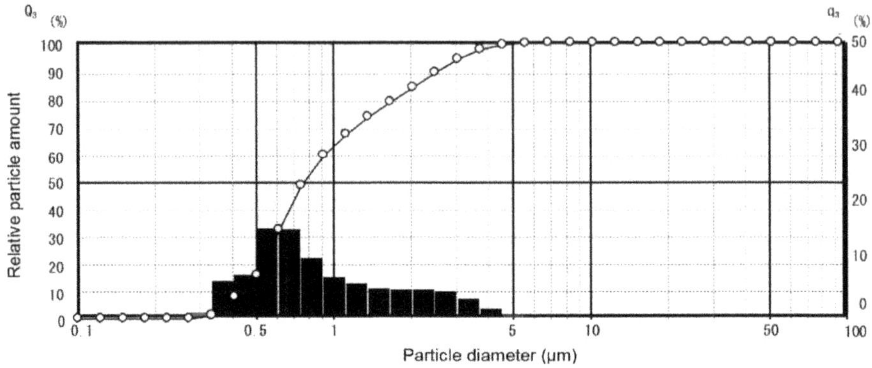

Fig. 7.9 Distribution of particle diameter of carbon paste.[7] Reproduced by permission of CMC Publishing Co. Ltd

this mixture with the high-dispersion carbon paste mentioned before produces a uniform and stable paint for the positive electrode (Fig. 7.8). This paint can be applied to aluminum foil and allowed to dry to obtain a smooth electrode with a surface roughness (Ra) of about 0.7 μm. Use of PVD binder produces an electrode with an Ra of about 1.0 μm at its best.

The dispersion state of the binder on the electrode surface has been observed with SEM (Fig. 7.10). The dispersion of the conductive carbon on the active material surface depends on the binder type, i.e., PVDF or BM-500B. It has been found that the BM-500B binder provides a more uniform dispersion, which is suspected to greatly influence battery properties. However, the optimum dispersion state for batteries has not been clarified yet.

Flexibility is another important characteristic of the electrodes, since it determines how easily electrodes can be rolled or how easily cracks are induced, which is of primordial importance in the making of prismatic batteries thinner. BM-500B is an elastomer with a glass transition temperature of –40°C that provides flexibility to electrodes even with a small additive amount. Figure 7.11 shows the results of flexibility evaluation of electrodes using a loop stiffness tester. Figure 7.12 shows the correlation between the thicknesses of the electrode ring obtained with the equipment shown in Fig. 7.11 (stiffness of aluminum collector is subtracted). The flexibility of the electrodes can be determined with ring thickness and stiffness at an electrode –break point. No cracks in the electrode were observed when BM-500B was used, even if when the electrode was pressed to 1 mm. This result shows stiffness is small (soft). Figure 7.13 shows the correlation between the amount of binder additive and the electrode flexibility. Note that compared with PVDF, BM-500B provides flexibility to electrodes with a minimal amount of additive.

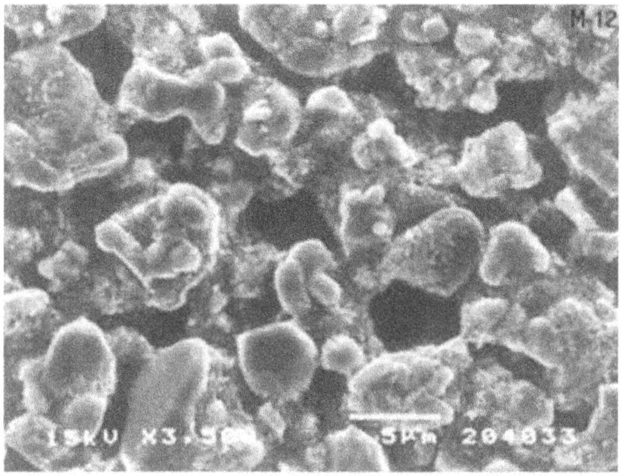

Using BM-500B

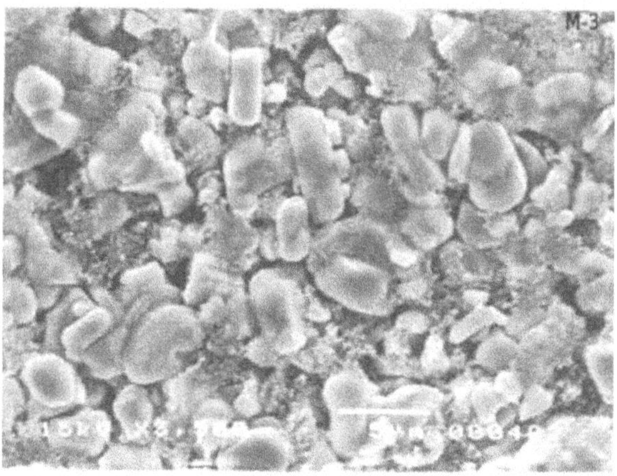

Using PVDF

Fig. 7.10 SEM image of positive electrode.[7] Reproduced by permission of CMC Publishing Co. Ltd

7.5 Summary

High-performance secondary batteries better than the lithium-ion secondary batteries have not been developed yet. We think that the lithium-ion secondary battery will become widely available and its application expanded. In addition, demands for the improvement of battery performance will drive the creation of miniaturized, thinner, higher capacity, and safer batteries. These performances are translated into, for example, facilitation in manufacturing electrode paint, speedup of electrode manufacture, high-speed impregnation of electrolytes to electrodes, and high-speed

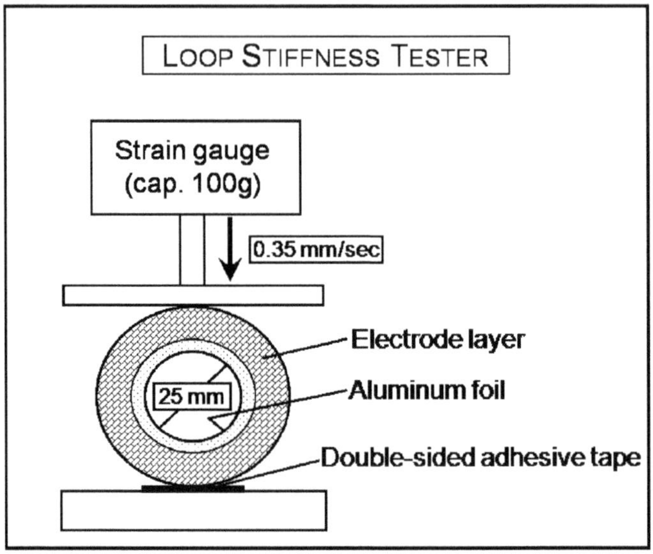

Fig. 7.11 Flexibility measurement using loop stiffness tester.[7] Reproduced by permission of CMC Publishing Co. Ltd

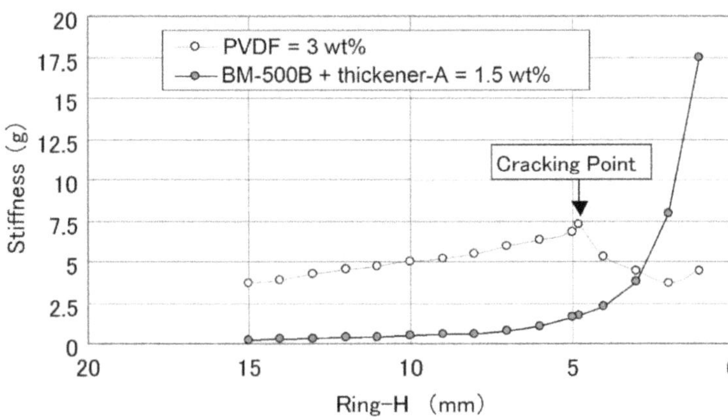

Fig. 7.12 Flexibility of an electrode.[7] Reproduced by permission of CMC Publishing Co. Ltd

rolling of electrodes. As expected, the binder will play an important role in these productivity improvements. Traditional development of electrode material has focused on active material, electrolyte, and separators.[6] We hope that the development and improvement of binders will be regarded as equally important as these traditional factors.

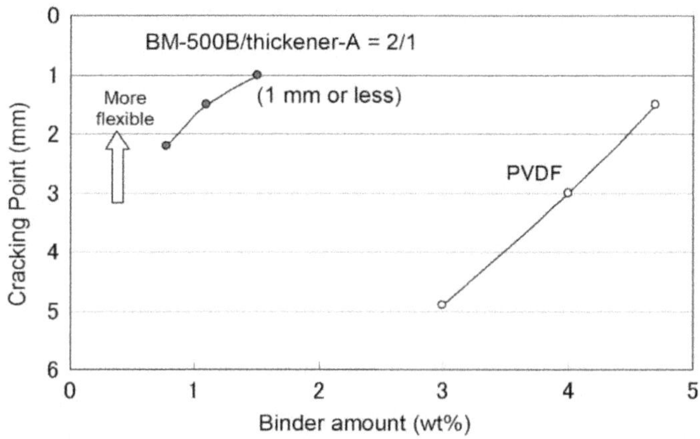

Fig. 7.13 Correlation between binder addition amount and flexibility of an electrode.[7] Reproduced by permission of CMC Publishing Co. Ltd

References

1. *Development of Material for Li-ion Battery and Market*, CMC, Minneapolis, MN, 1997, p.114.
2. For example, AH4-51459, AH5-74461, AH10-814519, A2001-15116, A2001-76731, A2002-75373, A2002-75377, A2002-75458.
3. For example, A2000-195521, A2001-28381, A2001-35496, A2001-256980, A2001-332265, A2002-56896, A2002-110169.
4. A. Kurihara, A. Nagai, Proceedings of 39th Battery Symposium in Japan, Japan, 1998, pp. 309–310.
5. S. Ikeda, A. Kozawa, *Kogyo Zairyo*, **49** (6) (2001) 65–68.
6. M. Yoshio, A. Kozawa, *Li-ion Secondary Battery* (2nd ed.), The Nikkan Kogyo Shinbun, Osaka, 2000.
7. *Secondary Battery Material: Recent Ten Years and Hereafter*, CMC, Tokyo, 2003, Ch. 13.

Chapter 8
Production Processes for Fabrication of Lithium-Ion Batteries

Kazuo Tagawa and Ralph J. Brodd

8.1 Introduction

This chapter is intended to provide an overview of the various aspects of manufacturing lithium-ion (Li-Ion) cells. While the basic principles of cell design and manufacture are well known, each manufacturer maintains proprietary, specific details of their cell designs and assembly and the equipment used in cell fabrication. Nonetheless, the overall principles and processes involved are detailed below. Figure 8.1 shows a schematic of the components of a cell (battery).

The International Electrotechnic Commission (IEC) has established a common nomenclature for describing the various cell sizes and chemistry.[1] For instance, ICR18650 translates into: I is for Li-Ion technology, R is for a round cell, and C is for cobalt cathode, 18 is for the cell diameter in millimeters and 650 is the cell height in tenths of a millimeter. For a prismatic cell, IMP366509, I for Li-Ion, M is for manganese, P for prismatic, 36 mm wide, 65.0 mm long, and 9 mm thick.

8.2 Cell Design

Figure 8.2 depicts the various components of a commercial cell. Each segment and boundary represents unique considerations for good cell operation. For instance, the boundary of the current collector with the active mass involves the conductivity of the active mass and the current distribution (collection) within the porous electrode structure of the active mass. These are intertwined with the electrolyte being common to all components and a key consideration in developing a safe high-performance commercial cell. Safe cell designs alone are not sufficient criteria for a safe commercial cell. The productions operations must not introduce defects that

K. Tagawa
Hohsen Corporation, 10-4-601 Minami Semba 4-chome,
Chuo-ku, Osaka, 542-0081, Japan

M. Joshio et al. (eds.), *Lithium-Ion Batteries*,
DOI 10.1007/978-0-387-34445-4_8

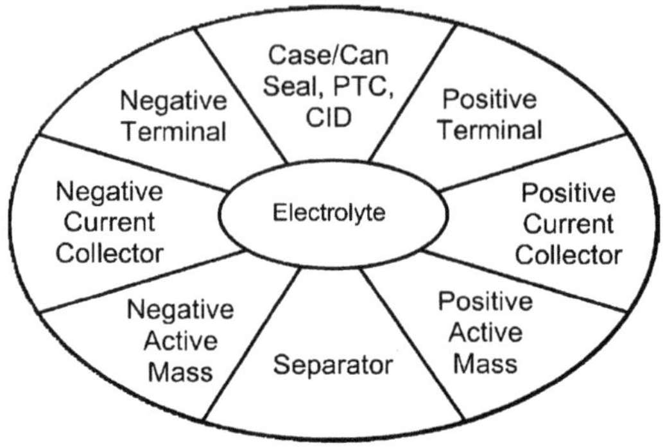

Fig. 8.1 Depiction of a cell showing the components that make up a battery. Each *line* represents an interface between components and an area of concern in cell operation. The electrolyte is a common denominator and contacts all of the cell components. Different phenomena occur at each interface

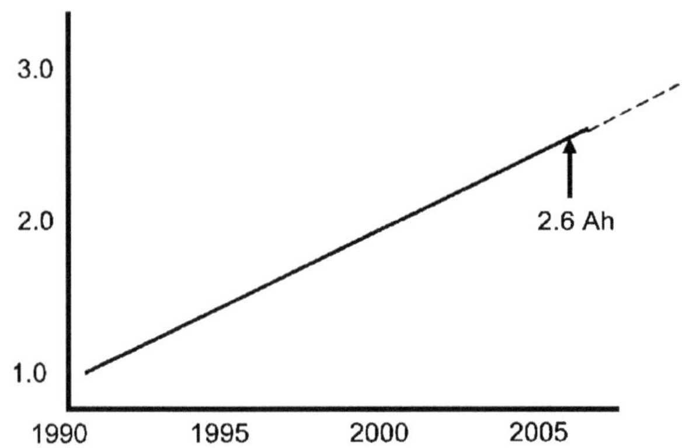

Fig. 8.2 Capacity increase in an ICR18650 cell as a function of time (smoothed)

create safety and performance problems. With this in mind, every new piece of equipment also must be qualified to ensure that it does not introduce defects in the cell that can cause safety problems. Once a cell design is finished, prototype cells are constructed, and comprehensive, statistically valid performance and safety testing are carried out to confirm cell performance. The improvement in cell ampere hour (Ah) capacity since the commercial introduction of the Li-Ion by Sony is given in Fig. 8.2.[2,3]

It is good practice to develop a computer program to predict the performance based on sound chemical engineering principles.[4,5] These programs are capable of closely approximating actual cell performance. Important cell characteristics include current distribution within the electrode structure, reactivity of the active mass (ex-change current), electrode thickness and porosity, ratio of conductive diluent to active powder, the balance of anode to cathode capacity, and the electrolyte conductivity among others. The goal is to have uniform current density across the geometric surface of the electrode strip as well as uniform current distribution inside the electrode coating. Once the computer program is written, it can be tested and updated using actual cell performance.

In addition to the materials and reactions that produce current in the cell, most Li-Ion cell designs incorporate safety devices such as:

- Shutdown separator undergoes a phase change that closes the pores and increases the internal resistance of the cell to reduce/stop current flow to stop the cell operation.
- Positive temperature coefficient (PTC) resistor operates by a phase change in the conductive polymer placed internally to terminal connection which increases its resistance to minimize current flow to the cell terminals when either the current flow exceeds the design point or the cell internal temperature exceeds the set point.
- A current interrupt device (CID) disconnects the electrodes from the cell terminal to stop current flow when the internal cell pressure reaches a predetermined pressure, usually the result of a high internal cell temperature.

8.3 Fabrication of Cylindrical and Prismatic Cells

An outline of the Li-Ion battery manufacturing process is shown in Fig. 8.3. The Li-Ion battery is manufactured by the following process: coating the positive and the negative electrode-active materials on thin metal foils, winding them with a separator between them, inserting the wound electrodes into a battery case, filling with electrolyte, and then sealing the battery case. The manufacturing process for the Li-Ion battery can be divided roughly into the five major processes:

1. Mixing, kneading, coating, pressing, and slitting processes of the positive electrode and negative electrode materials.
2. Winding process of the positive electrode, negative electrode, and separator.
3. Insertion of the wound cell core and electrolyte injection into the battery case.
4. Cell closing or sealing process.
5. Formation, aging, and cell selection

8.4 Mixing and Coating

In the electrode fabrication process, the active materials are coated onto metal foils and calendered. The positive electrode consists of active material, such as $LiCoO_2$, $LiNiO_2$, or Li_2MnO_4; a carbon-conductive agent such as acetylene black, Ketjen

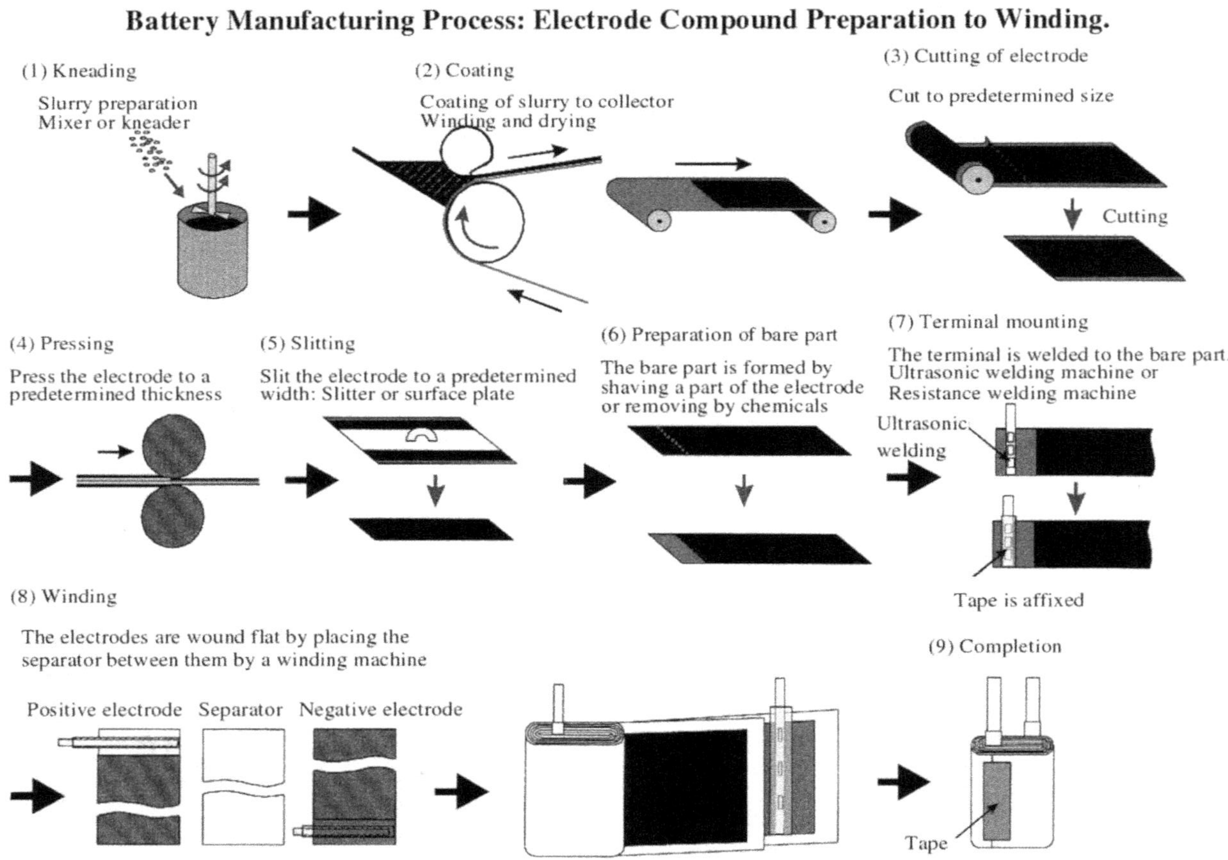

Fig. 8.3 General overview of the various processes that constitute the processes for manufacturing Li-Ion cells, starting with the coating operations for preparing the electrode stock. More specific manufacturing operations follow for the assembly of the cylindrical, prismatic and flat plate/laminate cell constructions

black, and graphite; and a binder, such as polyvinylidene difluoride (PVdF), or ethylene-propylene-diene methylene linkage (EPDM). The graphite and oxide are dry mixed in a multipurpose mixer (conventional impeller blade-type mixer). When different kinds of solids (except binder) are mixed, they first should be thoroughly mixed under dry solid-state conditions. The dry mixed solids are fed into a ball mill along with the previously prepared solution of PVDF dissolved in *N*-methyl pyrrolidone (NMP), and then thoroughly stirred. The ball mill contains ceramic balls, e.g., glass, zirconia, and so forth, about 2–3 mm in diameter, to assist in mixing. The mixing condition in the ball mill can strongly influence the performance of the battery.

The process for the negative electrode follows essentially that of the positive electrode but with different materials. Carbon or graphite is used for the negative electrode-active material. PVDF, carboxymethylcellulose (CMC), or styrene butadiene rubber (SBR latex), and so forth, are used for the binder. Polyimide may be added in some cases. Depending on the binder, solvents such as NMP for PVDF and water for EPDM, and so forth, are used for coating. PVDF has a bonding effect for the carbon material and the metal collector (copper foil). The wet mixing for the negative electrode is carried out using a planetary mixer for best results. The planetary mixer is an apparatus having two to three impeller blades on different axes so as to uniformly mix the side wall and the center side of the mixing vessel. While one may mix all the materials of a predetermined solid fraction in the laboratory, a better result can be obtained by first kneading the materials into a stiff paste and then adjusting the viscosity by adding the solvent so that it can be applied easily in the coating process. The electrode slurries (the viscosity is from 10,000 to 20,000 cps) are uniformly coated onto both sides of the current collector (aluminum foil for the positive and copper foil for the negative) of 15–20-μm thickness. Proper mixing techniques result in a uniform distribution of the components of the active mass during the coating operation.

Coating operations can use a slot die, reverse roll coating, or doctor blade coating equipment.[6] Each can produce satisfactory electrode stock. Close control of coating thickness is essential to ensure that all the parts fit into the can during final assembly. Usually, a second coating is applied onto the opposite side to produce a double-sided electrode foil. The record of each roll coating operation is archived for each roll as a quality control measure. The coating thickness for different cell designs can vary from 50 to 300 μ, depending on the cell design. The coating may be interrupted at regular intervals that correspond to the length of the electrode for winding into the cell core in preparation for winding the cell core. Next, the dried electrode is compressed with a roller press machine (calendered) to provide accurate control of the electrode thickness and to increase the density of the electrode mass. The load and the speed of the calender vary with each manufacturer. If the calendering process is not carried out properly, the yield from the winding process will decrease. After calendering, the master roll is slit to the width specified for cell construction and wound onto a roll for the winding operations.

8.5 Cylindrical-Cell Fabrication

A schematic of the assembly of cylindrical cells is shown in Fig. 8.4. The next step in cell fabrication is the winding operation to produce the cell core. The reels of the slit anode, cathode, and separator stock are mounted on the winding machine. The strips have been sized for length, width, and thickness that match the cell design. Once the foils are threaded into the machine, the machine starts and continues to operate automatically until the reels are used up. This process of mounting the reels is repeated throughout the workday. An aluminum tab (about 0.08–0.15-mm thickness) is affixed to the bare part of the aluminum foil positive electrode by ultrasonic welding before winding. Likewise, a nickel tab (about 0.04–0.1-mm thickness) is affixed to the copper negative electrode collector by ultrasonic welding. If the electrode foils were not coated with an intermittent pattern, the coated foil is cleaned before the welding process.

The winding machine then combines the two electrodes and separator strip on a mandrel and winds the combination in jellyroll fashion into a tight bobbin or cell core. Cylindrical cells are wound on a round mandrel while the prismatic cells use a flat paddle. Winding requires constant tension on the coil as it is formed and grows to the final size. Any irregularity leads to a gap between the separator and electrode, resulting in a nonuniform current distribution that can lead to a malfunction or shortened cycle life. Vision and X-ray systems are used to continuously monitor critical processes and ensure precise placement of components. The wound coil is checked for internal shorts before being inserted into the can with a "hi-pot" or impedance tester. In the last step, the cell core is taped to keep it tightly wound before inserting it into the cell case. It is important that neither peeling or sloughing off of the active material nor twisting of the separator occur during the winding process. Early rejection of potential cell faults is an economy and prevents investing more work on bad cells.

Following insertion into the cell case, a tubular mandrel with a serration may be inserted in the winding core. The function of this mandrel is to improve the integrity of the cell core and for safety of the battery. When the internal pressure rises, gas has free passage through the hollow center of the mandrel and is released through the vent. When the cell is crushed, the mandrel causes both electrodes to short-circuit and instantaneously discharge. A welding electrode is inserted through the hole of the mandrel to weld the anode tab to the cell case.

Any moisture contamination of the cell has a deleterious effect on its operation. Therefore, all cell assembly operations are usually carried out in a dry room or dry box. Alternatively, the cell then may be put in a heated vacuum oven for 16–24 h in order to extract residual water from the cell core before electrolyte filling. In the next process, the cell is filled with the electrolyte using a vacuum injection apparatus. Electrolyte is added to the cell by a precision pump and then vacuum filled to ensure that the electrolyte permeates and completely fills the porosity in the separator and electrode structures. Precision pumps meter the exact amount of electrolyte needed for good cell operation. The electrolyte salt usually is $LiPF_6$ dissolved in a

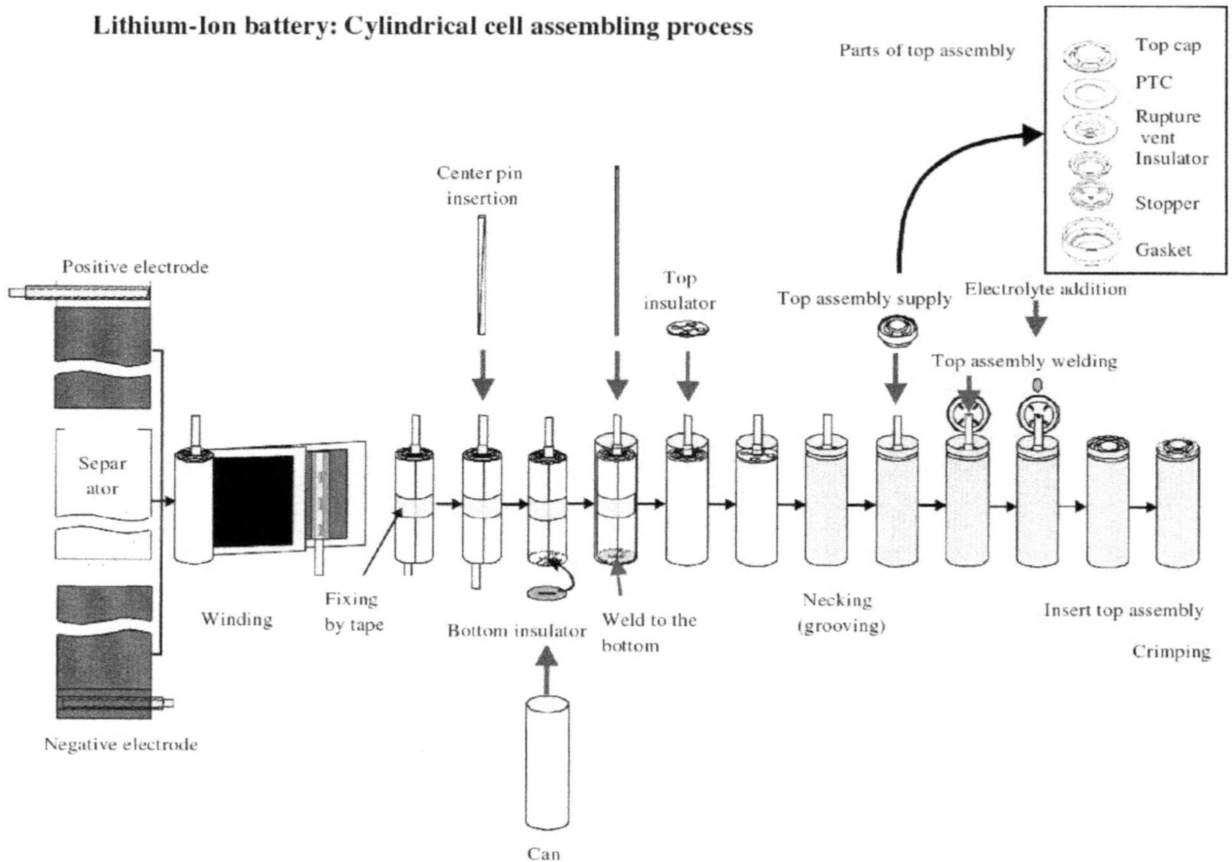

Fig. 8.4 General schematic for cylindrical cell assembly. Several operations, such as beading a ledge near the top of the can to seal the cell cap, the cap assembly itself, and the vacuum for electrolyte addition are not shown here. The polymer binder may be PVDF of SBR, depending on the design specification

mixture of organic carbonate solvents. The exact composition varies with each manufacturer.

After filling, the cap is placed and a polymer compression seal is applied to close the cell from. The cell cap contains the vent, PTC, and CID safety devices. Both the CID and PTC are safety devices designed to activate and prevent dangerous temperatures and pressures from developing internal to the cell. The PTC function is to stop current flow when the current or cell temperature exceeds a set value. The current and temperature are checked for each lot of devices before use. The CID is designed to interrupt the current when the internal cell pressure exceeds the set value before the cell reaches the vent pressure. Afterward, both the cylindrical cell and the prismatic cell are completely washed with isopropyl alcohol or acetone containing a small amount of water in order to remove any adhering electrolyte. An electrolyte leakage test is done using a smell sensor apparatus in order to confirm a good seal.

The internal construction of the cell is inspected by X-ray for proper top cap installation, misalignment of the winding from processing, the improper bending of the tabs, and so forth, which potentially could cause an internal short circuit or cell malfunction. It is common to print a cell number and other information such as production line number, date, and so forth, to serve as identification for future reference. The number can be used to trace the source and day of manufacture, to identity the assembly line, and the identity of all cell components, materials, electrolyte, separator, and so on. It also may be jacketed (labeled). Detailed quality assurance checks become a part of the archived data on the materials, processing, and conditions for the fabrication of each cell.

Finally, the cell is subjected to an aging process to select out cells with microshorts and to categorize cells by capacity for assembly into battery packs. The temperature, length of storage, and cell checking methodology vary. This process is designed to identify cells with internal cell faults and microshorts that are not found during cell fabrication.

8.6 Prismatic Cell Fabrication

Assembly of prismatic cells shown in Fig. 8.5 follows the general pattern for assembly of coating and winding for cylindrical cells until the process step of applying the top. The process then deviates because of the characteristics of the prismatic geometry as opposed to the cylindrical cell geometry. Prismatic cells use an aluminum cell case in order to decrease its weight, and an aluminum-laminated material is used in order to decrease the thickness (4 mm or less) of the battery. An insulating material is placed on the bottom of a cylindrical case and the wound coil is placed in the case using an insertion apparatus. For the cylindrical cell, the negative electrode tab is welded to the bottom of the can. For the prismatic cell with an aluminum case, the positive electrode tab is welded to the aluminum case. For the

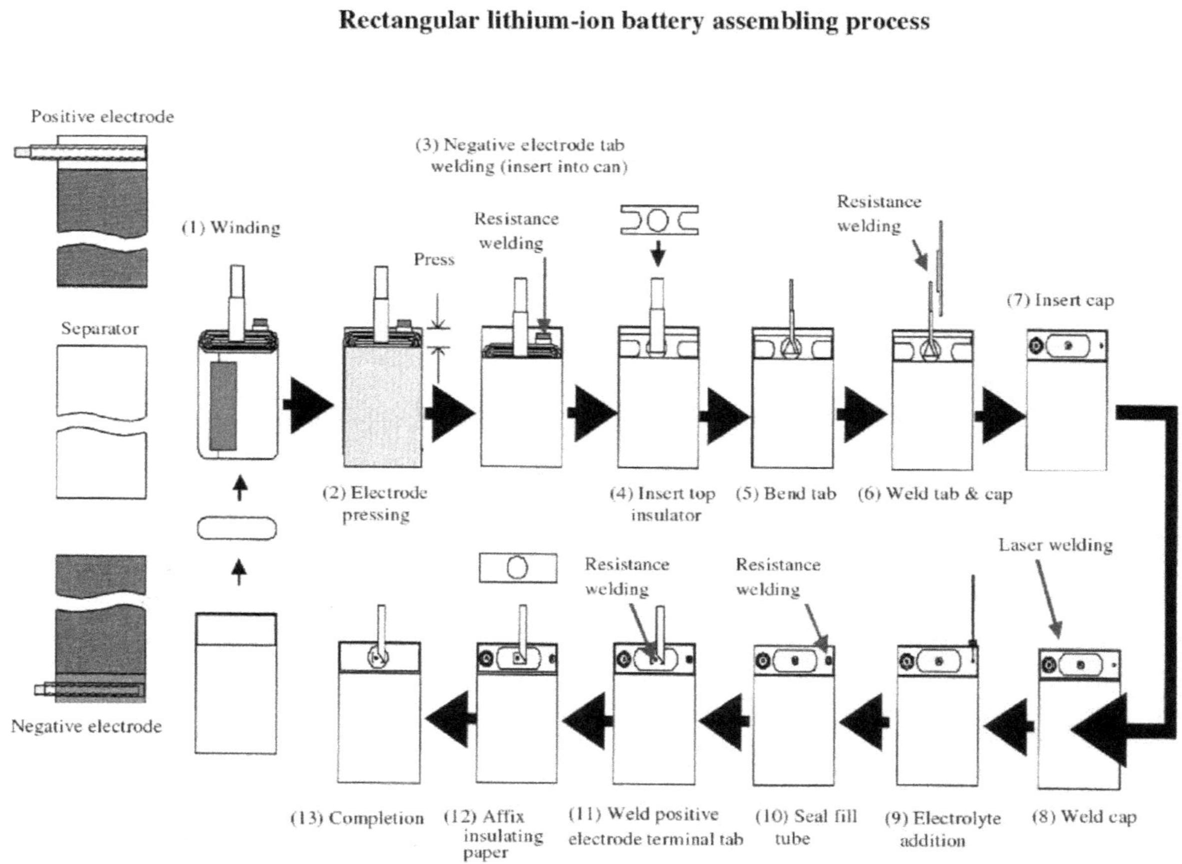

Fig. 8.5 General schematic for prismatic cell assembly. Several operations, such as beading a ledge near the top of the can to seal the cell cap, the cap assembly itself, and the vacuum for electrolyte addition are not shown here. The polymer binder may be PVDF of SBR, depending on the design specification

prismatic cell, a top cap and a tab usually are laser-welded to the cell case. If previous processes are not carried out in a dry room, the cells are dried under vacuum, overnight or for 24 h, in order to extract any water in the cells. The cap is pushed into the case with a cap inserting apparatus, etc., and the cap is temporarily welded to the case using a laser-welding machine; finally, the surroundings of the cap and the case are completely laser-welded. Next, the electrolyte fill tube is sealed using a laser-welding machine.

8.7 Li-Ion Flat Plate and Polymer Cell Fabrication

Li-Ion polymer and flat plate cells are produced in small sizes for cellular phones (about 0.5 Ah and higher) and large sizes (up to 200 Ah) for energy storage and motive power applications. A common characteristic is that the polymer (adhesive) holds the electrolyte and forms a physical barrier between the anode and cathode to prevent internal shorting. Most polymer cells are flat plate (prismatic) construction. There is no one general cell fabrication process for polymer cells as there is for the liquid electrolyte cylindrical and prismatic cells. Each manufacturer has a slightly different process flow. Cell designs are essentially laminated constructions produced by several different fabrication processes examples are shown in Figs. 8.6 and 8.7. In general, cell fabrication follows the same description of the prismatic cells. The active mass may be coated onto foils or expanded metal strips, depending on cell design. Cell assembly follows the general outline for cylindrical and polymer cells. The rolls of the anode and cathode electrode stock and polymer electrolyte are "cookie-cuttered" in the specified dimensions and stacked in layers with the polymer film, as depicted in Fig. 8.8. Alternatively, a Z-fold construction may be used with the electrodes bonded to the separator.[7] The stacks then are bonded together using ultrasonic or heat to ensure uniform bonding and dimensional control.

In the following section, the manufacturing process of a lithium polymer battery and a lithium-ion battery, which use a laminated film as the exterior case, will be briefly explained. The methods of coating the positive electrode and the negative electrode are the same as previously described. The following methods are now being used for making the cell core or electrode stack:

- The positive electrode, the negative electrode, and the separator are wound into a coil and then heated and pressed flat.
- The positive electrode, the negative electrode, and the separator are weaved using a Z-fold or the W weaving (Thuzuri-Ori) method.
- Stacking the positive electrode, the negative electrode, and the separator (repeatedly layering the positive electrode, the separator, and the negative electrode).

The cell case is formed from an aluminum-polymer laminated film. The film is folded in half, and the folded film is cut into pieces. The insulator film is fusion bonded, the electrode stack is inserted, and the exterior case is fused by a heat or

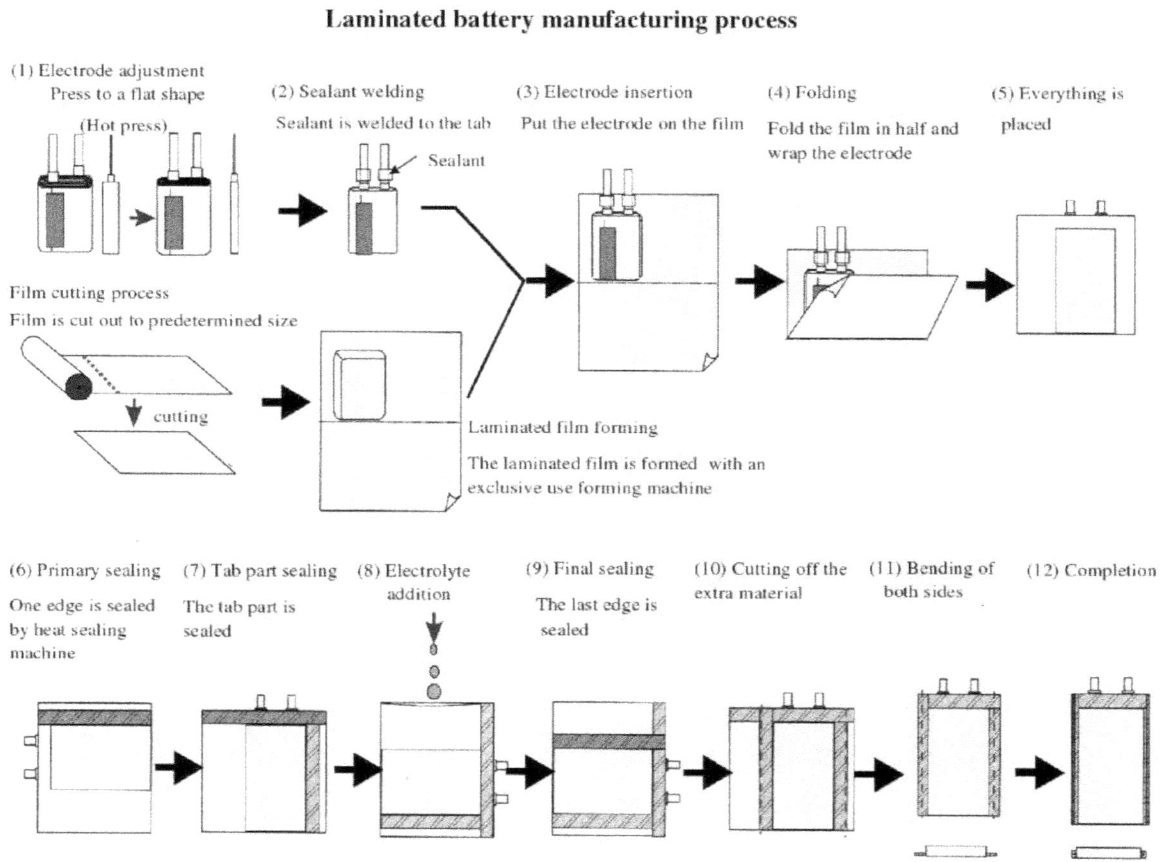

Fig. 8.6 General schematic for flat plate/laminated cell assembly. Several operations, such as beading a ledge near the top of the can to seal the cell cap, the cap assembly itself, and the vacuum for electrolyte addition are not shown here

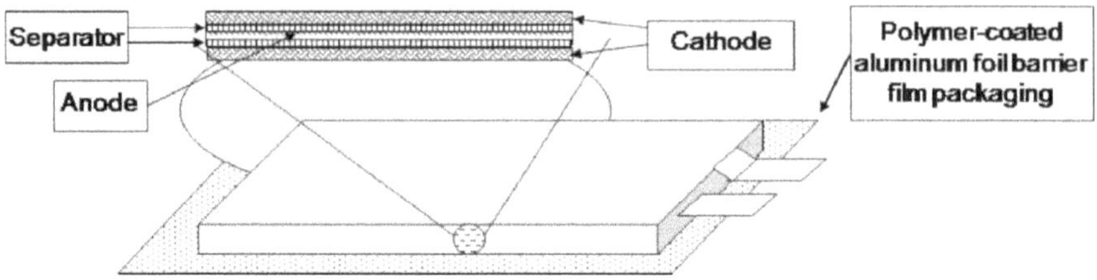

Fig. 8.7 A depiction of the flat cell construction. A "bicell" configuration is shown

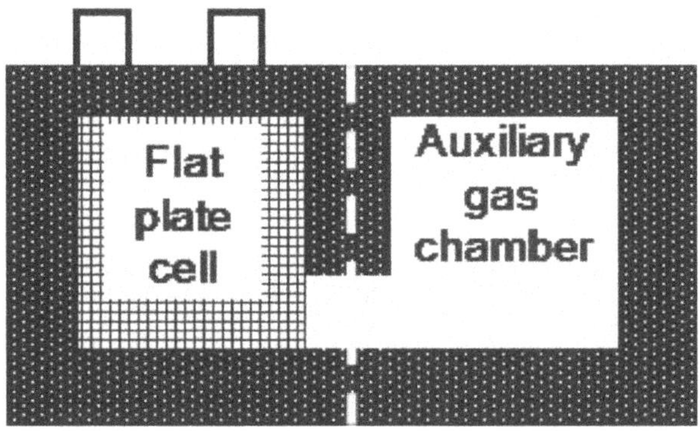

Fig. 8.8 The flat plate/polymer cells may use a polymer pouch construction with an aluminum barrier layer. An auxiliary chamber may be used to contain the gases generated during formation. After formation the auxiliary chamber is penetrated with a needle and any gas formed during formation is evacuated. The opening between the two chambers is then sealed

an ultrasonic sealing apparatus. The assembly then is checked for internal shorts, the moisture is removed by heating in a vacuum dryer, the electrolyte (polymer) is injected, and the cell is vacuum-sealed in a two-chamber configuration. After formation, any gas formed is removed by a vacuum, the cell resealed, and the gas chamber discarded. Finally, the cell is subjected to top sizing, both side bending, leakage tested with an odor sensor, cleaning and drying, inspection of any coil winding misalignment using an X-ray apparatus, lot number printing, storage, charging and discharging, and then final inspection.

8.8 Formation and Aging

Once the cell assembly process is complete, the final step in the overall production process shifts to the formation and aging of the cells. This applies to cylindrical, prismatic, flat plate, and polymer cell constructions. Li-Ion cells are assembled in the discharged condition and must be activated by charging. The first charge is called "formation," which activates the active materials in the cells and establishes their ability to function. The first charge typically starts at a lower current to properly form the protective solid electrolyte interface (SEI) layer on the graphite/carbon anode and then increases to the normal current at about 30% into the charge period. The cell voltage is measured after the first charge and stored (aged) for a set time period. The storage time and temperature vary from one manufacturer to another. The voltage and capacity measurements are stored for use in the cell-selection/matching process. These will be used later to sort cells out with internal microshorts or other manufacturing defects. Differences in voltage at the start and end of the storage period can be used to identify problem cells with low voltage and low capacity. Cells with low voltage result from "soft" or "micro" internal shorts and are discarded. The exact first charge regimen and cell selection procedure may vary with manufacturer. Some manufacturers give the cell one or two cycles after formation to check the capacity for cell matching in assembly of battery packs. Others use the voltage measurement after first charge.

8.9 Safety

Safety is a key issue for Li-Ion batteries. The U.S. Department of Transportation (DOT) and the United Nations classify Li-Ion and Li-Ion polymer batteries as hazardous materials for shipping.[8,9] The DOT grants exemptions for shipping small Li-Ion cells, provided that the cells/battery with limited "lithium-equivalent" content can pass certain safety-testing protocols. Other organizations that follow these regulations include the International Maritime Organization (IMO) and the International Air Transport Association (IATA). The International Civil Air Organization Association (ICOA) also follows the United Nations regulations.

The IEC,[10] the Underwriters Laboratory (UL)[11,12] and the Battery Association of Japan[13] have developed a series of standardized test protocols for verifying the safety of a particular cell design. Typical tests include internal shorting, overcharge, overdischarge, shock, vibration, temperature variations, and other situations that can be commonly encountered in normal and abuse environments. Safety tests should be carried out on a regular basis on production cells as well as during the qualification of all new cell designs and equipment changes.

Li-Ion batteries have the built-in capability to self-destruct under abuse conditions. If all the energy in a 2.4-Ah cell is released internally by an internal short circuit, it can heat itself to over 700°C in a matter of minutes. This places extreme pressure on the cell design and cell manufacturing processes to produce defect-free cells that meet the design criteria. The safety issues of Li-Ion center on preventing thermal runaway inside the cell. As the cell temperature increases, the protective layer at the surface of the anode (SEI layer) becomes unstable at around 125–130°C. If the temperature continues to increase, the reaction of the cathode with the electrolyte is an energetic reaction with an autocatalytic characteristic. Nickel and cobalt cathode materials are good catalysts for this reaction that begins at around 180–250°C. The reaction with manganese cathode materials begins somewhat higher at around 300°C. The phosphate cathode-electrolyte reaction is less energetic and begins at temperatures above 500°C

Li-Ion batteries require electronic controls to prevent cell voltage from exceeding predetermined high- and low-voltage limits. Voltage excursions outside these limits can damage the cell and may create safety issues in the cell. The electronic controls also determine the state-of-charge and stop-cell operation.

References

1. *Secondary Cells and Batteries Containing Alkaline or Other Non-acid Electrolytes – Secondary Lithium Cells and Batteries for Portable Applications*, IEC 61960 Edition 1.0.
2. Y. Nishi, Prologue, this volume.
3. T. Nagaura, T. Tozawa, *Progr. Batt. Solar Cells*, **10** (1991) 218.
4. E. Cohen, E. Gutoff, *Modern Coating and Drying Technology*, Wiley-VCH, New York, 1992.
5. V. Srinivasan, J. Newman, *J. Electrochem. Soc.*, **151** (2004) A1530.
6. M. Doyle, J. Newman, A. S. Gozdz, C. N. Schmutz, J.-M. Tarascon, *J. Electrochem. Soc.*, **143** (1996) 1890.
7. J. Kim, J. J. Hong, S. Koh, Proceedings of the 42nd Power Sources Conference, Philadelphia, PA, June 2006.
8. DOT 49CFR 173.185.
9. *Recommendations on the Transport of Dangerous Goods*, Manual of Tests and Criteria, 3rd rev. ed., United Nations, New York and Geneva, 2002.
10. *Secondary Cells and Batteries Containing Alkaline or Other Non-acid Electrolytes – Safety Requirements for Portable Sealed Secondary Cells, and for Batteries Made from Them, for Use in Portable Applications*, IEC 62133 Edition 1.0.
11. Underwriters Laboratories Inc., UL1642.3, 1995.
12. Underwriters Laboratory Inc. UL 2054, Household and Commercial Batteries, 2003.
13. *Guidance for Safe Usage of Portable Lithium-Ion Rechargeable Battery Pack*, 1st ed., Battery Association of Japan, Tokyo, 2003.

Chapter 9
Polyanionic Cathode-Active Materials

Shigeto Okada and Jun-ichi Yamaki

9.1 First-Generation 4-V Cathodes

In the 1980s, the layered rock salt types $LiCoO_2$[1] and $LiNiO_2$[2] and spinel-type $LiMn_2O_4$[3] were successively proposed as 4-V class cathode-active materials by Goodenough's group. Ten years later, the rocking chair-type battery using a $LiCoO_2$ cathode and carbonaceous anode was marketed by a Japanese battery maker under the brand name "lithium-ion (Li-Ion) battery." Although the energy density was limited by use of an insertion-type carbon anode instead of a lithium metallic anode, it demonstrated an excellent advantage in regard to safety and cyclability. The Li-Ion battery has been widely applied in various portable electronic products such as cellular phones, notebook PCs, and camcorders due to the strong downsizing demands in this market. The Li-Ion battery has made the following three significant impacts in regard to technology, the electronics market, and future development:

1. It is the first commercialized device based on intercalation reaction.
2. It has the highest energy density and longest cycle life of any battery in the market.
3. It is a one of the strongest power source candidates for electric vehicles.

After the success of the Li-Ion battery, the research for cathode-active materials has been concentrating on lithium-containing first-row transition-metals oxide with 4 V-class high electromotive force, because it can serve as a lithium source to a carbon-negative electrode. Unfortunately, all 4 V-class rechargeable cathodes, $LiCoO_2$, $LiNiO_2$, and $LiMn_2O_4$, have the essential problems of cost and environmental impact, because these cathodes commonly include rare metals as redox center. As shown in Table 9.1, these problems become serious especially for $LiCoO_2$ along with the further rapid growth of the market for electric vehicles expected in the near future.

S. Okada (✉) and J.-i. Yamaki
Institute for Materials Chemistry and Engineering, Kyushu University, 6-1 Kasuga Koen, Kasuga-shi, Fukuoka 816-8580, Japan
s-okada@cm.kyushu-u.ac.jp

M. Yoshio et al. (eds.), *Lithium-Ion Batteries*,
DOI 10.1007/987-0-387-34445-4_9

Table 9.1 Cost, deposits,[4] and environmental regulation value[5] of transition metals

	Fe	Mn	Ni	Co
Market price of metal ($/kg)	0.23	0.5	13	25
Atomic contents in crust (ppm)	50,000	950	75	25
Permissible amount in air (mg/m^3)	10	5	1	0.1
Permissible amount in water (mg/l)	300	200	13.4	0.7

Furthermore, the unusual valence state of Co^{4+} or Ni^{4+} on fully charged $LiCoO_2$ or $LiNiO_2$ and Jahn-Teller-unstable Mn^{3+} ($3d^4$:$t^3_{2g\uparrow}e^1_{g\uparrow}$) high-spin state on the discharged $LiMn_2O_4$ are factors of concern regarding the thermal and chemical stability of these cathodes. Especially, solutions to the problems of economical efficiency and safety are indispensable for the use of Li-Ion batteries in electric vehicles or for load leveling.

In order to solve the problems described below, recent research trends regarding the next generation cathode are introduced in this section.

1. Raw material cost and environmental impact of large-scale cells and mass production.
2. Production cost of solid-state synthesis using high and long heating process.
3. Oxygen release and heat generation from the cathode in a fully charged state.
4. Sensitivity of safety for charge cutoff voltages.
5. Sensitivity of cathode performance for stoichiometry.
6. Low practical capacity of the cathode being half that of a carbonaceous anode.

9.2 Second-Generation Cathodes

The followings transition metals are attractive as the redox center of the next generation cathodes:

1. Most abundant is iron, with stable trivalent state.
2. Second most abundant is titanium, with stable tetravalent state.
3. Vanadium, with wide valence change (V^{2+}–V^{5+}).
4. Molybdenum, with wide valence change (Mo^{4+}–Mo^{6+}).

Actually, these redox couples have been used as central metals of the polyanionic cathodes in recent years, as shown in Table 9.2. It can be expected that the three-dimensional framework of the oxygen-closed pack framework is sufficiently stable to tolerate repeated lithium intercalation/deintercalation reaction, because it does not included weak van der Waals bonding in the matrix.

On the other hand, it is difficult to secure a sufficient diffusion path for Li-Ion conduction in the matrix. As a solution, it is possible to replace oxygen by a larger polyanion in order to expand the bottleneck for lithium diffusion. Of course, the introduction of $(XO_n)^{m-}$ oxyanion larger than O^{2-} reduces theoretical capacity. Thus, this idea has attracted little interest from cathode researchers.

Table 9.2 Typical polyanionic cathode-active materials

Structure		Polyanionic cathodes
NASICON	$M^{3+}_2(X^{6+}O_4)_3$	Monoclinic $Fe_2(SO_4)_3$,[6] rhombohedral $Fe_2(SO_4)_3$,[7] $Fe_2(MoO_4)_3$[8,9]
	$LiM^{3+}_2(X^{6+}O_4)_2(X^{5+}O_4)$	$LiFe_2(SO_4)_2(PO_4)$[10]
	$Li_3M^{3+}_2(X^{5+}O_4)_3$	Monoclinic $Li_3Fe_2(PO_4)_3$,[11,12] rhombohedral $Li_3Fe_2(PO_4)_3$,[13] monoclinic $Li_3V_2(PO_4)_3$,[11,12] rhombohedral $Li_3V_2(PO_4)_3$,[14] $Li_3Fe_2(AsO_4)_3$[13]
	$LiM^{4+}_2(X^{5+}O_4)_3$	$LiTi_2(PO_4)_3$[15]
	$Li_2M^{4+}M^{3+}(X^{5+}O_4)_3$	$Li_2TiFe(PO_4)_3$,[16] $Li_2TiCr(PO_4)_3$[17]
	$LiM^{5+}M^{3+}(X^{5+}O_4)_3$	$LiNbFe(PO_4)_3$[16]
	$M^{5+}M^{4+}(X^{5+}O_4)_3$	$NbTi(PO_4)_3$[16]
Pyrophosphate		$Fe_4(P_2O_7)_3$,[18] $LiFeP_2O_7$,[18] TiP_2O_7,[19] $LiVP_2O_7$,[19,20] MoP_2O_7,[21] $Mo_2P_2O_{11}$[22]
Olivine		$LiFePO_4$,[18,23] Li_2FeSiO_4[24]
Amorphous $FePO_4$		$FePO_4 \cdot nH_2O$[25,26] $FePO_4$[27]
$MOXO_4$	$M^{5+}OX^{5+}O_4$	α-$MoOPO_4$,[28] β-$VOPO_4$,[28,29] γ-$VOPO_4$, 30δ-$VOPO_4$,[30] ε-$VOPO_4$,[31] β-$VOAsO_4$[30]
	$LiM^{4+}OX^{5+}O_4$	α-$LiVOPO_4$[31]
	$M^{4+}OX^{6+}O_4$	β-$VOSO_4$30,[32]
	$Li_2M^{4+}OX^{4+}O_4$	Li_2VOSiO_4[33]
Brannerite		$LiVMoO_6$[34]
Borate		Fe_3BO_6,[35] $FeBO_3$,[35,36] VBO_3,[36] $TiBO_3$[36]

However, adjoining XO_n oxoanion units tend to connect through corner-sharing linkages to release the electrostatic repulsive force between them, because the hetero-atom X of the XO_n oxoanion has a high oxidation state. The crystal structure having this corner-sharing matrix provides large and generally unoccupied sites for intercalants, and these sites are three-dimensionally connected to each other through large bottlenecks. Therefore, we can expect that such compounds have an ability to act as intercalation hosts not only for Li^+ but also for bigger and cheaper Na^+ or multivalent Mg^{2+} and Ca^{2+}.

9.2.1 Nasicon

One of the typical examples is NASCION compound (Fig. 9.1), an acronym for *Na* *s*uper *i*on *c*onductor.[37] Before attracting interest as a cathode-active material, NASICON has a long research history of over 20 years as a solid electrolyte. Since Delmas reported[38,39] that while $LiTi_2(PO_4)_3$ works as a lithium intercalation host and $NaTi_2(PO_4)_3$ works as a Na intercalation host, many attractive cathodes have been proposed based on the NASICON material group.

The features of the NASICON cathodes can be summarized as follows:

1. A shared three-dimensional corner framework(-O-X-O-M-O-X-O-)
   - high diffusivity through the large bottleneck $[Na_{1+x}Zr_2(PO_4)_{3-x}(SiO_4)_x]$[37]

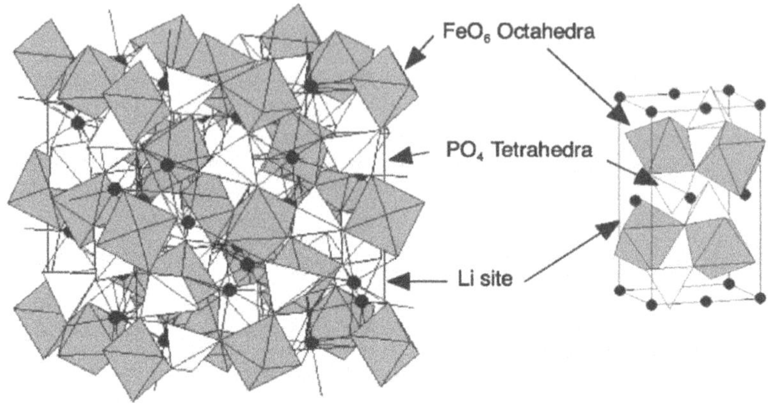

Fig. 9.1 Crystal structure of NASICON and olivine-type cathode active materials

2. Large multiple guest accommodation sites
   - promising candidate as a cathode for post-Li-Ion battery ($Na_xTi_2(PO_4)_3$,[39] $Mg_xTi_2(PO_4)_3$,[40] $Na_xFe_2(SO_4)_3$,[41] $Ca_xFe_2(SO_4)_3$,[42] $Na_xFe_2(MoO_4)_3$,[8,43] and $Na_xFe_2(WO_4)_3$[43]
3. Quick and easy synthesis process
   - rendering unnecessary the mixing process of raw materials by means of the pyrolysis method or reactant grinding process by means of the precipitation method [$Fe_2(SO_4)_3$][11]
4. Chemical stability of both end members
   - thermal stability at fully charged and fully discharged states [$Fe_2(SO_4)_3$][11]
5. High voltage by inductive effect
   - charge-discharge reaction without unusual multivalent state [$Fe_2(SO_4)_3$][6,7]
6. Flat voltage plateau by two-phase reaction
   - rendering unnecessary the use of a DC/DC converter [$Fe_2(SO_4)_3$][11]
7. Diversity in material design by element substitution of redox couple and countercation
   - voltage compatibility with the 3.6-V commercialized Li-Ion battery [$Fe_2(SO_4)_3$][11]

9.2.2 Olivine

As shown in Table 9.2, the various polyanionic cathodes are reported in addition to NASICON. Among them, phosphates provide the largest cathode-active material group, because condensate salts such as pyrophosphate $(P_2O_7)^{4-}$, tripolyphosphate

$(P_3O_{10})^{5-}$, or polymetaphosphate $(P_nO_{3n})^{n-}$ can be easily produced. Iron phosphates such as olivine $LiFePO_4$, pyrophosphate $LiFeP_2O_7$, and $Fe_4(P_2O_7)_3$ that show similar discharge plateaus at 3 V are attractive as rare metal free cathodes. Especially, olivine $LiFePO_4$ has the highest theoretical capacity (170 mAh/g) in the iron-based polyanionic cathodes.

The discharged $LiFePO_4$ of the triphylite structure and the fully charged phase $FePO_4$ of the heterosite structure have the same space group, *P*nma, and it shows a 3.3 V flat discharge profile corresponding to the two-phase reaction. In the first paper concerning the olivine cathode, the capacity was restricted to 120 mAh/g in spite of the low rate of 0.05 mA/cm^2, caused by the low lithium diffusivity of the interface between the two phases and the low electrical conductivity through the long metal atom distance in olivine matrix.

However, the poor rate capability, which is the largest weakness of $LiFePO_4$, is being gradually resolved by carbon nanocoating using organic precursors,[44–46] substitutional doping to iron,[47] and minimizing the particle size.[48] According to the patent of Hydro-Québec,[49] the addition of carbon precursors such as polypropylene (PP) and cheaper saccharides on calcinating $LiFePO_4$ raw material at 700°C in argon apparently improves the electric contact through carbon coating on $LiFePO_4$ particles and prevents oxidation from Fe^{2+} to Fe^{3+}. As a consequence, it shows a high capacity close to the theoretical limit of 170 mAh/g at the operating temperature of 80°C. Moreover, Chung[47] has reported that the bulk electronic conductivity can be enhanced from 10^{-9} to 10^{-1} S/cm by 1 atom% substitutional doping to iron. It allowed improvement to the conductivity of $LiFePO_4$, which is larger than that of $LiMn_2O_4$ (10^{-5} S/cm) and $LiCoO_2$ (10^{-3} S/cm). On the other hand, Yamada[48] succeeded in increasing the large capacity up to 160 mAh/g at room temperature at a rate of 0.12 mA/cm^2 by low-temperature synthesis less than 600°C in order to suppress particle growth.

In addition, various synthesis approaches such as hydrothermal synthesis,[50] microwave synthesis,[51] carbothermal reduction method,[52] and high-temperature quick-melting method[53] also have been developed to reduce the production cost, as shown in Table 9.3.

Furthermore, reports that support the predominance of $LiFePO_4$ for the first-generation 4-V class cathodes in regard to the chemical and thermal stability are still published in the literature.[55–57] Currently, $LiFePO_4$ rather than $LiMn_2O_4$ is regarded as the most likely cathode candidate for the large-sized Li-Ion batteries for electric vehicles, as shown in Table 9.4. Actually, in the BATT (batteries for the advanced transportation technology) program,[58] started by the Department of Energy of United States in 2,000, the large-scale Li-Ion batteries using $LiFePO_4$ seem to be the main point of focus.

The following features of the improved $LiFePO_4$ have drawn attention for its use as a next-generation cathode candidate:

- Perfectly rare metal-free
- Larger gravimetric [Ah/g] and volumetric [Ah/cc] capacity than those of $LiMn_2O_4$

Table 9.3 Typical composition conditions of $LiFePO_4$

Starting materials				
Fe source	Li source	P source	Heating condition	Ref.
$FeC_2O_4{\cdot}2H_2O$	$LiOH{\cdot}H_2O$	$(NH_4)_2HPO_4$	800°C, 6 h in N_2	54
$Fe_3(PO_4)_2{\cdot}8H_2O$	Li_3PO_4 with PP(3w/o)		350°C, 3 h→700°C, 7 h in Ar	49
$(CH_3COO)_2Fe$	CH_3COOLi	$NH_4H_2PO_4$	350°C, 5 h→700°C, 10 h with 15 w/o sol-gel carbon in N_2	45
$FeC_2O_4{\cdot}2H_2O$	Li_2CO_3	$(NH_4)_2HPO_4$	320°C, 12 h→800°C, 24 h with 12 w/o sugar in Ar	46
$FeC_2O_4{\cdot}2H_2O$	Li_2CO_3	$NH_4H_2PO_4$	600–850°C with 1 atm% dopant in Ar	47
$(CH_3COO)_2Fe$	Li_2CO_3	$NH_4H_2PO_4$	320°C, 10 h→550°C, 24 h in N_2	48
$FeSO_4$	LiOH	H_3PO_4	Hydrothermal synthesis at 120°C 5 h in teflon reactor	50
$(NH_4)_2Fe(SO_4)_2{\cdot}2H_2O$	LiOH	H_3PO_4	A few minutes microwave heating with 5 w/o CB in air	51
Fe_2O_3	LiH_2PO_4		Carbothermal reduction at 750°C, 8 h with carbon in Ar	52
FeO	$LiOH{\cdot}H_2O$	P_2O_5	High temp, quick melting synthesis at 1,500°C in Ar	53

Table 9.4 Comparison of the major candidates for cathodes in rechargeable large-sized lithium-Ion batteries

Cathode	Redox couple	Gravimetric capacity (Ah/kg)	Volumetric capacity (Ah/l)	Discharge voltage (V)	Material cost ratio
$Li_{1-x}CoO_2$	Co^{3+}/Co^{4+}	140	722	4.0	1
$Li_{1-x}NiO_2$	Ni^{3+}/Ni^{4+}	200	956	3.7	1/2
$Li_{1-x}Mn_2O_4$	Mn^{3+}/Mn^{4+}	125	535	4.0	1/4
$Li_{1+x}FePO_4$	Fe^{2+}/Fe^{3+}	160	547	3.3	1/10

- Excellent thermal and chemical stability at both end members
- No Jahn-Teller ion such as high-spin Mn^{3+} ($3d^4{:}t^3_{2g\uparrow}e^1_{g\uparrow}$) or low-spin Ni^{3+} ($3d^7{:}t^6_{2g\uparrow}e^1_{g\uparrow}$) during cycling
- No disproportionation reaction such as Mn^{3+}- > Mn^{2+}+ Mn^{4+}during cycling
- Easily protected from over charging by monitoring cell voltage
- Flat discharge profile which does not need a DC/DC converter
- Full-chargable up to 4 V without concern for electrolyte oxidization decomposition

9.2.3 Pyro- and Other Phosphates

Pyro- and other phosphates containing $(PO_4)^{3-}$ or $(P_2O_7)^{4-}$ oxoanion units can provide multiple large, empty sites for the gest in the matrix (Fig. 9.2). Thus, an unexpectedly large capacity may be possible, if multivalent transition metals such as vanadium or molybdenum are selected as redox couples. For example, $Mo_2P_2O_{11}$ shows a large reversible capacity of about 200 mAh/g not only for lithium, but also for sodium, as shown in Figs. 9.3 and 9.4. This means that the structure is suitable for a larger guest cation than Li^+.

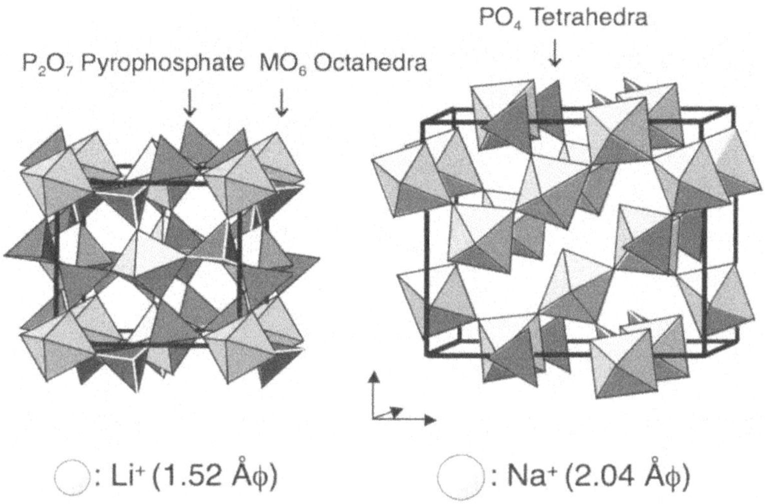

Fig. 9.2 Crystal structures of MoP_2O_7 (*left*) and $Mo_2P_2O_{11}$ (*right*)

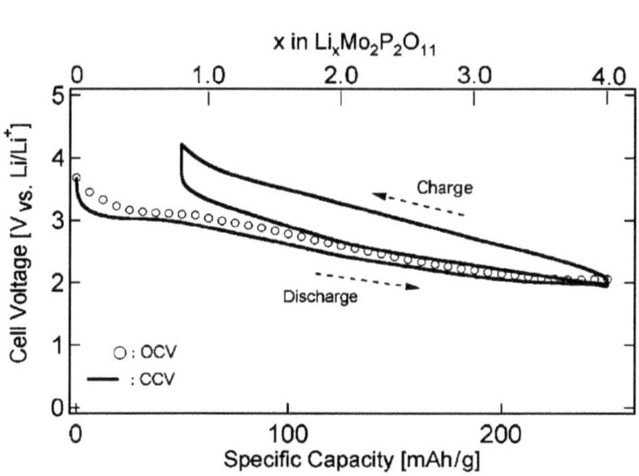

Fig. 9.3 Charge-discharge profile of $Li/Mo_2P_2O_{11}$

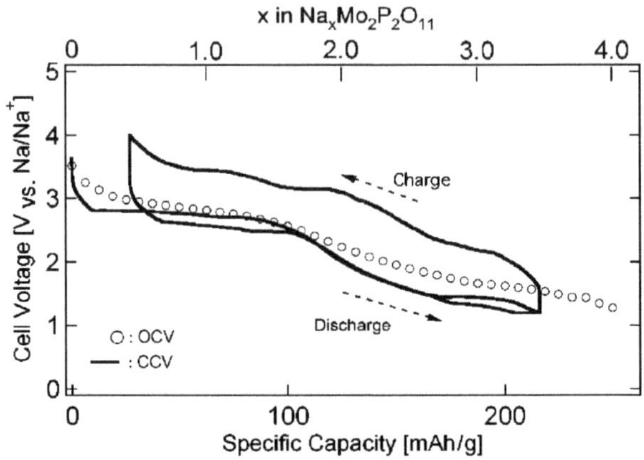

Fig. 9.4 Charge-discharge profile of $Na/Mo_2P_2O_{11}$

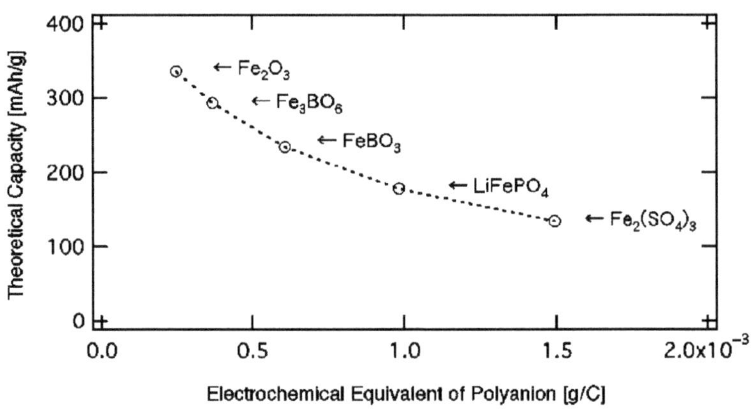

Fig. 9.5 Theoretical capacities of various iron-based cathode active materials

9.2.4 Borates

Ferric borates, having the lightest weight borate polyanions, are attractive for application in various iron-based polyanionic cathodes or anode materials (Fig. 9.5). The theoretical capacity of calcite $FeBO_3$ estimated by Fe^{3+}/Fe^{2+} redox reaction is 234 mAh/g (856 mAh/cc) for calcite. The volumetric capacity of $FeBO_3$ is equal to that of a graphite anode (855 mAh/cc). The 1.5-V mean voltage on lithium

intercalation into ferric borates is unfortunately too high for anodic use in Li-Ion batteries. However, the mean charge-discharge voltage of MBO_3 can be tuned by V^{3+} or Ti^{3+} substitution in place of Fe^{3+} on $FeBO_3$ based on the analogy of the 3.6-V NASICON cathode, $Fe_2(SO_4)_3$, and 2.6-V NASICON cathode, $V_2(SO_4)_3$ (Fig. 9.6).

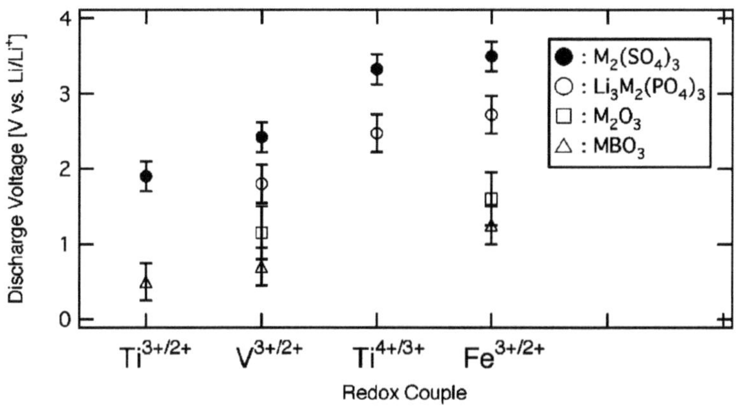

Fig. 9.6 Redox potential map of various polyanionic cathodes

Table 9.5 Post-lithium candidates

Anode	Li^+	Na^+	Mg^{2+}	Ca^{2+}
Potential (V vs. SHE)	–3.04	–2.71	–2.66	–2.87
Ionic radius (Å)	0.96	1.16	0.86	1.14
Theoretical capacity (Ah/g)	3.86	1.17	2.21	1.34
Theoretical capacity (Ah/cc)	2.05	1.13	3.84	2.07
Clarke number (%)	0.006	2.63	1.93	3.39

Discovery of the polyanionic anodes leads us to expect the ultimate realization of an "all polyanionic solid-state battery" composed of a polyanionic cathode, an electrolyte, and an anode. The interface contact between both the electrodes and solid-state electrolyte should be improved by tightening the connection through the covalent polyanionic matrix. On the other hand, studies of the "multivalent battery," which is composed of a rare metal-free polyanionic cathode and abundant multivalent cation such as Mg^{2+} or Ca^{2+} (Table 9.5), also have been started as part of the development of a post-Li-Ion battery.

References

1. K. Mizushima, P. C. Jones, P. J. Wisemann, J. B. Goodenough, *Mater. Res. Bull.*, **15**, 783 (1980).
2. M. G. S. R. Thomas, W. I. F. David, J. B. Goodenough, P. Groves, *Mater. Res. Bull.*, **20**, 1137 (1985).
3. M. M. Thackeray, P. C. Johnson, L. A. de Picciotto, J. B. Goodenough, *Mater. Res. Bull.*, **19**, 179 (1984).
4. Ullmann's Encyclopedia of Industrial Chemistry, Wiley-VCH, New York, 1989.
5. R. P. Pohanish, *Handbook of Toxic and Hazards Chemicals and Carcinogens*, Noyes, Berkshire, UK, 1985.
6. A. Manthiram, J. B. Goodenough, *J. Solid State Chem.*, **71**, 349 (1987).

7. S. Okada, H. Ohtsuka, H. Arai, M. Ichimura, *Proc. Symp. New Sealed Rechargeable Batteries Super Capacitors*, **93-23**, 431 (1993).
8. A. Nadiri, C. Delmas, R. Salmon, P. Hagenmuller, *Rev. Chim. Mine.*, **21**(4), 537 (1984).
9. S. Okada, T. Takada, M. Egashira, M. Tabuchi, H. Kageyama, T. Kodama, R. Kanno, *Proc. HBC,* **99**, 325 (1999).
10. A. K. Padhi, V. Manivannan, J. B. Goodenough, *J. Electrochem. Soc.*, **145**, 1518 (1998).
11. K. S. Nanjundaswamy, A. K. Padhi, J. B. Goodenough, S. Okada, H. Ohtsuka, H. Arai, J. Yamaki, *Solid State Ionics*, **92**, 1 (1996).
12. S. Okada, H. Arai, K. Asakura, Y. Sakurai, J. Yamaki, K. S. Nanjundaswamy, A. K. Padhi, C. Masquelier, J. B. Goodenough, *Progr. Batteries Battery Mater.*, **16**, 302 (1997).
13. C. Masquelier, A. K. Padhi, K. S. Nanjundaswamy, J. B. Goodenough, *J. Solid State Chem.*, **135**, 228 (1998).
14. J. Gaubicher, C. Wurm, G. Goward, C. Masquelier, L. Nazar, *Chem. Mater.*, **12**, 3240 (2000).
15. C. Delmas, A. Nadiri, J. L. Soubeyroux, *Solid State Ionics*, **28–30**, 419 (1988).
16. A. K. Padhi, K. S. Nanjundaswamy, C. Masquelier, J. B. Goodenough, *Proceedings of the 37th Power Sources Conference*, Palisades Institute for Research Services, New Jersey, June (1996).
17. M. Sato, S. Hasegawa, K. Yoshida, K. Toda, *Denki Kagaku*, **66**, 1236 (1998).
18. A. K. Padhi, K. S. Nanjundaswamy, C. Masquelier, S. Okada, J. B. Goodenough, *J. Electrochem. Soc.*, **144**, 1609 (1997).
19. Y. Uebou, S. Okada, M. Egashira, J. Yamaki, *Solid State Ionics*, **148**, 323 (2002).
20. G. Rousse, C. Wurm, M. Morcrette, J. R. Carvajal, J. Gaubicher, C. Masquelier, *Int. J. Inorg. Mater.*, **3**, 881 (2001).
21. Y. Uebou, S. Okada, J. Yamaki, *Electrochemistry*, **71**(5), 308–312 (2003).
22. Y. Uebou, S. Okada, J. Yamaki, *J. Power Sources*, **115**, 119–124 (2003).
23. A. K. Padhi, K. S. Nanjundaswamy, J. B. Goodenough, *J. Electrochem. Soc.*, **144**, 1188 (1997).
24. A. Nytén, A. Abouimrane, M. Armand, T. Gustafsson, J. Thomas, *Electrochem. Commun.*, **7**, 156–160 (2005).
25. Y. Song, S. Yang, P. Y. Zavalij, M. S. Whittingham, *Mater. Res. Bull.*, **37**, 1249–1257 (2002).
26. C. Masquelier, P. Reale, C. Wurm, M. Morcrette, L. Dupont, D. Larcher, *J. Electrochem. Soc.*, **149**, A1037–A1044 (2002).
27. S. Okada, T. Yamamoto, Y. Okazaki, J. Yamaki, M. Tokunaga, T. Nishida, *J. Power Sources*, **146**, 570–574 (2005).
28. N. Imanishi, K. Matsuoka, Y. Takeda, O. Yamamoto, *Denki Kagaku*, **61**, 1023 (1993).
29. J. Gaubicher, T. Mercier, Y. Chabre, J. Angenault, M. Quarton, *Abstract of IMLB-9*, Poster II, Thur 30 (1998).
30. N. Dupré, J. Gaubicher, J. Angenault, G. Wallez, M. Quarton, *J.Power Sources*, **97–98**, 532 (2001).
31. T. A. Kerr, J. Gaubicher, L. F. Nazar, *Electrochem. Solid-State Lett.*, **3**, 460 (2000).
32. J. Gaubicher, T. Mercier, Y. Chabre, J. Angenault, M. Quarton, *J. Electrochem. Soc.*, **146**, 4375 (1999).
33. M. Orichi, Y. Katayama, T. Miura, T. Koshi, *Abstr. Electrochem. Soc., Japan*, 2H06, **130** (2000).
34. J. B. Goodenough, V. Manivannan, *Denki Kagaku*, **66**, 1173 (1998).
35. J. L. C. Rowsell, J. Gaubicher, L. F. Nazar, *J. Power Sources*, **97–98**, 254 (2001).
36. S. Okada, T. Tonuma, Y. Uebou, J. Yamaki, *J. Power Sources*, **119–121**, 621–625 (2003).
37. J. B. Goodenough, H. Y.-P. Hong, J. A. Kafalas, *Mater. Res. Bull.*, **11**, 203 (1976).
38. C. Delmas, A. Nadiri, J. L. Soubeyroux, *Solid State Ionics*, **28–30**, 419 (1988).
39. C. Delmas, F. Cherkaoui, A. Nadiri, P. Hagenmuller, *Mater. Res. Bull.*, **22**, 631 (1987).
40. K. Makino, Y. Katayama, T. Miura, T. Kishi, *J. Power Sources*, **97–98**, 512 (2001).
41. S. Okada, H. Arai, J. Yamaki, *Denki Kagaku*, **65**, 802 (1997).
42. K. Akuto, H. Ohtsuka, M. Hayashi, Y. Nemoto, *NTT R&D*, **50**, 592 (2001).
43. P. G. Bruce, G. Miln, *J. Solid State Chem.*, **89**, 162 (1990).

44. N. Ravet, J. B. Goodenough, S. Besner, M. Simoneau, M. Hovington, M. Armand, *The 196th Electrochemical Soc. Meeting Abstracts*, No.127, The Electrochemical Society, New Jersey, USA, 1999.
45. H. Huang, S.-C. Yin, L. F. Nazar, *Electrochem. Solid-State Lett.*, **4**, A170 (2001).
46. Z. Chen, J. R. Dahn, *J. Electrochem. Soc*, **149**, A1184 (2002).
47. S.-Y. Chung, J. T. Bloking, Y.-M. Chiang, *Nature Mater.*, **1**, 123 (2002).
48. A. Yamada, S. C. Chung, K. Hinokuma, *J. Electrochem. Soc.*, **148**, A224 (2001).
49. N. Ravet, S. Besner, M. Simoneau, A. Vallee, M. Armand, J.-F. Magnan, Patent No.CA2307119.
50. S. Yang, P. Y. Zavalij, M. S. Whittingham, *Electrochem. Commun.*, **3**, 505 (2001).
51. K. S. Park, J. T. Son, H. T. Chung, S. J. Kim, C. H. Lee, H. G. Kim, *Electrochem. Commun.*, **5**, 839 (2003).
52. J. Barker, M. Y. Saidi, J. L. Swoyer, *Electrochem. Solid-State Lett.*, 6, A53 (2003).
53. S. Okada, Y. Okazaki, T. Yamamoto, J. Yamaki, T. Nishida, *Meeting Abstract of 2004 ECS Joint International Meeting at Hawaii*, No. 584, The Electrochemical Society, New Jersey, USA, 2004.
54. M. Th. Paques-Ledent, *Ind. Chim. Belg.*, **39**, 845 (1974).
55. D. D. MacNeil, Z. Lu, Z. Chen, J. R. Dahn, *J. Power Sources*, **108**, 8 (2002).
56. J. Jiang, J. R. Dahn, *Electrochem. Commun.*, **6**, 39 (2004).
57. N. Iltchev, Y. Chen, S. Okada, J. Yamaki, *J. Power Sources*, **119–121**, 749 (2003).
58. http://berc.lbl.gov/BATT/BATT.html

Chapter 10
Overcharge Behavior of Metal Oxide-Coated Cathode Materials

Jaephil Cho, Byungwoo Park, and Yang-kook Sun

10.1 Introduction

Commercial lithium-ion (Li-ion) cells usually operate at a maximum temperature of 60°C, and protective devices are used to control the cell-operating voltage. These protective devices consist of a positive temperature coefficient (PTC) material and protective circuits that block overcharging above 4.35 V, overdischarging below 3 V, and the overcurrent above 1 C. Despite this protection, many accidents (fire and explosion) associated with the cell itself or malfaction of the protective devices have been reported.[1] Such accidents are the result of the thermal runaway of the cell. Thermal runaway occurs when heat generation exceeds heat dissipation because the rate of heat generation increases exponentially with the increasing cell temperature, while the rate of heat transfer to a cool environment increases only linearly.[2]

Hence, controlling the degree of heat generation is critical to prevent thermal runaway, which is due mainly to the violent exothermic reaction of Li_xCoO_2 with the flammable electrolyte, resulting in oxygen evolution from the cathode.[3–7] In order to minimize the reaction, previous studies have focused on reducing the flammable nature of the electrolytes by adding phosphorus-based additives and cosolvents to the electrolytes or redox shuttle additives.[8–11] However, Cho et al. reported a more fundamental approach to improving the thermal stability of the Li_xCoO_2 cathode. They coated the cathode with $AlPO_4$ nanoparticles prepared from water.[12] The $AlPO_4$ coating improved the thermal stability of the cathode, as opposed to using either Al_2O_3 or ZrO_2 coatings derived using the sol–gel method.[13–15] Despite this, some metal oxides have been reported to avoid the unwanted surface reaction and protect the bulk.[16–26]

J. Cho (✉), B. Park and Y.-k. Sun
Department of Applied Chemistry, Hanyang University, Ansan 426–791, South Korea
jpcho@hanyang.ac.kr

M. Yoshio et al. (eds.), *Lithium-Ion Batteries*,
DOI 10.1007/987-0-387-34445-4_10

10.2 Overcharge Reaction Mechanisms in Cathode Materials

Figures 10.1 and 10.2 compare the 12-V overcharge reaction divided into four regions, and provide a diagram of the factors affecting the increase in the cell temperature during a portion of the overcharge tests. Regions III and IV should be

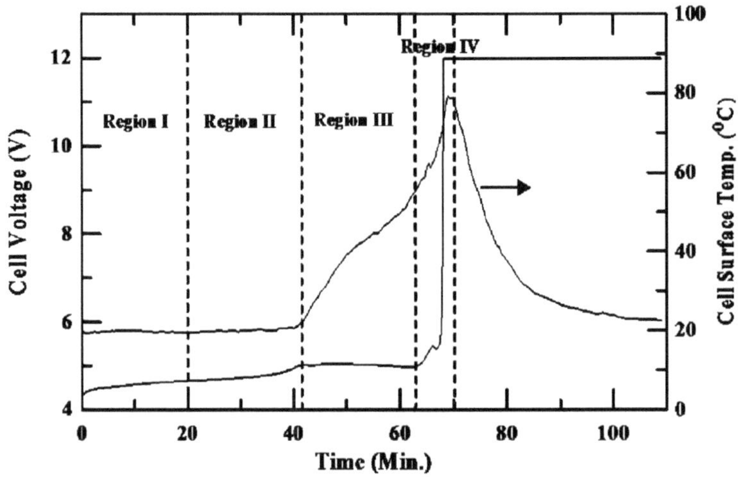

Fig. 10.1 Plot of the cell voltage and cell-surface temperature as a function of time in the course of overcharging to 12 V

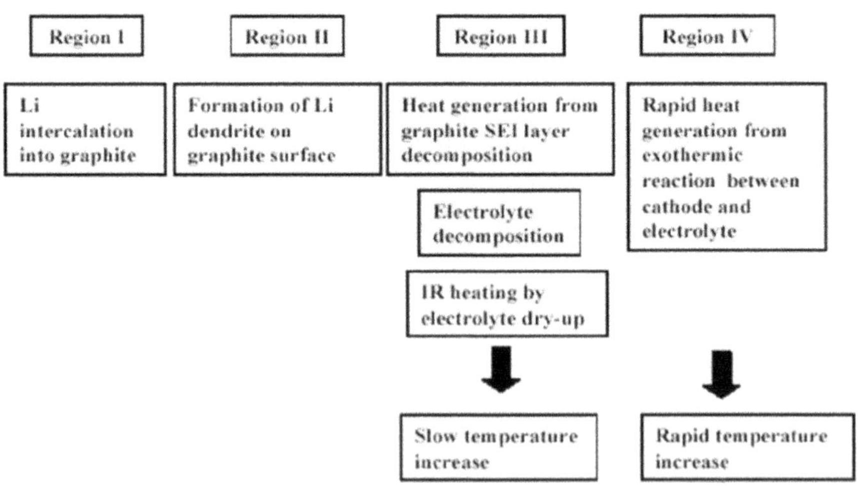

Fig. 10.2 Diagram of the reactions governing the cell temperature increase in the four regions shown in Fig. 10.1

particularly noted where the cell-surface temperatures increased faster than in the other regions. The cell-surface temperature in region III rather slowly increases mainly due to the decomposition of the electrolyte, anode, and its solid electrolyte interface (SEI) layer decomposition. However, the cell-surface temperature rises faster in region IV as a result of the exothermic decomposition of the cathode within a short time, producing oxygen. This reaction plays a dominant role in determining the cell safety, and no explosion of the cell would be expected despite the short-circuiting as long as the delithiated cathode remains stable. In terms of the cell geometry, when the heat dissipation rate is higher than the rate of heat accumulation, the cell will not explode.

10.3 Dependence of $AlPO_4$ Coating Thickness in $LiCoO_2$ Cathode

Figure 10.3 shows the voltage curves and the cycle life of the bare and coated cathodes in the coin-type half-cells between 3 and 4.6 V. Even though the initial discharge capacities of the samples before or after the coating are similar to each other, the cycle life performance was drastically improved as a result of the coating. A small peak around 4.1 V was observed only in the $AlPO_4$ 1 wt% and $AlPO_4$ 3.2 wt% coated cathodes and was reported to be due to a phase transition between the hexagonal and monoclinic phase in $LiCoO_2$. At this time, the origin of such an enhanced phase transition of the P30 and P100-coated cathodes to bare cathode is not known.

Figure 10.4 shows the differential scanning calorimetry (DSC) scans of the Li_xCoO_2 electrodes at 4.7 V at a heating rate of 3°C/min. Since x in Li_xCoO_2 is approximately 0.2 at this voltage, and the oxidation state of Co approaches 4+, the Li_xCoO_2 turns into a very strong oxidant. Hence, any reaction with the electrolyte leads to the violent evolution of oxygen from the cathode. The onset temperature for this oxygen release from bare Li_xCoO_2 is ~190°C, which accompanies a significant amount of oxygen evolution (substantial amounts of heat generation). The baseline of the bare $Li_{0.2}CoO_2$ was higher than that of the coated cathodes, which may be indicative of the beginning of an exothermic reaction around 100°C. The quantity oxygen evolved can be calculated by integrating the dimensions of the exothermic peaks between 100 and 300°C. The calculated total heat evolved from the 0.3, 1, and 3.2 wt%-coated $LiCoO_2$ was estimated to be ~35, ~4, and ~0.1 W/g, respectively, while the bare one was ~320 W/g from 100 to 300°C.

However, the onset temperatures of the $AlPO_4$-coated samples increased to 217°C, which was accompanied by a drastic decrease of oxygen evolution. Even though the initiation temperature of oxygen release remains almost constant, irrespective of the coating thickness, heat generation from the reaction depends strongly on the coating thickness. That is, the heat generation rapidly decreased with increasing coating thickness, and it appeared to disappear in the 3.2 wt%-coated sample. This suggests that the $AlPO_4$ protecting layer is effective in

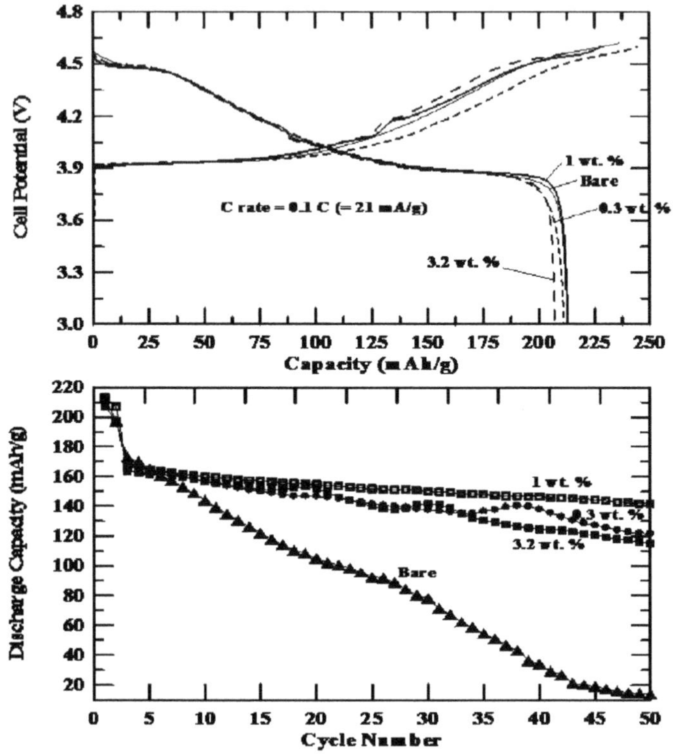

Fig. 10.3 Plots of (**a**) the initial charge and discharge curves of the cells containing the bare and $AlPO_4$-coated $LiCoO_2$ with a different coating thickness at 0.1 C rate (=21 mA/g) between 4.6 and 3 V, and (**b**) the cycle life results of the cells with the identical cells in (**a**). The cycle test was carried out initially at 0.1 C for 1 cycle, which was increased to 0.5 and 1 C rates for 1 and 48 cycles, respectively, for a total of 50 cycles

retarding the reaction between the $LiCoO_2$ and the electrolytes, leading to a decrease in heat generation. Such superior thermal stability over the bare sample is due to the strong covalent bonding of $(PO_4)^{3-}$ with Al.[27] Oxides with $(PO_4)^{3-}$ bonding were reported to have greater thermal stability than λ-MnO_2 even at the fully delithiated state.[27,28]

Figure 10.5 shows the cell voltage and cell-surface temperature profiles of the bare and $AlPO_4$-coated cathodes up to 12 V in the Li-ion cells. As the cell voltage was increased to 12 V, the cell temperature also was increased as a result of joule heat (i^2R). In addition, electrolyte decomposition around 5 V (see the plateau in the bare $LiCoO_2$) accompanying simultaneous oxygen decomposition from the Li_xCoO_2 accelerates heat generation, thereby accelerating the temperature rise. At around 6 V, because of the rapid increase in the cell internal resistance, a steep voltage upsurge to 12 V was observed. Even though the cell components, such as the anode and binders, also aid to the increase in cell temperature, the most contributing factor was the exothermic reaction of the cathode with the electrolytes.[4–6]

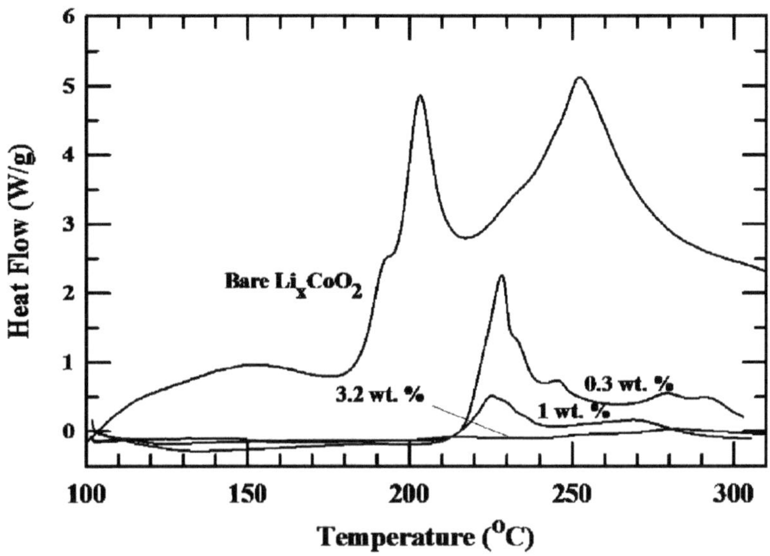

Fig. 10.4 DSC scans of the charged cathodes containing the bare and $AlPO_4$-coated $LiCoO_2$ with a different coating thickness at 4.7 V

When the internal temperature exceeds the melting temperature of the separator (<140°C), the separator shuts down and the temperature begins to decrease as long as an internal short circuit does not occur. Note that the temperature difference between the internal and the surface of the cell case was approximately 100°C.[4] All these phenomena can be observed in Fig. 10.4. However, it should be noted that the electrolyte decomposition plateau at 5 V continues to reduce with increasing coating thickness. Accordingly, such a plateau appears to disappear in the P100-coated sample. Moreover, the cell-surface temperature decreases as the coating thickness increases, and the P100-coated-$LiCoO_2$ showed the lowest surface temperature of 60°C (cell-surface temperature of the bare material shows 100°C on reaching 12 V). The results suggest that a 5 V plateau plays a key role in controlling the heat generation of the cell. This result is well correlated with the DSC result.

10.4 Comparison of Al_2O_3- and $AlPO_4$-coated $LiCoO_2$

A comparison of the transmission electron microscopy (TEM) images between the Al_2O_3- and $AlPO_4$-coated $LiCoO_2$ particles is shown in Fig. 10.6. In both cases, the aluminum or phosphorus elements are distributed over the $LiCoO_2$ surfaces. The possible formation of a solid solution from a reaction between the coating materials and lithium (or even cobalt) during the heat treatment is not ruled out.

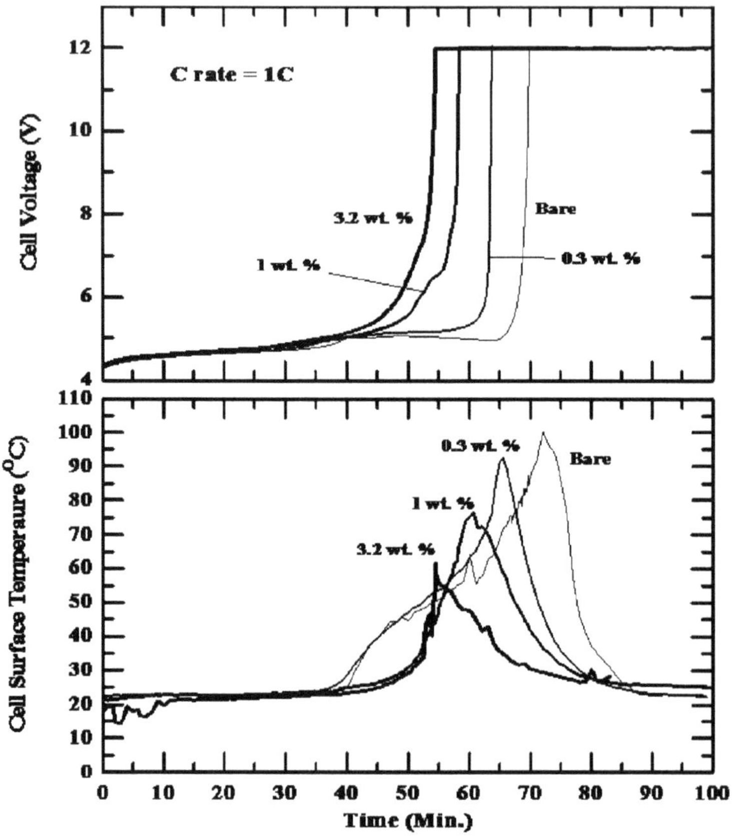

Fig. 10.5 Plots of (**a**) the 12-V overcharge curves of the bare and $AlPO_4$-coated $LiCoO_2$ with a different coating thickness in the Li-ion cells. Before overcharging, all the fresh cells were initially charged to 4.2 V at a 1 C rate (=900 mA), and (**b**) the corresponding cell-surface temperature profiles to (**a**) during overcharging

X-ray photoelectron spectroscopy (XPS) was used to compare the bonding nature of the Al_2O_3- and $AlPO_4$-coated cathodes, as shown in Fig. 10.7. The binding energies of the Al 2*p* in the bulk Al_2O_3 and $AlPO_4$ were reported to be observed at ~74.7 eV and ~74.5 eV, respectively.[29, 30] A peak in the Al_2O_3-coated $LiCoO_2$ at ~71 eV agrees with the metallic nature of aluminum. The variation in the binding energies of aluminum in the coated cathodes may be related to a lithium (or even cobalt) reaction with the coating layer, and further study aimed at understanding the detailed microstructures of the nanoscale coating layer currently is underway.

The initial capacity and cycle-life performance of the Al_2O_3-coated cathode are similar to those in the $AlPO_4$-coated samples at 4.6-V cycling. However, increasing the charge voltage from 4.6 to 4.8 V leads to a drastic difference between these two cathodes, as shown in Fig. 10.8. Even though the charge capacities of both cathodes

(a)

(b)

Fig. 10.6 TEM images of (**a**) Al_2O_3- and (**b**) $AlPO_4$-coated $LiCoO_2$. Energy dispersive spectroscopy confirms the existence of either aluminum or phosphorus in the surface, and cobalt in the interior of the powder

are similar to each other (244 and 247 mA h/g for Al_2O_3- and $AlPO_4$-coated $LiCoO_2$, respectively), the discharge capacity is obviously different: Al_2O_3- and $AlPO_4$-coated $LiCoO_2$ show 220 and 233 mA h/g, respectively. Cobalt dissolution into the solution is coupled with the release of lithium and oxygen, resulting in structural degradation.[31] The cobalt dissolution rate in the Al_2O_3-coated cathodes was four times higher than that in the $AlPO_4$-coated cathodes after the initial 0.1 C rate cycling (between 4.8 V and 3 V), and the concentrations were ~160 and ~40 ppm for Al_2O_3- and $AlPO_4$-coated $LiCoO_2$, respectively. This affected the cycling stability of the coated cathode.[32] The $AlPO_4$-coated $LiCoO_2$ exhibits superior capacity retention after 46 cycles (at a 1 C rate), showing ~82% capacity retention, while that of the Al_2O_3-coated cathode shows ~68%. This is due to the structural degradation of the $LiCoO_2$ from cobalt dissolution, which was ~450 and ~160 ppm for the

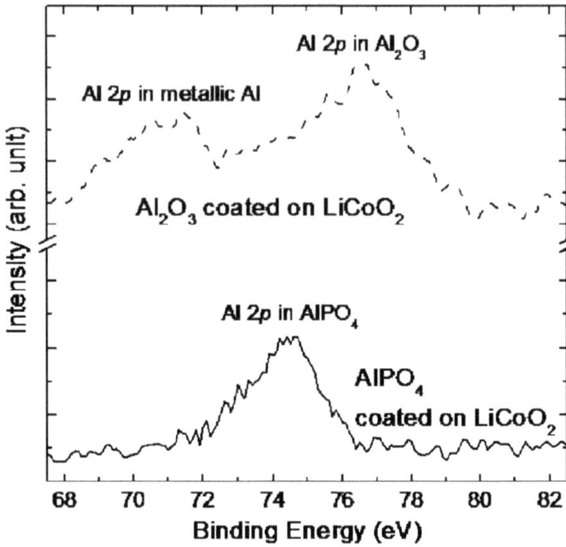

Fig. 10.7 XPS spectra of the Al 2*p* binding energies from the Al_2O_3 (*dashed line*) and $AlPO_4$ (*solid line*) nanoscale-coating layer on $LiCoO_2$

Al_2O_3- and $AlPO_4$-coated cathodes, respectively. These results confirm that the $AlPO_4$-coating layer is chemically more stable at the 4.8 V electrochemical window compared to the Al_2O_3 coating.

This is supported by the cycle-life performance in the Li-ion cells (with carbon as an anode) containing Al_2O_3- and $AlPO_4$-coated $LiCoO_2$ cathodes. As shown in Fig. 10.9 with a charge-cutoff voltage of 4.5 V at the 1 C rate, the capacity-fading rate of theAl_2O_3-coated $LiCoO_2$ is similar to that of the $AlPO_4$-coated $LiCoO_2$ for the initial ~40 cycles. However, its capacity rapidly decreases to ~27% after 160 cycles (between 4.5 and 2.75 V), while the capacity-fading rate from the $AlPO_4$-coated cathode is much slower, showing ~75% after 160 cycles. The amount of cobalt dissolution after 40 cycles in the cell containing the Al_2O_3-coated cathode is similar to that in the $AlPO_4$-coated cathode, showing ~70 ppm. However, the amount of cobalt dissolution from the Al_2O_3-coated cathode increased to ~3,200 ppm after 160 cycles. (The cobalt dissolution in $AlPO_4$-coated cathode was ~300 ppm.) The result suggests that the Al_2O_3-coating layer is relatively stable during the initial cycles, but chemically less stable either for higher voltage or for a long-time exposure to the electrolyte. Actually, the dissolved amount of aluminum into the electrolyte in the Al_2O_3-coated cathode was ~980 ppm after 160 cycles compared to the ~100 ppm dissolution of Al in the $AlPO_4$-coated one.

Figure 10.10a shows the rate capability of bare and Al_2O_3- and $AlPO_4$-coated $LiCoO_2$ in Li-ion cells at the rates of 0.2, 0.5, 1, and 2 C between 4.2 and 3 V. Capacity-retention rate of the $AlPO_4$-coated $LiCoO_2$ is enhanced, compared to the

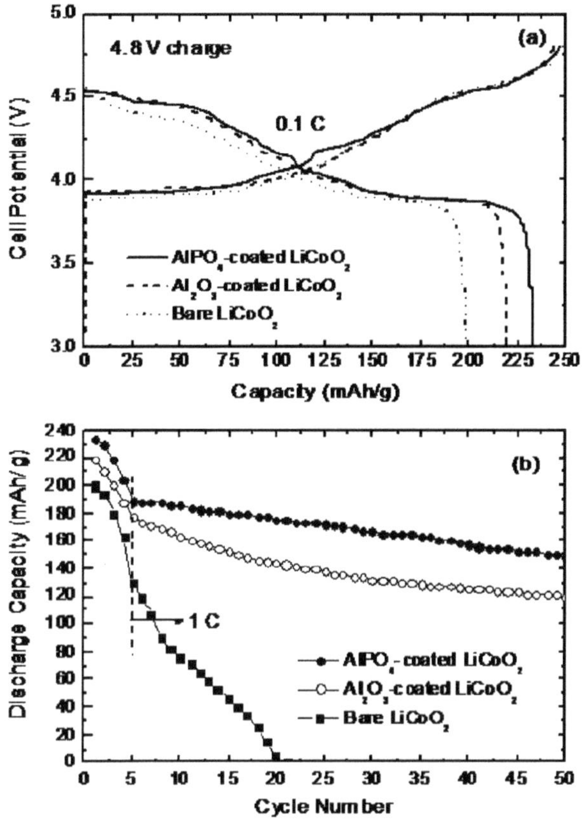

Fig. 10.8 (**a**) Voltage profiles and (**b**) capacity retention of bare and Al_2O_3- and $AlPO_4$-coated $LiCoO_2$ cathodes in coin-type half cells (Li as an anode) with a charge voltage of 4.8 V (precycled at the rates of 0.1 C for the initial 2 cycles, and 0.2 and 0.5 C each for the next two cycles)

Al_2O_3-coated $LiCoO_2$ with an increasing current rate the former showing 8% larger capacity retention than the latter at a 2 C rate. Figure 10.10b shows the voltage profiles of bare and Al_2O_3- and $AlPO_4$-coated $LiCoO_2$ at 0.2, 0.5, 1, and 2 C rates between 4.2 and 3 V as a function of the normalized capacity. Since other factors such as the carbon anode and electrolytes are fixed in the Li-Ion cells, the change in the voltage profile should be affected by the coating layer. The enhanced voltage profiles of the $AlPO_4$-coated cathode compared to Al_2O_3-coated one at different C rates are believed to be due to the higher lithium diffusivity in the $LiCoO_2$ powder by the $AlPO_4$ coating.

The DSC scans of the bare, Al_2O_3-, and $AlPO_4$-coated $LiCoO_2$ electrodes are shown in Fig. 10.11. The exothermic peak area indicates the amount of heat generation (related to oxygen generation) from the decomposed cathode after a reaction

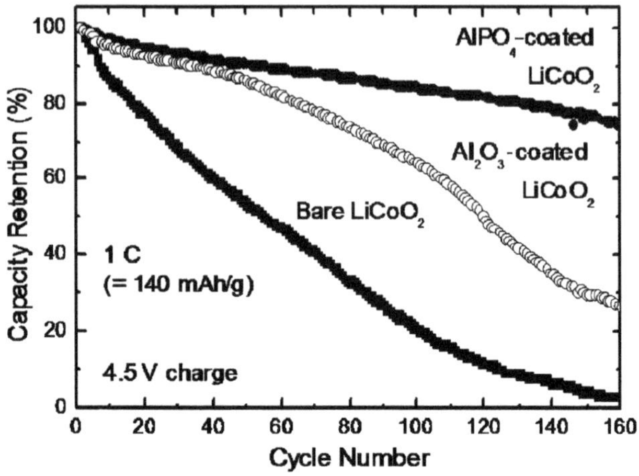

Fig. 10.9 Capacity retention of bare and Al_2O_3- and $AlPO_4$-coated $LiCoO_2$ cathodes in the Li-ion cells (with carbon anode) with a charge-cutoff voltage of 4.5 V

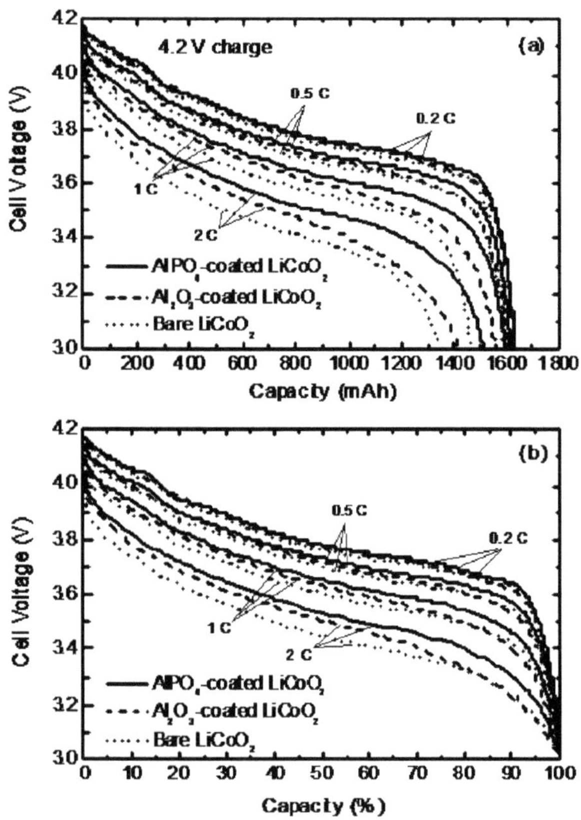

Fig. 10.10 Plots of (**a**) rate capabilities and (**b**) normalized capacity of bare and Al_2O_3- and $AlPO_4$-coated $LiCoO_2$ cathodes in Li-ion cells (with carbon anode) with a charge-cutoff voltage of 4.2 V

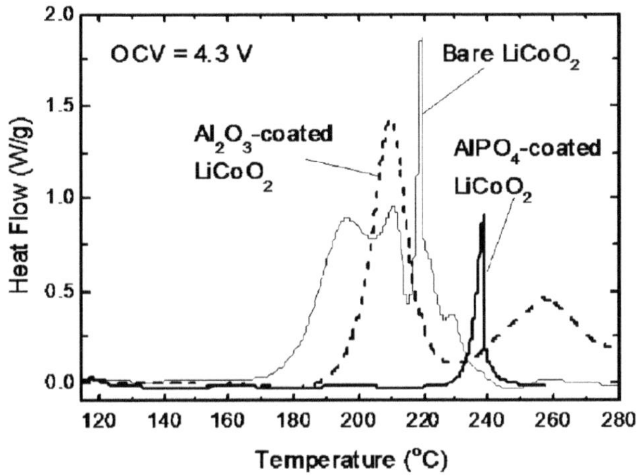

Fig. 10.11 DSC scans of bare and Al_2O_3- and $AlPO_4$-coated $LiCoO_2$ after 4.3 V charge. The scan rate was 3°C/min

with the electrolyte. The onset temperature of oxygen evolution from the cathode in the bare electrode is ~170°C, but improved to ~190°C in the Al_2O_3-coated cathode. However, the $AlPO_4$-coated electrode shows the highest onset temperature among the other electrodes (~230°C), indicating that the $AlPO_4$-coating layer effectively retards the initiation of the oxygen generation from the cathode, with a much lower amount of exothermic-heat release.

In order to investigate the 12-V overcharge behavior of the Al_2O_3- and $AlPO_4$-coated $LiCoO_2$, the Li-ion cells containing these cathodes are overcharged to 12 V and held at that voltage until the applied current decreases to ~30 mA. Previously, it was suggested that the short circuit at 12 V resulted from the direct contact between the anode and cathode due to the separator shrinkage, accompanying a rapid temperature upsurge, which induces a thermal runaway according to the stability of the cathode.[4] Figure 10.12 compares the 12-V overcharge behaviors of the Li-ion cells containing the bare and Al_2O_3- and $AlPO_4$-coated $LiCoO_2$ cathodes. When a separator melts at ~120°C, it triggers a large heat output induced by an internal short circuit.

After spiking to 12 V, the cell containing the Al_2O_3-coated cathode shows thermal runaway with the surface temperature of the cell over ~500°C. In this condition, the cell was burnt off completely. This behavior is quite similar to that observed in the bare cell.[4] Even though the cell with an $AlPO_4$ nanoparticle-coated $LiCoO_2$ cathode has a short circuit, the temperature increases to only ~60°C without burning the cell. All these are related to the cobalt dissolution, the cycle-life performance, and the DSC results. Another interesting feature observed during the 12-V overcharging is that the cell voltage recovers to 12 V, followed by another internal short circuit after showing the first internal short circuit, as shown in

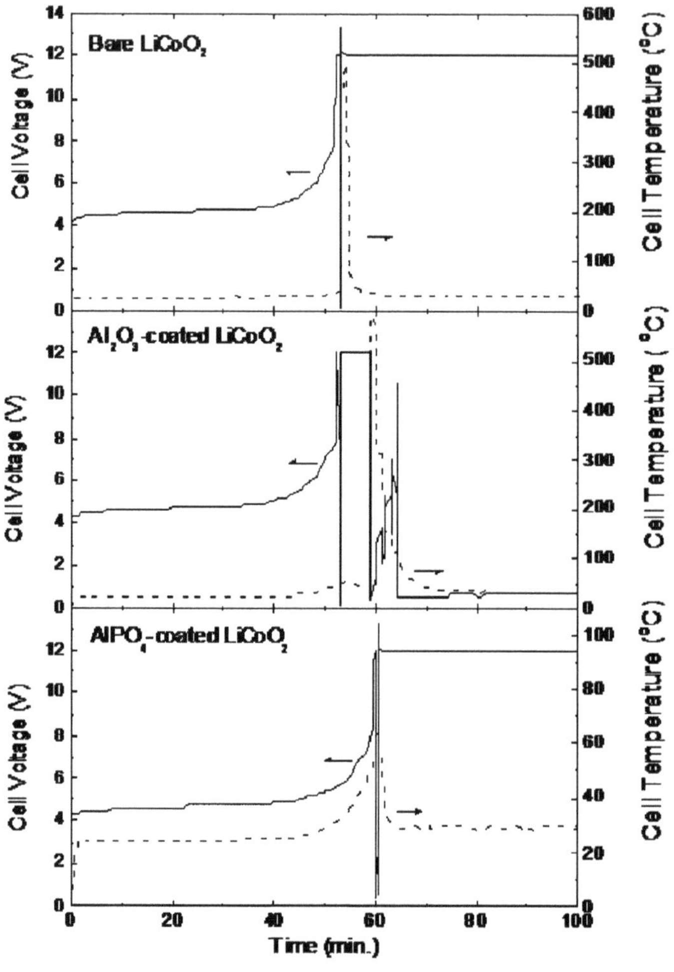

Fig. 10.12 Voltage (*solid lines*) and temperature (*dotted lines*) profiles of the cells with bare and Al_2O_3- and $AlPO_4$-coated $LiCoO_2$ cathodes as a function of time, showing the breakdown of the cells by one full short circuit. All the cells were first charged to 4.2 V, overcharged to 12 V at a rate of 1 C, and then held at that voltage for ~50 min

Fig. 10.13. In this case, the cell containing the Al_2O_3-coated $LiCoO_2$ cathode exhibits the maximum surface temperature of ~500°C. However, $AlPO_4$-coated $LiCoO_2$ shows a maximum temperature of only ~60°C, indicating that the $AlPO_4$-coating layer is still stable after the successive short circuits in contrast to the Al_2O_3-coating layer.

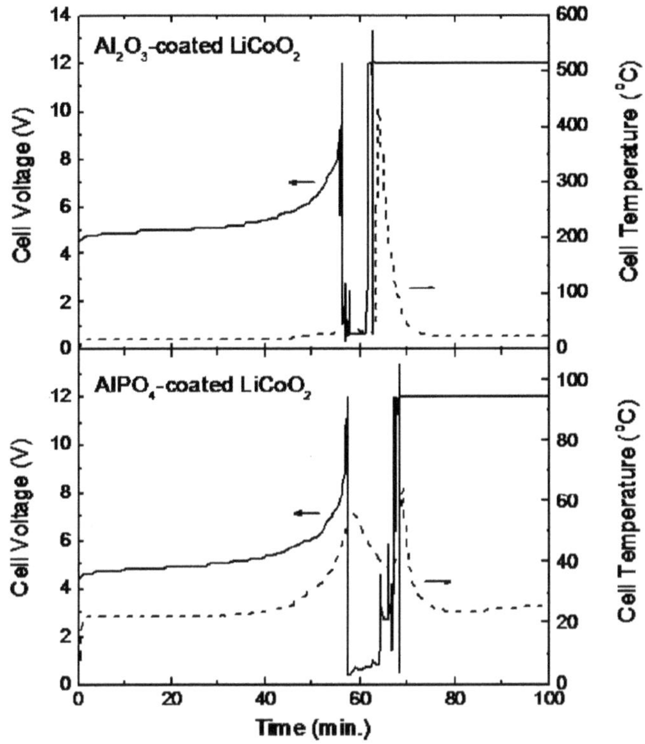

Fig. 10.13 Voltage (*solid lines*) and temperature (*dotted lines*) profiles of the cells with the Al_2O_3- and $AlPO_4$-coated $LiCoO_2$ cathode as a function of time, showing breakdown of the cells by another internal short circuit after the first internal short circuit. The sequential local short-circuit phenomenon was not observed in bare cathode materials. The charging conditions were same as in Fig. 20.9

10.5 Comparison of $AlPO_4$-coated $LiCoO_2$ and $LiNi_{0.8}Co_{0.1}Mn_{0.1}O_2$ Cathodes

Along with the improvement in the electrochemical properties of nickel-based cathode materials ($LiNi_{1-x}M_xO_2$ where M = metals), thermal instability with respect to Li_xCoO_2 is a great concern for use in Li-ion cells.[33–44] Nickel-based cathode materials have not met the safety guidelines that require no explosions, fire, and smoke during the nail penetration test at a 4.35-V overcharged state in Li-ion cells.[45] In terms of the thermal stability, the violent exothermic reaction of the cathode with the electrolyte with accompanying substantial heat generation should be avoided. Otherwise, it causes thermal runaway exhibiting fire, sparks, and explosions.

In order to minimize oxygen evolution (heat generation), the addition of magnesium, titanium, or aluminum in $LiNiO_2$ was reported to be effective in reducing the exothermic reaction, but it greatly sacrificed the discharge capacity.[46],[47] For example, a $LiNi_{0.7}Ti_{0.15}Mg_{0.15}O_2$ cathode showed 190 mA h/g even at the charged voltage of 5 V, and its specific capacity at a 4.3 V cutoff is smaller than $LiCoO_2$.[46] To overcome the thermal instability of nickel-based cathode materials, some additives (for example, γ-butyrolactone) were used to reduce the direct reaction of the cathode with the electrolyte at the charged states. This solvent was reported to decompose into the organic products, thereby encapsulating the cathode, blocking any direct reaction with the electrolyte.[45] As a consequence, Li-ion cells containing this solvent did not explode during the nail penetration (at 4.35 V) and overcharge tests to 12 V. However, such an additive has raised concerns as its compatibility with the anode and cathode, which therefore still needs to tailor its electrochemical properties.

$LiCoO_2$ (100 g) with an average particle size of ~10 μm was added to the coating solution and thoroughly mixed for 5 min. The slurry then was dried in an oven at 120°C for 6 h and heat-treated in a furnace at 700°C for 5 h. TEM confirmed that the aluminum and phosphorus elements were confined within ~10 nm from the surface.[48] The $Ni_{0.8}Co_{0.1}Mn_{0.1}(OH)_2$ starting material consisting of spherical particles (~13 μm in diameter) was prepared by co-precipitation from a solution containing stoichiometric amounts of nickel and cobalt nitrates and manganese nitrate by the addition of NaOH and NH_4OH solutions in the reactor.[34] The $LiNi_{0.8}Co_{0.1}Mn_{0.1}O_2$ was prepared by mixing stoichiometric amounts of LiOH and $Ni_{0.8}Co_{0.1}Mn_{0.1}(OH)_2$, followed by heat treatment at 800°C for 20 h in a stream of dried air:

$$0.8\text{Ni}^2+(\text{aq})+0.1\text{Co}^2+(\text{aq})+0.1\text{Mn}^{2+}(\text{aq})+x\text{NH}_4\text{OH(aq)}$$
$$\rightarrow[\text{Ni}_{0.8}\text{Co}_{0.1}\text{Mn}_{0.1}(\text{NH}_3)_{0.1}]^2+(\text{aq})+0.1\text{H}_2\text{O}+(x-0.1)\text{NH}_4\text{OH(aq)} \quad (10.1)$$

and

$$[\text{Ni}_{0.8}\text{Co}_{0.1}\text{Mn}_{0.1}(\text{NH}_3)_{0.1}]^2+(\text{aq})+y\text{OH}^-+z\text{H}_2\text{O}\rightarrow\text{Ni}_{0.8}\text{Co}_{0.1}\text{Mn}_{0.1}(\text{OH})_2(\text{s})$$
$$\downarrow+z\text{NH}_4\text{OH}+(0.1-z)\text{NH}_3+(y-0.1)\text{OH}^- \quad (10.2)$$

An excess amount of lithium (1.02 mol) was used to compensate for the loss of lithium during firing. The resultant powders showed a Brunauer–Emmett–Teller (BET) surface area of 1 m^2/g. The process for $AlPO_4$ coating on the cathode is identical to the $LiCoO_2$ described above.

The $Ni_{0.8}Co_{0.1}Mn_{0.1}(OH)_2$ particle size was controlled by the reaction time in order to enhance the BET surface area. A longer reaction time in the solution resulted in a larger particle size. The co-precipitated particles continued to be rotated with 1,000 rpm in the reactor at 50°C. The pH was maintained at 11.5 by controlling the amount of NaOH. Co-precipitation methods are industrially well known. Ohzuku et al.[49] used commercial hydroxide precursors. On the other hand, Dahn's group[50] used the method by controlling with LiOH, and their recipe could not yield high density. However, we obtained spherical powders with high density through the co-precipitation method by controlling with NH_4OH and NaOH. Scanning electron microscopy images of the bare, $AlPO_4$-coated $LiCoO_2$ and $LiNi_{0.8}Co_{0.1}Mn_{0.1}O_2$ cathode particles are shown in Fig. 10.1. The surface morphologies of both coated particles are distinctly different to those of the bare ones, and that of the coated $LiNi_{0.8}Co_{0.1}Mn_{0.1}O_2$ changed from a rough to smooth shape after the coating (Fig. 10.14). This suggests that the $AlPO_4$ nanoparticles at least reacted at the surface of powders.

The cathodes for the battery test cells were made from the cathode material, super phosphorus carbon black, and polyvinylidene fluoride (PVDF) binder (Kureha Company) in a weight ratio of 96:2:2. The electrodes were prepared by

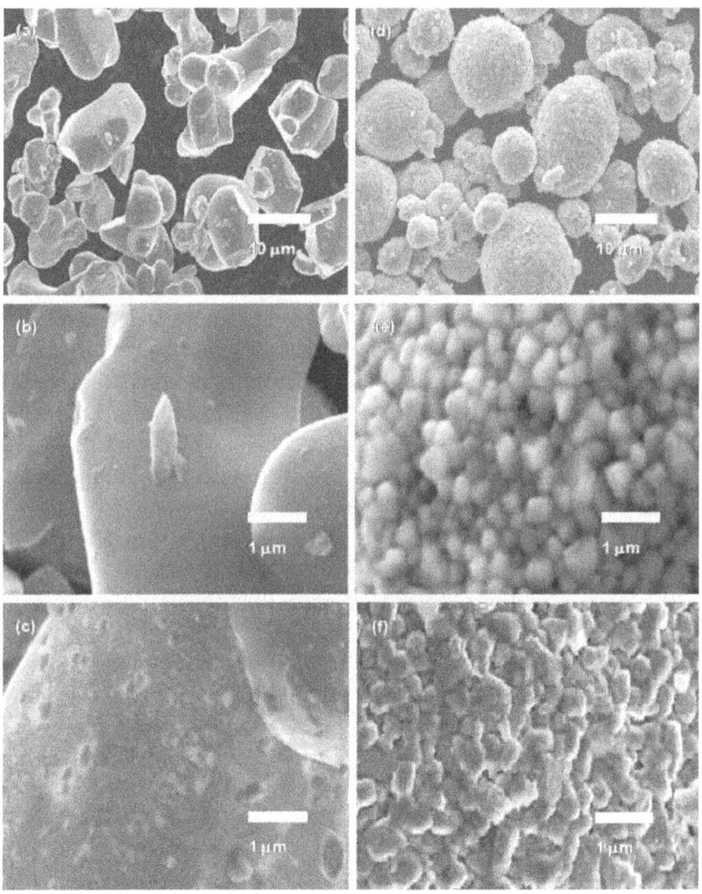

Fig. 10.14 SEM micrographs of (**a**) and (**b**) for bare $LiCoO_2$, and (**c**) for $AlPO_4$-coated $LiCoO_2$. Micrographs of (**d**) and (**e**) are from bare $LiNi_{0.8}Co_{0.1}Mn_{0.1}O_2$, while (**f**) is from $AlPO_4$-coated $LiNi_{0.8}Co_{0.1}Mn_{0.1}O_2$

coating a cathode slurry onto an aluminum foil followed by drying it at 130°C for 20 min. The slurry was prepared by thoroughly mixing a *N*-methyl-2-pyrrolidone (NMP) solution of PVDF, carbon black, and the powdery cathode material. The coin-type battery test cells (size 2,016) prepared in helium-filled glove box contained a cathode, a lithium metal anode, and a microporous polyethylene separator. The electrolyte was 1 M $LiPF_6$ with ethylene carbonate/diethylene carbonate/ethylmethyl carbonate (EC/DEC/EMC) (30:30:40 vol%). We used standardized coin-cell parts (2016R-type), and normally the amount of the electrolyte was ~0.1 g in each cell. Each test cathode contained approximately 25 mg of either $LiCoO_2$ or $LiNi_{0.8}Co_{0.1}Mn_{0.1}O_2$. The test cells were aged at room temperature for 24 h prior to the electrochemical test. Cells with a dimension [3.4 × 40 × 62 mm^3 (thickness × length × width)] were used for the 12-V overcharging experiments and the anode material was synthetic graphite. The dimensional ratio of anode to cathode (matching ratio) was 1.08:1 for the Li-ion cells. The overcharge test used the guidelines reported elsewhere.[51] The cell-surface temperature was monitored using a K-type thermocouple placed on the center of the largest face in the cell can, and the thermocouple was tightly glued with insulating tape.

To compare the capacity of the $AlPO_4$-coated $LiCoO_2$ with $LiNi_{0.8}Co_{0.1}Mn_{0.1}O_2$ cathode, coin-type half-cells were used with a charge cutoff voltage of 4.3 V at the charge and discharge rates with a constant current of 0.1 C (14 mA/g and 18 mA/g, respectively). The cycling tests of the Li-ion cells containing $AlPO_4$-coated $LiCoO_2$ and $LiNi_{0.8}Co_{0.1}Mn_{0.1}O_2$ were carried out by charging and discharging the cells at a 1 C rate for 200 cycles at 21°C. The rate-capability tests of the Li-ion cells containing $AlPO_4$-coated $LiCoO_2$ and $LiNi_{0.8}Co_{0.1}Mn_{0.1}O_2$ were carried with a discharge rate of 0.5, 1, and 2 C (while the charge rate was fixed at 1 C).

The Li-Ion cells for the DSC tests were charged to the predetermined cell voltages, 4.2 and 4.6 V for bare cathodes, and 4.2, 4.4, 4.6, and 4.8 V for the coated cathodes at the constant-current mode of 1 C (710 and 820 mA for Li_xCoO_2 and $Li_xNi_{0.8}Co_{0.1}Mn_{0.1}O_2$, respectively), which was followed by holding these voltages until the currents decreased to 30 mA. The charged cathodes were then disassembled from the cells in a dry room. The cathode electrode typically contained ~20 wt% electrolyte, ~25 wt% aluminum foil, a ~5 wt% combined binder and carbon black, and a ~50 wt% cathode material. Approximately 10 mg of the cathode was cut and hermetically sealed in an aluminum DSC sample pan. Only the cathode material was used to calculate the specific heat flow. The heating rate of the DSC tests was 3°C/min. Due to the high reactivity of the electrolyte with moisture, its content in the dry room was maintained below 50 ppm. Since the dry room was used for dissembling the cells for the DSC experiments, compressed DSC pans may contain oxygen. However, our DSC experiments up to 300°C showed negligible effects: pans assembled in the dry room were identical to those assembled in a glove box.

Figure 10.15 shows the X-ray diffraction (XRD) patterns of $AlPO_4$-coated $LiCoO_2$ and $LiNi_{0.8}Co_{0.1}Mn_{0.1}O_2$ powders. Both materials are indexed to the hexagonal-type space group $R\bar{3}m$. The lattice constants, *a* and *c*, for the $AlPO_4$-coated $LiCoO_2$ material are a = 2.815 ± 0.004 Å and c = 14.051 ± 0.043 Å (with

2.816 ± 0.004 Å and 14.034 ± 0.043 Å, respectively, for bare samples), and those for the coated $LiNi_{0.8}Co_{0.1}Mn_{0.1}O_2$ material are a = 2.874 ± 0.001 Å and c = 14.217 ± 0.008 Å (with 2.877 ± 0.001 Å and 14.219 ± 0.003 Å, respectively, for bare samples). The XRD patterns of the coated $LiCoO_2$ or $LiNi_{0.8}Co_{0.1}Mn_{0.1}O_2$ powders are identical with the bare samples, even though the possible formation of nanophases or interdiffusion could not be ruled out.

To examine the electrochemical properties, coin-type half-cells with the coated $LiCoO_2$ and $LiNi_{0.8}Co_{0.1}Mn_{0.1}O_2$ were charged between 4.3 and 3 V at a rate of 0.1 C (14 mA/g and, respectively). The first discharge curves in Fig. 10.16 show that

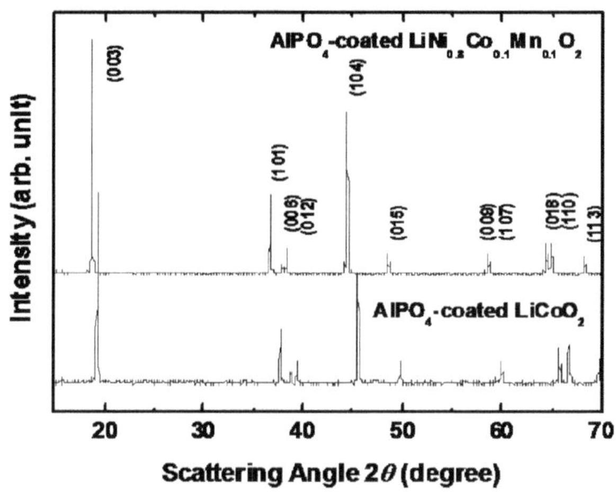

Fig. 10.15 XRD patterns of the $AlPO_4$-coated $LiCoO_2$ and $LiNi_{0.8}Co_{0.1}Mn_{0.1}O_2$ cathodes

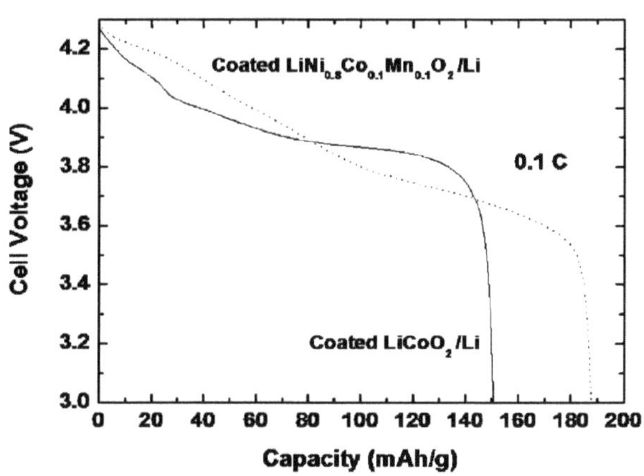

Fig. 10.16 Plots of the first discharge curves from the $AlPO_4$-coated $LiCoO_2$ and $LiNi_{0.8}Co_{0.1}Mn_{0.1}O_2$ cathodes in the coin-type half-cells between 4.3 and 3 V at 0.1 C rate (lithiumas an anode), while the 1 C rate corresponds to 140 mA/g and 180 mA/g, respectively

the initial voltage profiles of the coated $LiNi_{0.8}Co_{0.1}Mn_{0.1}O_2$ cathodes above 4 V are higher than that of the coated $LiCoO_2$. The discharge capacity of the coated $LiNi_{0.8}Co_{0.1}Mn_{0.1}O_2$ is 188 mA h/g, while that of the coated $LiCoO_2$ is 150 mA h/g. These values are identical to those of the bare cathodes. The larger discharge capacity of the $LiNi_{0.8}Co_{0.1}Mn_{0.1}O_2$ cathode than the coated $LiCoO_2$ is expected to increase the nominal capacity of the Li-ion cell.

The rate capabilities of the coated $LiCoO_2$ and $LiNi_{0.8}Co_{0.1}Mn_{0.1}O_2$ in the Li-ion full cells (between 4.2 and 3 V) are shown in Fig. 10.17a at a discharge rate of 0.5, 1, and 2 C (with the charge rate of 1 C). With relatively good capacity retention, the overall voltage profiles of the coated $LiNi_{0.8}Co_{0.1}Mn_{0.1}O_2$ appear enhanced as the C rate increased from 0.5 to 2 C, compared to the coated $LiCoO_2$. For the cycle-life performance, the coated $LiNi_{0.8}Co_{0.1}Mn_{0.1}O_2$ exhibits better capacity retention than the $LiCoO_2$ cathodes after 200 cycles, as shown in Fig. 10.17b.

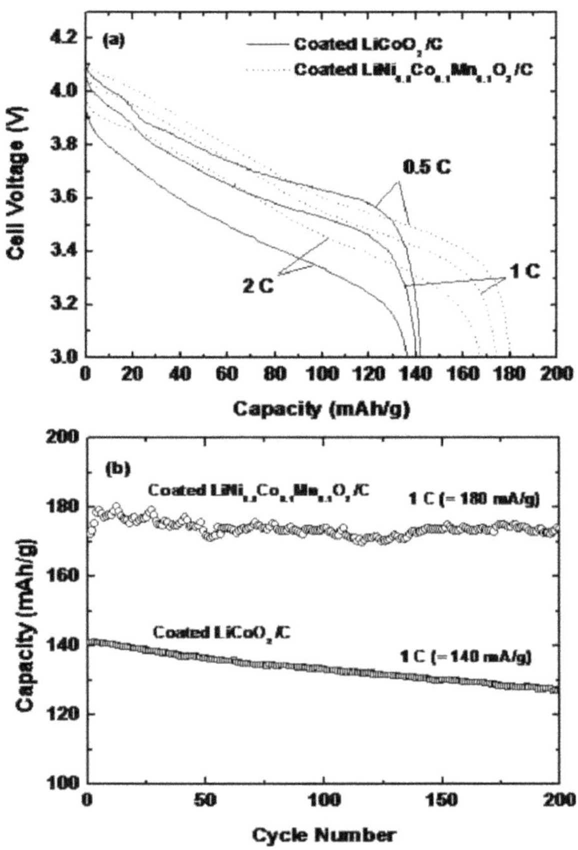

Fig. 10.17 Plots of (**a**) the rate capabilities of the $AlPO_4$-coated $LiCoO_2$ and $LiNi_{0.8}Co_{0.1}Mn_{0.1}O_2$ cathodes in the Li-Ion cells at 0.5, 1, and 2 C rates between 4.2 and 3 V and (**b**) the discharge capacity as a function of the cycle number at 1 C rate

Figure 10.18 compares the XRD patterns of the charged $AlPO_4$-coated Li_xCoO_2 and $Li_xNi_{0.8}Co_{0.1}Mn_{0.1}O_2$ cathode electrodes at 4.2, 4.4, 4.6, and 4.8 V in Li-ion cells (with carbon anode). In both cases, the (003) lines shift toward lower diffraction angles during the charge up to 4.2 V, and then shift back to the higher diffraction angles as the lithium ions are further removed. At 4.8 V, the XRD pattern of the coated Li_xCoO_2 is indexed to the hexagonal unit cell in the space group of *P3m1* (CdI_2 or O1 type). Because 4.8 V in the Li-ion cell corresponds to ~4.9 V vs. lithium metal, the lithium content in the Li_xCoO_2 is zero at 4.8 V. These results agree well with Ohzuku et al.[52] and Amatucci et al.[53] On the other hand, the XRD patterns of the coated $Li_xNi_{0.8}Co_{0.1}Mn_{0.1}O_2$ are indexed to the $R\bar{3}m$ ($CdCl_2$ or O3 type) space group even at a low lithium content. Usually, the (101) and (104) peaks for a

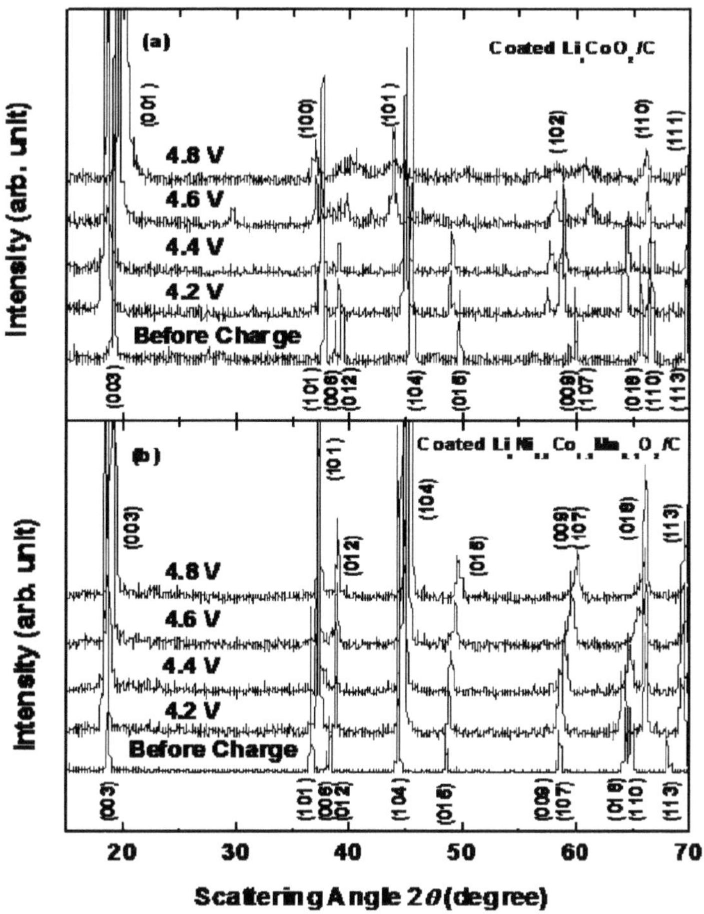

Fig. 10.18 XRD patterns of (**a**) coated Li_xCoO_2, and (**b**) coated $Li_xNi_{0.8}Co_{0.1}Mn_{0.1}O_2$ cathode electrodes as a function of the charge voltage during the first charge

hexagonal lattice observed for $LiNiO_2$ were split into two peaks in the monoclinic lattice (*C2/m*).[20] However, these are not observed in the XRD patterns of the $Li_xNi_{0.8}Co_{0.1}Mn_{0.1}O_2$, which is similar to those of $Li_xNi_{0.5}Co_{0.5}O_2$ having a homogenous hexagonal phase transition.[54]

Ohzuku et al. and Amatucci et al. reported that delithiated Li_xNiO_2 to $x \sim 0$ maintained the $R\bar{3}m$ structure without a monoclinic distortion at a low lithium content, which is unlike $LiCoO_2$.[52], [53] Similarly, the XRD patterns of $Li_xNi_{0.8}Co_{0.1}Mn_{0.1}O_2$ show that its structure at $x \sim 0$ (at 4.8 V) remains as $R\bar{3}m$.

A comparison of the $AlPO_4$-coating effect on the cathode materials was investigated using DSC. Figure 10.19 shows the DSC scans of the $AlPO_4$-coated and the bare Li_xCoO_2 cathodes in the Li-ion cells at different charge voltages. It shows continuous heat generation of cathodes beginning at approximately 150°C. As the charge voltage increases, the exothermic peak above 150°C, which is indicative of oxygen evolution from cathode decomposition, increases in both cathodes. The peak area of the coated $LiCoO_2$ cathode is approximately two to three times lower than that of the bare $LiCoO_2$.

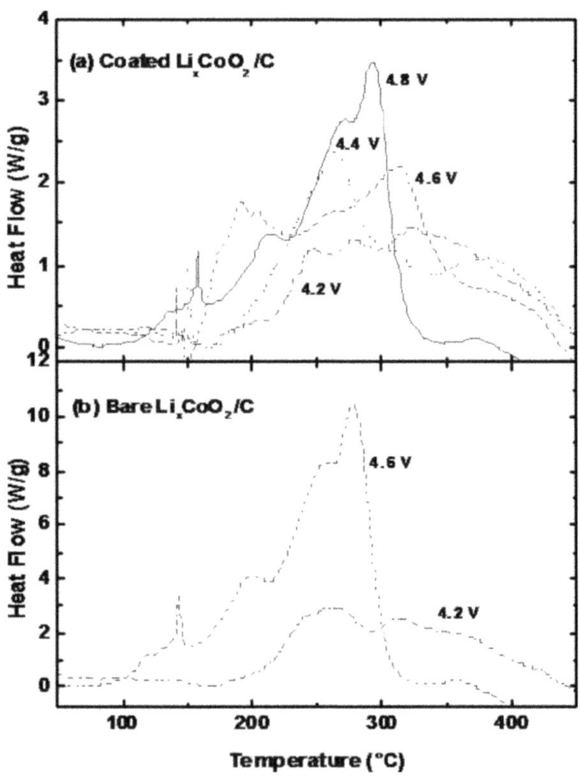

Fig. 10.19 DSC scans of the coated and bare Li_xCoO_2 cathodes at different charge voltages. The cathodes were extracted from the Li-ion cells, and the scan rate was 3°C/min

Figure 10.20 compares the DSC scans of the $AlPO_4$-coated and the bare $Li_xNi_{0.8}Co_{0.1}Mn_{0.1}O_2$ cathodes in the Li-ion cells at the different charge voltages. The coated $Li_xNi_{0.8}Co_{0.1}Mn_{0.1}O_2$ shows two sets of exothermic peaks at ~75°C and ~200°C, which is similar to the bare ones. The peak at ~75°C may be related to the decomposition of organic compounds residing at the particle surface.[55,56] Andersson et al.[55] reported that the delithiated $LiNi_{0.8}Co_{0.2}O_2$ cathode surface contained a mixture of organic polycarbonates species, LiF or $Li_xPF_yO_z$-type compounds, and electrolyte species due to high surface reactivity of the cathode powder. An SEI layer on delithiated electrodes containing various organic and inorganic electrolyte-decomposition products showed a mild heat generation below 100°C.[55,56]

The DSC results suggest that the $AlPO_4$-nanoparticle coating reduces heat generation during the electrochemical reactions. It should be noted that the peak size of the bare cathodes is significantly diminished after the $AlPO_4$ coating (Figs. 10.19 and 10.20). The BET surface area of all the samples was measured and values from the $LiCoO_2$ and $LiNi_{0.8}Co_{0.1}Mn_{0.1}O_2$ powders were 0.6 and 1 m^2/g, respectively, whether the samples were coated or not. However, the DSC results showed significant

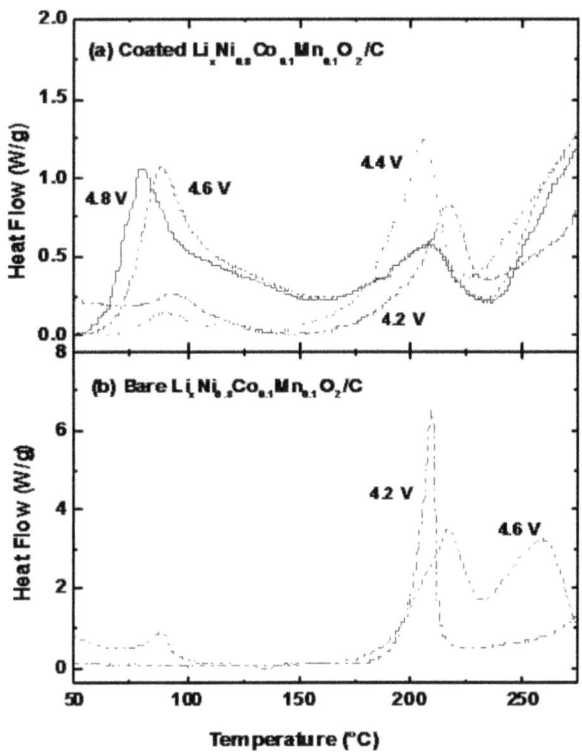

Fig. 10.20 DSC scans of the coated and bare $Li_xNi_{0.8}Co_{0.1}Mn_{0.1}O_2$ cathodes at different charge voltages. The cathodes were extracted from the Li-ion cells, and the scan rate was 3°C/min

heat reduction after the $AlPO_4$ coating, indicating that the crucial factor governing the exothermic reaction is the interfacial reaction with the electrolyte and cathode. With the surface reaction minimized by the coating, the oxygen generation inside the powders can be reduced. It also is believed that the covalent-bond nature of $(PO_4)^{3-}$ with the aluminum cation contributes to a strong resistance to the reaction with the electrolyte.[57, 58]

The coated $Li_xNi_{0.8}Co_{0.1}Mn_{0.1}O_2$ cathode exhibits more diffuse peaks than the coated Li_xCoO_2 with increasing cell voltage, suggesting possibly slower reaction kinetics. This indicates that Li_xCoO_2 has a higher chance of undergoing thermal runaway compared to the $Li_xNi_{0.8}Co_{0.1}Mn_{0.1}O_2$, with the competing heat-generation vs. heat-dissipation rates in Li-ion cells. To evaluate the effect of heat generation on the bare and $AlPO_4$-coated $LiCoO_2$ and $LiNi_{0.8}Co_{0.1}Mn_{0.1}O_2$ in the Li-ion cells during the overcharge to 12 V, the cells were charged at three different current rates of 1, 2, and 3 C to 12 V (Figs. 10.21 and 10.22). At higher C rates, the cell-surface temperature increases rapidly due to the rapid increase in the joule heat, electrolyte oxidation, cathode decomposition, and other cell-component reactions. Among these, a violent exothermic reaction of the cathode with the electrolyte was reported to trigger thermal runaway of the cell.[3, 59] The Li-ion cell containing $AlPO_4$-coated $LiCoO_2$ exhibits increasing cell-surface temperature with an increasing C rate: the cell-surface temperature reaches ~170°C at a overcharging rate. Since the difference between the internal and external temperature is approximately 100°C above 6 V,[3] the cell-internal temperature is estimated to be ~270°C.

The $AlPO_4$-coated $LiNi_{0.8}Co_{0.1}Mn_{0.1}O_2$ exhibits lower maximum cell-surface temperatures than the coated $LiCoO_2$. This is due to faster heat dissipation than heat accumulation, which is consistent with the DSC results. A common feature of the

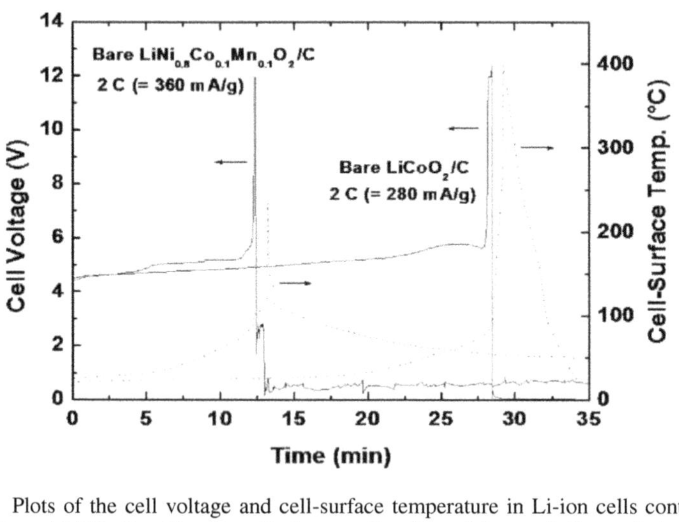

Fig. 10.21 Plots of the cell voltage and cell-surface temperature in Li-ion cells containing the bare $LiCoO_2$ and $LiNi_{0.8}Co_{0.1}Mn_{0.1}O_2$ cathodes as a function of time at 2 C rate during the 12-V overcharge test

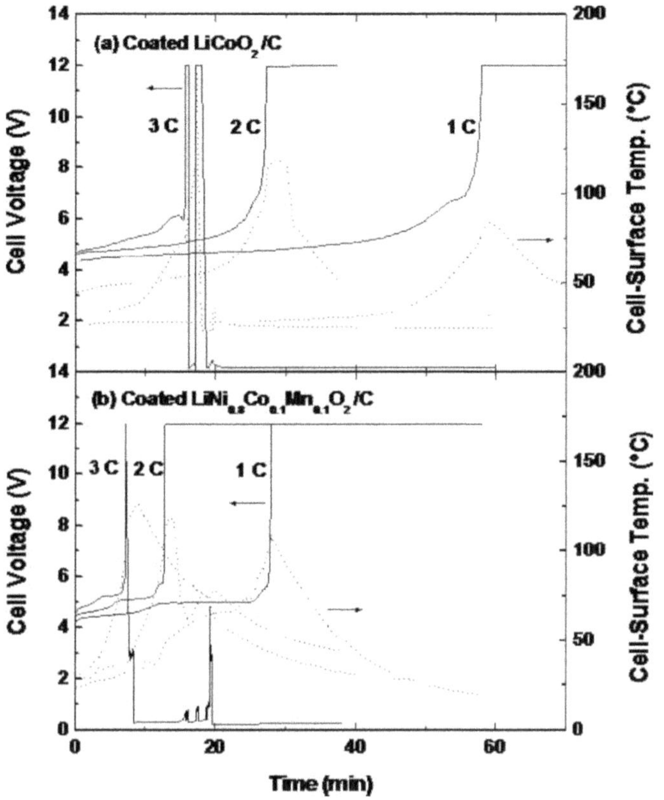

Fig. 10.22 Plots of the cell voltage and cell-surface temperature in Li-ion cells containing $AlPO_4$-coated $LiCoO_2$ and $LiNi_{0.8}Co_{0.1}Mn_{0.1}O_2$ cathodes at 1, 2, and 3 C rates during the 1–2 V overcharge test

coated $LiCoO_2$ and $LiNi_{0.8}Co_{0.1}Mn_{0.1}O_2$ is that a short circuit occurred after shooting to 12 V at a 3 C rate charging. With the exothermic reactions of the cell components occurring within ~10 min, any direct contact between the cathode and the anode electrode can cause a short circuit, which usually results in the thermal runway of the cell containing a bare cathode. However, the surface temperature of the cell containing the coated $LiNi_{0.8}Co_{0.1}Mn_{0.1}O_2$ is limited to ~125°C, lower than that containing the coated $LiCoO_2$. Also, the $AlPO_4$-coated $LiNi_{0.8}Co_{0.1}Mn_{0.1}O_2$ cathode reaches 12 V approximately 10–30 min earlier than $LiCoO_2$. This may be related to the different decomposition behaviors of the cathodes.

When the first heat generation from the cathode particle induces a separator shutdown before the second heat generation begins at higher temperatures, the internal resistance of the cell increases greatly. Eventually, it causes a decrease of current, resulting in a decrease in temperature. However, the coated $LiCoO_2$ cathode exhibits continuous heat generation beginning at ~150°C with a slimmer peak shape. Since this behavior causes faster heat accumulation than heat dissipation, the maximum cell-surface temperature at 12 V increases. This is well supported by the fact that the cell temperature of the bare $LiCoO_2$ cathode increases abruptly to ~400°C at 12 V upon short circuiting, as shown in Fig. 10.8. However, a Li-ion cell containing bare $LiNi_{0.8}Co_{0.1}Mn_{0.1}O_2$ exhibits a maximum temperature of ~225°C, accompanying only smoke from the bottom rupture of the can.

10.6 ZnO Coating on $LiNi_{0.5}Mn_{1.5}O_4$ Spinel Cathode

Spinel $LiMn_2O_4$ is one of the most promising cathode materials due to low cost, abundance, and nontoxicity.[60] It is well known that the spinel $LiMn_2O_4$ electrode shows good stability during cycling at room temperature, but a poor cycling behavior at elevated temperature, e.g., 50–80°C, which prevents its wider use as cathode materials for lithium secondary batteries.[61]

Recently, some research groups have reported that transition metal-substituted spinel materials ($LiM_xMn_{2-x}O_4$, M = Cr, Co, Fe, Ni) showed a higher voltage plateau at around 5 V.[62–65] These research groups also reported that the 5-V plateau originates from oxidation of substituted transition metals (M), while the 4-V plateau is associated with Mn^{3+} oxidation to Mn^{4+}. Furthermore, it is well known that the discharge capacity and the 5-V plateau in $Li/LiM_xMn_{2-x}O_4$ cells strongly depend on the kind of transition metals and their concentration.

The major factor responsible for the capacity loss for the spinel $LiMn_2O_4$ electrode in the 4-V region at elevated temperatures is ascribed to:

1. A dissolution of the MnO into the electrolyte due to the HF formation resulting from the reaction of fluorinated anions with residual H_2O.
2. A phase transition from cubic to tetragonal symmetry due to Jahn–Teller distortion.
3. Oxygen deficiency.[66–68]

The poor cycling performance could be improved by cation and anion substitution and surface treatment of electrode active materials.[69–71] Surface passivation treatments of $Li_{1.05}Mn_{1.95}O_4$ particles coated with lithium borate glass and acetylacetone were found to improve the performance of the material at 55°C, which has been attributed to the minimization of the $LiMn_2O_4$/electrolyte interface.[72] However, there is not yet explicit experimental data for the structural degradation of the coated spinel particle surface after cycling to confirm many researcher's observation and the suggestion that is associated with capacity loss.$LiNi_{0.5}Mn_{1.5}O_4$ powders were prepared by a sol–gel method using glycolic acid as a chelating agent. $Li(CH_3COO){\cdot}H_2O$, $Ni(CH_3COO)_2{\cdot}4H_2O$, and $Mn(CH_3COO)_2{\cdot}4H_2O$ (cationic ratio of Li:Ni:Mn = 1:0.5:1.5) were dissolved in the distilled water and added dropwise to a continuously stirred aqueous solution of glycolic acid. The pH of the solution was adjusted to 8.5–9.0 using ammonium hydroxide. The resultant solution was evaporated at 70–80°C until a transparent sol and gel were obtained. The resulting gel precursorswere decomposed at 500°C for 10 h in air and calcined at 850°C in air for 10 h. To coat ZnO on the $LiNi_{0.5}Mn_{1.5}O_4$ powders, $Zn(CH_3COO)_2{\cdot}2H_2O$ was first dissolved in the distilled water. The synthesized $LiNi_{0.5}Mn_{1.5}O_4$ powders were poured into the solution, and then mixed for 4 h at room temperature. The amount of Zn in the coating solution corresponded to 1.5 wt% of the $LiNi_{0.5}Mn_{1.5}O_4$ powders. After drying, the ZnO-coated $LiNi_{0.5}Mn_{1.5}O_4$ powders at 120°C were further calcined at 400°C for 1 h in air.

X-ray diffraction patterns of the as-prepared $LiNi_{0.5}Mn_{1.5}O_4$ and ZnO-coated $LiNi_{0.5}Mn_{1.5}O_4$ powders are shown in Fig. 10.23. It was confirmed that two materials were well-defined spinel phases with the space group *Fd3m*, although there were small amount of NiO peaks as impurity near the (311), (400), and (440) peaks. Even if ZnO-related phases do not appear on the XRD pattern of the ZnO-coated $LiNi_{0.5}Mn_{1.5}O_4$ powders, it is speculated that a small amount of ZnO might exist on the surface (cf Fig. 10.25). The lattice constants *(a)* of the as-prepared $LiNi_{0.5}Mn_{1.5}O_4$ powders were calculated by Rietveld refinement from the X-ray diffraction data to be 8.160, which is much lower than that of stoichiometric spinel.[73]

Figure 10.24 shows SEM images for the as-prepared and ZnO-coated $LiNi_{0.5}Mn_{1.5}O_4$ powders. The particles of the $LiNi_{0.5}Mn_{1.5}O_4$ powders have a well-developed (100) planes and a size distribution ranging from 0.5 to 2 μm. For the ZnO-coated $LiNi_{0.5}Mn_{1.5}O_4$ powders, ZnO with the order of 10 nm or less is uniformly distributed over the surface of $LiNi_{0.5}Mn_{1.5}O_4$ powders as shown in Fig. 10.24b. In order to confirm the component of the coated material on the surface of the $LiNi_{0.5}Mn_{1.5}O_4$ powders, EDAX analysis was performed, as shown in Fig. 10.25. It clearly revealed that zinc existed on the surface of the $LiNi_{0.5}Mn_{1.5}O_4$ particles.

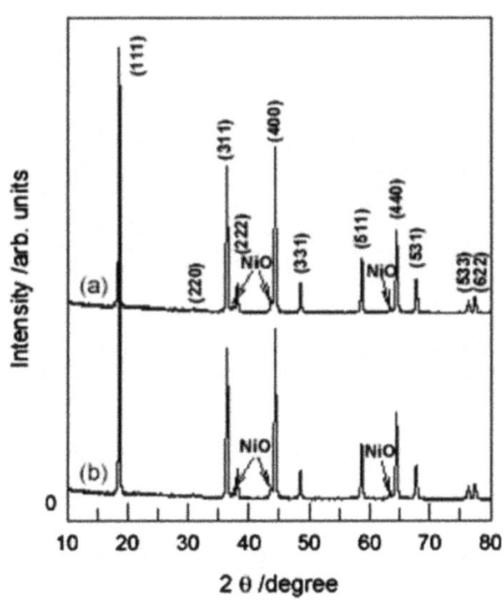

Fig. 10.23 X-ray diffraction patterns of (**a**) the as-prepared $LiNi_{0.5}Mn_{1.5}O_4$ and (**b**) ZnO-coated $LiNi_{0.5}Mn_{1.5}O_4$ powders

Fig. 10.24 SEM of (**a**) the as-prepared $LiNi_{0.5}Mn_{1.5}O_4$ and (**b**) ZnO-coated $LiNi_{0.5}Mn_{1.5}O_4$ powders

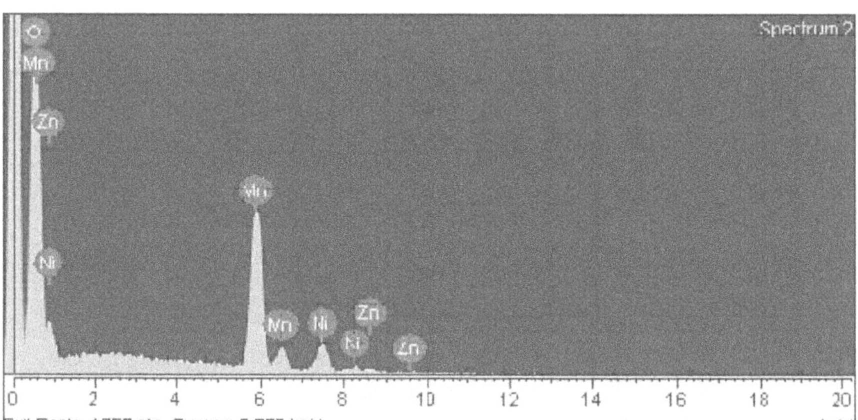

Fig. 10.25 The EDAX spectrum of the ZnO-coated $LiNi_{0.5}Mn_{1.5}O_4$ powders

Figure 10.26 shows charge–discharge curves for the as-prepared and ZnO-coated $LiNi_{0.5}Mn_{1.5}O_4$ electrodes as a function of the cycle number at 25°C. The Li/$LiNi_{0.5}Mn_{1.5}O_4$ cell shows a very flat discharge plateau at 4.75 V vs. Li/Li$^+$. When comparing voltage profiles of as-prepared $LiNi_{0.5}Mn_{1.5}O_4$ electrode with that of the ZnO-coated $LiNi_{0.5}Mn_{1.5}O_4$ electrode, we could not find any difference between the two. The as-prepared and the ZnO-coated $LiNi_{0.5}Mn_{1.5}O_4$ electrodes cycled at 25°C delivered an initial capacity of 140 and 137 mA h/g, respectively. The capacity retention was 92% for as-prepared $LiNi_{0.5}Mn_{1.5}O_4$ and 94% for ZnO-coated $LiNi_{0.5}Mn_{1.5}O_4$ electrodes after 50 cycles.

However, the cycling characteristics for the as-prepared and ZnO-coated $LiNi_{0.5}Mn_{1.5}O_4$ electrodes at 55°C are quite different. Figure 10.27 shows charge–discharge curves for the as-prepared and various amounts of ZnO-coated $LiNi_{0.5}Mn_{1.5}O_4$ electrodes at 55°C. The polarization (voltage difference between the charge and discharge curves) of the as-prepared $LiNi_{0.5}Mn_{1.5}O_4$ electrode dramatically increased with increasing cycle number, indicating the structural degradation of the $LiNi_{0.5}Mn_{1.5}O_4$ electrode material. On the other hand, the polarization of ZnO-coated electrodes decreased with increasing amounts of the ZnO coating. For the 1.5 wt%-coated $LiNi_{0.5}Mn_{1.5}O_4$ electrode (Fig. 10.27d), the measured charge–discharge curves were unchanged during cycling and the electrode showed a very flat discharge plateau at b4.75 V vs. Li/Li$^+$. The ZnO-coated spinel electrodes

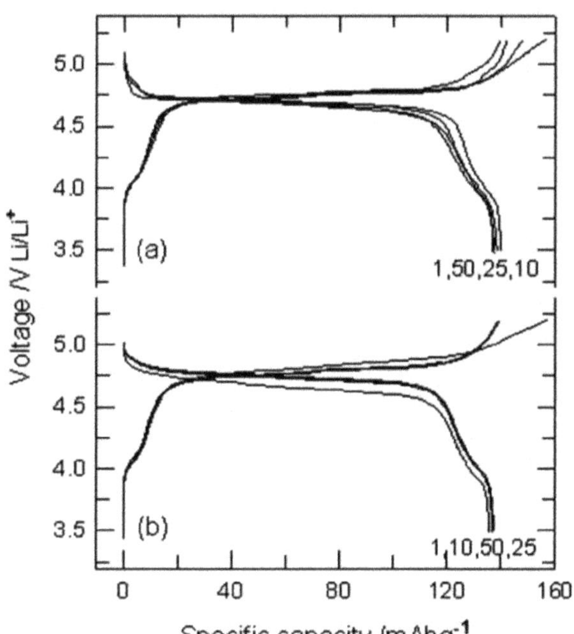

Fig. 10.26 Charge–discharge curves for (**a**) the as-prepared $LiNi_{0.5}Mn_{1.5}O_4$ and (**b**) ZnO-coated $LiNi_{0.5}Mn_{1.5}O_4$ electrodes at 25°C at a constant current densities of 0.4 mA/cm^2

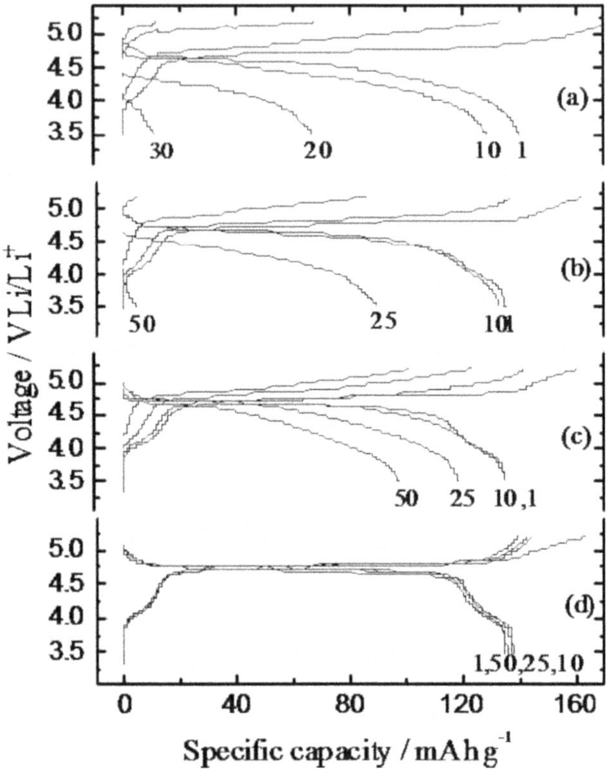

Fig. 10.27 Charge–discharge curves (**a**) the as-prepared, (**b**) 0.1%, (**c**) 0.5%, and (**d**) 1.5% ZnO-coated $LiNi_{0.5}Mn_{1.5}O_4$ electrodes

clearly show an improved cycling behavior compared with the as-prepared uncoated spinel (Fig. 10.27a). The initial capacities (137 mA h/g) of different spinel materials are almost identical, but the capacity retention among the as-prepared and ZnO-coated $LiNi_{0.5}Mn_{1.5}O_4$ electrodes is significantly different. The capacity retentions of the as-prepared, 0.1, 0.5 wt% ZnO-coated spinel electrodes are 8%, 29%, and 88% after 30 cycles, respectively. However, 1.5 wt% ZnO-coated $LiNi_{0.5}Mn_{1.5}O_4$ electrode delivers an initial capacity of 137 mA h/g and retains excellent cycling behavior without any capacity loss even after 50 cycles 55°C and a C/3 rate. This cycling behavior of the ZnO-coated electrode material shows clearly the impact of ZnO coating in protecting the surface of $LiNi_{0.5}Mn_{1.5}O_4$ against hydrogen fluorine (HF) attack.

In order to explore the difference in cycling behavior observed with the as-prepared and ZnO-$LiNi_{0.5}Mn_{1.5}O_4$ electrodes, the electrodes were removed after 50 cycles at 55°C in the discharged state and characterized using XRD and high-resolution transmission electron micrograph (HRTEM). The as-prepared $LiNi_{0.5}Mn_{1.5}O_4$ electrode, characterized by large capacity loss, exhibit the emergence of peaks related to tetragonal phase ($Li_2Mn_2O_4$), rock salt phase (Li_2MnO_3), and a spinel phase, shown in Fig. 10.28. On the other hand, 1.5 wt% ZnO-coated $LiNi_{0.5}Mn_{1.5}O_4$ electrode maintains its original spinel phase even after cycling at 55°C. It is commonly accepted that the capacity loss for the spinel $LiMn_2O_4$ electrode could be attributed largely to slow dissolution of MnO at the spinel particle into electrolyte due to the formation of HF resulting from the reaction of fluorinated anions with residual H_2O.[70,74] Recently, the author reported[75] that the degradation mechanism of the spinel electrode in the 4-V region at elevated temperature was attributed to the formation of rock salt Li_2MnO_3 resulting from the dissolution of MnO from $Li_2Mn_2O_4$ due to HF presence in the electrolyte. Therefore, it is very important to decrease the HF content at the electrode surface. ZnO is amphoteric oxide and could react with HF at the cathode surface. In order to elucidate the role of ZnO, we examined the reactivity of ZnO with HF in electrolyte solution. Various powder samples of ZnO, as-prepared, and ZnO-coated $LiNi_{0.5}Mn_{1.5}O_4$ were immersed in the electrolyte at both 25 and 55°C. After 1-week aging in the electrolyte, the residual HF content in the electrolyte was analyzed and the results are shown in Table 10.1. In the case of an experiment carried out at 25°C, the initial amount of HF content in the electrolyte was 274 ppm. The large excess of HF concentration used in this

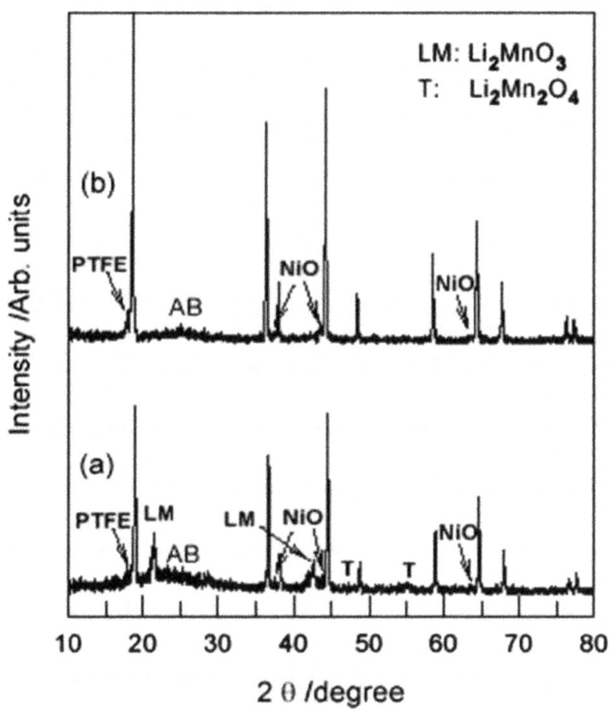

Fig. 10.28 X-ray diffraction patterns for (**a**) the as-prepared $LiNi_{0.5}Mn_{1.5}O_4$ and (**b**) 1.5% ZnO-coated $LiNi_{0.5}Mn_{1.5}O_4$ electrodes after 50 cycles at 50°C

Table 10.1 HF contents in the 1-M $LiPF_6$ in EC/DMC (1:2) after treated various sample for 1 week at 25 and 55°C

Temperature (°C)	Samples	Residual HF (ppm)
25	Bulk electrolyte (1 M $LiPF_6$ in EC/DMC)	274
	As-prepared $LiNi_{0.5}Mn_{1.5}O_4$ powders	274
	1.5% ZnO-coated $LiNi_{0.5}Mn_{1.5}O_4$ powders	7.3
	Only ZnO powders	0
55	Bulk electrolyte (1 M $LiPF_6$ in EC/DMC)	68
	As-prepared $LiNi_{0.5}Mn_{1.5}O_4$ powders	68
	0.5% ZnO-coated $LiNi_{0.5}Mn_{1.5}O_4$ powders	53
	1.5% ZnO-coated $LiNi_{0.5}Mn_{1.5}O_4$ powders	32
	Only ZnO powders	0

case was necessary to quantity the role of ZnO in getting HF. The result from Table 10.1 shows that the residual HF content in the electrolyte decreases dramatically with increasing amount of ZnO coating. It is noticeable that the residual HF contents in the electrolyte treated with only ZnO powders were 0.0 ppm regardless of temperature. These results strongly suggest that ZnO reacts with HF and prevent manganese from dissolving into the electrolyte. Another reason for improving electrochemical performance by coating $LiNi_{0.5}Mn_{1.5}O_4$ powders with ZnO may be attributed to suppression of surface reaction between cathode surface and electrolyte. Figure 10.29a shows the TEM image of the ZnO-coated $LiNi_{0.5}Mn_{1.5}O_4$ powder in [110] zone. At this resolution, no visible surface phases can be seen on the surface of the particle, which suggests the ZnO is covering the particle surface in subnanometer thickness. The observation is in agreement with the energy dispersive X-ray spectroscopy (EDS) analysis near the edge of the particle through which no trace of zinc was found within the EDS detection limit. Shown in Fig. 10.29b is the bright field TEM image of the uncoated $LiNi_{0.5}Mn_{1.5}O_4$ particle after 50 cycles at 55°C. As indicated by arrows, the surface of the particle was covered with carbonaceous phases. In higher magnification shown in Fig. 10.29c, entangled graphitic planes with its characteristic 002 plane spacing of 0.34 nm can be clearly seen. In contrast, the powder coated with ZnO does not show such graphitic phases on the surface of the particle after cycling at the elevated temperature (Fig. 10.29d). It is not clear how the carbon phase has nucleated on the particle surface during cycling and to our best knowledge no such surface phase formation has been previously reported. We speculate that the surface phase is related to the breakdown of the electrolytic solution at 5 V and cation radicals generated from the electrolyte oxidation could provide the source for forming the graphitic phase on the charged cathode surface. Regardless of the formation mechanism, the graphitic phase on the particle surface, having a rather entangled structure, would certainly hinder the lithium intercalation process and lead to capacity loss of the material. In comparison, the ZnO coating appears to prevent the nucleation of the carbon phase and subsequent capacity degradation observed with the $LiNi_{0.5}Mn_{1.5}O_4$ without the ZnO surface coating.

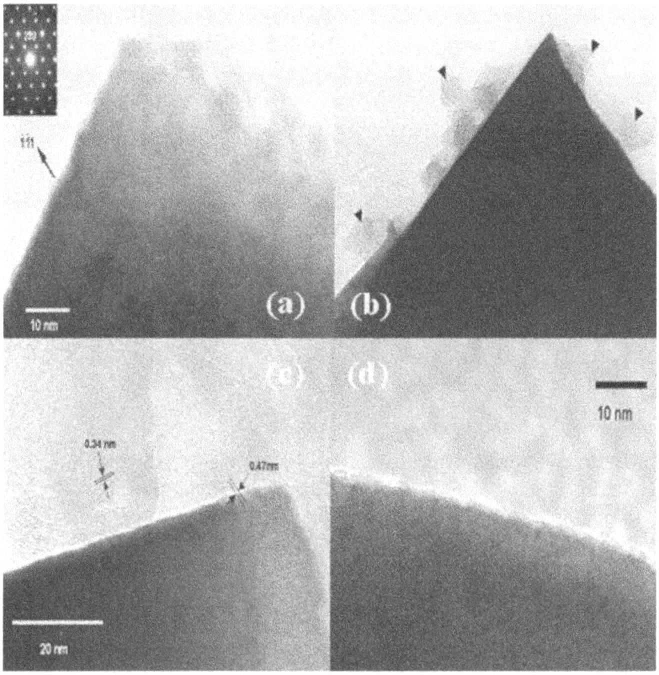

Fig. 10.29 TEM images of (**a**) the ZnO-coated $LiNi_{0.5}Mn_{1.5}O_4$ particle in [110] zone, (**b**) uncoated $LiNi_{0.5}Mn_{1.5}O_4$ particle after cycling at 55°C (*arrows* indicate the carbonaceous phases), (**c**) uncoated $LiNi_{0.5}Mn_{1.5}O_4$ particle after cycling at 55°C at a higher magnification (002 graphitic and 111 spinel spacings are shown for comparison), and (**d**) ZnO-coated $LiNi_{0.5}Mn_{1.5}O_4$ particle after cycling showing no carbonaceous phases on the particle surface

Acknowledgment This work was supported by ITRC research center project and the Korea Science and Engineering Foundation (KOSEF) grant funded by the Korea government (MEST) (NO. R01-2008-000-20142-0).

References

1. *Laptop Batteries Are Linked to Fire Risk*, New York Times, March 15, 2001; US Consumer Product Safety Commission (http://www.cpsc.gov/cpscpub).
2. S. C. Levy, P. Bro, *Battery Hazards and Accident Prevention*, Plenum Press, New York, NY,, 1994.
3. R. A. Leising, M. J. Palazzo, E. S. Takeuchi, K. J. Takeuchi, *J. Electrochem. Soc.*, **148**, (2001) A838.
4. H. Maleki, S. A. Hallaj, J. R. Selman, R. B. Dinwiddie, H. Wang, *J. Electrochem. Soc.*, **146**, (1999) 947.
5. A. Duaquier, F. Disma, T. Bowmer, A. S. Gozdz, G. G. Amatucci, J.-M. Tarascon, *J. Electrochem. Soc.,* **145**, (1998) 472.
6. H. Maleki, G. Deng, A. Anani, J. Howard, *J. Electrochem. Soc.*, **146**, (1999) 3224.

7. N. Takami, H. Inagaki, H. Ishii, R. Ueno, R. M. Kanda, IMLB11 – *11th International Meeting on Lithium Batteries*, Monterey, CA, USA, June 23–28, 2002.
8. K. Xu, M. S. Ding, S. Zhang, J. Allen, T. R. Jow, *J. Electrochem. Soc.*, **149**, (2002) A622.
9. S. C. Narang, S. C. Ventura, B. J. Dougherty, M. Zhao, M. S. Smedley, G. Koolpe, US Patent 5830660.
10. M. Adachi, K. Tanaka, K. Sekai, *J. Electrochem. Soc.*, **146**, (1999) 1256.
11. X. Wang, E. Yasukawa, S. J. Kasuya, *J. Electrochem. Soc.*, **148**, (2001) A1058.
12. J. Cho, Y. W. Kim, J.-G. Lee, B. Kim, B. Park, *Angew. Chem. Int. Ed.*, **42**, (2003) 1618.
13. J. Cho, Y. J. Kim, T.-J. Kim, B. Park, *Chem. Mater.*, **13**, (2001) 18.
14. J. Cho, Y. J. Kim, B. Park, *Chem. Mater.*, **12**, (2000) 3788.
15. J. Cho, J.-G. Lee, B. Kim, B. Park, *Chem. Mater.*, **15**, (2003) 3190.
16. M. Mladenov, R. Stoyanova, E. Zhecheva, S. Vasslev, *Electrochem. Commun.*, **3**, (2001) 410.
17. J. Cho, Y. J. Kim, B. Park, *Chem. Mater.*, **12**, (2000) 3788.
18. J. Cho, Y. J. Kim, T.-J. Kim, B. Park, *Angew. Chem. Int. Ed.*, **40**, (2001) 3367.
19. J. Cho, C.-S. Kim, S.-I. Yoo, *Electrochem. Solid-State Lett.*, **3**, (2000) 362.
20. Z. Wang, C. Wu, L. Liu, F. Wu, L. Chen, X. Huang, *J. Electrochem. Soc.*, **149**, (2000) A466.
21. Z. Chen, J. R. Dahn, *Electrochem. Solid-State Lett.*, **5**, (2002) A213.
22. Z. Wang, L. Liu, L. Chen, X. Huang, *Solid State Ionics*, **148**, (2002) 335.
23. J. Cho, T.-J. Kim, Y. J. Kim, and B. Park, *Electrochem. Solid-State Lett.*, **4**, (2001) A159.
24. D. Rahner, W. Plieth, M. Kloss, *Modified Electrode Material and its Use*, US Patent No. 6348259, March 30, 1999.
25. T. Miyasaka, *Non-Aqueous Lithium Ion Secondary Battery*, US Patent No. 6037095, March 30, 1998.
26. H.-J. Kweon, G.-B. Kim, D.-G. Park, *Positive Active Material for Rechargeable Lithium Battery and Method Of Preparing Same*, US Patent No. 6183911, October 29, 1999.
27. A. K. Padhi, K. S. Nanjundaswamy, J. B. Goodenough, *J. Electrochem. Soc.*, **144**, (1997) 1188.
28. S. Okada, S. Sawa, M. Egashira, J.–I. Yamaki, M. Tabuchi, H. Kageyama, T. Konishi, A. Yoshio, *J. Power Sources*, **97**, (2001) 430.
29. *Handbook of X-ray Photoelectron Spectroscopy*, J. Chastain, R. C. King, Jr. (Eds.), Physical Electronics, Inc., Minnesota, 1995.
30. J.A. Rotole, P.M.A. Sherwood, *Surf. Sci. Spec.*, **5**, (1998) 60.
31. G.G. Amatucci, J.M. Tarascon, L.C. Klein, *Solid State Ionics*, **83**, (1996) 167.
32. Y.J. Kim, J. Cho, T.-J. Kim, B. Park, *J. Electrochem. Soc.*, **150**, (2003) A1723.
33. K.-K. Lee, W.-S. Yoon, K.-B. Kim, *J. Electrochem. Soc.*, **148**, (2001) A1164.
34. J. Cho, G. Kim, Y. Park, S. Kim, US Patent # 6241959 (2001).
35. Y. Nishida, K. Nakane, T. Satoh, *J. Power Sources*, **68**, (1997) 561.
36. H. Watanabe, T. Sunagawa, H. Fujimoto, N. Nishida, T. Nohma, *Sanyo Tech. Rev.*, **30**, (1998) 84.
37. Y. Sato, T. Koyano, M. Mukai, K. Kobayakawa, Meeting Abstracts of the 192nd Electrochem. Soc. Meeting, Paris, France, Aug. 31–Sep. 5, 1997.
38. M. Yoshio, H. Noguchi, J.-I. Itoh, M. Okada, T. Mouri, *J. Power Sources*, **90**, (2000) 176.
39. C. Nayoze, F. Ansart, C. Laberty, J. Sarrias, A. Rousset, *J. Power Sources*, **99**, (2001) 54.
40. J. Cho, T.-J. Kim, Y. J. Kim, B. Park, *Electrochem. Solid-State Lett.*, **4**, (2001) A159.
41. J. Cho, H. Jung, Y. Park, G. Kim, H. Lim, *J. Electrochem. Soc.*, **147**, (2000) 10.
42. J. Cho, G. Kim, H. Lim, *J. Electrochem. Soc.*, **146**, (1999) 3571.
43. W. Li, J. C. Currie, *J. Electrochem. Soc.*, **144**, (1997) 2773.
44. H. Arai, M. Tsuda, K. Saito, M. Hayashi, Y. Sakurai, *J. Electrochem. Soc.*, **149**, (2002) A401.
45. N. Takami, H. Inagaki, R. Ueno, M. Kanda, IMLB11–11th International Meeting on Lithium Batteries, Monterey, CA, USA, June 23–28, 2002.
46. Y. Gao, M. V. Yakovleva, W. B. Ebner, *Electrochem. Solid-State Lett.*, **1**, (1998) 117.
47. I. Yoshiyuki, K. Kochichi, Y. Shuji, K. Motoya, Meeting Abstracts of the 202nd Electrochem. Soc. Meeting, Salt Lake City, UT, USA, Oct. 20–24, 2002.
48. J. Cho, *Electrochem. Commun.*, **5**, (2003) 146.

49. N. Yabuuchi, T. Ohzuku, *J. Power Sources*, 119–121, (2003) **171.**
50. Z. Lu, D. D. MacNeil, J. R. Dahn, *Electrochem. Solid-State Lett.*, **4**, (2001) A200.
51. *Guideline for the Safety Evaluation of Secondary Lithium Cells*, Japan Battery Association, Tokyo, 1997.
52. T. Ohzuku, A. Ueda, M. Nagayama, *J. Electrochem. Soc.*, **140**, (1993) 1862.
53. G. G. Amatucci, J. M. Tarascon, L. C. Klein, *J. Electrochem. Soc.*, **143**, (1996) 1114.
54. A. Ueda, T. Ohzuku, *J. Electrochem. Soc.*, **141**, (1994) 2010.
55. A. M. Andersson, D. P. Abraham, R. Haasch, S. MacLaren, J. Liu, K. Amine, *J. Electrochem. Soc.*, **149**, (2002) A1358.
56. J.-I. Yamaki, H. Takatsuji, T. Kawamura, M. Egashira, *Solid State Ionics*, **148**, (2002) 241.
57. A. K. Padhi, K. S. Nanjundaswamy, J. B. Goodenough, *J. Electrochem. Soc.*, **144**, (1997) 1188.
58. S. Okada, S. Sawa, M. Egashira, J.-I. Yamaki, M. Tabuchi, H. Kageyama, T. Konishi, A. Yoshino, *J. Power Sources*, **97**, (2001) 430.
59. J. R. Dahn, E. W. Fuller, M. Obrovac, U. von Sacken, *Solid State Ionics*, **69**, (1994) 265.
60. D. Guyomard, J.-M. Tarascon, *J. Electrochem. Soc.*, **139**, (1994) 222.
61. Y. Xia, Y. Zhou, M. Yoshio, *J. Electrochem. Soc.*, **144**, (1997) 2593.
62. C. Sigala, D. Guyomard, A. Verbaere, Y. Piffard, M. Tournoux, *Solid State Ionics*, **81**, (1995) 167.
63. H. Kawai, M. Nagata, H. Kageyama, H. Tsukamoto, A.R. West, *Electrochim. Acta*, **45**, (1999) 315.
64. H. Shigemura, H. Sakaebe, H. Kageyama, H. Kobayashi, A.R. West, R. Kanno, S. Morimoto, S. Nasu, M. Tabuchi, *J. Electrochem. Chem.*, **148**, (2001) A730.
65. K. Amine, H. Tukamoto, H. Yasuda, Y. Fujita, *J. Electrochem. Soc.*, **1607**, (1996) 143.
66. D. H. Jang, Y. J. Shin, S. M. Oh, *J. Electrochem. Soc.*, **143**, (1996) 2204.
67. A. D. Pasquier, A. Bylr, P. Courjal, D. Larcher, G. Amatucci, B. Gerand, J.-M. Tatscon, *J. Electrochem. Soc.*, **146**, (1999) 48.
68. Y. Xia, T. Sakai, T. Fujieda, X.Q. Yang, X. Sun, Z.F. Ma, J. McBreen, M. Yoshio, *J. Electrochem. Soc.*, **148**, (2001) A723.
69. Y. Xia, N. Kumada, M. Yoshio, *J. Power Sources*, **90**, (2000) 135.
70. G. G. Amatucci and J.-M. Tarascon, US Patent No. 5,674,645 (1997).
71. Y.-K. Sun, G.-S. Park, Y.-S. Lee, M. Yoshio, K. S. Nahm, *J. Electrochem. Soc.*, **148**, (2001) A994.
72. G. G. Amatucci, A. Bylr, C. Siagala, P. Alfonse, J.-M. Tarascon, *Solid State Ionics*, **104**, (1997) 13.
73. A.D. Pasquier, A. Blyr, P. Courjal, D. Larcher, G. Amatucci, B. Gerand, J.-M. Tarascon, *J. Electrochem. Soc.*, **146**, (1999) 428.
74. Y. Xia, N. Kumada, M. Yoshio, *J. Power Sources*, **90**, (2000) 135.
75. Y.-K. Sun, C.S. Yoon, C.K. Kim, S.G. Youn, Y.-S. Lee, M. Yoshio, I.-H. Oh, *J. Mater. Chem.*, **11**, (2001) 2519.

Chapter 11
Development of Metal Alloy Anodes

Nikolay Dimov

11.1 Introduction

Since the beginning of the 1990s we have witnessed the development of lithium-ion batteries (LIB), which currently are the power source of choice for most portable electronic devices, such as cell phones, digital cameras, and laptop computers. Although these batteries offer the highest specific energy density, the demand for greater electronic performance is continuing to place increasing pressure on their storage capabilities. Recently, many ongoing research activities have focused on the development of second-generation cathode and anode materials with enhanced capacity or cyclic performance. Although the specific capacity (in mA h g^{-1}) of the cathode material is often lower (about half that of anode material), it is unlikely that significant advancement can be made in the near future. On the contrary, materials with theoretical capacities many times higher than that of carbonaceous anodes exist. Advanced LIB therefore can be designed using high-capacity anode material to accommodate for low-capacity cathode material in the limited volume of a battery interior.

The advantage of using higher-capacity anode material can be easily quantified. Considering the capacity (C_C) of a given cathode material, this may vary from ca. 140 ($LiCoO_2$, spinels) to ca. 200 mA h g^{-1} ($LiMnO_2$ and its derivatives); there is a simple relationship between the total capacity of the LIB material as a function of anode specific capacity (C_A). Qualitative expression of total specific capacity with the variation in C_A can be expressed as

$$\text{Total} = \frac{C_C}{1 + \dfrac{C_C}{C_A}} = \frac{C_A C_C}{C_A + C_C} \quad (\text{mA h g}^{-1}) \tag{11.1}$$

Equation (11.1) shows that for a fixed cathode-specific C_C, the total specific capacity does not increase linearly with the linear increase in C_A. A simulation based on

N. Dimov
Department of Applied Chemistry, Saga University,
1 Honjo, Saga 840-8502, Japan

M. Yoshio et al. (eds.), *Lithium-Ion Batteries*, 241
DOI 10.1007/978-0-387-34445-4_11 © Springer Science + Business Media LLC 2009

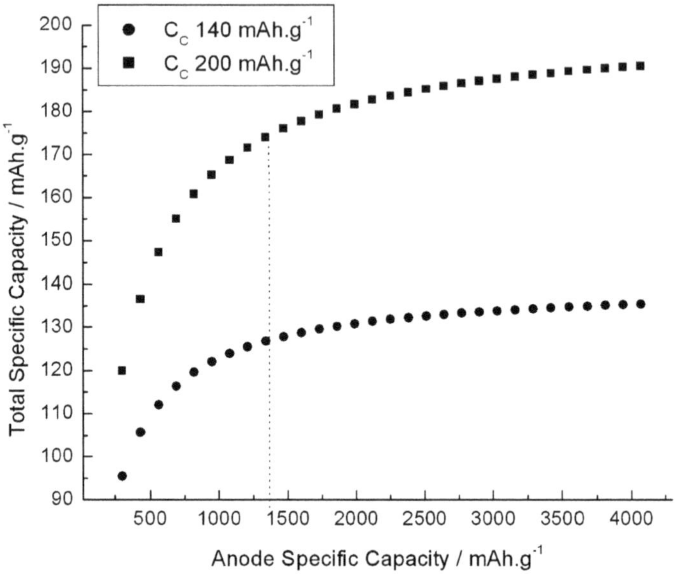

Fig. 11.1 Total capacity of LIB material presented as a function of anode capacity. Two possible cathodes with capacities of 140 and 200 mA h g^{-1}, respectively, are considered. As shown, a comparatively fast increase in the total capacity with an anode-specific capacity (C_A) of 300–1,200 mA h g^{-1}, is followed by a "tail" with a smaller slope when C_A exceeds ca. 1,200 mA h g^{-1}

(11.1) is plotted in Fig. 11.1; total capacity increases quickly with an initial increase in C_A then a plateau-like region appears. This behavior can be further confirmed by exploring the limit of $C_A \to \infty$

$$\frac{d(\text{Total})}{d(C_A)} = \frac{(C_C)^2}{(C_C + C_A)^2} \to 0 \text{ as } C_A \to \infty \tag{11.2}$$

The rate at which the total capacity increases will depend on the value of C_C. As seen in the figure, there are two patterns of increase: A comparatively fast increase in total capacity with increasing C_A from 300–1,200 mA h g^{-1} followed by a "tail" with a smaller slope when C_A exceeds ca. 1,200 mA h g-^{1}. Therefore, the most noticeable improvement in LIB will be seen if the presently used carbonaceous anode is replaced with one having a capacity in the order of 1,000 mA h g^{-1}.

The driving force of such studies is based on the fact that the potential of the anode vs. Li^+ should be close to 0 V. Hence, the electrochemical reaction at the anode site should not necessarily be based on an intercalation type of reaction. Alloying of Li^+ with certain metals is very attractive, because the Li:M mole ratio in the Li_xM alloy at the end of charge might be much higher than in the case of intercalation hosts which generally cannot accommodate and release large amounts of Li^+ in order to maintain a stable crystal structure over the cycles. Numerous metals – Si, Al, Sn, Sb, Ge, Pb, Ag, etc – are able to alloy electrochemically with lithium when polarized to a sufficiently negative potential in a Li_x-containing liquid organic electrolyte. The respective solid phases Li_xM that form during this electrochemical alloying exhibit high theoretical capacity. From a practical point of view tin, silicon, and their alloys are the most attractive candidates, because they not only offer high theoretical capacity, but also are abundant and environmentally friendly. Unfortunately, most high-capacity anodes based on the aforementioned elements are intrinsically unstable during cycling and their advantages often are undermined by their short cycle life. The lack of cycling performance is due primarily to mechanical stress induced in the material during repetitive charge and discharge operations where large changes in the specific volume of the material occur. The goal of this chapter is to give an overview on silicon-based anode materials for ambient temperature rechargeable LIB. A brief chronological survey of lithium alloying anodes will introduce the principle concepts to overcome the problems with the dimensional instability of the metallic host materials. Recent work on promising (composite and thin film) silicon-based lithium-alloying electrodes will be highlighted.

11.2 Historical Remarks on the Lithium Alloys in Room Temperature Lithium Batteries

This review deals with solid metals in liquid organic electrolytes at room temperature, which is the common system of the currently used LIB. However, the first systematic works on lithium alloys were connected with their use in molten salt electrolytes, which operate at high temperature (~400°C). These pioneering works on Li–Al,[1–6] Li–Sn,[7–9] Li–Mg,[10] Li–Sb,[11–13] and Li–Si[14–18] served as a basis for further, still ongoing research. It was revealed that at high temperature, reconstitution reactions with the formation of respective well-defined crystal solid-state phases take place. Between the late 1970s and early 1990s much effort has been put into the replacement of the metallic lithium anode in rechargeable lithium cells by such lithium alloys.[19–29] The objective of this work was to overcome the performance and safety problems of the lithium electrode. The first use of lithium alloy anodes in commercial cells was the employment of an alloy of bismuth, lead, tin, and cadmium (Wood's metal) in button-type cells developed by Matsushita-Panasonic.[30–37] The main drawback of these metal electrodes was that reasonable cycling life could be reached only when the cycling depth of a lithium alloy anode was limited to very thin reacting layers (approximately; 10% depth of discharge). It turned out that only such a "shallow" cycling mode could keep the mechanical stresses caused by the successive alloying and dealloying sufficiently small. As a result, the specific charge density of the first commercial alloying electrodes was an order of magnitude lower than that expected for lithium-rich alloys observed at high temperatures. The first anode material that allowed deep charge–discharge cycling is coke/graphite.[38] In 1985, Asahi Chemical Industry first patented an LIB[39] commercialized by Sony in the beginning of 1990s.

After this invention, research and development focused on the new carbonaceous anodes. Lithium alloys again attracted much attention in 1995, when Fuji Photo Film Celltech Co. (Japan) announced its Stalion® lithium-ion cell,[40–43] which has a higher specific energy and energy density than conventional lithium-ion cells with carbon anodes. The Stalion® cell employed an amorphous tin-based composite oxide (TCO) as the anode material. It is formed in situ according to the following reaction scheme

$$SnO_x + 2xLi^+ + 2xe^- \rightarrow xLi_2O + Sn \text{ irreversible conversion step} \quad (11.3)$$

$$xLi_2O + Sn + yLi^+ + ye^- \rightleftarrows xLi_2O + Li_ySn \text{ reversible cycling} \quad (11.4)$$

This material also contained B_2O_3, Al_2O_3, and tin phosphates. The latter components added to the SnO precursor to form a glassy electrochemically inactive "matrix" during the initial cycle (conversion step), which acts as a supporting media for the in situ formed tin nanodomains. Thus, despite the fact that the anode contains some parasite mass (the inactive matrix), it can be deeply cycled, which results in much higher total capacity than all previously studied alloying anodes. For example, TCO exhibited a stable capacity of 475 mA h g $^{-1}$, although the theoretical value for fully lithiated SnO (i.e., up to Li_2O + $L_{4.4}Sn$) is 875.7 mA h g^{-1}. However, a common problem of all tin oxide-based materials is the severe irreversible capacity loss during the initial conversion cycle. Attaching a thin lithium-metal layer to the convertible anode may solve the problem on the expense of increased complexity of battery production.

To overcome this problem without using a metallic lithium layer, an active–active concept has been proposed. It also strove to avoid the presence of inactive solid phases that considerably reduce the overall anode capacity. Such an example is the multiphase system – Sn/SnSb – proposed by Besenhard's research group.[44] It was revealed that this alloy reacts with Li^+ according to the following scheme[45–48]

$$SnSb + 3Li^+ + 3e^- \rightleftarrows Li_3Sb + Sn \text{ reversible} \quad (11.5)$$

$$Sn + yLi^+ + ye^- \rightleftarrows Li_ySn \text{ reversible} \quad (11.6)$$

Due to different alloying potential ranges of both tin and tin-antimony, these reactions proceed sequentially. After the complete formation of Li_3Sb domains, the remaining tin domains react further with lithium, yielding various Li_ySn phases. Separation of two distinctive solid phases from the initial SnSb results in a more favorable morphology, as the original SnSb domain segregates into a nanostructured intimate mixture of Li_3Sb and Sn domains. During lithium removal, the SnSb phase is restored both in the first and in the latter cycles. It is believed that continuous phase separation and restoration during cycling could counteract an aggregation of the finely dispersed tin into large tin regions, which is considered to be the main reason for electrode failure. It also was found that the cycle life depends on the morphology and the electrode fabrication techniques. Sn/SnSb can be used as a composite powder electrode as well as in the form of directly electroplated, binder-free electrode. More than 150 cycles in the case of the electroplated multiphase hosts and more than 200 cycles in the case of the Sn/SnSb powders have been obtained at 360 mA h g $^{-1}$. However, higher charging levels were followed by a rapid decrease of the achievable cycle number, indicating that the still existing large volume changes lead to fast electrode deterioration.[44]

Last, but not least, work on tin-based alloying anodes has concentrated on materials with lower irreversible capacity losses. J. Dahn's group has performed considerable work in this direction. They investigated various intermetallic phases of tin and iron, such as Sn_2Fe, SnFe, Sn_2Fe_3, and Sn_3Fe_5.[49–54] During lithiation, the tin acts as a reactant forming Li_xSn, which is surrounded by the released elemental iron. The fine iron (~10 nm) is electrochemically inactive and acts as a matrix, supporting the intergrain electronic contact. The best results have been obtained with composites of active Sn_2Fe and almost inactive $SnFe_3C$, having grain sizes in the range of 10–20 nm. Even at deep charge–discharge cycling (up to $Li_{4.4}Sn$) the composite showed quite stable specific capacity of about 200 mA h g $^{-1}$. Only about 20% of the charging capacity in the first cycle is irreversibly lost (~50 mA h g $^{-1}$). However, this stability is obtained at the expense of severe reduction of the capacity due to the presence of a large amount of inactive buffering media. A summary of all above considered methods is presented in Table 11.1.

The possibility for successful deep cycling of anodes based on an alloying type of reaction initiated further research on metal and intermetallic anodes. Si-based anode materials were the most logical choice for further studies. The reason is that Si is 4.2 lighter than Sn (atomic weights: Si 28.086, Sn 118.690) and their alloys with maximum lithium uptake (at elevated temperatures) have the same stoichiometry: $Li_{22}M_5$. However, due to the lighter atomic weight of silicon, composite materials containing the same amount of Si instead of Sn will have a factor of 4.2 higher overall capacity for the same final Li_xM alloy stoichiometry. Thus, even small amounts of Si (~20%) in a composite material would lead to impressive results.

11.3 Mechanism of the Electrochemical Reaction Between Silicon-Containing Materials and Lithium

As mentioned above, early investigations focused on lithium–silicon alloys as anodes for molten salt electrolyte batteries, in which silicon forms well-defined lithium–silicon crystal phases.[14–18] Later, the same alloying reaction was studied in low-temperature organic solvents and it was found that upon electrochemical alloying silicon–lithium systems form a set of Zintl (i.e., amorphous) phases depending on the conditions.[55–64] Since their volume is considerably larger than that of the parent silicon phase, conventional electrodes with an active powder mass suffer from poor reversibility, typical of all alloying anodes that undergo large volumetric variations; the latter are considered to be responsible for microstructural cracks in the bulk of the silicon particles, leading to fast electrode deterioration.

Table 11.1 Overview of composite tin-based alloying anodes[a]

Anode material	Maximum lithiation	Active phase	Matrix	Capacity (mA h g $^{-1}$)	Cycle life	Ref.
Graphite (C_6)	LiC_6	C	–	372.3	Good	38
Sn	$Li_{4.4}Sn$	Sn	–	993.7	Poor	44
SnO	$Li_{4.4}Sn/Li_2O$	Sn	Li_2O	875.7	Fair	40–43
SnO_2	$Li_{4.4}Sn/2Li_2O$	Sn	Li_2O	782.7	Fair	40–43
$SnO/B_2O_3/Al_2O_3/Sn_2P_2O_7$ (TCO)	$Li_{4.4}Sn/Li_2O/B_2O_3/Sn_2P_2O_7/Al_2O_3$	Sn	$Li_2O/Sn_2P_2O_7$ B_2O_3/Al_2O_3	475.8	Good	40–43
Sn_2Fe	$2Li_{4.4}Sn/Fe$	Sn	Fe	804.5	Good	49–54
$Sn_2Fe/SnFe_3C$ (25:75 w/w)	0.25 ($Li_{4.4}SnFe$)/0.75 $SnFe_3C$	Sn	$Fe/SnFe_3C$	201.1	Good	49–54
Sn/SnSb as "$Sn_{0.88}Sb_{0.12}$"	0.88 $Li_{4.4}Sn/0.12Li_3Sb$	Sn, SnSb	Sn, Li_3Sb	953.7	Fair	44

[a] Theoretical specific capacities calculated with respect to the masses of the fresh anode materials. The practical cycle life of these composites is very sensitive to the amount and the type of conductive additives and binders, and therefore this is a very rough estimation. For details, see the references

The reaction of lithium with binary silicon alloys also was considered. Particularly Mg_2Si, which has an antifluorite structure, has been pointed out as a promising candidate, and experiments revealed that its reaction with lithium is very attractive; it has a large capacity within a useful potential range at elevated temperatures.[65–69] Experiments on the reaction of lithium with Mg_2Si at ambient temperatures also have been performed.[58] The reported capacity is over 400 mA h g $^{-1}$ with an average potential of about 0.35 V vs. lithium. A later study also showed that this material could deliver a stable capacity of 100 mA h g $^{-1}$, when cycled over a range of 50–225 mV.[61] A maximum discharge capacity of 830 mA h g $^{-1}$ can be observed by cycling over a wider potential window of 5–650 mV vs. lithium, but in the latter case, the capacity fade is rapid. The decrease in charging potential leads to the formation of binary lithium alloys and incomplete reformation of Mg_2Si. There also has been interest in amorphous negative electrode materials containing silicon, for example, silicon oxides and lithium silicate glasses and other silicon-containing oxides, such as amorphous SiO–SnO mixtures prepared by means of mechanical milling.[70–73] Netz et al. have shown that the initial lithiation of several disilicides (SiO and SiB_3) results in the formation of a lithium-containing product from which lithium can be extracted and reinserted.[62] They have observed that the compositional dependency of the potential during the first extraction and second insertion cycles and showed that all these materials have approximately the same potential profiles (Fig. 11.2). This is an indication that they have comparable thermody-

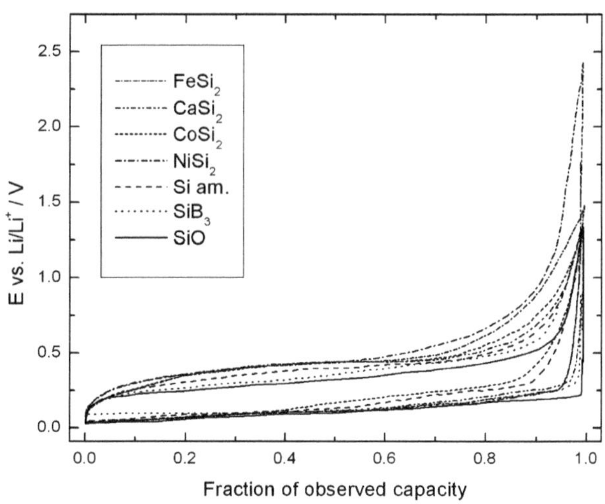

Fig. 11.2 Potential profiles recorded during the first Li^+ extraction and second Li^+ insertion processes for binary silicon containing compounds and amorphous silicon. Galvanostatic cycling performed at a current density of 0.1 mA cm^{-2} in a voltage range from 1.5 V to 25 mV. Data are plotted as a function of the fraction of observed capacities. Reprinted from Ref. 62, copyright (2003), with permission from Elsevier

Table 11.2 Electrochemical properties of the silicides given in Fig. 11.2

Precursor	Molar weight	First Li extraction (Li mol $^{-1}$)	Extraction capacity (mA h g $^{-1}$)
$CoSi_2$	115.11	0.25	58
$FeSi_2$	112.03	0.25	60
$NiSi_2$	114.87	0.85	198
$CaSi_2$	96.26	1.15	320
SiB_3	60.52	1.0	443
SiO	44.09	1.1	669
a-Silicon	28.09	1.05	1,002

namic and kinetic properties. X-ray observations showed that the products of the first Li^+ insertion of these materials are amorphous. Moreover, experiments performed with amorphous silicon show that its measured capacity and potential profile correspond quite well with the results found with other binary silicon-containing compounds after the first Li^+ insertion step. This is shown in Fig. 11.2, in which the potential profiles are normalized by their respective capacities (data obtained from Ref. 73). These similarities confirmed the assumption that these reactions proceed as electrochemically driven amorphization rather than intercalation in a well-defined crystal lattice. Therefore, at higher Li^+ insertion levels most of the electrochemically active silicides studied so far behave simply as a mixture of silicon and the corresponding metal.

Although the data presented in Fig. 11.2 are useful for a fundamental understanding of silicon-containing materials, considering the practical applications, the capacity per unit weight is the most important parameter. As can be seen clearly in Table 11.2, most of the silicides can hardly compete with modern carbonaceous anodes. Therefore, a technically interesting choice is to use composite materials based on elemental silicon or SiO.

11.4 Technical Aspects of Preparing Silicon-Based Anodes

11.4.1 The Composite Electrode Concept

In the composite electrode concept, the electrode material consists of at least two separate, intimately mixed, solid phases. Most research efforts have been directed toward preparation of composites containing small silicon particles uniformly distributed within an electrochemically active or inactive solid phase and acting as a mechanical and electrochemical buffer. The latter is necessary for preventing electrochemical sintering of the silicon particles; this takes place during the alloying step with Li^+. The fundamental reason explaining the importance of particle grain

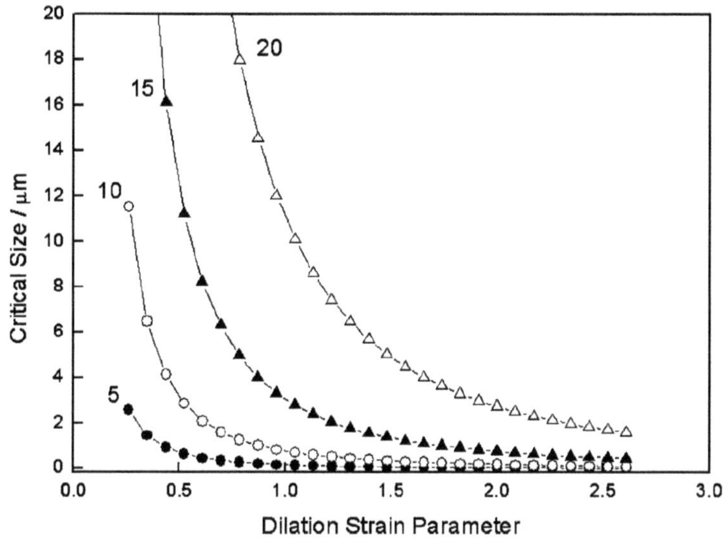

Fig. 11.3 Influence of the dilation strain parameter, upon a critical size below which a fracture will not occur (from Ref. 73). Different *numbered curves* stand for different fresh solid phases with various fracture toughness parameters. The dilation strain parameter accounts for the volume mismatch between the pristine and lithiated phases

size in the case of electrodes that undergo large volumetric variations was derived by Huggins et al.[73] Their final results are plotted in Fig. 11.3. It implies that each material has its own critical particle size, which depends on the mechanical toughness of the pristine metal phase as well as the volume mismatch between the fresh and lithiated phases.

In the case of pure silicon, it is difficult to make a reasonable estimation based only on the above theory, because there are different possible phases that might form on the surface of the silicon particles during the first Li^+ insertion. It also is possible that some of the assumptions made during derivation of the model (presented in Fig. 11.3) do not hold true in practice. Nevertheless, we believe that such a trend always exists and represents a fundamental property of such systems, and therefore it is a good starting point when designing silicon-containing composite materials.

To the best of our knowledge, the silicon grain particle size that allows preparation of morphologically stable composites lies within the range of 0.5–1.0 µm, which is about twofold higher than the typical "nano"-range, within which quantum confinement effects appear.

The literature describes numerous methods for practically synthesizing silicon-containing composite material. The difference between them is in the choice of buffering media and the method of composite formation. However, whatever the preparation method, the capacity of each composite can be easily estimated by means of (11.3). The total capacity of each silicon-containing composite material is represented by the following relationship:

$$\text{Cap(tot)[mAh g}^{-1}] = (B[\text{wt\%}]B[\text{mAh g}^{-1}] + \text{Si[wt\%] Si[mAh g-1]})/100 \quad (11.3)$$

where B stands for the buffering media used in the composite material.

On the other hand, the capacity with respect to the pure silicon phase is represented by:

$$\text{Si[cap]} = \frac{n \times 26{,}800}{28.09} (\text{mA h g}^{-1}) \quad (11.4)$$

where n stands for the number of lithium atoms in the final Li_xSi alloy, 26,800 is the Faraday divided by 3.6, and 28.09 the atomic weight of silicon. Combining (11.3) and (11.4) allows estimation of composite capacity as a function of silicon content and the mole fraction of lithium in the Li_xSi alloy after charging. Such a simulation is graphically shown in Fig. 11.4 for the cases where the buff-

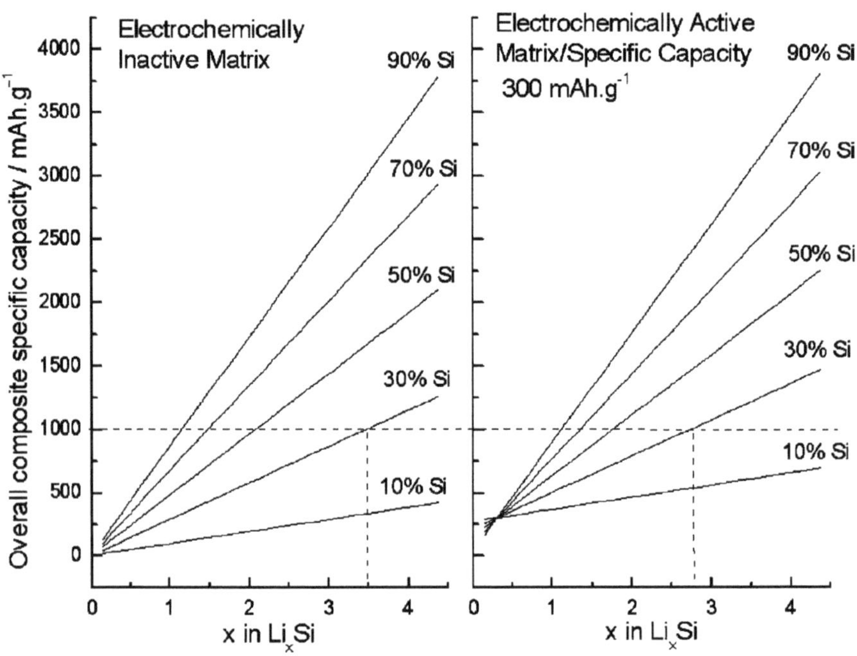

Fig. 11.4 Estimation of the theoretical capacity of Si-containing composite material as a function of the Li^+ mole fraction in the alloy Li_xSi. Amount of Si in the composite is indicated on each line. If electrochemically active matrix is used, Li_xSi alloy with lower Li mole fractions could be used in order to reach certain predefined capacity. This effect is more pronounced when the silicon content is lower as exemplified in the figure

ering media is electrochemically inactive ($B = 0$ mA h g $^{-1}$) and when it has a specific capacity of 300 mA h g^{-1}(e.g., natural/synthetic graphytes). As shown, there is a considerable difference between the cases. As a practical example, considering a composite with a desirable capacity of 1,000 mA h g $^{-1}$ and containing 30 wt% silicon, the theoretical composition of the alloy after charging should be $Li_{3.5}Si$ and $Li_{2.8}Si$, respectively. This difference cannot be neglected, since the cycling stability generally depends on the degree of silicon-phase lithiation over the first charge.

Therefore, carbonaceous phases are the logical choice of buffering media for silicon-containing composites, because they are not only electrochemically active, but also have good electrical conductivity and permeability for Li^+. Mixtures of silicon and different blends of graphite were previously studied in the author's laboratory and it was found that graphite is important in improving cycling performance.[74–77] The composites were typically prepared by means of milling the ingredients (silicon and graphite powders) for different time intervals. In addition, some samples were coated with a layer of hard carbon deposited from a gaseous phase by means of the thermal vapor deposition (TVD) technique. The volume resistance of the silicon powder was 1,500 Ω cm, while that of the carbon-coated composite was only around 100 mΩ cm. Composites containing comparatively large silicon particles (>1 μm) could cycle up to 50 cycles when the insertion capacity is limited (<800 mA h g $^{-1}$). These improved characteristics are ascribed mainly to the continuous electric networks around the silicon particles. In contrast, particulate silicon anodes, containing only silicon and PVDF, do not show considerable reversibility and deteriorate completely in just a few cycles.

Examination of the voltage profiles (Fig. 11.5) reveals that there are no well-defined plateaus, which might correspond to specific phase transformations. In situ XRD spectra of silicon-containing composite is shown in Fig. 11.6.[78]

No new Bragg reflections corresponding to lithium–silicon alloy with a new crystal structure can be observed during the initial Li^+ insertion. The most notable changes in Fig. 11.6 are the significant peak broadening and intensity reduction of (111), (220), and (311) reflections in the spectra. These data suggest that on the first insertion, the primary process is an irreversible breakdown of the crystal structure of the pristine silicon solid phase, which is in agreement with other previous studies (e.g., Li et al.[79]).

A semiquantitative analysis carried out by integrating the silicon-related peak intensities revealed that at the beginning of the first lithiation Li^+ preferentially saturated the graphite phase (intercalation), and later, silicon became involved in the reaction (alloying). However, due to the milder alloying conditions, in the presence of the carbonaceous phase, the formation of lithium–silicon alloys with high lithium content on the surface of the alloying silicon particles has been suppressed. Therefore, the carbonaceous media led to a more uniform distribution of Li^+ within the bulk of the active silicon particles.

Differential capacity plots and cyclic voltammograms (Figs. 11.7 and 11.8) provide further information for the electrochemical properties of the silicon–carbon composite. Differential capacities dQ/dE were recorded for the carbon-coated

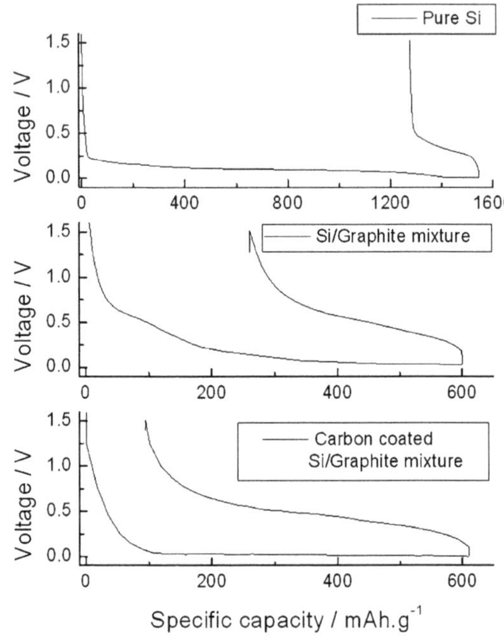

Fig. 11.5 Voltage profiles over the first cycle of pure silicon, a mixture of silicon–graphite, and a carbon-coated silicon–graphite. In all cases, the composite electrode contained 10 wt% PVDF and the silicon particle grain size was in the range of 1–5 μm. Degrees of silicon lithiation for samples presented in the figure were roughly the same and could be estimated by the simulation given in Fig. 11.4.

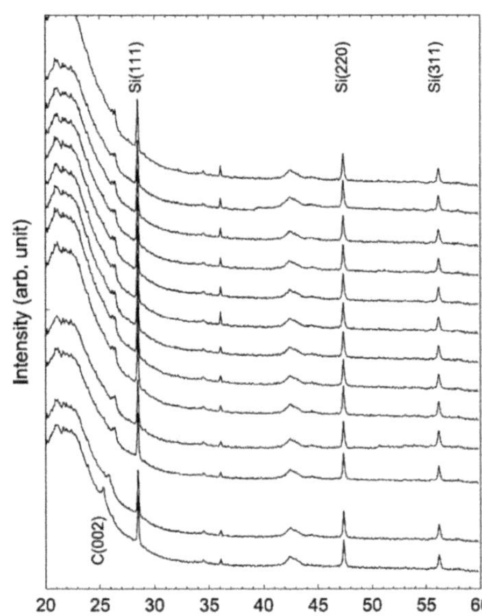

Fig. 11.6 Selected in situ synchrotron-based XRD patterns of a carbon-coated silicon electrode during the first Li^+ insertion; current rate ~50 mA g^{-1}. The 2θ angles have been converted to those corresponding to Cu Kα radiation (λ = 1.54 Å).

silicon composite electrode cycled between 1.5 V and 5 mV vs. Li/Li^+. Only extraction peaks are given in this case because during the insertion after some cycles the potential was held at 5 mV, thus excluding the possibility to define dQ/dE values. It is seen (Fig. 11.7) that during the first extraction there is only one peak

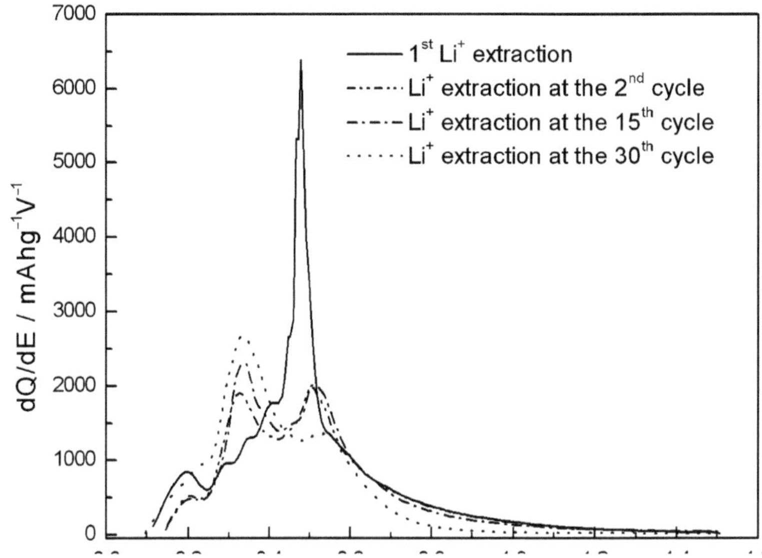

Fig. 11.7 Differential capacity plots of the carbon-coated composite silicon anode in 1-M $LiPF_6$ EC:DMC (1:2). Only extraction *peaks* are presented in the graph

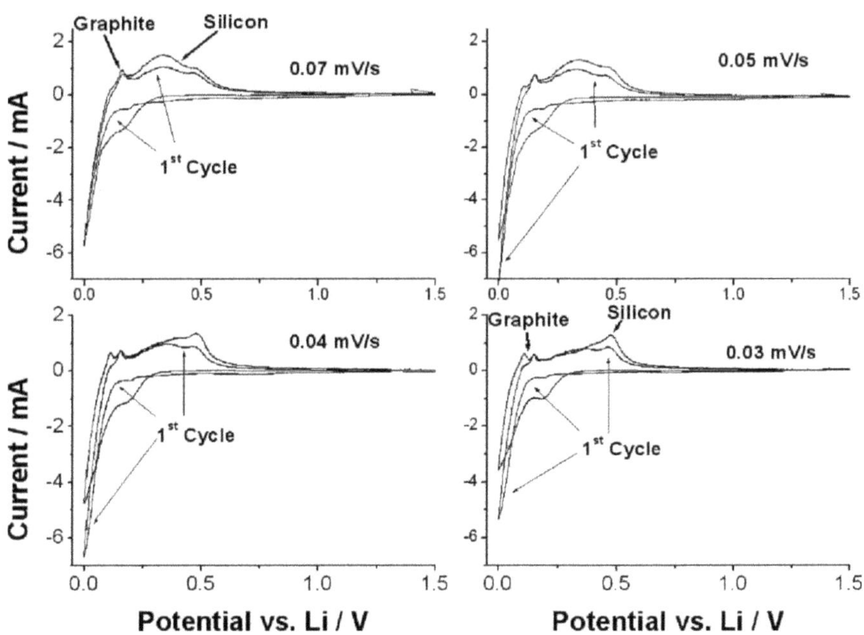

Fig. 11.8 Cyclic voltammograms of the carbon-coated silicon electrode at slow sweep rates indicated on the *curves*. Data recorded by means of three-electrode cell with lithium used as a reference and counter-electrode in 1-M $LiPF_6$ dissolved in EC:DMC (1:2)

at 480 mV. From the second cycle, there are two peaks located at ca. 380 and 510 mV, roughly corresponding to the cyclic voltammetry (CV) data. While CV gives the transient response of the electrode at a low sweep rate, the differential capacity was directly derived from the cycling test at a high current density (1.5 mA cm^{-2}or approximately 0.35mA $mg^{1)}$ and is useful to monitor the changes of the electrode with the cycles. Increase in the cycle number leads to a gradual decrease of peak area (extraction capacity) (11.7). The peak area ratio also changes. The area of the peak located at 380 mV increases while the area of the peak located at 510 mV decreases with cycles. The peak located at 510 mV can be assigned to the lithium–silicon dealloying reaction. Since pure carbon component normally shows a peak below 200 mV, the peak at 380 mV is considered to be a contribution from the carbon and silicon components of the anode.

The CV plots show a typical transient response, which is in accordance with the profiles shown in Fig. 11.5. As seen from the figure, the relative contribution of the silicon phase increases during the second scan.

Having all above data in mind, the following likely mechanism could be summarized. During the first insertion, Li^+ preferentially enters the carbon component. After its saturation, the electrochemical alloying begins with the silicon component, thus leading to the formation of amorphous lithium–silicon alloy. At the beginning of the first extraction, there are two coexisting solid phases: Li^+-saturated carbon component and freshly formed amorphous lithium–silicon alloy. From the second cycle, Li^+ insertion goes concurrently into the carbon and amorphous silicon component. In terms of the anode material, this results in different spatial distribution of Li^+ between silicon and carbon components during the first and subsequent cycles, seen as different differential capacity and CV peak patterns. Such electrochemical behavior is characteristic for most silicon-containing composites, although they may contain different components as summarized in Table 11.3.

A similar result was obtained by Liu et al.[80] In this case, inclusion of a high-energy mechanical milling (HEMM) step between the two thermal pyrolysis reactions of PVDF resulted in a silicon/disordered carbon composite, in which the active silicon "cores" were homogeneously distributed within the pyrolyzed carbonaceous matrix. The composite offered a reversible capacity of ca. 900 mA h g^{-1} within 40 cycles and a relatively high coulombic efficiency of 80% over the initial cycle.

It is notable that in most of the references given at the end of this chapter, the authors emphasized that the further increase of the cycling performance of the silicon-containing composites is closely related to the decrease of the active silicon particle grain size down to nanometer level. However, it is not possible to perform this by means of simple mechanical milling. Moreover, methods described in the literature, such as laser ablation[79] for silicon nanoparticle preparation, are not suitable for practical applications. In addition, pure silicon nanopowder is unstable in air, because the amount of oxide in powders with comparatively large grain size (in the order of few micrometers) is negligible; however, the oxide layer cannot be neglected in the case of nanoscale powders. Such powders also have low tab density. Therefore, an appropriate method that relies on the nanoconcept should embody preparation of silicon nanoclusters directly embedded within the matrix.

Table 11.3 Overview of composite silicon-based alloying anodes[a]

Anode material	Maximum lithiation	Active phase	Matrix	Capacity (mA h g $^{-1}$)	Cycle life	Ref.
Graphite	LiC_6	C	–	372.3	Good	38
Si	$Li_{4.4}Si$	Si	–	~4,100	Poor	79
SiO	$Li_{1.7}Si/Li_2O$	Si	Li_2O	~500	Satisfactory	81,82
$Si/Al_2O_3/C$	–	Si	Al_2O_3/C	~550	Satisfactory	86
$Si/SiO_2/C$	–	Si,C	SiO_2/C	~700	Satisfactory	85
Si/TiN	~LiSi	Si	TiN	~300	Satisfactory	105
Ni(Fe)Si	–	Si	Ni(Fe)	600–1,000	Poor	95
Si/TiC	–	Si	TiC	~350	Poor	103
Si/TiB_2	–	Si	TiB_2	~500	Poor	102
$Li_2O/Al_xO/Si$	–	Si	Li_2O/Al_xO		Poor[b]	87
"$Ni_{20}Si_{80+}$"/C	–	Si,C	$NiSi_2/C$	~550	Satisfactory	104
Si/C	–	Si,C	C	500–900	Satisfactory[c]	74–78

[a] Theoretical specific capacities calculated with respect to the masses of the fresh composite materials. Li_xSi composition omitted for some of the samples, because the exact composite content was not given. For details, see the references

[b] Poor cycle life of this sample probably is due to the presence of residual amounts of $LiAlH_4$ and/or LiH, which are not compatible with the electrolyte

[c] As discussed in the text, capacity and cycle life particularly of silicon–carbon composites depend on both silicon content and degree of lithiation of the silicon phase during the first cycle

Such a method would be applicable without the need for isolation of silicon nanoparticles as a separate solid phase.

The simplest example of direct silicon nanocluster formation is the electrochemical reaction of SiO_x with Li^+, which proceeds in much the same way as tin convertible oxides (TCOs). Li_2O formed in situ is known to be an appropriate inactive matrix for accommodating active metal particles, and it is a good ionic conductor for Li^+. However, "convertible silicon oxides" have worse kinetic properties, because silicon is a semiconductor and the bulk conductivity of the resulting nanocomposite is several orders of magnitude less than in case of SnO_x

$$SiO + 2Li^+ + 2e \rightarrow Li_2O + Si \tag{11.5}$$

Nevertheless, in the presence of an appropriate additive, the electric conductivity can be enhanced and the corresponding material employing the Li_2O matrix might exhibit an attractive energy density and satisfactory rate performance. For example, Yang et al.[81] succeeded in cycling a composite consisting of SiO_x ($0.9 < x < 1.1$), with a grain size of 50 nm or less, a large specific surface area, and short ion diffusion path. In this case, however, the large irreversible capacity loss over the first cycle remained. One possible approach for avoiding this limitation is to use a mixture of SiO_x and Li^+-supplying material, consumed over the first cycle. An addition of a calculated amount of $Li_{2.6}Co_{0.4}N$ or stabilized lithium powder, which has been developed recently, might be an elegant solution to this problem and a new engineering approach to developing lithium ion cells.[82–84]

The remaining methods that appear in the literature focus on the idea of preparing composites that resemble the structure of TCOs with the conversion step performed ex situ. The advantage of this method is that the properties of such composites can be tailored, and in addition the initial reversible capacity loss can be greatly reduced.

The following describes several approaches for creating nanodispersed silicon composites. Although this list is not exhaustive, it gives a rough idea of the current trends in this rapidly changing field. Morita and Takami proposed an original preparation method for the material embedding nanosilicon clusters.[85] They employed a disproportionate reaction of SiO. Fine SiO powder was initially mixed with furfuryl alcohol and graphite. The alcohol was polymerized and the polymeric matrix was decomposed at 1,000°C for 3 h under argon atmosphere. The final product consisted of a SiO_2–C matrix, with silicon nanoclusters (ca. 20 nm) dispersed within it. It showed an initial capacity of 700 mA h g $^{-1}$, 80% of the capacity was retained at the 200th and 100th cycles when cycled vs. lithium metal and $LiCoO_2$, respectively. The electrochemical reaction in this case can be expressed as

$$(x+y)\mathrm{Li}^+ + \mathrm{Si}-\mathrm{SiO}_2-\mathrm{C}+(x+y)\mathrm{e}^- \rightleftarrows \mathrm{Li}_x\mathrm{Si}+\mathrm{SiO}_2-\mathrm{Li}_y\mathrm{C} \qquad (11.6)$$

The cycling stability was attributed to the nanosilicon clusters embedded in the SiO_2 matrix.

Another example for ex situ silicon nanocluster formation was proposed by Lee et al.[86]; here the authors used a mechanochemically driven reaction

$$\mathrm{SiO}_x + \mathrm{Al} \rightarrow \mathrm{Si} + \mathrm{AlO}_x \qquad (11.7)$$

Since the activation energy is caused by mechanical stress, it is distributed irregularly into small domains rather than uniformly as in the case of thermal activation, which prevents the growth of large silicon domains. The resulting material subsequently was carbon-coated and a final product with a stable reversible capacity of ca. 550 mA h g $^{-1}$ was obtained.

Another interesting high-temperature, hydrogen-driven, solid-state reaction was proposed by Reilly et al.,[87] who employed $LiAlH_4$ as a reducing agent, added in excess to the silicon oxide precursor. The following set of reactions was considered to occur

$$6\mathrm{LiAlH}_4 + \mathrm{SiO}_2 \rightarrow 6\mathrm{LiAlH}_2 + 6\mathrm{H}_2 + \mathrm{SiO}_2(\approx 150^\circ\mathrm{C}) \qquad (11.8)$$

$$6\mathrm{LiAlH}_2 + \mathrm{SiO}_2 \rightarrow 6\mathrm{LiH} + 6\mathrm{Al} + \mathrm{SiO}_2 + 3\mathrm{H}_2(\approx 300^\circ\mathrm{C}) \qquad (11.9)$$

$$6\mathrm{LiH} + \mathrm{SiO}_2 + 6\mathrm{Al} \rightarrow \mathrm{Li}_x\ \text{nanocomposite} + 3\mathrm{H}_2(\approx 500^\circ\mathrm{C}) \qquad (11.10)$$

$$x\mathrm{LiH} + \text{nanocomposite} \leftrightarrow \mathrm{Li}_x\ \text{nanocomposite} + x/2\,\mathrm{H}_2(\approx 500^\circ\mathrm{C}) \qquad (11.11)$$

Thermal treatment of the mixture resulted in formation of a nanocomposite material made of tiny grains with a diameter of ca. 20 nm. Hydrogen was considered to be the key in forming the nanocomposite material in this reaction. The hydrogen atoms were successively absorbed and desorbed from the lithium–silicon alloy many times, which makes the method more effective than physical grinding alone. Although the so-formed nanocomposite showed rapid capacity decay, this method is notable because nanocompsite formation was induced while avoiding HEMM. However, a practical obstacle of this method is the fact that residual hydride is a strong condensing agent for compounds, containing a carbonyl group; that is, most common electrolytes used in LIBs now. Therefore, composites prepared using LiH or its derivatives could be used if a suitable method is developed for removing traces of the residual hydride trapped within the composite.

There are indeed some other methods for producing silicon-based nanocomposites that appeared in the literature. Considerable work has been performed by J. Dahn's group and the decomposition of silane or polysilane pitch[88–93] within carbonaceous matrices has been widely explored. Other methods based on HEMM of electrochemically inactive phases such as iron, nickel, titanium–nickel, titanium–carbon, silicon–carbon, and so on also have been examined[94–106] using silicon powder with an initial particle grain size within either micrometric or nanometric ranges. The common problem in all these cases is that it is difficult to predict theoretically the appropriate matrix/silicon particles combination, i.e., there is no simple guidance rule in order to make reasonable predictions. An overview of all the above-described methods is summarized in Table 11.3.

The discussion so far has focused on the material properties only, neglecting the effect of the binding system. However, in the case of alloying elements, this factor seems to play a crucial role. Although it is not easy or safe to generalize in this rapidly changing field, one might expect there to be a trade-off between the composite material properties, particularly the active particle grain size, and binder properties as schematically shown in Fig. 11.9.

At this point the fundamental difference between intercalation and alloying type materials becomes obvious. With the intercalation hosts, studied over the past decades, the researchers simply "follow" nature; that is, the crystal lattice that is slightly modified in order to make it more favorable with respect to particular properties of interest,

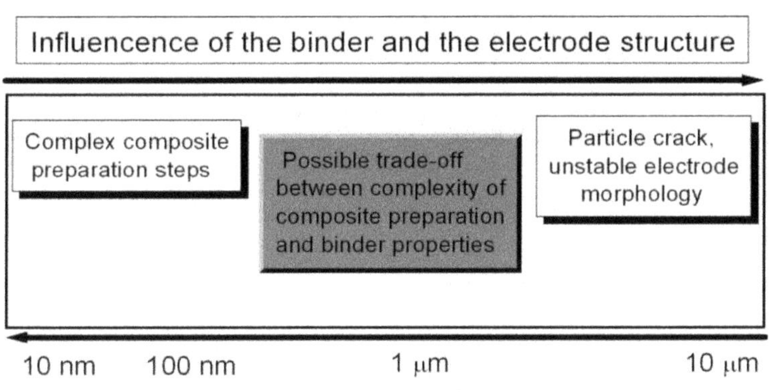

Fig. 11.9 Composites, containing Si particles larger than ca. 2–3 μm tend to deteriorate quickly. A possible trade-off is expected at the particle grain size of about 0.5–1.0 μm: an order of magnitude larger size than the typical nanorange, which is not easy to reach by means of mechanical milling

for example, rate performance, cycle life, capacity, and so forth. In other words, there is an "obvious" strategy followed by the researchers dealing with practically all types of intercalation hosts. All studies dealt with the crystal structure modification, particularly aiming to make it stable and at the same time able to accommodate as much as possible Li^+ with minimal volumetric variations. In the case of carbonaceous hosts, the main focus was making the crystal structure stable by means of protective layers (against solvent cointercalation) formed both in situ or ex situ, i.e., the main objective was focused again on the crystal structure and its stability with the cycles.

Unlike intercalation hosts, alloying type materials always exhibit large volumetric variations. Therefore, they are intrinsically unstable and the research is performed "against" nature, i.e., the electrode should be stable despite the unavoidable volumetric variations. Therefore, in this case there is no "obvious" strategy to follow, except for keeping the active "cores" as small as possible as explained above (cf Fig. 11.3).

Actually, all the methods described above for tailoring the properties of silicon-containing composites, particularly by sizing down the active cores, were performed in order to mimic the classic intercalation hosts, despite the fact that active working materials are not based on intercalation but on the alloying type of the reaction. The latter allows application of the classic electrode fabrication techniques currently adopted in the manufacturing of LIBs. However, as could be seen, this active particle sizing down is performed at the expense of increased complexity of preparation, which is not favorable regarding the practical applications. The recent trend in this field of study therefore is closely related to the development of new binding materials. Chen et al., who used a novel elastomeric binder,[107,108] achieved substantial improvement in the cycling performance of a-$Si_{0.64}Sn_{0.36}$ electrode. Similarly, the cycle performance of carbon-coated silicon samples has improved when an elastomeric SBR-CMC binder was used.[109]

It is interesting to note that both the aforementioned materials have different properties. The amorphous a-$Si_{0.64}Sn_{0.36}$ contracts and expands reversibly, while the reaction of silicon powder is irreversible with initial destruction of the crystal lattice as explained above. Despite of this difference, cycling performance appears to be very sensitive to the binding system. These results suggest that the electrochemical mechanism itself is not the main factor that should be considered when designing anodes with large volumetric variations.

One might suspect that the mechanical and morphological stability of the electrode would play a key role in the further improvement of such anodes, and since this research is still at the early stages, further substantial improvement can be expected.

11.4.2 The Thin-Film Concept

A thin film might be prepared by means of either ion sputtering[110–123] or vacuum deposition[124–130] techniques. In both cases, the thin-film electrode is binder free; that is, there is no parasite mass lowering its capacity. Moreover, the silicon layer is strongly attached to the copper substrate.

It is believed that the thin-film-forming method results in good adhesion between the silicon film and current collector due to the existence of an intermediate layer where the crystal lattices of the bulk copper and silicon phases interpenetrate. Such electrodes have shown remarkable cycling performance, which might be considered sufficient for practical LIB. An important feature of these films with excellent cycling performance is their self-organized structure. This structure can be formed on a specially roughened surface, which results in a higher contact area between both solid phases. During the first cycle of these electrodes, a columnar structure is formed as shown in Fig. 11.10.

Over the subsequent Li^+ insertion/extraction cycles, each "column" remains attached to the current collector and at the same time does not crack along its vertical axis, which explains the excellent cycling behavior of these electrodes. It also is possible to produce silicon-metal multilayer structures by means of sequential sputtering of silicon and other metals such as cobalt, iron, and zirconium.[131–133] However, such thin multilayer electrodes might be useful only in thin-film microbatteries, because the thickness of these films is in the order of 120 nm, which means a capacity of about 0.05 mA h cm^{-2}. Consequently, the working current density is around 30 μA cm^{-2}.

Despite the attractive features of thin-silicon-film electrodes, to the best of our knowledge, they are yet to be commercialized. The following reasons might be a partial explanation why:

1. The film-forming method is not commonly used in industries. Practical scaling-up therefore might not be commercially feasible, except for special applications where the cost is not a limitation.
2. The kinetic properties (power density) of these films are very sensitive to thickness.
3. Cycling life is also a function of electrode thickness and films thicker than ca. 5 μm tend to deteriorate.

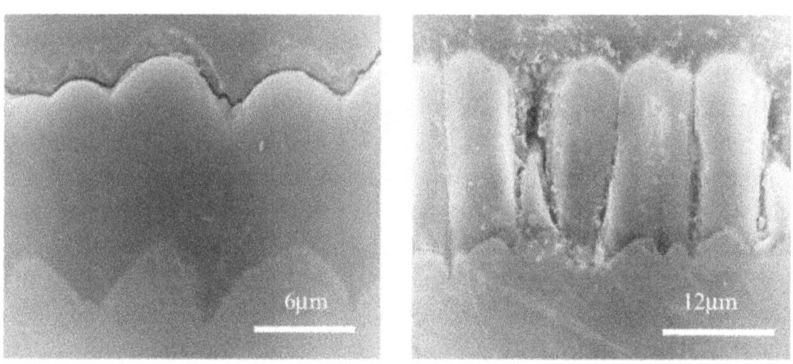

Fig. 11.10 Structure of a fresh thin silicon film (*left*) and after cycling (*right*). Note the roughened surface of the current collector, which causes macroscopic anisotropy of the film structure. The latter is considered the reason for the concentration of stress in the valleys of the irregularities, producing a stable structure over the first expansion

4. The thin silicon film electrodes are an order of magnitude thinner than cathodes, which might cause technological problems.[134]
5. Solid electrolyte interface (SEI) film formation on the moving phase boundary has yet to be studied in detail[135] and might highlight potential complications for real LIB employing such electrodes, especially during long-term cycling.

Nevertheless, the above results suggest that the alloying reaction itself is sufficiently reversible for commercialization, provided the electrode remains mechanically stable during operation.

11.5 Silicon–Carbon Composite Electrode

The preceding paragraphs reveal that creating reversible electrodes based on dimensionally unstable alloying components is not an easy task. In order to understand the fundamental reason causing these difficulties, it is essential to emphasize once more the difference between dimensionally stable (intercalation) and dimensionally unstable (alloying) lithium hosts. In both cases research strategies are likely to be different.

Dimensionally stable intercalation hosts allow easy electrode preparation with predictable properties that completely rely on the properties of their crystal lattices. Crystal lattice and its appropriate modification in this case is the natural and reliable starting point of the research.

Volumetrically unstable hosts represent the next level of complexity where entirely new properties appear, and the understanding of new behaviors requires new research strategies. In fact, integral properties of the entire electrode layer in this case become dominant. Actually this is a situation where knowledge of the underlying physical mechanism responsible for the life cycle of the electrode is partly known or missing. In such a complex situation it is almost impossible to find *the true model*, which would be able to predict their properties in detail. When there are a large number of potentially influential variables, the initial experiments therefore should be used to identify the most important variables. For example, cycle life of composite silicon-based electrodes is known to depend on many process variables including the amount of silicon in the composition and its particle grain size, binder properties, and electrode fabrication conditions. The goal of the study in this case is to identify the most important parameters as well as their interactions, which are responsible for the cycle performance of the electrode. To perform this task, two-level factorial designs are extremely useful. These designs are also known as screening designs. Such a design recently applied to composite silicon–carbon electrode directly revealed the importance of composite electrode structure and binder properties on electrode life cycle.[136] It also showed that a simple variation of fabrication properties such as slurry dry content and electrode drying temperature do not influence considerably the cycle performance of the composite silicon–carbon electrode.[136] Therefore, the electrode structure and its mechanical properties should be tailored in order to make its cycle life sufficiently long. Taking into account these factors allowed preparation of

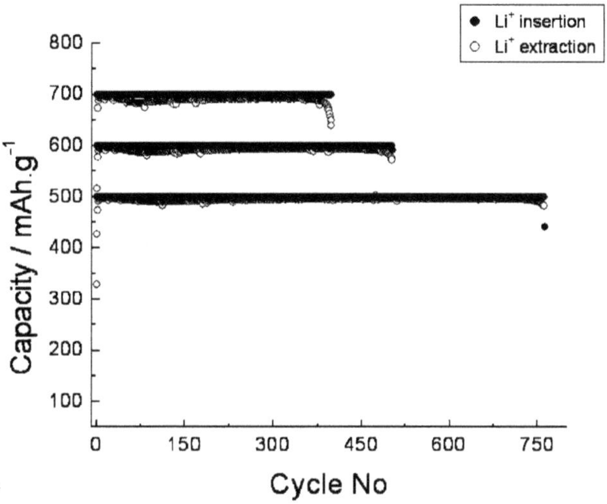

Fig. 11.11 Cycle life of porous composite electrode consisting of *p*-Si (1 μm) + MCMB 6–28 (1:1 by weight). Cycling performed at constant capacities 500, 600, and 700 mA h g^{-1}. Load density approximately 1 mA cm^{-2}, amount of the binder (SBR + CMC mass) 16 wt%, Coulombic efficiency over the first cycle around 70%, for the next cycles 98–99% vs. lithium metal, electrolyte 1-M $LiPF_6$ in MEC/EC 3:7

electrodes with a life cycle in the order of several hundreds of cycles, as shown in Fig. 11.11. Porous electrode structures containing large amounts of elastomeric styrene–butadiene rubber (SBR) copolymer mixed with sodium carboxymethyl cellulose (CMC) binder was created. No special efforts were made to improve the efficiency over the first cycle. Although the load density (approximately 1 mg cm^{-2}) in this case is about half of that required for practical LIB, it is expected that it can be doubled or tripled, since appropriate electrode structure and binder can buffer volume variations more effectively than in the case of thin-silicon films.

Finally, it should be noted that cycle life and reversibility of these electrodes may depend strongly on the properties of the electrolyte. Since no special efforts have been put into pursuing this direction, further substantial improvement may be possible.

11.6 Summary and Outlook

This short review focused on the fundamental aspects of electrochemical reaction between silicon and silicon-containing materials with Li^+. It was shown that crystal structure preservation is not mandatory to obtain a high capacity.

The main approaches used to date for investigating silicon-based anodes were summarized; that is, the composite approach and thin-film approach. The composite approach relies on the use of silicon in the form of tiny silicon "cores," uniformly distributed within an electrochemically active or inactive buffering media, which prevents silicon-silicon interaction and the formation of large clusters. To design such composites one should consider many factors and account for their spatial interaction. It seems that the critical silicon particle size that allows the formation of morphologically stable silicon-containing composites lies within the range of 0.5–1.0 μm, which allows the use of comparatively simple preparation techniques.

The idea of the thin-film approach is to deposit an active silicon layer as a continuous media on a specially roughened metal current collector. The film formation method, which results in strong adhesion between the current collector and silicon layer, leads to the formation of a stable, self-organized columnar structure during the first volumetric expansion. This structure remains stable over cycles, which is the reason for the remarkable cycling stability of such films. However, in spite of numerous research efforts, details on the SEI composition and growth kinetics on the moving-phase boundary remain controversial. Moreover, the film preparation method does not seem to be easily scalable, and therefore the use of such films might be limited to cases where price is not a limiting factor, such as military or spacecraft applications.

References

1. C.E. Johnson, M.S. Foster, *J. Electrochem. Soc.*, **116**, 1612 (1969).
2. N.P. Yao, L.A. Heredy, R.C. Saunders, *J. Electrochem. Soc.*, **118**, 1039 (1971).
3. E.C. Gay, D.R. Vissers, F.J. Martino, K.E. Andersen, *J. Electrochem. Soc.*, **124**, 1160 (1977).
4. C.A. Melendres, *J. Electrochem. Soc.*, **124**, 651 (1969).
5. J.R. Selman, D.K. DeNuccio, C.J. Sy, R.K. Steunenberg, *J. Electrochem. Soc.*, **124**, 1161 (1977).
6. C.A. Melendres, C.C. Sy, *J. Electrochem. Soc.*, **125**, 727 (1978).
7. M.S. Foster, C.E. Crouthamel, S.E. Wood, *J. Phys. Chem.*, **70**, 3042 (1966).
8. C.J. Wen, R.A. Huggins, *J. Solid State Chem.*, **35**, 376 (1980).
9. C.J. Wen, R.A. Huggins, *J. Electrochem. Soc.*, **128**, 1181 (1981).
10. C.S. TedmonJr., W.C. Hagel, *J. Electrochem. Soc.*, **115**, 151 (1968).
11. W. Weppner, R.A. Huggins, *J. Electrochem. Soc.*, **124**, 1569 (1977).
12. W. Weppner, R.A. Huggins, *J. Solid State Chem.*, **22**, 297 (1977).
13. W. Weppner, R.A. Huggins, *J. Electrochem. Soc.*, **125**, 7 (1978).
14. S. Lai, *J. Electrochem. Soc.*, **123**, 1196 (1976).
15. R.A. Sharma, R.N. Seerfurth, *J. Electrochem. Soc.*, **123**, 1763 (1976).
16. R.N. Seefurth, R.A. Sharma, *J. Electrochem. Soc.*, **124**, 1207 (1977).
17. C.J. Wen, R.A. Huggins, *J. Solid State Chem.*, **37**, 271 (1981).
18. G. Boukamp, G. Lash, R. Huggins, *J. Electrochem. Soc.*, **128**, 725 (1981).
19. M. Winter, J.O. Besenhard, M.E. Spahr, P. Novak, *Adv. Mater.*, **10**, 725 (1998).
20. M. Winter, J.O. Besenhard, in: M. Wakihara, O. Yamamoto (Eds.), *Lithium Ion Batteries*, Kodansha/Wiley, Tokyo/Weinheim (1998).
21. M. Winter, J.O. Besenhard, in: J.O. Besenhard (Ed.), *Handbook of Battery Materials*, Wiley, Weinheim (1999).
22. R.A. Huggins, *J. Power Sources*, **22**, 341 (1988).
23. R.A. Huggins, *J. Power Sources*, **26**, 109 (1989).
24. R.A. Huggins, in: B. Scrosati, A. Magistris, C.M. Mari, G. Mariotto (Eds.), *Fast Ion Transport in Solids*, Kluwer Dordrecht, The Netherlands, (1993).
25. D. Fateaux, R. Koksbang, *J. Appl. Electrochem.*, **23**, 1 (1993).

26. R. Yazami, Z.A. Munshi, in: M.Z.A. Munshi (Ed.), *Handbook of Solid-State Batteries and Capacitors*, World Scientific, Singapore, p. 425 (1995).
27. J.O. Besenhard, *Materials science volumes*, in:J. Rouxel, M. Tournoux, R. Brec (Eds.), *Soft Chemistry Routes to New Materials*, Trans-Tech Publications, Switzerland, p. 13 (1994).Vols. 152–153
28. J.O. Besenhard, in: W. Muller-Warmuth, R. Schollhorn (Eds.), *Progress in Intercalation Research*, Kluwer, Dordrecht, The Netherlands, p. 47 (1994).
29. D. Rahner, S. Machill, K. Siury, *Solid State Ionics*, , **86–88** 925 (1996).
30. Y. Toyoguchi, S. Nankai, J. Yamaura, T. Matsui, T. Ijima, *The 24th Battery Symposium in Japan, Extended Abstracts*, Osaka, p. 205 (1983).
31. Y. Toyoguchi, J. Yamaura, T. Matsui, T. Ijima, *Third International Meeting on Lithium Batteries*, Extended Abstracts, Kyoto, p. 113 (1986).
32. Y. Toyoguchi, J. Yamaura, T. Matsui, T. Ijima, J.-P. Gabano, Z. Takehara, P. Bro (Eds.), *Primary and Secondary Ambient Temperature Lithium Batteries*, PV88–6, The Electrochemical Society, Pennington, NJ, p. 659 (1988).
33. JEC Battery Newsletter, **3**, 4 (1989).
34. K. Nishio, N. Furukawa, in: J.O. Besenhard (Ed.), *Handbook of Battery Materials*, Wiley, Weinheim, Part I, Chap. 2 (1999).
35. J. Yamaki, S. Tobishima, in: J.O. Besenhard (Ed.), *Handbook of Battery Materials*, Wiley, Weinheim, Part III, Chap. 3 (1999).
36. T. Nagaura, K. Tozawa, *Prog. Batt. Solar Cells*, **9**, 209 (1990).
37. T. Nagaura, *Prog. Batt. Solar Cells*, **10**, 218 (1991).
38. H. Ikeda, Narukawa, H. Nakashima, Japanese Patent 1769661 (June 18, 1981), Sanyo, Japan.
39. A. Yoshino, K. Jitsuchika, T. Nakashima, Japanese Patent H4–24831 (May 10, 1985), Asahi Chemical Industry, Osaka.
40. Y. Idota, US Patent, 5478671 (1995).
41. H. Tomyama, Japanese Patent, 07–029608 (1995).
42. *Nippon Denki Shimbun*, March 11 (1996).
43. Y. Idota, T. Kubota, A. Matsufuji, Y. Maekawa, T. Miyasaka, *Science*, **276**, 1395 (1997).
44. M. Winter, J.O. Besenhard, *Electrochim. Acta*, **45**, 31 (1999).
45. J. Yang, M. Winter, J.O. Besenhard, *Solid State Ionics*, **90**, 281 (1996).
46. J.O. Besenhard, J. Yang, M. Winter, *J. Power Sources*, **68**, 87 (1997).
47. M. Winter, J.O. Besenhard, J.H. Albering, J. Yang, M. Watchler, *Prog. Batt. Batt. Mater*, **17**, 208 (1997).
48. J.O. Besenhard, M. Watchler, M. Winter, J. Yang, J. Albering, *Small Particle Size Li-alloy Anodes Li-alloy Anodes for Lithium Ion Batteries*, The First Hawaii Battery Conference, Big Island of Hawaii, January 5–7, (1998).
49. O. Mao, R.L. Turner, I.A. Courtney, B.D. Fredericksen, M.I. Buckett, L.J. Krause, R. Dahn, *Electrochem. Solid State Lett.*, **2**, 3 (1999).
50. O. Mao, R.A. Dunlap, I.A. Courtney, J.R. Dahn, *J. Electrochem. Soc.*, **145**, 4195 (1998).
51. O. Mao, R.A. Dunlap, J.R. Dahn, *J. Electrochem. Soc.*, **146**, 405 (1999).
52. O. Mao, J.R. Dahn, *J. Electrochem. Soc.*, **146**, 414 (1999).
53. O. Mao, J.R. Dahn, *J. Electrochem. Soc.*, **146**, 423 (1999).
54. O. Mao, R.A. Dunlap, J.R. Dahn, *Solid State Ionics*, **118**, 99 (1999).
55. W. Weidanz, P.M. Wilde, M. Wohlfahrt-Mehrens, R. Oesten, R.A. Huggins, in: W.A. Adams, A.R. Landgrebe, B. Scrosati (Eds.), Proceedings of the Symposium on Exploratory Research and Development of Batteries for Electric and Hybrid Vehicles, *The Electrochemical Society Proceedings*, Vol. 96-14, p. 223 (1996).
56. J.O. Besenhard, J. Yang, and M. Winter, *J. Power Sources*, **68**, 87 (1997).
57. R.A. Huggins, *J. Power Sources*, **81**, 13 (1999).
58. W. Weidanz, M. Wohlfahrt-Mehrens, R.A. Huggins, *J. Power Sources*, **81**, 237 (1999).
59. A. Netz, R.A. Huggins, W. Weppner, *Ionics*, **7**, 433, (2001).
60. R.A. Huggins, *Solid State Ionics*, **152**, 61 (2002).

61. G.A. Roberts, E.J. Cairns, J.A. Reimer, *J. Power Sources*, **110**, 424 (2002).
62. A. Netz, R.A. Huggins, W. Weppner, *J. Power Sources*, **119**, 95 (2003).
63. P. Limthongkul, Y.-I. Jang, N.J. Dudney, Y.-M. Chiang, *J. Power Sources*, **119**, 604 (2003).
64. T.S. Hatchard, J.R. Dahn, *J. Electrochem. Soc.*, **151**(6), A838 (2004).
65. A. Anani, R.A. Huggins, in: J.-P. Gabano, Z. Takehara, P. Bro (Eds.), Proceedings of the Symposium on Primary and Secondary Ambient Temperature Lithium Batteries, *Electrochemical Society*, Vol. 88-6, p. 635 (1988).
66. A. Anani, R.A. Huggins, *J. Power Sources*, **38**, 351 (1992).
67. A. Anani, R.A. Huggins, *J. Power Sources*, **38**, 363 (1992).
68. R.A. Huggins, A. Anani, US Patent 4,950,566 (21 August, 1990).
69. C.-K. Huang, B.V. Ratnakumar, S. Surampudi, G. Halpert, in: S. Megahed, B.M. Barnett, L. Xie (Eds.), Rechargeable Lithium and Lithium-Ion Batteries, *The Electrochemical Society Proceedings*, Vol. 94-28, p. 361 (1994).
70. K. Tahara, H. Ishikawa, F. Iwasaki, S. Yahagi, A. Sakata, T. Sakai, European Patent 582173 A1 93111938.2 (1994).
71. H. Morimoto, M. Tatsumisago, T. Minami, *Electrochem. Solid State Lett.*, A16 (2001).
72. H. Huang, E.M. Kelder, L. Chen, J. Schoonman, *J. Power Sources*, **81**, 362 (1999).
73. R.A. Huggins, W.D. Nix, *Ionics*, **6**, 57 (2000).
74. M. Yoshio, H. Wang, K. Fukuda, T. Umeno, N. Dimov, Z. Ogumi, *J. Electrochem. Soc.*, **149**, A1598 (2002).
75. N. Dimov, K. Fukuda, T. Umeno, S. Kugino, M. Yoshio, *J. Power Sources*, **114**, 88 (2003).
76. N. Dimov, S. Kugino, M. Yoshio, *Electrochim. Acta*, **48**, 1579 (2003).
77. N. Dimov, S. Kugino, M. Yoshio, *J. Power Sources*, **136**, 108 (2004).
78. X.-Q. Yang, J. McBreen, W.-S. Yoon, M. Yoshio, H. Wang, K. Fukuda, T. Umeno, *Electrochem. Commun.*, **4**, 893 (2002).
79. H. Li, X. Huang, L. Chen, G. Zhou, Z. Zhang, D. Yu, Y.J. Mo, N. Pei, *Solid State Ionics*, **135**, 181 (2000).
80. Y. Liu, K. Hanai, J. Yang, N. Imanishi, A. Hirano, Y. Takeda, *Solid State Ionics*, **168**, 61 (2004).
81. J. Yang, Y. Takeda, N. Imanishi, C. Capiglia, J.Y. Xie, O. Yamamoto, *Solid State Ionics*, **152**, 125 (2002).
82. J. Yang, Y. Takeda, N. Imanishi, O. Yamamoto, *Electrochim. Acta*, **46**, 2659 (2001).
83. C.R. Jarvis, M.J. Lain, Y. Gao, M. Yakovleva, 12th International Meeting on Lithium Batteries, Nara, Japan, June 27-Luly 2, 2004,Abs#182.
84. R.C. Morrison et al.,US Patent 5567474, US Patent 5776369, and US Patent 5976407.
85. T. Morita, N. Takami in 206th ECS meeting, Hawaii, October 3–8 2004,, Abs#312.
86. H.-Y. Lee, S.-M. Lee, *Electrochem. Commun.*, **6**, 465 (2004).
87. J.J. Reilly, J.R. Johnson, T. Vogt, G.D. dzic, Y. Zhu, J. Mc Breen, *J. Electrochem. Soc.*, **148**(6), A636 (2001).
88. A.M. Wilson, J.N. Reimers, E.W. Fuller, J.R. Dahn, *Solid State Ionics*, **74**, 249 (1994).
89. A.M. Wilson, J.R. Dahn, *J. Electrochem. Soc.*, **142**, 326 (1995).
90. W. Xing, A.M. Wilson, K. Eguchi, G. Zank, J.R. Dahn, *J. Electrochem. Soc.*, **144**, 2410 (1997).
91. W. Xing, A.M. Wilson, G. Zank, J.R. Dahn, *Solid State Ionics*, **93**, 239 (1997).
92. A.M. Wilson, W. Xing, G. Zank, B. Yates, J.R. Dahn, *Solid State Ionics*, **100**, 259 (1997).
93. A.M. Wilson, G. Zank, K. Eguchi, W. Xing, J.R. Dahn, *J. Power Sources*, **68**, 195 (1997).
94. D. Larcher, C. Mudalige, A.E. George, V. Porter, M. Gharghouri, J.R. Dahn, *Solid State Ionics*, **122**, 71 (1999).
95. G.X. Wang, L. Sun, D.H. Bradhurst, S. Zhong, S.X. Dou, H.K. Liu, *J. Power Sources*, **88**, 278 (2000).
96. I.S. Kim, P.N. Kumta, G.E. Blomgren, *Electrochem. Solid State Lett.*, **3**, 493 (2000).
97. E.M. Huang, X. Kelder, J. Schoonman, *J. Power Sources*, **94**, 108 (2001).
98. J. Yang, B. Fang, K. Wang, Y. Liu, J. Xie, Z. Wen, *Electrochem. Solid State Lett.*, **6**, A154 (2003).

99. H. Dong, X.P. Ai, H.X. Yang, *Electrochem. Commun.*, **5**, 952 (2003).
100. X. Wu, Z. Wang, L. Chen, X. Huang, *Electrochem. Commun.*, **5**, 935 (2003).
101. Z.S. Wen, J. Yang, B.F. Wang, K. Wang, Y. Liu, *Electrochem. Commun.*, **5**, 165 (2003).
102. I.-S. Kim, G. Blomgren, P. Kumta, *Electrochem. Solid State Lett.*, **6**, A157 (2003).
103. P. Patel, I.-S. Kim, P.N. Kumta, *Mater. Sci. Eng.*, **B116**, 347 (2005).
104. H.-Y. Lee, Y.-L. Kim, M.-K. Hong, S.M. Lee, *J. Power Sources*, **141**, 159 (2005).
105. Y. Zhang, Z.-W. Fu, Q.-Z. Qin, *Electrochem. Commun.*, **6**, 484 (2004).
106. I.-S. Kim, G.E. Blomgren, P.N. Kumta, *J. Power Sources*, **130**, 275 (2004).
107. Z. Chen, L. Christensen, J.R. Dahn, *Electrochem. Commun.*, **5**, 919 (2003).
108. Z. Chen, L. Christensen, J.R. Dahn, *J. Electrochem. Soc.*, **150**(8), A1073 (2003).
109. W.-R. Liu, Z.-Z. Guo, D.-T. Shieh, H.-C. Wu, M.H. Yang, N.-L. Wu in *206th ECS meeting*, Hawaii, October 3–8 2004, Abs#310.
110. B.J. Neudecker, R.A. Zuhr, J.B. Bates, *J. Power Sources*, **81**, 27 (1999).
111. S. Bordearu, T. Brousse, D.M. Schleich, *J. Power Sources*, **81**, 233 (1999).
112. S.-J. Lee, J.-K. Lee, S.-H. Chung, H.-Y. Lee, S.-M. Lee, H.-K. Baik, J. Power Sources, **97** (2001).
113. K. Sayama, H. Yagi, Y. Kato, S. Matsuta, H. Tarui, S. Fujitani, *The 11th International Meeting on Lithium Batteries, Monterey, CA*, June 23–28, 2002, Abs#52.
114. T. Yoshida, T. Fujihara, H. Fujimoto, R. Ohshita, M. Kamino, S. Fujitani, *The 11th International Meeting on Lithium Batteries*, Monterey, CA, June 23–28, 2002, Abs#48.
115. J. Graetz, C.C. Ahn, R. Yazami, B. Fultz, *Electrochem. Solid State Lett.*, **6**, A194 (2003).
116. H. Jung, M. Park, Y.-G. Yoon, G.-B. Kim, S.-K. Joo, *J. Power Sources*, **115**, 346 (2003).
117. H. Jung, M. Park, S.H. Han, H. Lim, S.-K. Joo, *Solid State Commun.*, **125**, 387 (2003).
118. T. Hatchard, J.M. Topple, M.D. Fleischauer, J.R. Dahn, *Electrochem. Solid State Lett.*, **6**, A129 (2003).
119. S.-W. Song, K.A. Striebel, R.P. Reade, G.A. Roberts, E.J. Crains, *J. Electrochem. Soc.*, **150**(1), A121 (2003).
120. L.Y. Beaulieu, K.C. Hewitt, R.L. Turner, A. Bonakdarpour, A.A. Abdo, L. Christensen, K.W. Eberman, L.J. Krause, J.R. Dahn, *J. Electrochem. Soc.*, **150**(2), A149 (2003).
121. G.A. Roberts, E.J. Crains, J.A. Reimer, *J. Electrochem. Soc.*, **151**(4), A493 (2004).
122. Ye Zhang, Z.-W. Fu, Q.-Z. Qin, *Electrochem. Commun.*, **6**, 484 (2004).
123. K.-L. Lee, J.-Y. Jung, S.-W. Lee, H.-S. Moon, J.-W. Park, *J. Power Sources*, **129**, 270 (2004).
124. M. Green, E. Fielder, B. Scrosati, M. Watchler, J.S. Moreno, *Electrochem. Solid State Lett.*, **6**, A75 (2003).
125. T. Takamura, S. Ohara, J. Suzuki, K. Sekine, *The 11th International Meeting on Lithium Batteries, Monterey*, CA, June 23–28, 2002, Abs#257.
126. M. Uehara, J. Suzuki, K. Sekine, T. Takamura, *The 44th Battery Symposium in Japan, Sakai*, November 4–6 2003, Abs#1D08.
127. T. Shimokawaji, J. Suzuki, K. Sekine, T. Takamura, *The 44th Battery Symposium in Japan, Sakai*, November 4–6 2003, Abs#1D09.
128. J. Maranchi, A. Hepp, P. Kumta, *Electrochem. Solid State Lett*, **6**, A198 (2003).
129. M. Suzuki, J. Suzuki, K. Sekine, T. Takamura, *The 44th Battery Symposium in Japan, Sakai*, November 4–6 2003, Abs#1D10.
130. S. Ohara, J. Suzuki, K. Sekine, T. Takamura, *J. Power Sources*, **136**, 303 (2004).
131. Y.-L. Kim, H.-Y. Lee, S.-W. Jang, S.-H. Lim, S.-J. Lee, H.-K. Baik, Y.-S. Yoon, S.-M. Lee, *Electrochim. Acta*, **48**, 2593 (2003).
132. J.-B. Kim, H.-Y. Lee, K.-S. Lee, S.-H. Lim, S.-M. Lee, *Electrochem. Commun.*, **5**, 544 (2003).
133. S.-J. Lee, H.-Y. Lee, H.-K. Baik, S.-M. Lee, *J. Power Sources*, **119**, 113 (2003).
134. K. Zaghib, K. Kinoshita in *12th International Meeting on Lithium Batteries*, Nara, Japan, June 27–July 2, 2004, Abs#7.
135. I. Yonezu, H. Tarui, S. Yoshimura, S. Fujitani, T. Nohma in *12th IMLB*, Nara, Japan, June 27–July 2 2004, Abs#58.
136. N. Dimov, H. Noguchi, M. Yoshio, *J. Power Sources*, **156**, 567 (2006).

Chapter 12
HEV Application

Tatsuo Horiba

12.1 Introduction

In 1997, Toyota Motor Company launched electric hybrid vehicles into the world market, followed by Nissan and Honda a few years later. They were very popular because of the good fuel economy and the increasing interest in global environmental issue. Therefore, all of the car manufacturers worldwide accelerated their research and development for hybrid electric vehicle (HEV) technology. Although the batteries used for HEV by Toyota and Honda are nickel-metal hydride (Ni–MH) batteries and they are major batteries in the field at present, the pressure for development is put on not Ni-MH but lithium because of the superior characteristics in power, weight, heat generation, and so on to Ni-MH.

There are many varieties in HEV technology, for the power partition ratio between the engine and motor is adjustable and dependent only on the design concept. The distinction of the parallel hybrid electric vehicle (P-HEV) and the series hybrid electric vehicle (S-HEV) is a well-known rough classification for HEV: The former is powered mainly by the engine and partially assisted by motor, while the latter is, at any time, powered by the motor connected by the engine through a battery and a generator. As P-HEV is the major technology today, the term HEV will be used instead of P-HEV hereafter in this chapter. There are two types of HEV: High voltage systems of 300–144 V and low voltage system of 42–14 V. The former has a larger power share in power partition in driving than the latter, and the topics here are focused on the former.

In the United States, the national project named FreedomCAR for the car technology of high efficiency and low emission by the utilization, for example, of hydrogen fuel is now ongoing. Development of a high-power density battery is one of the development themes in the project. The project is sponsored by the U.S. Department of Energy (DOE) and many national laboratories, universities, and private companies

T. Horiba
Shin-Kobe Electric Machinery Co., Ltd, 2200 Oka, Fukaya, Saitama, 369-0297, Japan
t.horiba@shinkobe-denki.co.jp

M. Yoshio et al. (eds.) *Lithium-Ion Batteries*,
DOI 10.1007/978-0-387-34445-4_12

have joined it under the well-organized control of DOE.[1] In Japan, a 5-year national project for the development of high-power lithium batteries for fuel cell vehicle application started in 2002.

In the following sections, an overview of development status for practical large-sized lithium batteries for HEV application will be given briefly.

12.2 Outside of Japan

In the United States, a well-organized national project, integrating all of the national resources for the development of advanced HEV battery is now preceding according to the schedule. However, (SAFT) is the sole manufacturer responsible for the production of the HEV battery in the project. The most outstanding feature for the battery is the nickel-based positive electrode material. They reported the power density of 1,400 W/kg and a predicted calendar life longer than 15 years.[2,3] It is expected that their aggressive effort to commercialize this promising technology will become successful soon.

In Korea, Samsung and LG, major manufacturers of the lithium-ion (Li-Ion) battery for consumer product application, have extended their interest to vehicle applications and have presented several papers.[4,5] LG's paper on the HEV battery presents the battery as a polymer lithium battery of manganese-graphite chemistry.[5] It is remarkable that they also present a battery module design and its development is expected.[6]

12.3 Inside of Japan

Other than Hitachi/Shin-Kobe, Japan Storage Battery is very actively developing an HEV battery. They launched a Li-Ion battery for an HEV truck for Mitsubishi Motor Co. in 2002. The feature for their battery is the manganese-based material in the positive electrode. The members of the national project for fuel-cell vehicle application, other than Japan Storage Battery and Hitachi/Shin-Kobe as battery developers, are Panasonic, and national laboratories, universities, private companies, and so forth as the advanced technology developers. The synergistic energy of this project is expected to accelerate the development.

12.4 Hitachi/Shin-Kobe[7]

The electrode-active material used in the HEV battery is a lithium-manganese spinel of partially substituted manganese with lithium, for the positive electrode-active material, and hard carbon, which gives a gradually sloping voltage profile, for the negative electrode-active material. The load for the HEV batteries is not a continuous charge or discharge but frequent pulses of narrow width. The energy to drive a car is stored in a fuel tank; therefore, the battery should be not an energy battery of high energy density but a power battery of high power density. To cope with the demand for high power density performance, reduction of the cell's internal resistance is the most effective measure. It is the most effective to make the electrode thinner than that for other applications. The thinner electrode not only increases the electrode area to lessen the current density, but also makes the interelectrode distance shorter, which also lowers the resistance.

An outline of the single cell for HEV application is shown in Table 12.1 and the photograph is shown in Fig. 12.1. The cell is cylindrical and is 40 mm in diameter and 108 mm long. It weighs 300 g and has a capacity of 3.6 Ah in constant current-constant voltage (CC-CV) charge up to 4.1 V. The power density of 2,000 W/kg at 50% state of charge (SOC) and 25°C was calculated by extrapolation of the 5 s constant current discharges at different current values. The charge/discharge characteristics for a single cell are shown in Fig.12.2 . Although the voltage difference between the start and end of the discharge is rather large, the smooth decreasing voltage curve profile enables us to monitor the SOC with good accuracy by simply measuring the cell voltage in ordinary accuracy. The voltage changes from 4.1 to 2.7 V and the difference is 1.4 V, while that for a nickel-metal hydride (Ni-MH) cell is 0.4 V. Therefore, a manganese-based Li-Ion cell has a 3.5 times higher accuracy than a Ni-MH cell with the same voltage detection method of identical precision.

Table 12.1 Specifications of the HEV cell and the battery module

	Cell	Module
Dimensions (mm)	φ 40 × 108	541 × 260 × 160
Weight (kg)	0.3	20.2
Nominal voltage (V)	3.6	173
Capacity (Ah)	3.6	3.6
Output power density at 50% SOC (W/kg)	2,000	1,350
Cooling system	–	Compartment air suction

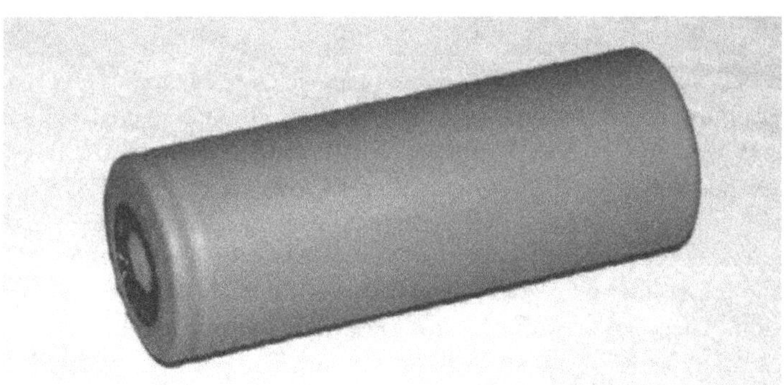

Fig. 12.1 Photograph of an HEV cell. Reproduced from Ref. [7], copyright (2003) with permission from Elsevier Ltd

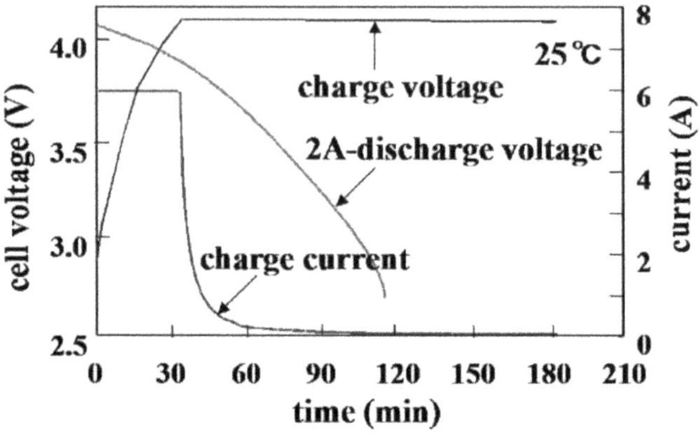

Fig. 12.2 Charge/discharge characteristics of the HEV cell (25°C). Reproduced from Ref. [7], copyright (2003) with permission from Elsevier Ltd

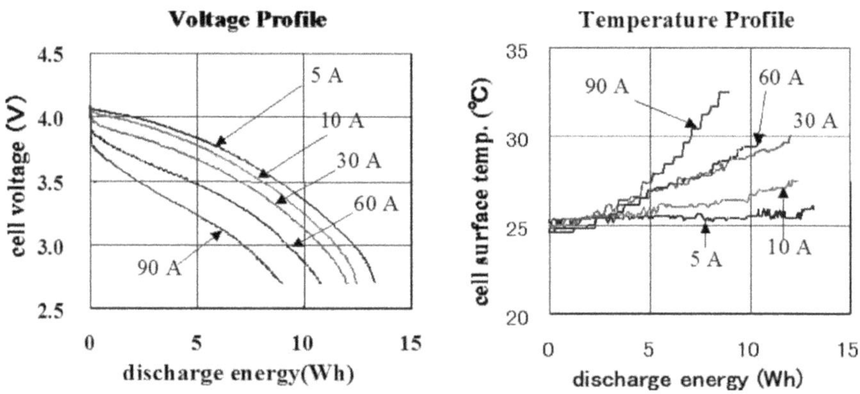

Fig. 12.3 Discharge rate characteristics of the HEV cell (25°C). Reproduced from Ref. [7], copyright (2003) with permission from Elsevier Ltd

Figure 12.3 shows the high-rate continuous discharge capability and temperature profile during discharge. A continuous discharge up to 90 A is possible with a cell-voltage drop of 0.6 V for a 5-A discharge. At a 10-A discharge (= 3 C), the cellsurface temperature increased by 2°C, while for a 90-A discharge (= 25 C), the surface temperature elevation was only 8°C. This result means the low heat generation in the charge/discharge process for the manganese-based Li-Ion cells, which is mainly caused by the low internal resistance as low as 4 mΩ/cell and low enthalpy for the cell reaction. These low resistance and low voltage drop values must support the low heat generation at frequent high-power input/output operations in the HEV application.

The capacity change by storage at 50% SOC and different temperatures showed a tendency that the higher the temperature, the faster the capacity changes.[8]

However, the capacity change showed the tendency of saturation at every temperature. The result of the internal resistances measured simultaneously showed the similar tendency of temperature dependency and saturation as the capacity. From these results and the estimated temperature distribution in a battery module during practical HEV use,[8] we evaluated the practical lifetime of the HEV battery as more than 5 years.

12.5 HEV Battery Module

We developed a 48-cell battery module to adapt the HEV single cells to an HEV application. The module consists of 48 cells connected in series, 6 cell controllers, a fuse, and a module case that contains all the parts. The cell controller is a printed circuit including a microcomputer to detect individual cell voltages and temperatures in the module box and to communicate the results to the upper-level computer in the vehicle that controls the cell SOC in a balanced state. The specifications for the module is already summarized and shown in Table 12.1 and its photograph is shown in Fig. 12.4. As mentioned above, the heat generation for our battery is small, therefore, the battery module can be cooled by air in spite of the frequent input/output during HEV applications. There are cooling air inlets and outlets on the top of the module case.

The input/output power at 25°C for a module calculated by extrapolation of I-V curves for a 5 s constant current charge/discharge is shown in Fig.12.5 . As the rated output/input power for the module is 12.5 kW and 8 kW, we can use the module between 15 and 90% SOC. This wide serviceable SOC range means the ability to supply or receive power at almost any time and allows the battery to have a large

Fig. 12.4 Photograph of the HEV battery module. Reproduced from Ref. [7], copyright (2003) with permission from Elsevier Ltd

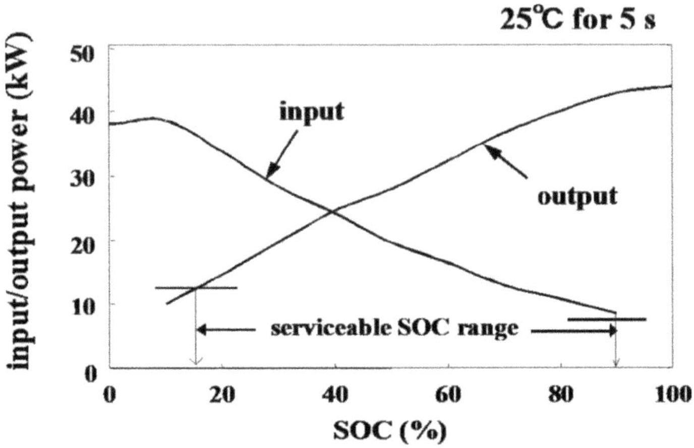

Fig. 12.5 Input/output power for the HEV module. Reproduced from Ref.[7], copyright (2003), with permission from Elsevier Ltd

capacity for power assistance and regeneration. Thus it is expected that the high fuel efficiency during HEV application must contribute to the solution of the global environmental issue, which is the main purpose of developing the HEV system for automobiles.

The developed manganese-based Li-Ion batteries of high-power density for HEV application are being used in the Nissan Tino Hybrid already released into the market.[9,10] If the technology can be further refined and production increases, the cost will be expected to drop, as the small-sized lithium batteries for consumer products proved, because of the lower price of the raw materials for the manganese-based lithium battery. It then will be promising to apply these batteries not only to HEV and other motor-assisting drive systems, but also to other power sources judging from the superior and flexible characteristics of lithium ion battery technology.

12.6 Future Prospects

Although the Li-Ion battery for HEV application still has little experience in the real market, research and development activity for it is very dynamic at present worldwide. This fact must surely reflect the common recognition that the technology is one of the most promising for the HEV battery of the coming generation. It is predicted that the HEV will be significantly widespread by 2010. The Li-Ion battery is expected to take a certain portion of the application market at that time. Other than the high-power Li-Ion battery applications mentioned above are the battery power-assisted bicycle by Panasonic released in 2002, the electric scooter by Yamaha also released in 2002, and idling stop/start system for Toyota Vitz released in 2003. Although the technology for these types of application is not exactly the same as that for HEV application, it is certain that the development of these applications will help to accelerate the evolution of the high-power Li-Ion battery technology along with the intensive activity for HEV applications mentioned above.

References

1. T. Q. Duong, R. A. Sutula, J. A. Barnes et al., Proc. 19th Intern. Electric Vehicle Symposium, 287 (2002).
2. P. Blanchard, L. Gaignerot, S. Herreyre et al., Proc. 19th Intern. Electric Vehicle Symposium, 321 (2002).
3. K. Nechev, M. Saft, G. Chagnon, A. Romero, Proc. 2nd Advanced Automotive Battery Conference, Las Vegas, NV, February 2002.
4. K. H. Kim, J. Y. Ryu, J. K. Kim et al., Proc. 19th Intern. Electric Vehicle Symposium, 1217 (2002).
5. J.-S. Yu, S.-W. Kim, H.-S. Choo, M.-H. Kim, Proc. 19th Intern. Electric Vehicle Symposium, 300 (2002).
6. M. L. Hinton, Proc. 19th Intern. Electric Vehicle Symposium, 1766 (2002).
7. T. Horiba, K. Hironaka, M. Matsumura , *J. Power Sources*, **119–121**, 4 (2003).
8. F. Saito, N. Hirata, S. Ogata et al., Proc. 18th Intern. Electric Vehicle Symposium, 1C-3 (2001).
9. T. Miyamoto, E. Oogami, M. Origuchi et al., Proc. SAE 2000 World Congress, 00PC-166 (2000).
10. M. Origuchi, N. Hirata, K. Suzuki et al., Proc. 17th Intern. Electric Vehicle Symposium, 3B-4 (2000).

Chapter 13
Flame-Retardant Additives for Lithium-Ion Batteries

Masashi Otsuki and Takao Ogino

13.1 Introduction

Lithium-ion (Li-Ion) batteries possess high-energy density compared with other secondary batteries such as nickel–metal hydride (Ni–MH) and nickel–cadmium (Ni–Cd) batteries. Since the Li-Ion battery, which consists of carbon anode (graphites, cokes, glassy carbons etc.) and lithiated transitional metal oxide cathode ($LiCoO_2$, $LiNiO_2$, $LiMn_2O_4$ etc.), provides higher voltage and longer life in smaller, lighter packages than the other batteries and the Li-Ion batteries also maintain their capacity well through repeated rechargings, it has become a common power source in mobile electronic devices such as digital still cameras, camcorders, laptop-type computers, cellular phones, and so forth. Small Li-Ion cells with a capacity of 2,000–2,400 mAh (18,650 size) are now commercially available for these electronic devices. Due to the above-mentioned inherent marvelous properties, Li-Ion batteries have attracted much interest in recent years as a power source for electric vehicles and/or electric-utility load leveling.[1–3]

In spite of success in improving the electrochemical properties, e.g., capacity and high-rate discharge character,[4–6] the stubborn drawback of the Li-Ion battery regarding safety concern remains an untouched issue. The safety issue is a serious hindrance in the widespread acceptance of Li-Ion battery with electric vehicle (EV) and hybrid electric vehicle (HEV) application. Under abusive operating conditions, e.g., overcharge, external impact, heat shock, Li-Ion battery undergoes thermal runaway,[7–9] producing exceedingly high temperature, smoke, explosion, and fire because fire-catchable organic solvent is used as an electrolyte in the high-energy cell.

Battery manufacturers and sellers employ external safety devices in the battery package to surmount these problems: an Negative Temperature Coefficient sensor (NTC) thermistor has been used as an overcurrent interrupter, and Charge Protection

M. Otsuki (✉) and T. Ogino
Central Research, Bridgestone Corporation, 3-1-1 Ogawa-higashi,
Kodaira, 187-8531 Tokyo, Japan
masashi.otsuki@bridgestone.co.jp

M. Yoshio et al. (eds.), *Lithium-Ion Batteries*,
DOI 10.1007/978-0-387-34445-4_13

Integrated Circuit with field effect transistor (FET-IC) has been used as an overcharge/overdischarge interrupter.[10] However, the safety margin of the Li-Ion cell is still too small for practical use.[11,12] Application of these external devices, however, is not an essential counterplan for the safety design.

Recently, safety improvement of cell materials has been extensively studied with powerful efforts. For example, thermally and/or electrochemically stable electrolyte supporting salt and electrolyte have been studied.[13–20] In these papers, Barthel et al. clarified that some thermally stable organoborates have a sufficiently high oxidation limit in carbonate solutions. Sasaki et al. also found out that lithium bis [2,3-naphthalenedioxalato(2-)-O,O′]borate, which does not hamper the electrochemical performance, is thermally stable above 320°C.[19] on Functional additives for safe use of Li-Ion battery also have been investigated. Sacken et al. reported that biphenyl compounds act as suppressors for overcharge of the cell.[21] Tobishima et al. also reported that cyclohexylbenzene and hydrogenated diphenyleneoxide provide better performance than biphenyl as an overcharging suppressor.[22] By using a nonflammable electrolyte and/or a flame-retardant, which can make conventional electrolytes nonflammable, also is an important approach for the safe design and safe operation of the Li-Ion battery. It can be assumed that nonflammable electrolyte technology is an indispensable one for inducing scale-up of Li-Ion battery. Battery manufacuturers have experimented with various additives to render the electrolyte in Li-Ion batteries nonflammable.[23] However, all previous additives have affected battery performance adversely, and none of them have been commercially successful. Prakash et al. worked to achieve a nonflammable electrolyte by modifying the conventional electrolyte with a novel flame-retardant additive. In his work, hexamethoxycyclotriphosphazene, a solid additive, was found to be an electrochemically stable additive and it induced thermal stability in a $Li/LiNi_{0.8}Co_{0.2}O_2$ cell.[24] We also have found that a liquid additive that consists of a phosphorus nitrogen compound, so-called phosphazene, can overcome a stubborn drawback of Li-Ion batteries. Mixed into the electrolyte at a concentration of 5–10 vol% of phosphazene, it renders the solution nonflammable without adversely affecting battery performance. It also reduces the generation of heat inside the battery. At the following section, we review a phosphazene additive, Phoslyte (registered Japanese trademark), which could make Li-Ion batteries a competitive alternative in EV, HEV, and in large-scale storage systems for electric power.[25–27]

13.2 Phosphazene Compounds

Compounds that contain alternating phosphorus-nitrogen double bonds in their skeleton are called "phosphazenes." As historical background, Liebig first reported the synthesis procedure of chlorophosphazene $(NPCl_2)_n$, prepared by the reaction between phosphorus pentachloride with ammonia in 1834.[28] Then Gladstone, Besson, Rosset, Couldridge, and Stokes furthered the fundamental chemistry of the phosphazene, i.e., substitution, hydrolysis, polymerization, and so forth, from 1850

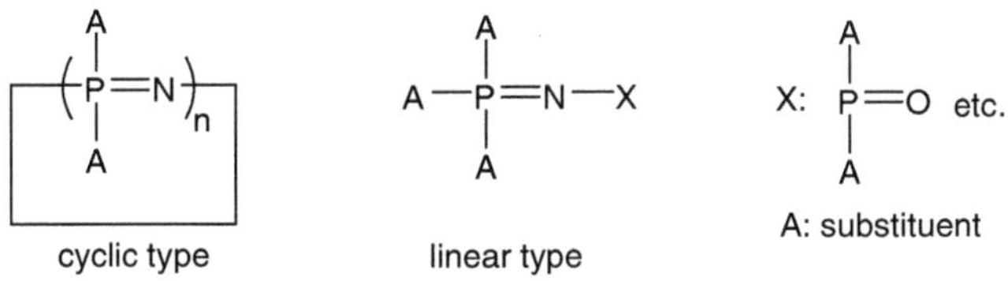

Fig. 13.1 Representative structures of phosphazene

to 1900.[29–47] Then, from the 1940s to the 1980s, structure, reaction, and their application were reviewed by Allcock, Audrieth, Gribova, Paddock, Shaw, Schmuldbach, and Kajiwara.[48–54] One of the applications of phosphazene compounds was flame-retardance in plastics, textiles, synthetic fibers, and so forth.

The structure of phosphazene material can be classified into two types, as shown in Fig. 13.1:

1. Linear type
2. Cyclic type

The substituent, A, can be halogen, pseudohalogen, or many kinds of organic groups such as alkoxy, aryloxy, alkyl, and so forth Some can be prepared by direct synthesis and others are obtained by appropriate substitution reactions by using a suitable precursor. The physical and chemical properties of phosphazene compounds can be varied by choosing a wide variety of substituents.

The reasons why authors directed their attention to this material as a flame-retardant additive for nonaqueous electrolyte are described as follows:

1. They grasped its high function as a flame-retardant for various resins.
2. They have an impression of its high-voltage resistance through their examination as a base oil of the electric rheological fluid (ERF).
3. Phosphazene polymers were widely examined as a solid electrolyte in the United States in 1980s.

From these accounts, they expected that phosphazene compounds would become a key material for safe design of Li-Ion battery.

13.3 Optimization of Phosphazene Structure as Additive Agents

The flame-retardant additives that can be applied to Li-Ion battery have to possess the following characteristics:

1. High contribution to incombustible with low amounts.
2. Mixing well with conventional electrolytes.
3. Low viscosity almost equal to that of the conventional electrolyte.

4. High boiling points at least more than that of the conventional electrolytes.
5. No affection to the ionic conductivity decreasing.
6. Wide electrochemical voltage window.
7. Less reactivity with cell materials.
8. Well-soaking into the electrode.

Although the solid substance also was considered to be an additive agent, it frequently could not accommodate the above-mentioned requirements.As a structure of the flame-retardant additive, the authors thought that the liquid phase substance is favorable to obtain both excellent flame-retardant ability and electrochemical performances. In order to synthesize the liquid-phase phosphazene conveniently, the authors examined the synthesis procedures, focusing on hexachlorocyclotriphosphazene $(PNCl_2)_3$ as a starting material, which was produced and commercialized in large amounts.

In making phophazene compounds low viscosity similar to that of conventional electrolytes, it was found that the substitution with fluorine in the phosphazene skeleton could drastically decrease its viscosity. However, the phosphazene in which all side chains were substituted with fluorine, hexafluorocyclotriphosphazene has a freezing point near room temperature. Thus, the boiling point was not acceptable; it was extremely low (52°C). Therefore, partially fluorinated organocyclotriphosphazene, as shown in Fig. 13.2, was adopted as a suitable structure, where the organic substituent is a monovalent organic group such as alkoxy, aryloxy, alkyl, aryl, and so forth. The optimized structure of phosphazene compounds regarding the flame-retardant additives for Li-Ion battery is called "Phoslyte," as a registered trademark. Phoslyte possesses following physical properties:

1. Low viscosity (0.8–2.0 mPa.s)
2. High boiling point (80 to more than 400°C)
3. Low freezing point (less than –20°C)
4. High contribution to incombustibility of the electrolyte

Figure 13.2 shows a representative structure of the additive (Phoslyte-A), which consists of fluorine and appropriate organic substituent. The additive has a viscosity and boiling point, 1.2 mPa.s and 194°C, respectively. Although the ionic conductivity of the Phoslyte added EC/DEC(1/1) electrolyte containing 1 M $LiPF_6$ (7.2 mS/cm) decreased a small amount compared with no additive electrolyte [EC/DEC (1/1, volume ratio)], 1 M $LiPF_6$, 7.6 mS/cm), viscosity of the electrolyte with Phoslyte is a little bit lower than that of the base electrolyte as shown in Fig. 13.3. Remarkably,

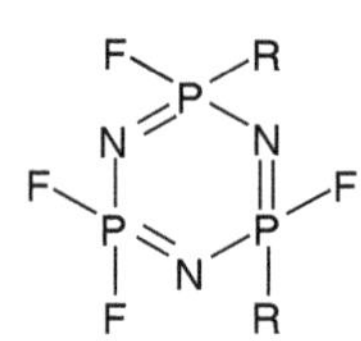

Fig. 13.2 Representative structures of the Phoslyte

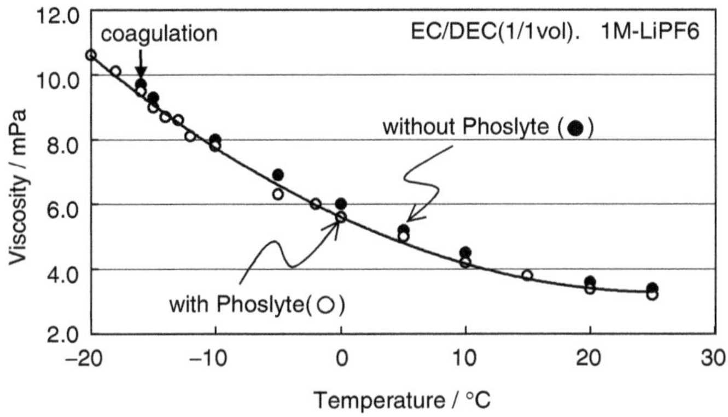

Fig. 13.3 Viscosity of the EC/DEC (1/1) electrolyte with or without Phoslyte: (open circle) with Phoslyte; (closed circle) without Phoslyte

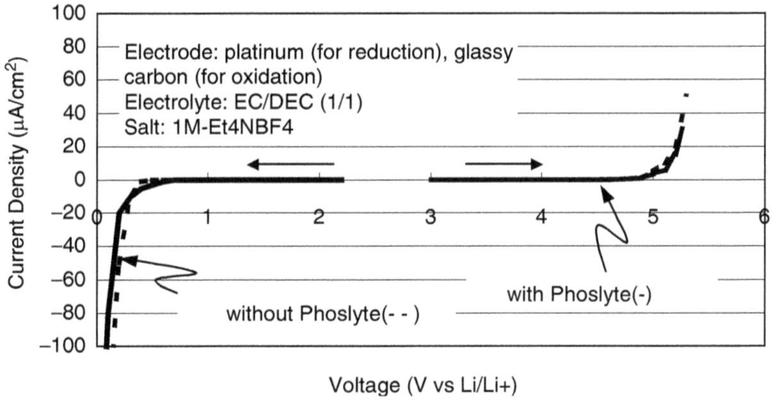

Fig. 13.4 Electrochemical stability of the EC/DEC (1/1) electrolyte with or without Phoslyte. *Solid line*: with Phoslyte; *broken line*: without Phoslyte

addition of Phoslyte into EC/DEC(1/1) electrolyte could make freezing point lower: although EC/DEC(1/1) containing 1-M $LiPF_6$ showed a freezing point near –15°C, the electrolyte that contains 10 vol% of Phoslyte could not be frozen at –20°C. It is noted that the freezing point of Phoslyte-A itself is lower than –50°C. Thus, the phenomenon is expected to contribute to the improvement of the low-temperature performances of the Li-Ion battery using the ethylene carbonate (EC) system.

Figure 13.4 shows cyclic voltammograms on a glassy carbon disk electrode (for oxidation) and on a platinum disk electrode (for reduction) with/without Phoslyte between 0.0 V and 5.2 V versus Li/Li^+. Both electrolytes were electrochemically stable within 0.3–5.0 V. Addition of Phoslyte did not affect the electrochemical voltage window of base electrolyte.

13.4 Method of Flammability Determination

In this section, flammability of the electrolyte is discussed. The flammability is determined by using several methods, i.e., burning test with UL94HB method, flash point test with ASTM D56 and ASTM D93 method, and limiting oxygen index test with JIS K 7,201 method. These test procedures sometimes are arranged with these requirements.

UL94HB horizontal burning test generally is considered to be the easiest test to pass and materials that pass any of the other tests such as V or VTM tests (vertical burning tests) usually will be accepted by Underwriting Laboratory. In the test, we use a 1″ × 5″ specimen held at one end in a horizontal position with marks at 1" and 4" from the free end. A flame is applied to the free end for 30 s or until the flame front reaches the 1" mark, as shown in Fig. 13.5. If combustion stops before the 4" mark, the time of combustion and the damaged length between the two marks are recorded. A set of three specimens are tested. A material that is less than 0.118" in thickness will be classified 94HB if it has a burning rate of less than 3" per minute or stops burning before the 4" mark. The HB-rated materials are considered to be "self-extinguishing." If one specimen from the set of three fails to comply, then a second set of three are tested. All three of this second set must comply. Thus, the test is a burning test for solid-state specimen. In order to evaluate the flammability of the electrolyte, we arranged as follows: 0.5 mm thickness of 1" x 5" silica fiber (nonflammable sheet) was used as a electrolyte holder. One milliliter of the examined electrolyte was dipped into the fiber, then the specimen was fixed in a horizontal position. The authers call this method "arranged UL94HB." A specimen with a burning rate of less than 3"/min or that stops burning before the 4" mark is considered to be "self-extinguishing." If burning stops within 1" mark, it is considered as "flame retardant." If the specimen could not catch fire, it was considered to be "nonflammable."

Oxygen index [the limiting oxygen index (LOI)], also called the critical oxygen index (COI) methods, which can describe the tendency of material to sustain a

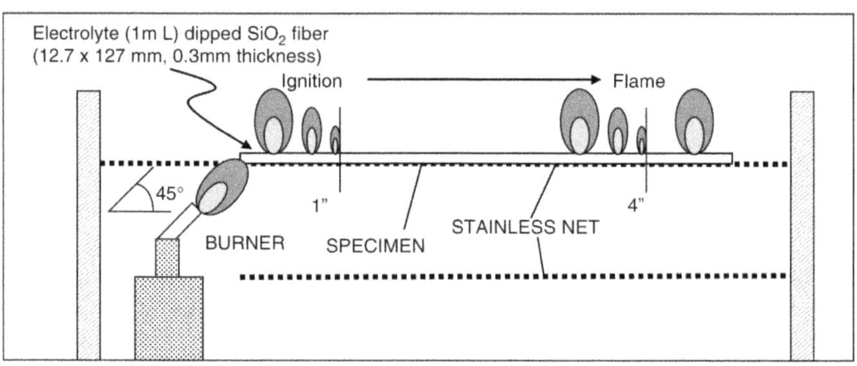

Fig. 13.5 Horizontal burning test for arranged UL-94HB classification

flame, are widely used as a tool to investigate the flammability of polymers. The method provides a convenient and reproducible means of determining a numerical measure of flammability. These methods have been used to systematically investigate flammability of fire retardants, frequently comparing the effectiveness of fire-retardants.

The oxygen-index method is described as follows: the sample which is vertically fixed into the gas-controlled cylindrical tube is burned under a controlled atmosphere. When the sustained burning is observed under an appropriate atmosphere, the value of the oxygen concentration is considered to be LOI. The effectiveness of a flame-retardant is evaluated from the comparison of these LOI values with/without additives.

Since the air comprises about 20.95% oxygen by volume, any materials with a LOI less than this value will burn easily in the air. Nelson investigated how the introduction of a fire-retardant changes the oxygen index of a material.[55] He classified these into two groups, i.e., LOI < 20.95 and LOI > 100. Nelson referred to these materials "flammable" and "intrinsically nonflammable," respectively. On the other hand, Horrocks et al. have suggested that materials with a LOI greater than 28% are generally "self-extinguishing."[56] Moreover, Nelson referred to materials that are between the 20.95 < LOI < 28, as being "slow-burning."[57]

Authors attempted to understand the flammability of the electrolyte quantitatively. Thus, some arrangement was attended: 0.5 mm thickness of 1" x 5" silica fiber, just as in the above-mentioned UL94-HB test, was used as an electrolyte holder. One milliliter of the examined electrolyte was dipped into the fiber and vertically fixed into the glass cylinder of the LOI instrument, and then LOI was measured according to the regulation (JIS K 7,201 method), as shown in Fig. 13.6.

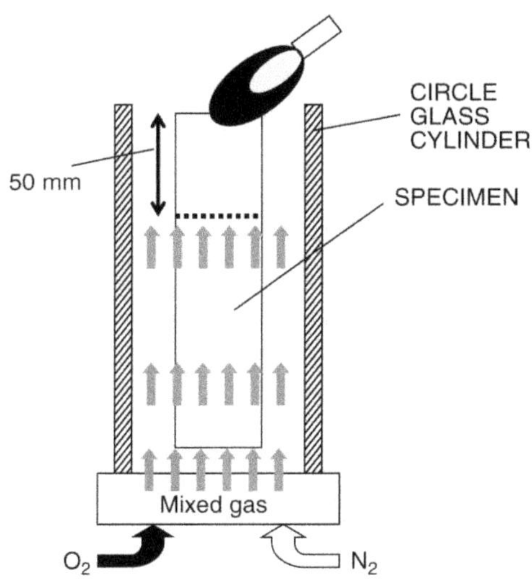

Fig. 13.6 Limiting oxygen index determination test for arranged JIS K 7,201 classification

Figure 13.7 shows the photographs of the arranged UL94-HB test of EC/DEC(1/1) in the presence or absence of Phoslyte-A (10 vol%). The electrolyte with Phoslyte did not catch fire, although the electrolyte without Phoslyte easily caught fire and the fire spread rapidly. Figure 13.8 shows results of flammability tests of PC/DME(1/1), EC/DEC(1/1), and EC/DEC(3/7) in the presence or absence of Phoslyte-A and B by using arranged UL94-HB and arranged LOI method. As a result, it is recognized that the optimal structure of the Phoslyte for each electrolyte can make the conventional electrolyte effectively nonflammable.

Moreover, the flash point also changes dramatically with addition of Phoslyte. Here, an explanation is added about the flash point. The flash point is the lowest temperature at which a liquid can form an ignitable mixture in air near the surface of the liquid. The lower the flash point, the easier it is to ignite the material. ASTM methods are available for determination of the flash point: liquid that has a kinematic viscosity of less than 5.5 mm^2/s at 40°C does not contain suspended solids and has no tendency to form a surface film while under testing. The standard method of test for flash point by tag closed cup tester (ASTM D56) method can be

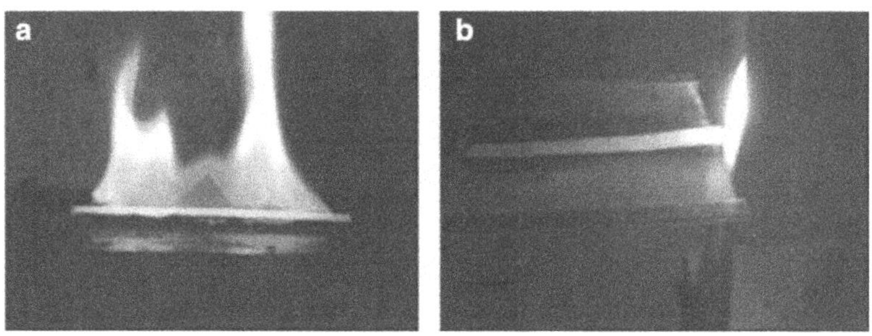

Fig. 13.7 Flammability of the electrolyte evaluated by the arranged UL94-HB method. (**a**) Conventional electrolyte; (**b**) the electrolyte with Phoslyte

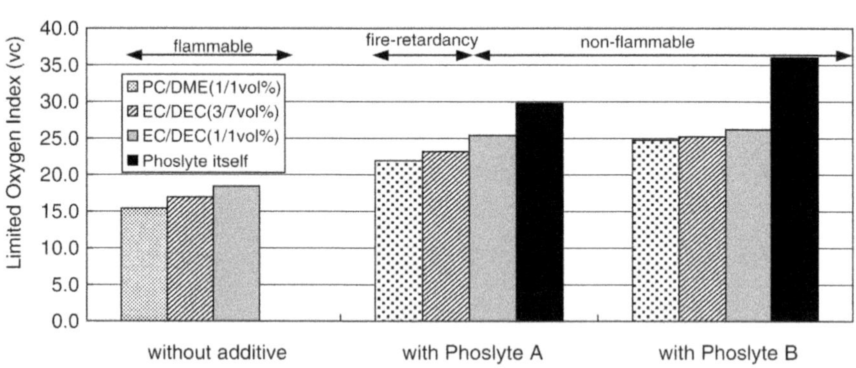

Fig. 13.8 Relationship between the arranged UL94-HB test and the limiting oxygen index. Phoslyte A and B consist of cyclotriphosphazene structure with a different substituent

available. If the above viscosity is more than 5.5 mm²/s at 40°C, the standard method of test for the flash point determination by Penski-Martens closed cup tester (ASTM D93) should be used. The detail of the flash point analysis can be seen in the ASTM method.

In the Japan Fire Code, dangerous materials are classified by their flash point. The substance that has no flash point can be classified as "noncorresponding substance for danger" by law. Since the storageable quantity, standard treatment, handling permission, and so on are restricted from the classification, improvement of the flash point might affect the large-scale battery production. Figure 13.9 shows the flash point of the carbonate system with or without Phoslyte. Addition of an appropriate kind of Phoslyte can shift the flash point to high temperature and it then becomes higher by increasing amounts of Phoslyte, finally achieving non-flash point electrolyte.

Authors speculate that the nonflammable effect might be caused by the following mechanisms through their fundamental studies of the gas analysis and pyrolytic analysis of the electrolytes:

1. The nonflammable gas (phosphate ester), which is generated by decomposition of Phoslyte, intercepts the continuous supplying oxygen.
2. The above-mentioned phosphate ester promotes surface carbonization on the electrode, producing carbide film (char). The char intercepts the continuous supplying oxygen.
3. Halogen and the phosphorus radical must catch the active radical in the flame field.

electrolyte	flammability	flash point/°C
Conventional electrolyte	flammable	15
10vol.% Phoslyte addition	non-flammable	32
15vol.% Phoslyte addition	non-flammable	none

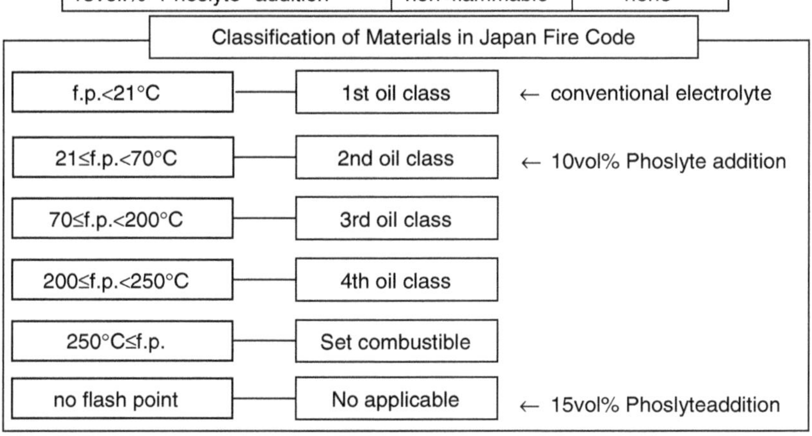

Fig. 13.9 Effect of Phoslyte on the flash point of conventional electrolyte (example)

13.5 Thermal Stability

As mentioned above, addition of Phoslyte contributes remarkably to making electrolytes nonflammable. It also has been discovered that the addition of Phoslyte provides thermal stabilization of the cell materials as well. Use of differential scanning calorimetry (DSC) and accelerating rate calorimetry (ARC) provide information regarding thermal behavior of the cell and cell materials.[3–7,58–66]

Figure 13.10 shows the thermal runaway behavior of the Li-Ion cell, which consists of $LiCoO_2$ and synthetic graphite with EC/EMC (1/2 volume) including $LiPF_6$ (1 M). As shown in this figure, onset temperature of the exothermal reaction of the fully charged $LiCoO_2$ (4.2 V) was shifted to higher (20–30°C) temperature by the addition of the Phoslyte. The total heat flow sometimes was decreased as well.

These phenomena also were observed in the case of the lithiated synthetic graphite. Prakash also reported the effect of the addition of hexamethoxycyclotriphosphazene as a solid retardant on thermal stability of the charged cathode. In his paper, addition of 1.68 wt% flame-retardant additives into the Li/$LiNi_{0.8}Co_{0.2}O_2$ cell reduced the overall heat generation produced due to the exothermic reaction.[24]

Figure 13.11 shows the ARC profile of the fully charged cell (2,032 coin cell) in the presence or absence of Phoslyte. The ARC data clearly provide the thermal behavior and thermal runaway conditions of full cell under adiabatic atmosphere. It clearly is indicated that the onset temperature of exothermic reaction of the cell including Phoslyte shifts to higher than that without Phoslyte. In addition, the onset temperature of the thermal runaway reaction of the cell which contains Phoslyte also shifts to higher heat compared with the cell without Phoslyte. The self-heat rate of the cell with Phoslyte shows one digit smaller than the cell without Phoslyte.

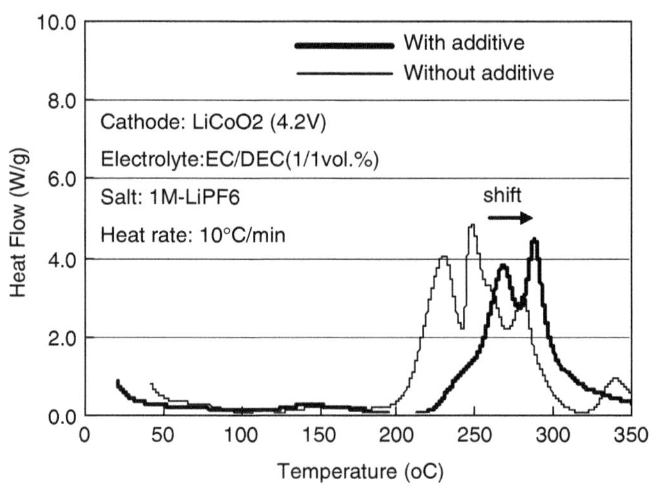

Fig. 13.10 Thermal stability of the fully charged cathode. The data were obtained by using 15th cycled materials (4.2 V)

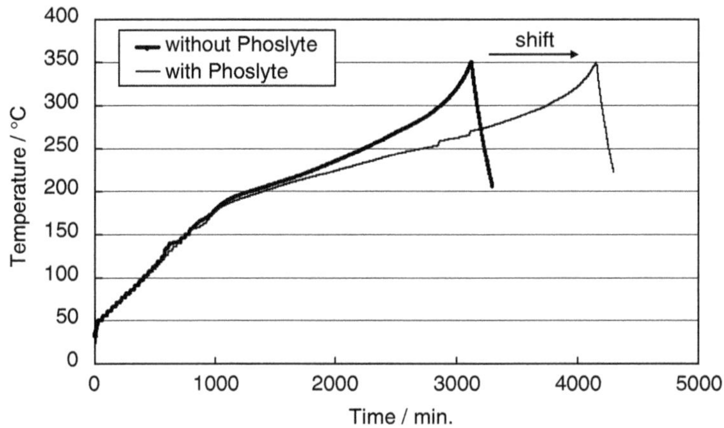

Fig. 13.11 Thermal runaway behavior of the $LiCoO_2$/graphite cell containing EC/EMC(1/2) and $LiPF_6$(1 M) with or without Phoslyte

13.6 Cell Performance

In this final section, cell performances, i.e., cyclability, temperature dependence of the capacity, and so forth are described. In general, additives sometimes give excellent function, but sometimes they bring negative effects on cell performance. For example, although trialkylphosphate effectively renders conventional electrolyte nonflammable, it reduces cell capacity and cyclability.[20] Phoslyte shows good performance as a flame-retardant additive without adversely affecting battery performance.

Figure 13.12 shows initial discharge capacity and the temperature dependence of the coin cell (2,032 type) which consists of $LiCoO_2$ cathode and synthetic graphite anode system by using EC/DEC(1/1) electrolyte. First discharge capacity was almost equal in both cases. Interestingly, low-temperature performance of the cell with additive surpasses that without additives. This was caused by the effect of additives on making low-coagulation temperature and low viscosity: the coagulation point of EC/DEC (–15°C) is shifted lower (less than –20°C) and viscosity is decreased from 3.6 mPa s to 3.2 at 0°C. Furthermore, Phoslyte endured high-temperature preservation and gave chemical stability under nonmoisture condition. Figures 13.13 and 13.14 show a cyclability characteristic and a high temperature preservation characteristic. Above-mentioned results suggest that Phoslyte considerably can improve cell performances.

The authors would emphasize some side effects of Phoslyte. It also has been found that cyclability of the cell with Phoslyte is better than that without Phoslyte when the cell is fabricated by using metal oxide cathode and lithium metal anode. It might be described that Phoslyte suppresses the formation of dendrites through cycling of charge and discharge.

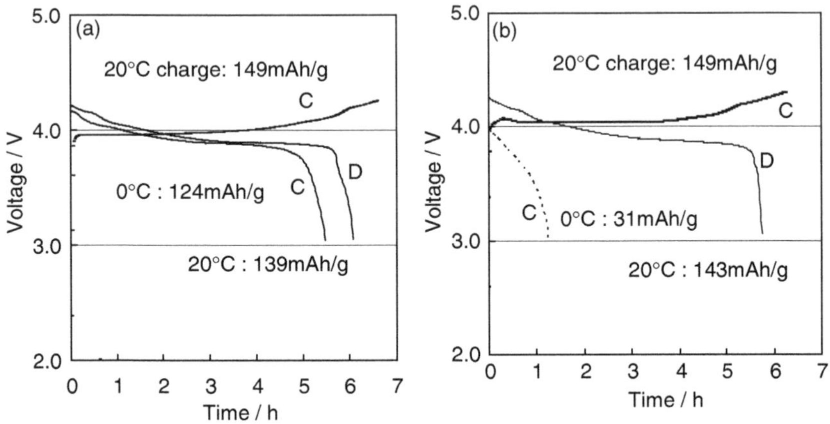

Fig. 13.12 Temperature dependence of the capacity. (**a**) With Phoslyte; (**b**) without Phoslyte; (**c**) charge curve; (**d**) discharge curve

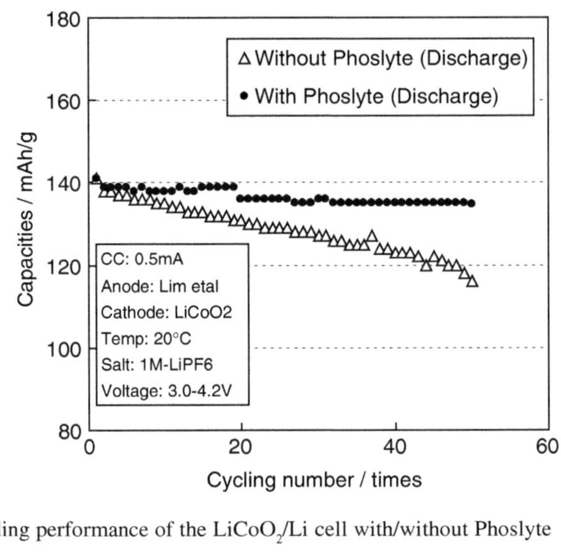

Fig. 13.13 Cycling performance of the $LiCoO_2$/Li cell with/without Phoslyte

Figure 13.15 shows wettability improvement of the separator by addition of Phoslyte. The wettability was analyzed by contact angle, base diameter, and volume change monitoring system. In general, conventional electrolytes have less ability for wettability to their electrode materials and separator. Phoslyte additive renders electrolytes more wettable to cell materials. Improvement of wettability by addition of Phoslyte was observed for cathode and anode as well.

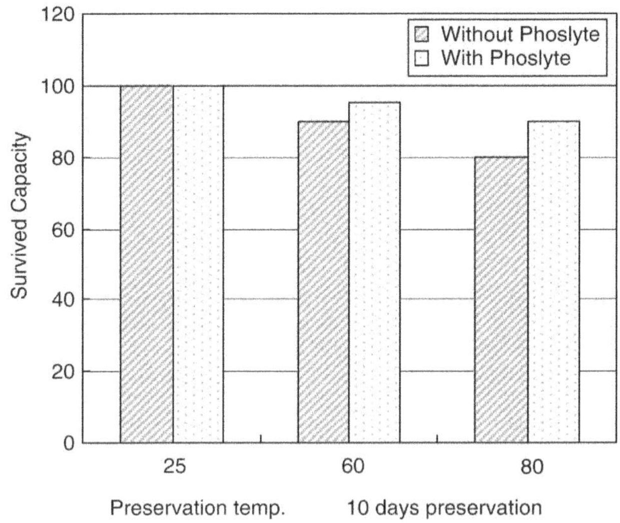

Fig. 13.14 High-temperature preservation performance of the lithium-ion cell in the presence or absence of some kinds of Phoslyte

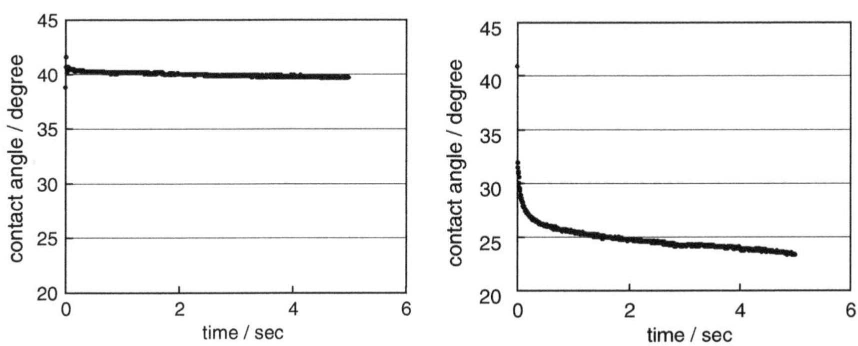

Fig. 13.15 Wettability study of the cell materials. Sequence of contact angle change of the electrolyte and separator: (**a**) without Phoslyte; (**b**) with Phoslyte

13.7 Conclusion

We have developed the high performance of flame-retardant additives, which consists of phosphazene skeleton. Partially fluorinated cyclophosphazene shows excellent ability not only for flame-retardance but for cell performance, dendrite suppressor, and wettability improvement as well. The additives show no or little stimulus for skin, low toxicity, and no carcinogenicity.

The additives eliminate the fire danger in lithium-ion battery without affecting battery performance adversely even though previous additives have contributed to bad battery performance. By mixing the electrolyte with a concentration of 5–10% as volume, it renders the solution nonflammable. It also reduces the generation of the heat inside the battery under abusive operating conditions. Phoslyte is a promising material for resolving the stubborn drawback of the lithium battery.

References

1. J. M. Tarascon, D. Guyomard, *Electrochim. Acta*, **38** (1993) 1221.
2. S. Megahed, B. Scrosati, *Electrochem. Soc. Interface*, **3**(4) (1994) 34.
3. Q. Wu, W. Lu, J. Prakash, *J. Power Sources*, **88** (2000) 237.
4. W. Lu, C. W. Lee, R. Venkatachalapathy, J. Prakash, *J. Appl. Electrochem.*, **30** (2000) 1119.
5. W. Lu, V. S. Donepudi, J. Prakash, J. Liu, K. Amine, *Electrochim. Acta*, **47** (2002) 1601.
6. S. Al Hallaj, J. Prakash, J. R. Selman, *J. Power Sources*, **87** (2000) 186.
7. P. Biensan, B. Simon, J. P. Peres, A. de Guibert, M. Broussely, J. M. Bodet, F. Perton, *J. Power Sources*, **81** (1999) 906.
8. Y. Saito, K. Kanari, K. Takano, *J. Power Sources*, **68** (1997) 451.
9. D. Fouchard, L. Xie, W. Ebner, S. Megahed, *ECS Proc.*, **94–28** (1994) 348.
10. Handbook of Battery Application, ed. By transistor Gijyutsu Hensyubu, QC publication in JAPAN, 2005.
11. S. Tobishima, J. Yamaki, *J. Power Sources*, **81–82** (1999) 882.
12. S. Tobishima, K. Takei, Y. Sakurai, J. Yamaki, *J. Power Sources*, **90** (2000) 185.
13. J. Barthel, M. Wuhr, R. Buestrich, H. H. Gores, *J. Electrochem. Soc.*, **142**(8) (1995) 2527.
14. X. Sun, H. S. Lee, X. Q. Yang, J. McBreen, *J. Electrochem. Soc.*, **146**(10) (1999) 3655.
15. J. Barthel, R. Buestrich, H. J. Gores, M. Schmidt, M. Wuhr, *J. Electrochem. Soc.*, **144**(11) (1997) 3866.
16. F. Kita, A. Kawakami, J. Nie, T. Sonoda, H. Kobayashi, *J. Power Sources*, **68** (1997) 307.
17. J. Barthel, M. Schmidt, H. J. Gores, *J. Electrochem. Soc.*, **145**(2) (1998) L17.
18. R. McMillan, H. Slegr, Z. X. Shu, W. Wang, *J. Power Sources*, **81–82** (1999) 20.
19. M. Handa, S. Fukuda, Y. Sasaki, K. Usami, *J. Electrochem. Soc.*, **144**(9) (1997) L235.
20. K. Xu, M. S. Ding, S. Zhang, J. L. Allen, T. R. Jow, *J. Electrochem. Soc.*, **149**(5) (2002) A622.
21. U. V. Sacken, *Abstract of the 9th International Meething on Lithium Batteries, Friday-93* (1998).
22. S. Tobishima, Y. Ogino, Y. Watanabe, *J. Appl. Electrochem.*, **33** (2003) 143.
23. X. Wang, E. Yasukawa, S. Kasuya, *J. Electrochem. Soc.*, **148**(10) (2001) A1058.
24. C. W. Lee, R. Venkatachalaphathy, J. Prakash, *Electrochem. Solid-State Lett.*, **3**(2) (2000) 63.
25. M. Otsuki, T. Ogino, K. Amine, *ECS Trans.*, **1**(24) (2006) 13–19. C.
26. M. Kajiwara, T. Ogino, T. Miyazaki, T. Kawagoe, Japanese Patent, 3055358 (2000).
27. M. Otsuki, S. Endo, T. Ogino, Japanese Patent, 83628 (2002).
28. M. Otsuki, S. Endo, T. Ogino, *IBA-HBC Proc.*, (2003) 39.
29. J. Liebig, *Ann. Chem.*, **11** (1834) 139.
30. J. H. Gladstone, *Ann.Chem.*, **76** (1850) 74.
31. J. H. Gladstone, *J. Chem. Soc. London*, **2** (1850) 121.
32. J. H. Gladstone, *J. Chem. Soc. London*, **3** (1851) 135.
33. J. H. Gladstone, *J. Chem. Soc. London*, **3** (1851) 353.
34. J. H. Gladstone, *Ann.Chem.*, **77** (1851) 314.
35. A. Besson, *C. R. Acad. Sci.*, **111** (1890) 972.
36. A. Besson, *C. R. Acad. Sci.*, **114** (1892) 1264.
37. A. Besson, *C. R. Acad. Sci.*, **114** (1892) 1479.
38. A. Besson, G. Rosset, *C. R. Acad. Sci.*, **143** (1906) 37.

39. A. Besson, G. Rosset, *C. R. Acad. Sci.*, **146** (1908) 1149.
40. W. Couldridge, *J. Chem Soc. London*, **53** (1888) 398.
41. W. Couldridge, *Bull. Soc. Chim. Fr.*, **2** (50) (1888) 535.
42. H. N. Stokes, *Am. Chem. J.*, **17** (1895) 275.
43. H. N. Stokes, *Chem. Ber.*, **28** (1895) 437.
44. H. N. Stokes, *Am. Chem. J.*, **18** (1896) 629.
45. H. N. Stokes, *Am. Chem. J.*, **18** (1896) 780.
46. H. N. Stokes, *Am. Chem. J.*, **19** (1897) 782.
47. H. N. Stokes, *Am. Chem. J.*, **20** (1898) 740.
48. H. N. Stokes, *Z. Anorg. Chem.*, **19** (1899) 36.
49. H. R. Allcock, *Phosphorus-Nitrogen Compounds*, Academic Press, New York and London, 1972.
50. L. F. Audrieth, R. Steimman, A. D. E. Toy, *Chem. Rev.*, **32** (1943) 104.
51. J. A. Gribova, U. U. Ban-Yuan, *Russ. Chem. Rev.*, **30** (1961) 1.
52. N. L. Paddock, H. T. Searle, *Quant. Rev. Chem. Soc.*, **1** (1959) 347.
53. R. A. Shaw, B. W. Fitzsimmons, B. C. Smith, *Chem. Rev.*, **62** (1962) 247.
54. P. Schmulbach, *Prog. Inorg. Chem.*, **4** (1962) 275.
55. M. Kajiwara, *Fine Chemical*, **14**(7) (1985) **5**; 14(9) (1985) 22.
56. M. I. Nelson, *Combust. Theor. Model.*, **5** (2001) 59.
57. A. R. Horrocks, M. Tunc, D. Price, *Textile Progr.*, **18**(1–3) (1989) 1.
58. C. P. Fenimore, *Flame-retardant Polymeric Materials, vol. 1*, M. Lewin, S. M. Atlas, E. M. Pearce (Eds.), Plenum, New York, (1975) 371.
59. M. I. Nelson, *Proc.Royal Soc. London*, **A454** (1998) 789.
60. H. Maleki, J. S. Hong, S. Al-Hallaj, J. R. Selman, *Electrochem. Soc. Meeting Abstr.*, **97** (1997) 143.
61. M. N. Richard, J. R. Dahn, *J. Electrochem. Soc.*, **146** (1999) 2068.
62. M. N. Richard, J. R. Dahn, *J. Electrochem. Soc.*, **146** (1999) 2078.
63. J. R. Dahn, E. W. Fuller, M. Obrovac, U. V. Sacken, *Solid State Ionics*, **69** (1994) 265.
64. Z. Zhang, D. Fouchard, J. R. Rea, *J. Power Sources*, **70** (1998) 16.
65. Y. Gao, M. V. Yakovleva, W. B. Ebner, *Electrochem. Solid State Lett.*, **13** (1998) 117.
66. F. Pasquier, T. Disma, J. M. Bowmer, *J. Electrochem. Soc.*, **145** (1998) 472.

Chapter 14
High-Energy Capacitor Based on Graphite Cathode and Activated Carbon Anode

Masaki Yoshio, Hitoshi Nakamura, and Hongyu Wang

Recently, electric double-layer capacitors (EDLCs) using activated carbon (AC) as polarizable electrodes have drawn extensive attention by virtue of the advantages such as high-power density, long cycle life, benignity to environment, and so forth. However, there is a disadvantage to be resolved before its further application in the future. That is its low-energy density. The storage of charge in the EDLCs is through the adsorption of ions at the interfaces between the electrolyte and both electrodes. The capacitance is given by $C = \varepsilon_0\varepsilon_r S/d$ (C, capacitance; ε_0, dielectric constant in vacuum; ε_r, specific dielectric constant of the solvent; S, surface area). The above equation implies that the capacitance is nearly proportional to the surface area of electrode materials. Thus numerous research efforts have been devoted to tailoring the porous structure of electrode material in order to increase the effective interfacial surface between electrodes and the electrolyte. Until now, this approach almost has met the saturation limit of capacitance versus specific surface area.[1] Moreover, the working voltage of EDLCs is limited to 2.7 V even in the stable organic electrolytes.[2] So the energy density of EDLCs ($E = 0.5CV^{2)}$ is heavily shackled nowadays (<4 Wh/kg in the practical application). As compared in the Ragone plot for different electric storage devices (Fig. 14.1), this type of capacitor can work safely at high voltages up to 3.5 V using stable organic electrolytes.

Conventional EDLC has high-power density but low-energy density, whereas batteries have high-energy densities but lack of adequate power densities. To get a good compromise between the above two systems, we attempted a novel megalo-capacitance capacitor (MCC) based on a graphitic carbon cathode and an AC anode, which possess both satisfactory power density and high-energy density.[3]

Figure 14.2 compares the typical X-ray diffraction (XRD) patterns of the carbon electrode materials in capacitors. AC (electrode materials in conventional EDLCs) shows very broad and weak diffraction peaks, which implies its low crystallinity. On the other hand, nanogate carbon demonstrates comparatively sharp diffraction peaks.[4] It is in fact an intermediate between highly crystalline graphite and disordered

M. Yoshio (✉) H. Nakamura, and H. Wang
Saga University, Saga 840-8501, Japan
yoshio@cc.saga-u.ac.jp

M. Yoshio et al. (eds.), *Lithium-Ion Batteries*,
DOI 10.1007/978-0387-34445-4_14

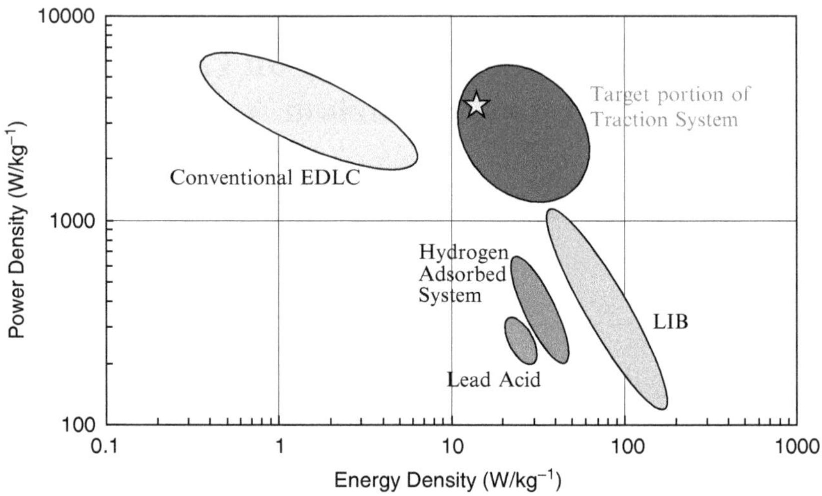

Fig. 14.1 Ragone plot of different electric storage devices

carbons. By contrast, the graphite cathode material applied in this study has the highest crystallinity among these three carbon materials. In MCC, we applied *highly crystalline graphite* (for example, KS6 from Timcal Co. Ltd.) as the cathode material and AC as the anode material.

Figure 14.3 compares the galvanostatic charge-discharge curves of a conventional EDLC and a MCC. The EDLC shows almost straight sloping curves in the voltage range from 0 to 2.7 V, but the MCC demonstrates bent curves. For instance, in the charge process, the cell voltage jumps up initially. From about 2 V, the rise-up trend of cell voltage starts to slow down. The curve roughly is composed of two portions with different slopes. To trace the respective potentials of graphite cathode and AC anode clearly, we introduced a reference AC electrode into the capacitor during the charge-discharge cycles. Figure 14.4 shows the potential profiles of graphite cathode and AC anode versus reference electrode in the first cycle of charge-discharge. The potential profiles of AC anode are simply linear curves. But those of graphite cathode are bent ones. In the charge process, the potential of graphite initially rise up swiftly to 1.35 V and then level off with time.

The charge-discharge curves of MCC actually equal the difference between the potential profiles of graphite and AC electrodes. By adjusting the relative weight ratios between cathode and anode materials, we can get very large discharge capacities for graphite. As shown in Fig. 14.5, the discharge capacity increases with the rise in the *AC/graphite* weight ratio. As the AC/graphite weight ratio is larger than 12, the discharge capacity becomes larger than 120 mAh/g.

Actually, the surface area of graphite is much smaller than that of AC, but can store more anions than AC. In the high-voltage branch of charge curve, the storage mechanism of anions is different from the common ion adsorption mechanism in

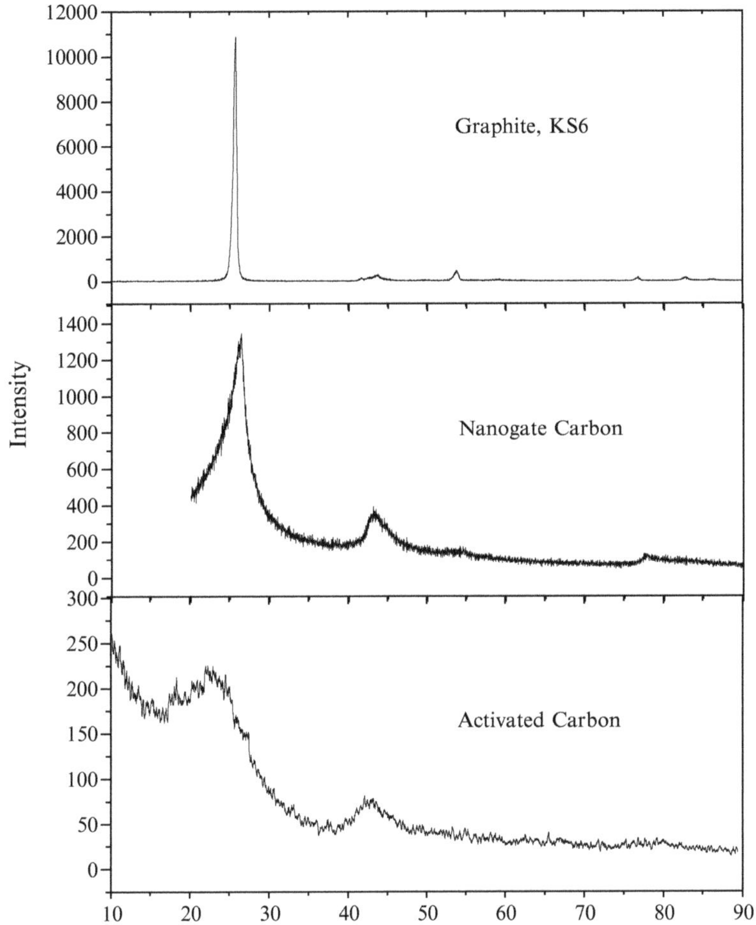

Fig. 14.2 Typical XRD patterns of the carbon materials in capacitors. Reproduced with permission from Ref. 6, copyright (2006), Electrochemical Society

conventional EDLCs. It is probably due to a strong interaction between the anion and graphite cathode. It is expected that the outer surface area of graphite is too small to accommodate so many anions. Thus the insertion of anions into the interlayer spaces between graphene layers of the graphitic carbon likely takes place at high voltage. Figure 14.6 shows the in situ XRD patterns of *KS6* cathode during the course of initial charge of an MCC [electrolyte, 1-M TEABF4-PC; tetraethylammonium (TEA)]. From OCV to 3.2 V, merely the (002) diffraction peak can be observed. Once the cell voltage exceeds 3.3 V, the (002) peak splits into small shoulder peaks at low Bragg angles. This result means the shallow insertion of anions into the KS6 cathode.[6] Deep intercalation into the electrode crystal lattice actually is not good for the performance of MCC since the drastic volume change

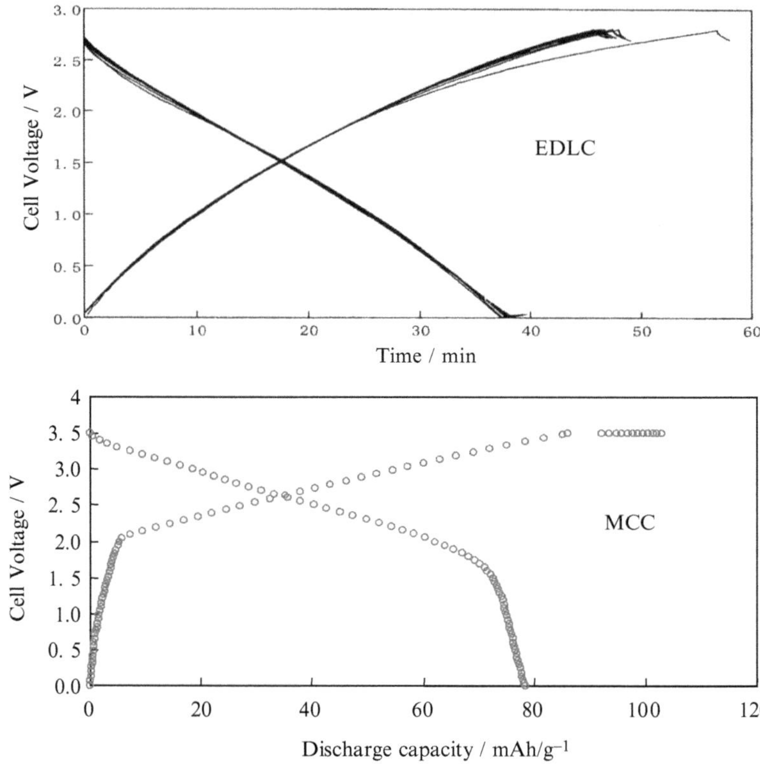

Fig. 14.3 Galvanostatic charge-discharge curves of a conventional EDLC (the AC/AC weight ratio 1; AC: stream activated carbon) and a MCC (the weight ratio of AC/KS6 is 4:1)

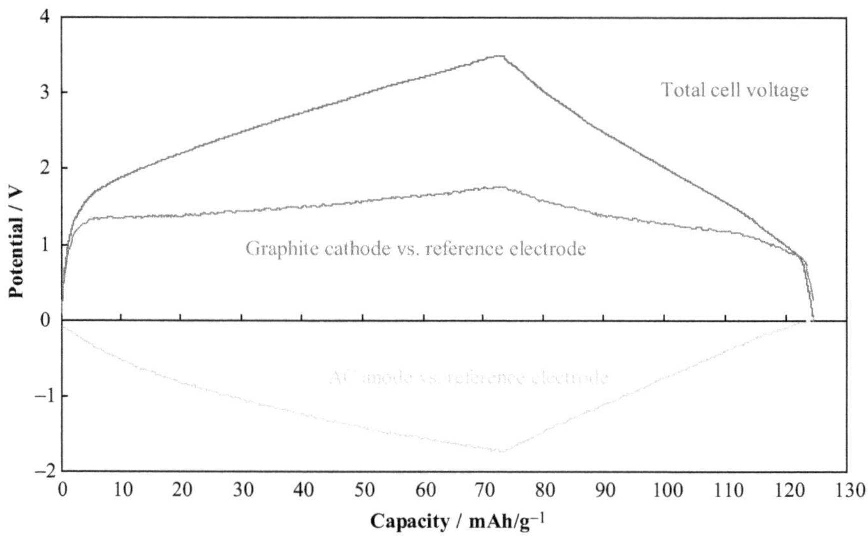

Fig. 14.4 Potential profiles of graphite cathode (KS6) and AC anode during the initial charge-discharge of KS6/AC capacitor (the weight ratio of AC/KS6 is 2:1)

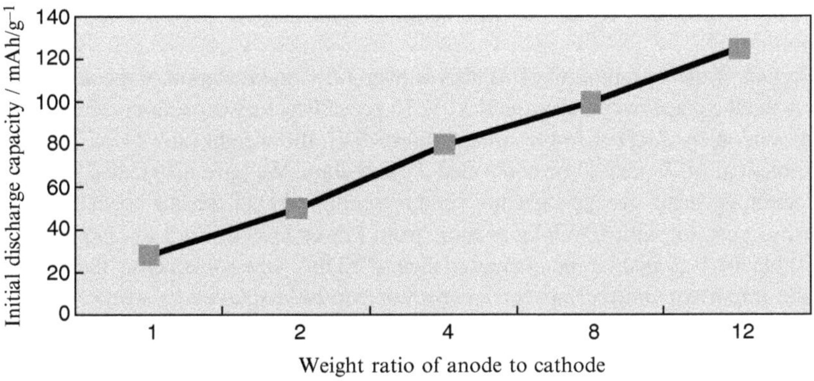

Fig. 14.5 Relationship between the weight ratio of AC anode to graphite cathode and initial discharge capacity of graphite cathode

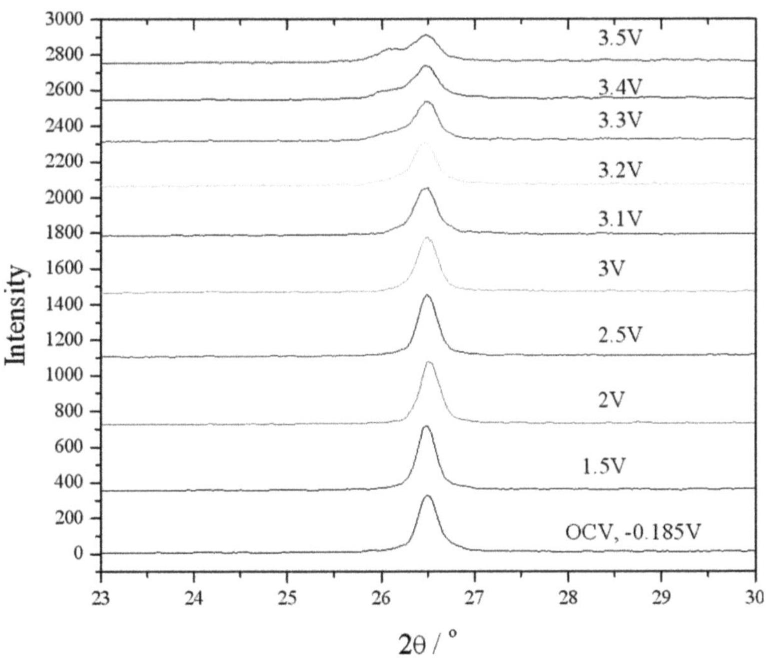

Fig. 14.6 In situ XRD patterns of KS6 cathode in KS6/AC capacitor in the initial charge process (weight ratio of AC/KS6 is 1). Reproduced with the permission from Ref. 6, copyright (2006), Electrochemical Society

during intercalation will deteriorate the cycle performance. Furthermore, the diffusion of anions inside the interlayer spaces is a rather slow process, which can retard the rate capability of capacitor. It is noted that quite a large part of the capacity is delivered in the voltage range lower than 3.3 V where no intercalation can be observed from in situ XRD. So the storage mechanism of anions at KS6 cathode is somehow different from the intercalation mechanism used to be in battery community.[6]

In fact, if the weight ratio of AC/KS is over 1.5:1, intercalation of the anion is very likely to take place over than around 3.0V. To get satisfactory capacitor performance (no intercalation by XRD at lower voltage than 3.5V), the weight ratio of AC/graphite in the practical MCC should be controlled around unity. We have also called this kind of capacitor as "nano-storage capacitor" and now commercialized as a novel high energy density capacitor with 10Wh/kg or more from Power System Co. Ltd., Japan.

Table 14.1 compares the characteristics of EDLC and nonaqueous batteries. One of the important fingerprints for a capacitor can be displayed by cyclic voltammograms (CV). As shown in Fig. 14.7, the CV of graphite cathode in MCC is quite different from the cathode material ($LiCoO_2$) in lithium-ion batteries. Unlike the sharp intercalation peaks in the CV of $LiCoO_2$, nearly rectangular waves can be seen in the CV of graphite. Moreover, the sweep rates in the CV for graphite are quite large, which implies the high rate capability of graphite cathode (high power).

Figure 14.8 shows the cycle performance of MCCs. The capacitance values can be kept to over 70% of their initial values. This result is comparable to that of conventional EDLCs.[7] Figure 14.9 shows the temperature dependence of MCCs. At temperatures lower than 0°C, the capacitance values drop drastically with the

Table 14.1 Some characteristics of EDLC and nonaqueous battery

	EDLC	Nonaqueous battery
Charge storage mechanism	Ion adsorption at surface	Ion intercalation into crystal lattice
Shape of charge-discharge curves	Straight sloping line	Plateau
Cycle life	Over tens of thousand cylces	About 300 cycles
Power density	>200 W/kg	About 100 W/kg
CV shape	Rectangular	Presence of peaks
In situ XRD	No shift or split of diffraction peaks	Shift or split of diffraction peaks
Theoretical capacity	Cannot be calculated	Can be calculated from chemical formula

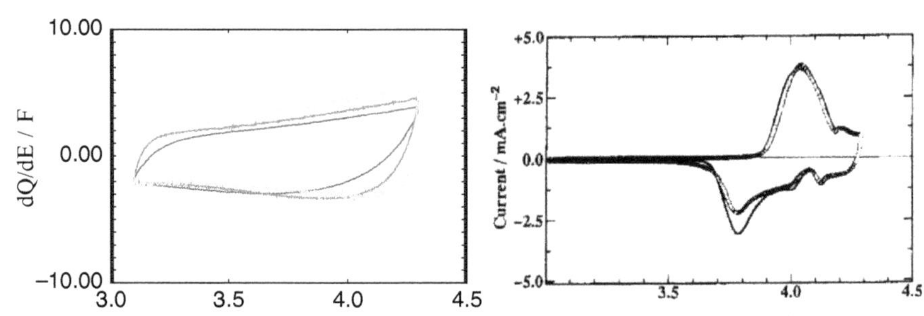

Fig. 14.7 Cyclic voltammograms of graphite cathode for MCC and $LiCoO_2$ cathode for lithium-ion battery

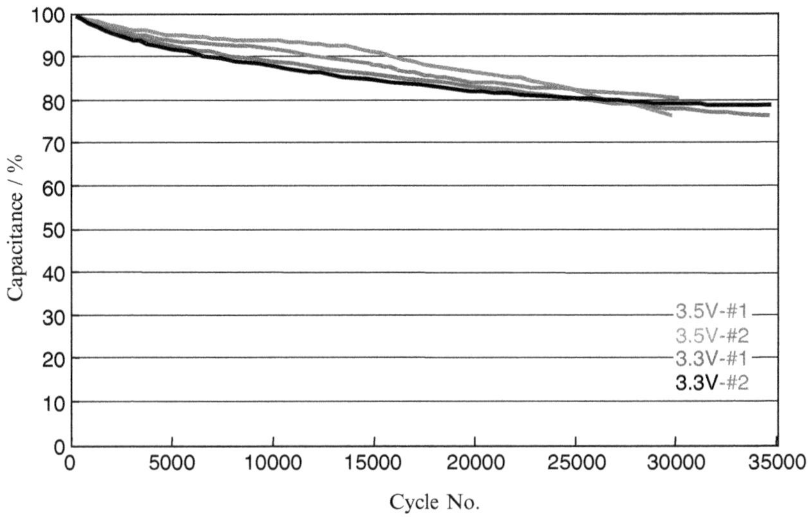

Fig. 14.8 Cycle performance of MCCs at 25°C (rate: 10 C). Reproduced with permission from Ref. 7, copyright (2006), Electrochemical Society

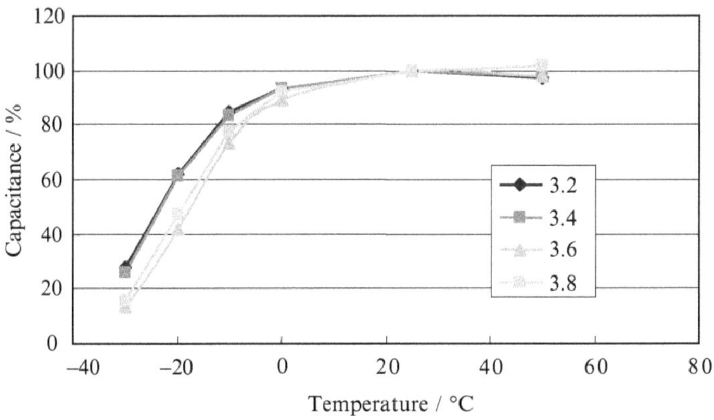

Fig. 14.9 Temperature dependence of MCCs

Table 14.2 Comparison between conventional EDLC and MCC

	Conventional	MCC	Unit
Capacitance	1,350	2,700	F
Voltage	2.7	3.5	V
Resistance	0.0015	0.006	ohm
Mass	0.21	0.23	kg
Volume	0.15	0.15	l
Energy density	6.5	19.5	Wh/kg
	9.1	30.0	Wh/l
Power density	5,786	2,219	W/kg

decrease of temperature. However, the low-temperature performance of MCCs is close to that of EDLCs.

Table 14.2 compares the features of MCC and EDLC. MCC has both a higher working voltage range and a higher density than EDLC. Accordingly, the energy density of MCC amounts to almost three times of that for EDLC. The MCC is a very promising electric storage system in the near future.

More recently, we have also explored the new type of electric energy storage devices with safety based on the graphite positive electrodes and metal oxides negative electrodes, such as Graphite/Nb_2O_5,[8]. Graphite/TiO_2 or Graphite/$Li_4Ti_5O_{12}$.[9] This kind of capacitors have more energy density and are more safe than LIB battery.

References

1. O. Barbieri, M. Hahn, A. Herzog, R. Kotz, *Carbon*, **43** (2005) 1303.
2. M. Hahn, A. Wursig, R. Gallay, P. Novak, R. Kotz, *Electrochem. Commun.*, **7** (2005) 925.
3. M. Yoshio, H. Nakamura, H. Wang, *Electrochem. Solid-State Lett.,* **9** (2006) A591.
4. M. Uchiyama, M. Okamura, *Jap. Pat. Kokai*, H11–317333 (1999).
5. M. Yoshio, H. Nakamura, H. Wang, *Electrochem. Solid-State Lett.,* **9** (2006) A561. Reproduced by permission of The Electrochemical Society.
6. J.A. Seel, J.R. Dahn, *J. Electrochem. Soc.*, **147** (2000) 892.
7. M. Yoshio, H. Wang, T. Ishihara, H. Nakamura, *ECS Trans.,* **3** (2006) 71.
8. A.K. Thapa, T. Ishihara, N. Moriyama, T. Kawamura, H. Nakamura. M. Yoshio, ECS Transactions, 16 (2008) 177.
9. Gum-Jae Park, Hiroyoshi Nakamura, Yun-Sung Lee, Masaki Yoshio, Koji Takano, ECS Transactions, 16 (2008) 169.

Chapter 15
Development of $LiCoO_2$ Used for Rechargeable Lithium-Ion Battery

Hidekazu Awano

15.1 Introduction

Currently, the lithium-ion battery is an indispensable battery for the power sources of mobile devices such as notebook-sized personal computers and mobile phones since the rechargeable lithium-ion battery was commercialized for the first time in 1990. In particular, the rechargeable lithium-ion battery and the mobile phone have both expanded their markets. The rechargeable lithium-ion battery has become widely used in the Asian countries in connection with the popularization of mobile phones in South Korea and China. The performance of the rechargeable lithium-ion battery (in particular, capacity) is improved every year. Initially, the number of manufacturers of rechargeable lithium-ion batteries was as high as 10–11 companies in Japan, including newcomer companies, while the number in China has not been accurately counted. On the other hand, the battery price is continuously decreasing. Currently, some manufacturers plan to withdraw from the rechargeable lithium-ion battery business, and there is the possibility that the existing manufacturers might be consolidated into a few companies in the near future. The manufacturing of a battery is not easily accomplished just by introducing new equipment; advanced know-how is required for its manufacturing. Therefore, it is assumed that the Japanese manufacturers still dominate the overseas manufacturers in this field. Moreover, when we pay attention to other energy devices, the developments of the capacitor and the fuel cell are also proceeding, and it is expected that they will compete or coexist with the rechargeable lithium-ion battery in the future. Nippon Chemical Industrial Co. Ltd., also is involved in the development so that we can survive as a supplier of the positive electrode-active material in the future.

In this chapter, we will introduce the manufacturing process and the characteristics of $LiCoO_2$, which is mainly used as the positive electrode-active material of

H. Awano
Nippon Chemical Industrial Co., Ltd., 9-11-1, Kameido, Koto-ku Tokyo 136-8515, Japan
hidekazu.awano@nippon-chem.co.jp

M. Yoshio et al. (eds.), *Lithium-Ion Batteries*,
DOI 10.1007/978-0-387-34445-4_15

the rechargeable lithium-ion battery. However, some technologies are confidential; therefore, we would like to ask you to overlook this area if not easily understood.

15.1.1 Background of Development

Nippon Chemical Industrial Co., Ltd. has been manufacturing and selling lithium salts at its establishment. The names of the products and their uses are listed in Table 15.1. $LiCoO_2$, which is used for the positive electrode-active material of the rechargeable lithium-ion battery, was selected in order to place a lithium salt product with a high additional value on the market. Its development began in 1991. The supply of samples to battery manufacturers began in 1992. The initial evaluation by the customer was finished in 1993, and several tens to hundreds of kilograms of product were experimentally manufactured using a pilot plant. The maximum production of 2 tons/month was recorded in 1994. The design of the mass production plant was started at the same time, and the mass production plant of 20 tons/month was set up in 1995. Afterward, the capacity of the plant was increased many times since then.

15.1.2 Why is $LiCoO_2$ Used?

As described in the first paragraph, $LiCoO_2$ is used mainly for the positive electrode-active material of the present rechargeable lithium-ion batteries. In the early stage of its development, it was expected that $LiCoO_2$ had problems with price and the resource of cobalt, and would soon be substituted by nickel-type or manganese-type material. Even in presentations in the journals, conferences, and so forth, at that time, examples of studying $LiCoO_2$ were few, although the development of the nickel-type and manganese-type were active. However, in actuality, the commercial rechargeable lithium-ion batteries using the nickel-type or the manganese-type material are still limited. Table 15.2 shows the features of each material. It can be understood that $LiCoO_2$ is a material that excels in the balance of the capacity,

Table 15.1 Lithium salts manufactured by Nippon Chemical Industrial Co., Ltd

Salt	Use
Lithium carbonate	Battery material; special glass
Lithium hydroxide	For carbon dioxide absorbent and grease
Lithium chloride	Aluminum welding material; humidity control material
Lithium bromide	For freezer and air conditioner
Lithium phosphate	Electronic and electric material
Lithium nitrate	Electronic and electric material

Table 15.2 Features of commercial rechargeable lithium-ion batteries

Material	Capacity[a]	Rate	High temp.	Safety	Cost	Process
$LiCoO_2$	B	B	B	B	C–B	A
$LiNi_{0.8}Mn_{0.1}Co_{0.1}O_2$	A	C	B	C	B	C
$LiNi_{1/3}Mn_{1/3}Co_{1/3}O_2$	B	C	C	B	B	B
$LiMn_2O_4$	C	C	C	A	B	C

[a]*A* excellent, *B* good, *C* fair

Fig. 15.1 $LiCoO_2$ powder

rate characteristics, cycling characteristics, high/low temperature characteristics, safety, and so on, when the battery characteristics of the rechargeable lithium-ion batteries are considered. Moreover, the manufacturing of $LiCoO_2$ can be carried out easily by mixing cobalt oxide with a lithium salt in a certain ratio and baking it at 700°C or more, as will be described later. Thus, the simplicity of its manufacturing process is considered to be one of the reasons why $LiCoO_2$ is used for the positive electrode-active material of rechargeable lithium-ion batteries.

15.1.3 Physical Properties of $LiCoO_2$

The main physical properties of $LiCoO_2$ are indicated in the following section (mainly for the $LiCoO_2$ manufactured by our company). A photograph of $LiCoO_2$ is shown in Fig. 15.1.

15.1.4 Resource and Price of Cobalt

In the past, cobalt generally has been recognized as an expensive raw material. The reason for this is that the cobalt is a speculation commodity, and its price always changes. Figure 15.2 shows the changes in the market price of cobalt since 1997. Thus, a very large change is observed. The price fell below $7 in April, 2002. (It is generally said that if the price becomes below $10, it does not pay the mining costs at the mine.) Afterward, it gradually increased, then rapidly increased to $28 in November, 2003. A possible reason for this increase is an increase in the amount of $LiCoO_2$ used for the positive electrode-active material because the rechargeable lithium-ion battery market is expanding quite rapidly. Moreover, as another reason for this increase, an increase in the concern of cobalt raw material also can be considered because each battery manufacturer announced his policy to expand the production volume of the rechargeable lithium-ion battery one after another. (The effect of the speculator is one of these factors.) Therefore, the battery manufacturers could not help raising the price of a cell by 8–10%. However, the transfer to the cell price does not appear to be occurring.

The Congo, Zambia, Canada, Morocco, and New Caledonia are considered the main places of production of the cobalt ore. The main use of the cobalt ore was for aircraft superalloys, hard metal tools, magnets, catalysts, and pigments before the rechargeable lithium-ion battery was developed. Therefore, the cobalt oxide that could be used for batteries was low. However, the number of manufacturers producing the cobalt oxide recently has increased due to the expansion of the market. In our company, various cobalt raw materials produced by 6–7 companies are always being evaluated.

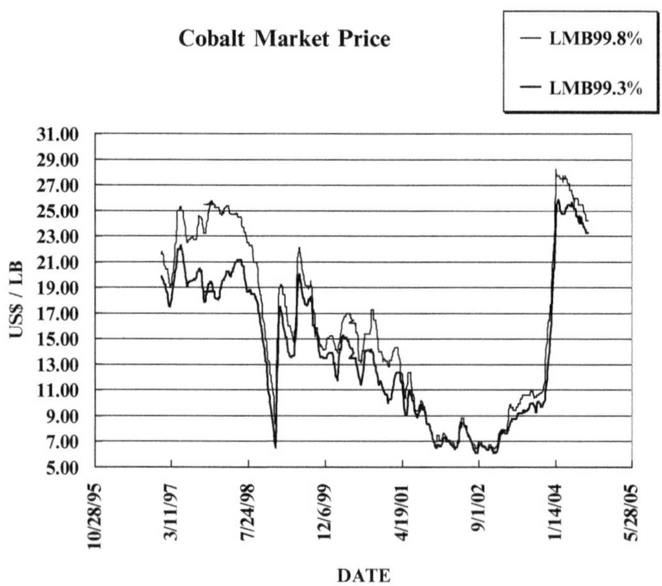

Fig. 15.2 Price fluctuation of cobalt

15.1.5 Trend of $LiCoO_2$ in the Future

Presently, the market price of cobalt is high; therefore, the application of $LiCoO_2$ to the positive electrode active material is depressed, and it is shifting to the nickel-type, manganese-type, and iron-type materials. However, it is expected that the price will settle to a certain level because a new cobalt project is scheduled to start after 2007. Moreover, it is considered that it will not become zero in volume because it is a material widely acknowledged in the market as a positive electrode-active material for rechargeable lithium-ion batteries. Furthermore, although mobile devices such as mobile phones and notebook-sized personal computers were the principal uses of the rechargeable lithium-ion battery, it is expected that the market for new uses (for power tools using a medium- to large-sized battery) will expand in the future. However, it is necessary to continue the effort to decrease the price of the material and to improve its characteristics. It is considered that there still remains enough room for improvement in the characteristics of $LiCoO_2$, and is described in detail in Chap. 3.

15.2 Manufacturing Method and Quality of $LiCoO_2$

Various synthesis methods were proposed at the beginning of development of the lithium-ion secondary battery. It is understood that the $LiCoO_2$ is obtained using a cobalt compound and a lithium compound as the raw materials and synthesizing it by heating as shown in Table 15.3. It is postulated that cobalt carbonate was first used as the cobalt compound. After large-scale production started, cobalt oxide (Co_3O_4) as the cobalt compound, and lithium carbonate (Li_2CO_3) as the lithium compound began to be mainly used due to the stability of their quality and supply,

Table 15.3 Examples of synthesis method of $LiCoO_2$

References	Synthesis method
Mizushima et al.[1]	Pellets consisting of a mixture of lithium carbonate and cobalt carbonate were calcined and then heat-treated at 900°C in air for 20 h. As a result of the analysis, the composition was assumed to be $Li_{0.99}Co_{1.01}O_2$
Molenda et al.[2]	Stoichiometric quantities of lithium carbonate and cobalt oxide were mixed and pressed into a pellet of 0.8-cm diameter, 0.1-cm high. It was heated at 1,170 K for 4 days
Reimer et al.[3]	Stoichiometric quantities of $LiOH{\cdot}H_2O$ and $CoCO_3$ were heated in air at 850°C
Gummow et al.[4]	The product (LT-$LiCoO_2$) synthesized in low temperature (400°C) exhibited low crystallinity and was more stable in the electrolyte compared to the product synthesized at high temperature (900°C)
Ohzuku et al.[5]	Stoichiometric quantities of Li_2CO_3 and $CoCO_3$ were calcined at 650°C in air for 12 h and then heat-treated at 850°C for 24 h
Gupta et al.[6]	Li_2CO_3 and Co_3O_4 were calcined at 550°C in air for 5 h, and then heat-treated at 850°C for 24 h

Table 15.4 Features of typical synthetic reaction of cobalt compound and lithium carbonate

	Co content (%)	Required amount (g) for synthesizing 1 kg of $LiCoO_2$		Reaction by-product (g) for synthesizing 1 kg of $LiCoO_2$		Preservation stability[a]
		Co salt	Li_2CO_3	CO_2	H_2O	
Cobalt oxide	73	820	380	225	–	B
Cobalt carbonate	49	1,220	380	675	–	D
Cobalt hydroxide	63	950	380	225	180	C

[a] *B* good, *C* fair, *D* bad

the easiness of handling, and so forth. Table 15.4 shows the features of the typical synthesizing reactions between the cobalt compound and the lithium carbonate. It can be understood that Co_3O_4, which produces fewer reaction by-products, is advantageous.

The reaction scheme for the $LiCoO_2$ is indicated by the following equation. It is a simple process:

$$2\ Co_3O_4 + 3\ Li_2CO_3 + 1/2\ O_2 \rightarrow 6\ LiCoO_2 + 3\ CO_2 \qquad (15.1)$$

Moreover, although the reaction of the cobalt oxide and the lithium carbonate is fundamental, it also is possible to synthesize $LiCoO_2$ by combining other cobalt salts and lithium salts. Formerly, it was possible to supply one grade of material to the customers of two to three companies. However, because other materials (electrolytic solution, electrolyte, negative electrode-active material, etc.), which are used in the rechargeable lithium-ion battery, are different and the required battery characteristics are different, we are now supplying a specific grade material for each customer or a specific grade material for each kind of battery (polymer type, rectangular type, and cylindrical type). Recently, a demand from our customers to improve the characteristics and to stabilize the quality has been gradually increasing.

15.2.1 *Manufacturing Method*

Figure 15.3 shows a flowchart of the manufacturing process of lithium cobaltate. As is understood from the flowchart, after mixing the cobalt oxide and the lithium carbonate at a certain Li/Co ratio, the mixture is baked. After it is baked, it becomes a block as shown in Fig. 15.4, so it is crushed and then packaged as a product. Hereafter, the most important point of each manufacturing process will be introduced.

15.2.1.1 Mixing Ratio

The molar ratio of lithium and cobalt is very important to the manufacturing of $LiCoO_2$. The particle size and the battery characteristics are significantly influenced

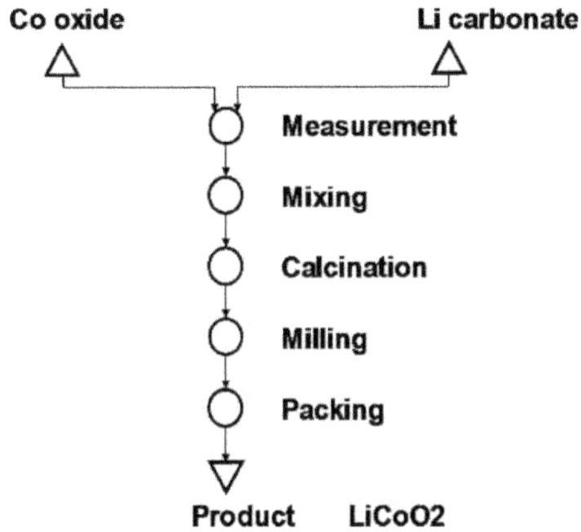

Fig. 15.3 Flowchart of manufacturing process of $LiCoO_2$

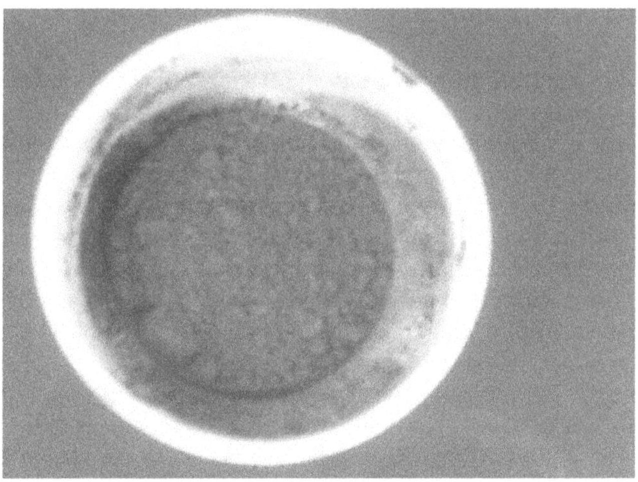

Fig. 15.4 Baked block

by changing the molar ratio. Therefore, it is a very important control point in the manufacturing process. In order to control this molar ratio within a predetermined range, controlling the purity of the lithium carbonate and the cobalt oxide raw materials is definitely required and also to accurately measure the weight of the raw materials. Moreover, it is important that the raw materials are accurately measured, and the entire quantity is placed in the mixer. When a strongly adhesive raw material is used, deviation in the blended charges from the predetermined rate might occur due to material remaining in the piping. This trouble is liable to be accidentally overlooked during examination in the laboratory.

15.2.1.2 Mixing Method

In the mixing process, it is important to uniformly mix the raw materials, which have been accurately measured. There are roughly two methods, i.e., the wet type and dry type. Generally, it seems that many manufacturers are adopting the dry type mixing for manufacturing $LiCoO_2$. An important point is that these two raw materials are uniformly mixed when using the dry process. The uniformity of the mixing is dependent on the kind of mixer. The selection of the mixer is based on the know-how of each company, although the details of the kind of mixer should be found in the manufacturer's literature. Moreover, the particle size, apparent density, and so forth of the lithium carbonate and cobalt oxide influence the homogeneity of the mixing state. Therefore, when the raw material is changed, it is important to optimize the mixing time, rotation speed, and so on. If the homogeneity worsens, the difference in a lot increases, the fraction of unreacted cobalt oxide and lithium carbonate increases, and the battery characteristics and so forth are negatively affected.

15.2.1.3 Baking Method

The blended charges are then baked in the baking furnace. This baking process significantly affects the quality of $LiCoO_2$. In particular, oxygen is consumed and carbon dioxide is released during the synthetic reaction as understood from the reaction equation. It is important to establish a baking method in which the supply of oxygen and discharge of carbon dioxide can be smoothly carried out. When running our study while considering this respect, it is necessary to select an electric furnace or gas furnace and to determine the firing atmosphere, firing time, and temperature rise pattern.[7] Moreover, various conditions also should be determined that include the kind of baking tray to be used for baking the blended charges, the amount of blended charges to be charged, and the thickness of the charging of the blended charges. In order to control the quality of the products, it becomes an important point as to how to control these points. The scanning electron microscopy (SEM) images of $LiCoO_2$ in various firing atmospheres are shown in Fig. 15.5 (under the conditions of the same firing temperature and Li/Co ratio).

It can be understood that the physical properties of $LiCoO_2$ changed depending on the baking atmosphere. The amount of charging of the blended charges also affects the quality of the $LiCoO_2$, although its effect is not remarkable as that of the baking atmosphere.

15.2.1.4 Milling

After baking, it becomes a lump as described in Sect. 15.2.1. It is necessary to mill this lump into particles of a certain size (in some cases, depending on the manufacturing method or the manufacturing condition, it might not become a lump). For $LiCoO_2$, strictly speaking, the expression "crushing" might be better than "milling." A typical SEM photograph of our $LiCoO_2$ is shown in Fig. 15.6.

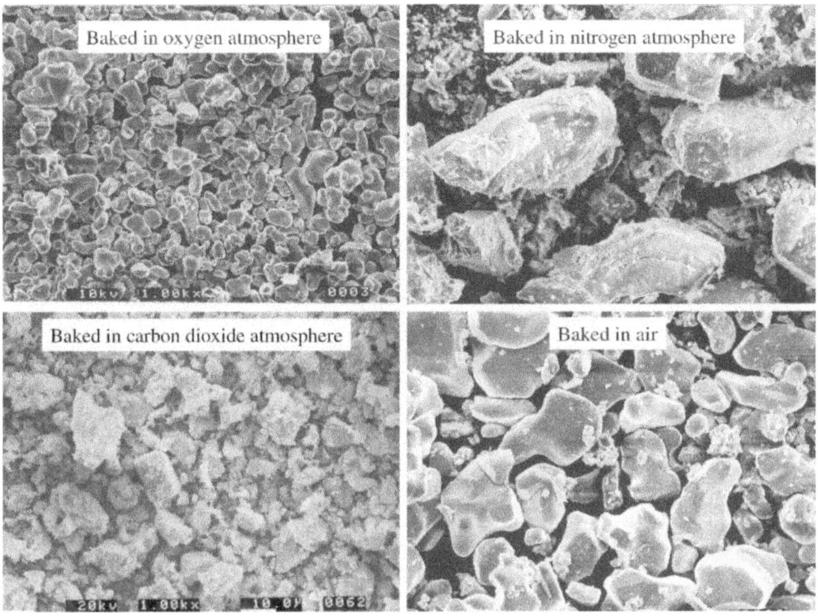

Fig. 15.5 Baking of $LiCoO_2$ in various atmospheres

Fig. 15.6 SEM photograph of $LiCoO_2$

During the milling process of our $LiCoO_2$, we do not mill the primary particle. It is confirmed in our test results that with excessive milling, which mills the primary particles, the number of fine powder particles increases and the battery characteristics, such as the charge and discharge cycle characteristics and so forth, deteriorate. Therefore, selection of the mill and optimization of the milling conditions are required.

15.2.2 *Qualities*

The physical properties listed in the following section are mainly enumerated as the quality of $LiCoO_2$:

- Average particle size, particle size distribution
- Specific surface area
- Water content
- Tap density
- Residual lithium content
- Crystallinity
- Particle shape
- Amount of impurities

All these items are assumed to influence the battery characteristics. Therefore, it recently has become important in manufacturing to stabilize these qualities. In order to stabilize these qualities, it is important to know how to control each manufacturing process as described in Section 2.1. In the following section, we will go into detail about the qualities, which are regarded as especially important.

15.2.2.1 Water Content

The battery characteristic deteriorates when the rechargeable lithium-ion battery is contaminated with water. Naturally, it is desired that the water content in $LiCoO_2$ used as the positive electrode-active material be controlled below a certain value. Moreover, currently a further decrease in the water content level is demanded by our customers. Although it may be considered that there is no problem because there is a process to make a slurry of the positive electrode-active material and to dry it, it is known that the water in $LiCoO_2$ consists of both physical adsorption and chemisorptions.[8] Because the water, which has been chemically adsorbed, cannot be removed during this drying temperature, it is important to stop the inclusion of moisture from the manufacturing process.

15.2.2.2 Residual Alkali

The lithium carbonate used as the raw material is alkaline. Basically, the ratio of Li/Co is 1. However, in our company, there are some grades where the lithium carbonate remains in the $LiCoO_2$ because they are baked using a blended charge having a composition with a slight excess of lithium. If the alkali remains in the product, there is a possibility of problems such that the viscosity of the slurry of the positive electrode-active material increases or the slurry becomes gelled [polyvinylidene fluoride (PVDF) is polymerized].[9] During the early stage of $LiCoO_2$ manufacturing, this problem frequently happened in the process of making the slurry. Moreover, it was reported that the content of silicon also is important for the effect of the residual alkali.[10] At present, a check of the alkali is carried out in the process and the quality problem such as gelation of the electrode slurry has not occurred.

15.2.2.3 Specific Surface Area

The specific surface area has a close relationship with the particle size described in the next paragraph. The surface is a place where the reaction with electrolyte occurs and the stabilization of the quality is desired. Our $LiCoO_2$ has a surface smoothness feature.[11]

15.2.2.4 Particle Size

The particle size control is a very important item for the quality of $LiCoO_2$ in our company. In general, the particle size distribution of our $LiCoO_2$ has a sharp peak[12]; however, an attempt to improve the performance by adjusting the particle size distribution also is being performed (described in Section 3.1). Our $LiCoO_2$ is being evaluated in the market as having a high safety for the rechargeable lithium-ion battery-positive electrode by controlling the particle size distribution and the particle shape.

15.2.2.5 Impurities

During the early stage of the development of $LiCoO_2$, we used materials that were as pure as possible. However, there is still room to examine what level of purity is required. Examination of the impurity level is required in order to decrease the price of $LiCoO_2$ in the future. However, it is important that metal impurities should be as low as possible. It is necessary to pay close attention to any contamination from the raw materials and contamination from the manufacturing facilities. Many metal parts are used in the manufacturing facilities of cobalt oxide as the raw material and $LiCoO_2$. The contamination from the process due to the wearing of the metal parts may cause an internal short circuit of the battery. In particular, $LiCoO_2$ has a high hardness and there is a significant possibility of wearing down the mill and transport facilities. The hardness of $LiCoO_2$, which was assumed by comparing the result of the pushing hardness measured by the micro-Vickers hardness test method with the correlation chart of Mohs hardness,[13] corresponds to 6–8 of Mohs hardness (cf Fig. 15.7b). It is almost the same hardness as that of quartz (Mohs hardness = 7). Consequently, it is important for manufacturing facilities to take measures against the wearing down of the metals. At present, a battery system of 2,000 mAh or more is the major design for the 18650 type and production of higher capacities, which is more than twice that of the early stages, is occurring. In order to secure the quality of batteries, decreasing the amounts of metal impurities as the capacity increases is required.

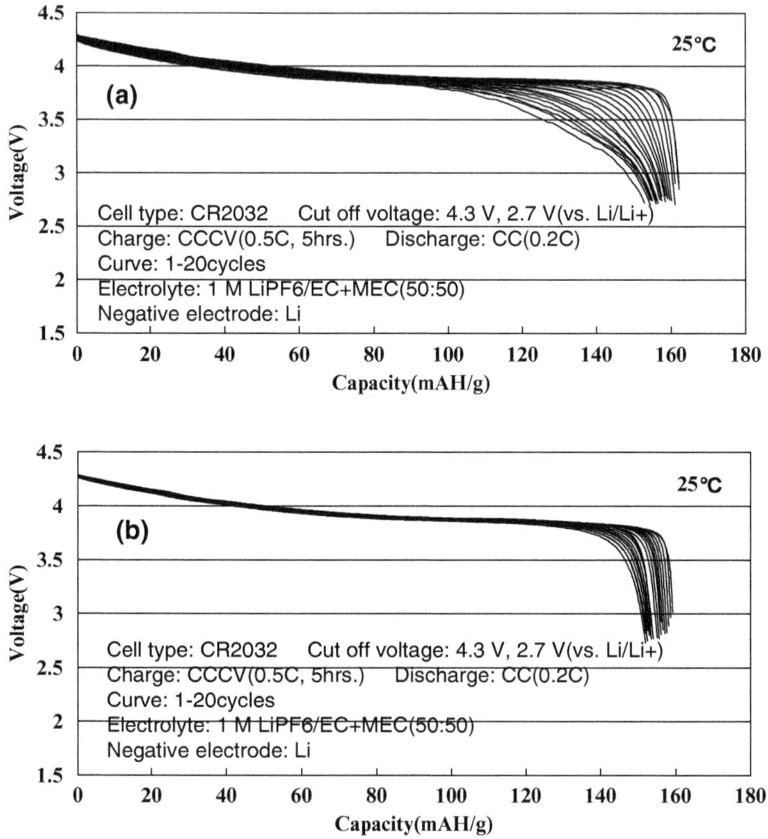

Fig. 15.7 (**a**) Cobalt oxide A and (**b**) cobalt oxide B

15.3 Further Improvement in the Characteristics of $LiCoO_2$

15.3.1 Particle Size Control

The shapes, the mean particle size, and the specific surface area of $LiCoO_2$ can be controlled by optimizing the mixing ratio, the mixing method, the baking method, and the milling conditions. The control of the particle size is comparatively easy compared with other nickel-type and manganese-type materials. Because $LiCoO_2$ is comparatively stable at high temperature (700–1,000°C), and when it is baked at high temperature the particle grows due to the sintering, a particle of any size can be obtained. On the other hand, for $LiNi_{0.8}Mn_{0.1}Co_{0.1}O_2$ and $LiMn_2O_4$, the temperature range where they are stable is narrower than that of $LiCoO_2$, and its grain growth cannot be easily performed. Therefore, the adjustments of the MnO_2 and co-precipitated hydroxide, which is the raw material of NiMnCo, are required. This easiness in treating might be one of the reasons why $LiCoO_2$ is favored. As described above, because the particle size and the specific surface area significantly affect the cell characteristics, their control is important.

15.3.2 Selection of Raw Material

As mentioned above, lithium carbonate is used as the raw material for lithium, and cobalt oxide is mainly used as the raw material for cobalt. However, there are many kinds of lithium carbonates and cobalt oxides, which have various powder properties. The characteristics of the synthesized $LiCoO_2$ is significantly different depending on the mean particle size, the liability to flocculate, and the impurities as shown in the example below. The examples are shown in Fig. 15.7. It can be understood that the two kinds of $LiCoO_2$, which have similar physical properties, exhibit different charge-discharge cycle characteristics.

15.3.3 Control of Impurities

There is an example where additives are positively used in order to satisfy the various needs of the battery characteristics at present. Many methods have been reported. They include methods to substitute another element (Ni, B, Al, etc.) for the cobalt site in order to reduce the change in the crystal lattice due to the occlusion and evolution of lithium,[13–16] and a method to substitute iron in order to reduce the cost.[17]

At present, various additives have been evaluated; however, some elements remarkably decreased the electric capacity. It is important to balance the kind and amount of additives, otherwise the total performance might decrease. On the other hand, the product quality might be negatively affected by other impurities. Because almost no contamination of impurities comes from the manufacturing process, controlling the raw material is important.

15.3.4 Surface Treatment

For the lithium-ion battery-positive electrode used for a high oxidation potential, it is considered that the control of the surface conditions will become the main point of development in the future. Recently, many attempts to improve the surface conditions of the lithium-ion battery electrode were carried out. A surface coating with an electric conductive material in order to improve the electric conductivity[18] and an improvement in the cycling characteristics by coating the oxide with Al_2O_3, ZrO_2, TiO_2, and MgO on the surface have been reported.[19] It is reported that a big difference was obtained when the cycle test was carried out under the conditions of the high potential of 4.5 V,[20] although no difference occurred with a low potential (4.2 V vs. Li/Li+). In connection with this phenomenon, there is a report suggesting that the cycling capability and the elution of cobalt into the electrolyte in the form of Co^{4+} can be correlated when the charging and discharging are repeated.[21] It will be important for the development in the future to know how to prevent these phenomena by surface modification and structure modification.

15.3.5 Producing Higher Capacity (Higher Density, Higher Voltage, Nickel-Type Material)

When the method for developing a higher capacity is discussed, the existence of the nickel-type material cannot be disregarded. The possibility of the nickel-type material had been discussed already before the lithium-ion battery was commercialized. Although the nickel-type material is being used for some batteries on the market, the amount being used is not very significant to replace $LiCoO_2$ yet. It appears to be a reality that the various characteristics, such as safety, residual lithium on the surface, and low electrical discharge potential, which have been discussed, have not yet been solved completely. The merits of $LiCoO_2$ are superior to those of the nickel-type material when considering the balance of the battery characteristics, even if there are some improved characteristics for the nickel-type. Therefore, it will be important to improve $LiCoO_2$ by making the material more attractive. When the method to increase the capacity of $LiCoO_2$ is considered, it seems that increasing the density of the electrode and the extent of raising the working voltage are the main problems. However, these two requirements are trade-offs. For instance, it is known that if the particle size is increased in order to increase the electrode density, the charging/discharging rate characteristics and the cycling characteristics become inferior, and if the operating voltage is increased, the safety and the cycling characteristics become insufficient although a high capacity is obtained. Therefore, the technology, which improves the total performance such as the cycling characteristics, the charging/discharging rate characteristics, the safety, and so forth, is required in order to obtain a high capacity.

15.4 Conclusion

With respect to $LiCoO_2$ of the positive electrode-active material for the lithium-ion secondary battery, some of the technological changes have been described. It is said that the improvement in the capacity of the lithium-ion secondary battery based on the positive electrode material is not proceeding as quickly compared with the improvements in negative electrode material. A significant reason is that an appropriate material that has a good balance, which can be used as a replacement of $LiCoO_2$, has not been found. However, a sudden rise in the cobalt price occurred in recent years, and the examination of $LiNi_{1\text{-}x\text{-}y}Mn_y$(or Al_y)Co_xO_2, $LiNi_{1/3}Mn_{1/3}Co_{1/3}O_2$, and $LiMn_2O_4$ is becoming active again as alternative materials. Although it is considered that these new electrode materials will take the place of $LiCoO_2$ in the future, the fact that $LiCoO_2$ is quantitatively dominating the market will be unchanged for a while.

References

1. Mizushima et al., *Mater. Res. Bull.*, **15**, 783 (1980).
2. J. Molenda et al., *Solid State Ionics*, **36**, 53 (1989).
3. J.N. Reimers et al., *J. Electrochem. Soc.*, **139**, 2091 (1992).
4. R.J. Gummow et al., *Mater. Res. Bull.*, **27**, 327 (1992).
5. T. Ohzuku et al., *J. Electrochem. Soc.*, **141**, 2972 (1994).
6. R. Gupta et al., *J. Solid State Chem.*, **121**, 483 (1996).
7. N. Yamazaki, K. Negishi, Japanese Examined Patent Application Publication NO.3274016.
8. N. Yamazaki, K. Negishi, H. Awano, Japanese Unexamined Patent Application Publication NO. 10–334919.
9. N. Yamazaki, K. Negishi, M. Kikuchi, Japanese Unexamined Patent Application Publication NO. 10–64518.
10. N. Yamazaki, H. Awano, K. Negishi, Japanese Unexamined Patent Application Publication NO. 11–162465.
11. N. Yamazaki, H. Awano, K. Negishi, Japanese Unexamined Patent Application Publication NO. 11–125325.
12. N. Yamazaki, K. Negishi, Japanese Examined Patent Application Publication NO. 3396076.
13. H. Tsukamoto et al., *J. Electrochem. Soc.*, **144**, 3164 (1997).
14. S. Levasseur et al., *Solid State Ionics*, **128**, 11 (2000).
15. T. Ohzuku, M. Kouguchi et al., The 35th Battery Symposium in Japan Abstract, 2C01, P129 (1994).
16. H. Mishima et al., The 35th Battery Symposium in Japan Abstract, 3C06, 175 (1994).
17. H. Tacbuchi et al., The 40th Battery Symposium in Japan Abstract, 1C02, 231 (1999).
18. M. Kadowaki et al., The 42th Battery Symposium in Japan Abstract, 1A01, 86 (2001).
19. G.T. Fey et al., IMLB 12th Meeting, Abstract No. 36 (2000).
20. H. Kurita et al., The 44th Battery Symposium in Japan, Abstract 1C05, 284 (2003).
21. G.G. Amatucci et al., *Solid State Ionics*, **83**, 167–173 (1996).

Chapter 16
Cathode Materials: $LiNiO_2$ and Related Compounds

Kazuhiko Kikuya, Masami Ueda, and Hiroshi Yamamoto

16.1 Introduction

The layered $LiNiO_2$ is expected to be a cathode material for the next generation of lithium-ion battery. Figure 16.1 shows the relationship between discharge capacity and voltage for cathode materials for the lithium-ion battery. It means that $LiNiO_2$ has a higher capacity compared with other materials. In spite of this, $LiNiO_2$ compounds have not been used in commercially available lithium-ion batteries. It is thought that $LiNiO_2$ presents the following potential problems:

- Cycle characteristic
- Thermal stability

Toda Kogyo Corporation is promoting the development for each cathode material for lithium-ion batteries depending on their use; for example, $LiCoO_2$ for notebook personal computers and cellular phones; $LiMn_2O_4$ for hybrid electric vehicles (HEV); and $LiNiO_2$ for high-capacity batteries. Special efforts are being made in the improvement of the above problems, i.e., cycle characteristics and thermal stability, for $LiNiO_2$, so that this compound can be included in the market trend of lithium ion batteries. This chapter presents the results of the process development to improve the above-mentioned problem.

16.2 Synthetic Process of $LiNiO_2$

Figure 16.2 shows the basic process of the production of $LiNiO_2$. There are three important steps in the preparation of $LiNiO_2$: the mixing process of the lithium salt and precursor, the synthesis of precursor with high-packing density, and the adding method of other elements to give thermal stability.

K. Kikuya (✉), M. Ueda, and H. Yamamoto
Toda Kogyo Corp., 1-26 Hibiki-machi, Wakamatsu, Kitakyushu, Fukuoka, 808-0021, Japan
kazuhiko_kikuya@todakogyo.co.jp

M. Yoshio et al. (eds.), *Lithium-Ion Batteries*,
DOI 10.1007/978-0-387-34445-4_16

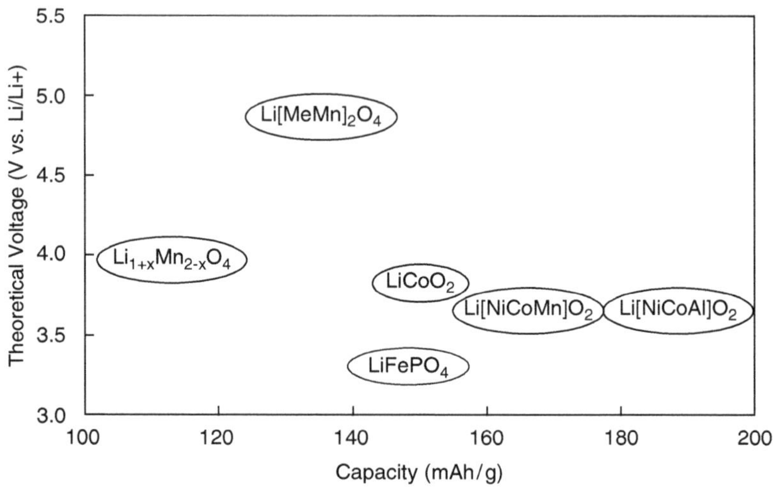

Fig. 16.1 Electrochemical potential of cathode material for lithium-ion battery

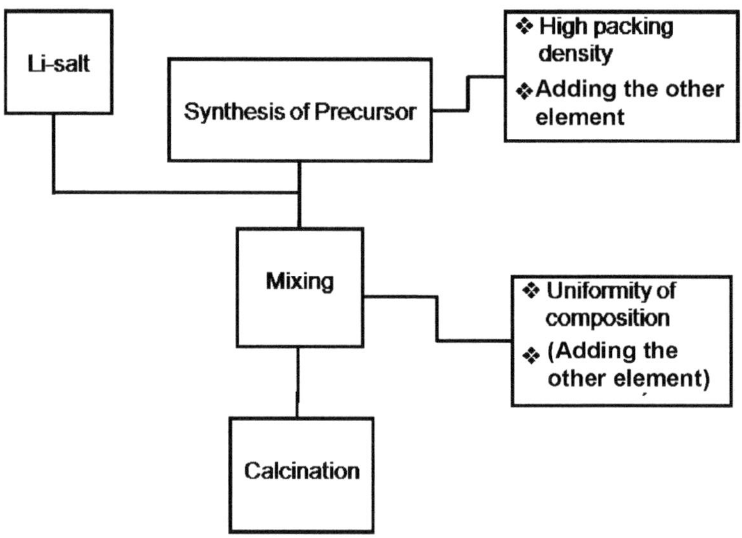

Fig. 16.2 General synthetic process of $LiNiO_2$

16.3 The Relationship Between Quality and Process

16.3.1 Synthesis of Precursor

Generally, $LiNiO_2$ is prepared using nickel hydroxide particles as a precursor, its high-packing density being one important characteristic of this compound. Toda Kogyo Corporation succeeded in the original development of a synthetic method to obtain a nickel hydroxide compound with highly dense particles. The $LiNiO_2$ with high density can be prepared by calcination after mixing the particles with the lithium salt. Table 16.1 shows the physical properties of the precursor and $LiNiO_2$, and Fig. 16.3 shows the scanning electron microscopy (SEM) image of $LiNiO_2$. In these results it was found that the particle size and tap density of $LiNiO_2$ strongly correlated to these properties of the precursor.

16.3.2 Addition of the Other Element

Extensive research has been devoted to improve the thermal stability of $LiNiO_2$ by the optimization and substitution of element such as cobalt, aluminum, and manganese in the nickel site and its amount.[1–4] There are two kinds of methods of aluminum doping: (1) aluminum doping by the wet process, and (2) aluminum doping by dry mixing. Figure 16.4 shows the X-ray diffraction (XRD) pattern of $LiNiO_2$ by both methods. The existence of extra diffraction in the wet mixing process of aluminum was not observed. On the other hand, the existence of Li_5AlO_4 was observed when the precursor and $Al(OH)_3$ were mixed by the dry method. That means that

Table 16.1 The comparison of physical properties between precursor and $LiNiO_2$

		Low-density particle	High-density particle
Precursor	TD (g/ml)	0.62	1.82
	PSD		
	D 10% (μm)	4.26	5.03
	D 50% (μm)	7.66	8.86
	D 90% (μm)	15.70	16.21
Products	TD (g/ml)	2.05	2.49
	PSD		
	D 10% (μm)	3.61	5.10
	D 50% (μm)	7.93	8.52
	D 90% (μm)	14.34	13.97

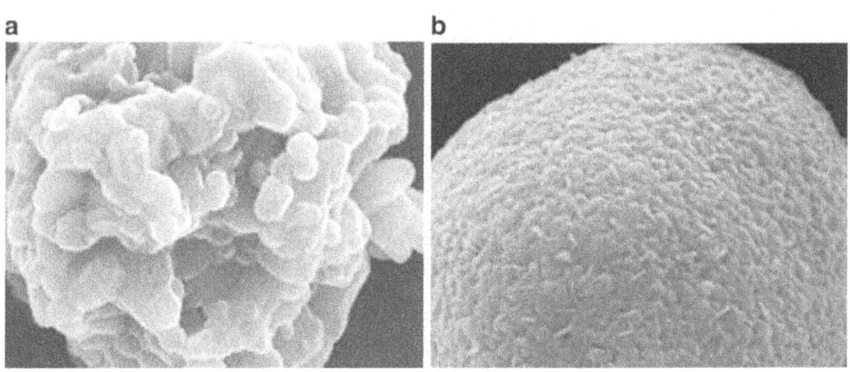

Fig. 16.3 Comparison of SEM images of low and high-density particles of $LiNiO_2$

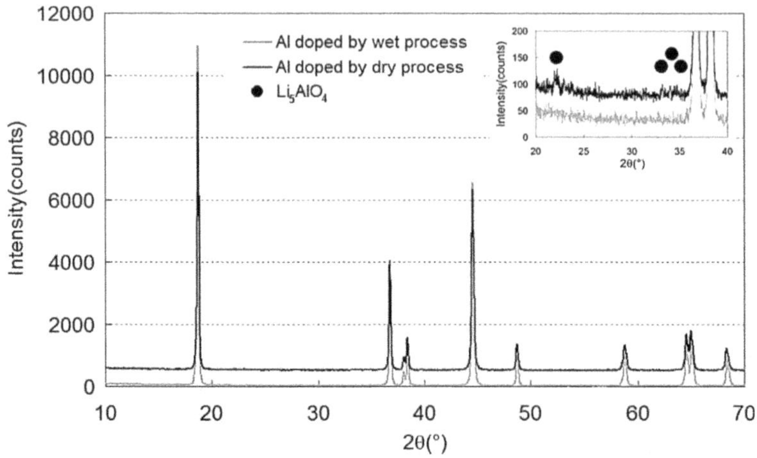

Fig. 16.4 Comparison of XRD patterns of $LiNiCoO_2$ by wet and dry aluminum doping processes

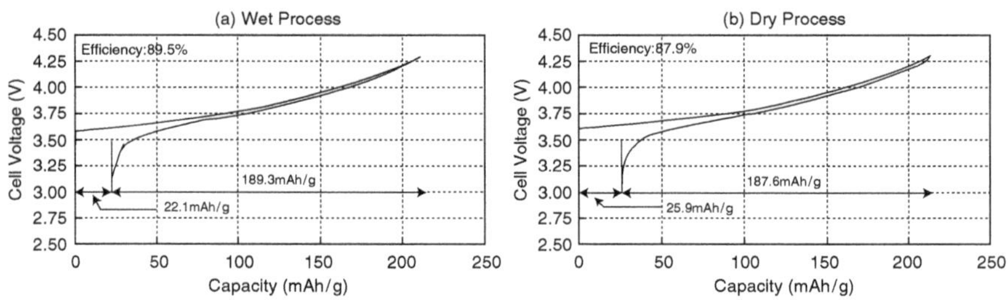

Fig. 16.5 Differences of electrochemical charge and discharge characteristics of $LiNiCoO_2$ by addition method of aluminum by **a** the wet process and **b** the dry process

aluminum did not dope uniformly. Figure 16.5 shows the results of the electrochemical characteristic of these materials. It seems that the initial cycle efficiency could be improved by the wet process. In addition, we analyzed differential scanning calorimetry (DSC) in the state of charge to confirm the thermal stability of these materials. Figure 16.6 shows that the exothermic peak point of the wet process material shifted to a higher temperature. These results showed that uniform doping of aluminum to nickel can improve initial efficiency and thermal stability, and such uniform doping can be obtained by wet mixing of aluminum.

16.3.3 Mixing Method of Precursor and Lithium Salt

A more uniform mixture is desirable, so that cation disorder may happen easily because of the ionic radius of Ni^{2+} and Li^+ may be approximated same in $LiNiO_2$.[5]

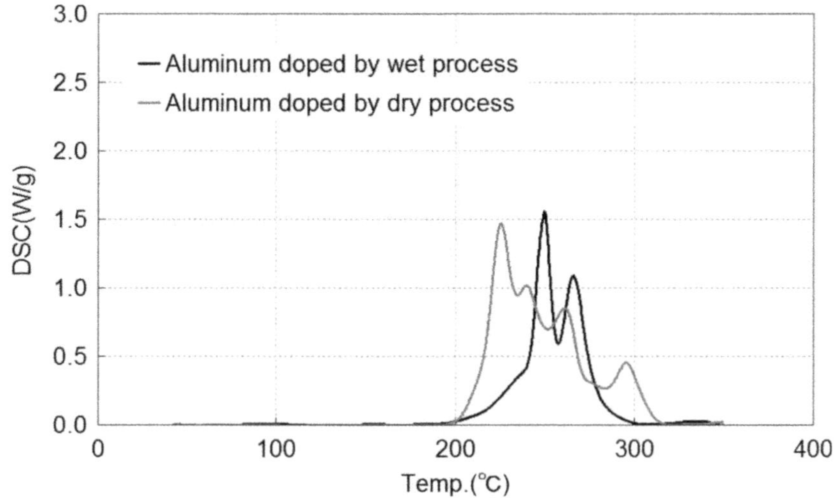

Fig. 16.6 DSC curves of $LiNiCoO_2$ in state of charge by different methods of adding aluminum

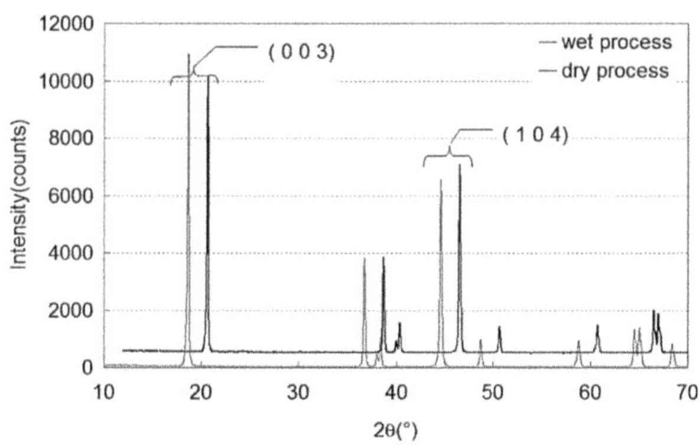

Fig. 16.7 The comparison of XRD patterns of $LiNiCoAlO_2$ by the wet and the dry lithium salt-mixing processes

The XRD pattern of calcinated $LiNiCoAlO_2$ is shown in Fig. 16.7 after mixing LiOH and precursor by the wet and dry processes. The 003/104 peak ratio was used as an indication of crystal structure for $LiNiO_2$.[6] The ratio of the wet mixing process was found to be higher compared with the dry mixing process (wet mixing = 1.39 and dry mixing = 1.08). This indicates that the crystallinity of the material by the wet LiOH mixing is better than dry mixing. Moreover, the electrochemical characteristics of these materials were evaluated. The material synthesized by the wet mixing process showed excellent cycle characteristics compared with those obtained by the dry mixing process (Fig. 16.8).

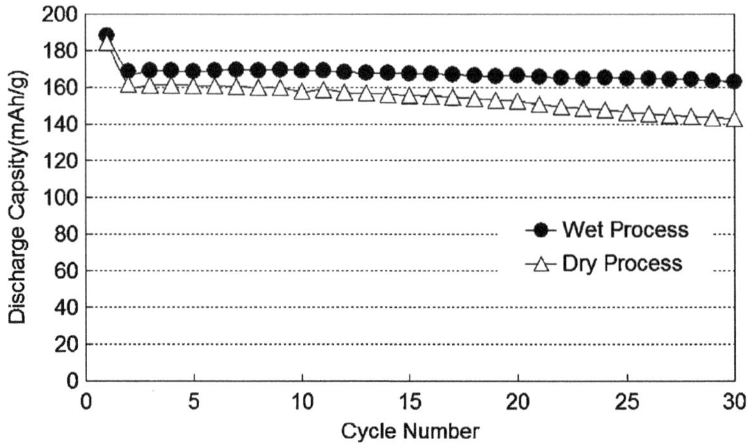

Fig. 16.8 The electrochemical characteristic of $LiNiCoAlO_2$ by the wet and the dry lithium salt-mixing processes

In addition, the ratio of Ni^{3+} by redox titration was examined.[7] The Ni^{+3} ratio of the wet process and the dry process were 99% and 95%, respectively. This can be attributed to cation disorder of Ni^{2+} and Li^{+}. These results indicate that the wet mixing process can uniformly mix the precursor and the lithium salt and it is effective in structural stabilization for $LiNiO_2$.

16.4 Summary

1. $LiNiO_2$ with high-packing density can be produced by using high-density precursor, and it should effectively improve the capacity of the lithium-ion battery.
2. The thermal stability of the lithium-ion battery can be more effectively improved by doping aluminum with the wet process instead of the dry process.
3. The structural stabilization can be achieved by the uniform mixture of the lithium salt and the precursor. It is expected that this can improve the cycle life of the lithium-ion battery.

These results indicate that the problems mentioned in the first section of this chapter can be resolved, and $LiNiO_2$ can be considered to be a cathode material of lithium-ion batteries.

References

1. T. Ohzuku, A. Ueda, M. Nagayama, Y. Iwakoshi, H. Komori, *Electrochem. Acta*, **38**(9), 1159–1167 (1993).
2. Q. Zong, U. von Sacken, *J. Power Sources*, **54**, 221–223 (1995).

3. T. Ohzuku, A. Ueda, M. Kouguchi, *J. Electrochem. Soc.*, **142**, 4033 (1995).
4. N. Takami , et al. *Electrochemistry*, **71**, 1162 (2003).
5. W. Li, J. N. Reimers, J. R. Dahn, *Am. Phys. Soc.*, **46**(9), 3236 (1992).
6. T. Ohzuku, A. Ueda, M. Nagayama, *J. Electrochem. Soc.*, **140**, 1862 (1993).
7. M. Yoshio, Y. Todorv, K. Yamato, H. Noguchi, J. I. Toh, M. Okada, T. Mouri, *J. Power Sources*, **74**, 46–53 (1998).

Chapter 17
Manganese-Containing Cathode-Active Materials for Lithium-Ion Batteries

Koichi Numata

17.1 Introduction

Manganese, which has a Clarke number of 0.06%,[1] is the tenth-most abundant element in the earth's crust, and has been utilized as a cathode-active material for manganese, alkaline-manganese, and lithium primary batteries. Mitsui Mining and Smelting (MMS) began the industrial production of electrolytic manganese dioxide (EMD) for such uses in 1949. Our company has been searching for new applications of manganese compounds. One major application is for use in lithium-ion batteries (LIB), for which we have been investigating two materials: lithium–manganese oxide spinel (LMO) and layer-structured material (LSM) containing manganese. These materials are expected to be more cost-effective than conventional lithium–cobalt oxide.

17.2 Lithium-Manganese Oxide Spinel

The stoichiometric composition of LMO is $LiMn_2O_4$. It has a cubic spinel structure with a lattice constant of 0.8248 nm. Lithium and manganese reside in the 8a tetrahedral site and 16d octahedral site in lattice positions of the Fd3m space group, respectively. The 8a site is located next to and shares a plane with the 16c octahedral site; it is assumed that lithium ions move along 8a, 16c, and 8a sites.[2]

The use of LMO in batteries was first reported by Hunter in 1982,[3] who prepared λ-MnO_2 by extracting lithium from LMO and applied it as a cathode-active material for lithium primary batteries. Application of LMO to a lithium secondary battery was first reported by Thackeray et al. in 1984.[4]

K. Numata
Corporate R&D Center, Mitsui Mining and Smelting Co. Ltd., 1333-2 Haraichi, Ageo, Saitama, 362-0021, Japan
k_numata@mitsui-kinzoku.co.jp

M. Yoshio et al (eds.), *Lithium-Ion Batteries*,
DOI 10.1007/978-0-387-34445-4_17

Since the commencement of mass production of LIB by Sony, a great deal of research and development effort has been performed to develop LMO for both consumer electronic devices and electric vehicles. LMO exhibit very different performance with changes in the preparation conditions, even if they have the same lithium/manganese ratio. LMO releases oxygen from about 700°C and decomposes to produce Li_2MnO_3 from about 950°C.[5] The released oxygen can be incorporated into the spinel framework by annealing at lower temperature. The oxygen deficiency is varied by changing the annealing conditions. As one of the major suppliers of EMD, we have been investigating the use of EMD as a raw material for LMO. EMD is a very dense material and can compensate for the small ideal capacity and the true density of LMO.

As mentioned above, LMO is expected to be cost-effective. However, there are some problems in that LMO has less capacity than lithium–cobalt oxide (LCO), and manganese dissolves from LMO into organic electrolytes, resulting in degradation of performance at high temperatures.[6] We have been attempting to resolve these problems by stabilization of the crystal structure and reduction of the specific surface area.

Our LMO production process is as follows. Lithium carbonate, EMD, and some other additives chosen to control the stability of material and the powder characteristics are used as raw materials. These materials are mixed thoroughly and heat-treated in a tunnel-type kiln at 700–900°C. Then, they are pulverized, sieved, and packed.

Table 17.1 shows some properties of our LMO. As the mean particle size of LMO is dependent on that of EMD used as a raw material, it is very important to choose EMD with the appropriate particle size, and it is sometimes necessary to control the particle size of EMD to meet the demands of the user.

The specific surface area should be small to reduce manganese dissolution. The smallest specific surface area of LMO is 0.2 m^2/g, and can be varied up to 1.5 m^2/g by changing the additives and heat treatment conditions.

Usually tapping density prepared from EMD is higher than that from chemical manganese dioxide (CMD). The typical tapping density of our product is 1.6 g/cm^3.

By adding foreign elements to the LMO system and adjusting the process parameters, we have improved the high-temperature performance of LMO. Table 17.1 shows some characteristics of our LMO with improved high-temperature characteristics. The control sample was a slightly lithium-rich spinel, Li/Mn = 0.54. Sample A contained more lithium and had a smaller specific surface area than the control. Samples B and C contained magnesium as a foreign element.

Table 17.1 LMOs with improved high temperature characteristics

	Control (Li rich)	A (Li rich)	B (Mg)	C (Mg)
Lattice constant (Å)	8.230	8.218	8.218	8.225
SSA (m^2/g)	1.0	0.2	0.2	0.5
Initial capacity (mAh/g)	120	110	105	115

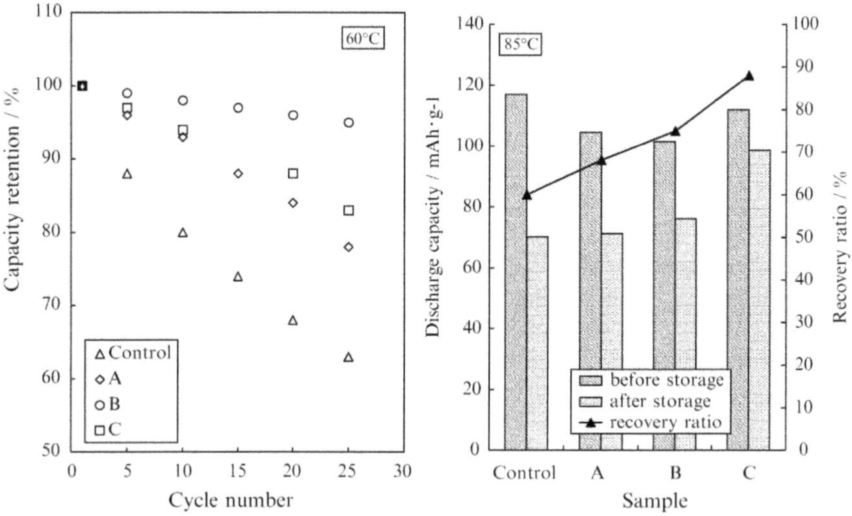

Fig. 17.1 High-temperature characteristics of Mitsui Mining and Smelting LMO. *Left*: Cycle performace at 60°C. *Right*: Storage performance at 85°C for 10 days in the charged state

CR2032-type coin cells with lithium anode and 1 M $LiPF_6$/EC + DMC (1:1) electrolyte were fabricated to evaluate the high-temperature cycle performance at 60°C and high-temperature storage performance at 85°C. The cycle performance was measured by repeated charging and discharging at 0.2 C rate. The storage performance was evaluated by charging the cell to 4.3 V, storing it in an oven at 85°C for 10 days, and then cycling at 20°C. Figure 17.1 shows the cycling performance of the improved LMO. Three samples showed better cyclability than the control, and among the three samples, sample B with magnesium and small specific surface area showed the best high-temperature cyclability. Sample C with magnesium and a higher capacity showed better storage performance. These effects were attributed to stabilization of the crystal lattice by the foreign element and the decrease in manganese dissolution with decreased specific surface area.

There is still a strong demand for higher capacity (e.g., for mobile equipment) and longer life (for hybrid electric vehicles), and we are continuing to improve LMO performance.

17.3 Layer-Structured Material Containing Manganese

There have been many attempts to improve both cycle life and safety of $LiCoO_2$ using Co–Ni,[7] Mn–Co,[8,9] and mixed oxides. Recently, the combination of manganese, cobalt, ad nickel as $LiMn_{1/3}Co_{1/3}Ni_{1/3}O_2$ mixed oxide (referred to as LSM) has attracted a great deal of interest from researchers in cathode materials for lithium and LIB,[10–13] from the viewpoints of reduced cobalt consumption and increased capacity.

We have investigated the Li/(Mn + Co + Ni) ratio dependence on the electrochemical properties of LSM.[14] The raw material of manganese–cobalt–nickel mixed hydroxide was prepared by the co-precipitation technique. The hydroxide was mixed thoroughly with lithium carbonate by ball milling. The Li/(Mn + Co + Ni) ratio was varied as 1.00, 1.05, 1.10, 1.15, and 1.20. The mixed powders were heat treated at 900°C for 20 h in air. The heat-treated samples were subjected to XRD measurement, and no other components than the layered materials were observed.

Battery tests were performed using CR2032 coin-type cells with the lithium anode described above. The batteries were cycled between 4.3 and 3.0 V vs. Li/Li^+. The constant current and constant voltage charge (CCCV charge) was performed galvanostatically at 0.4 C rate (160 mA/g was assumed to be the 1 C rate) and then potentiostatically at 4.3 V until the current dropped to less than 0.05 mA. The discharge current rate was 0.4 C. The rate performance was measured by varying only the discharge current at 0.4, 0.8, 1.2, 1.6, and 2.0 C rate.

Figure 17.2 shows the initial charge-discharge curves of the samples. The charge-discharge curves were not markedly different for samples with different Li/M molar ratios. However, the discharge capacity dropped to lower values when the Li/M ratio was over 1.10. The initial charge-discharge efficiency was between 87 and 93% and increased with increasing Li/M ratio.

Figure 17.3 shows the rate performance of samples with different Li/M molar ratios. The slope values (mAh/g C), calculated from a line drawn along the discharge capacity against discharge current (C), approached 0 with increasing

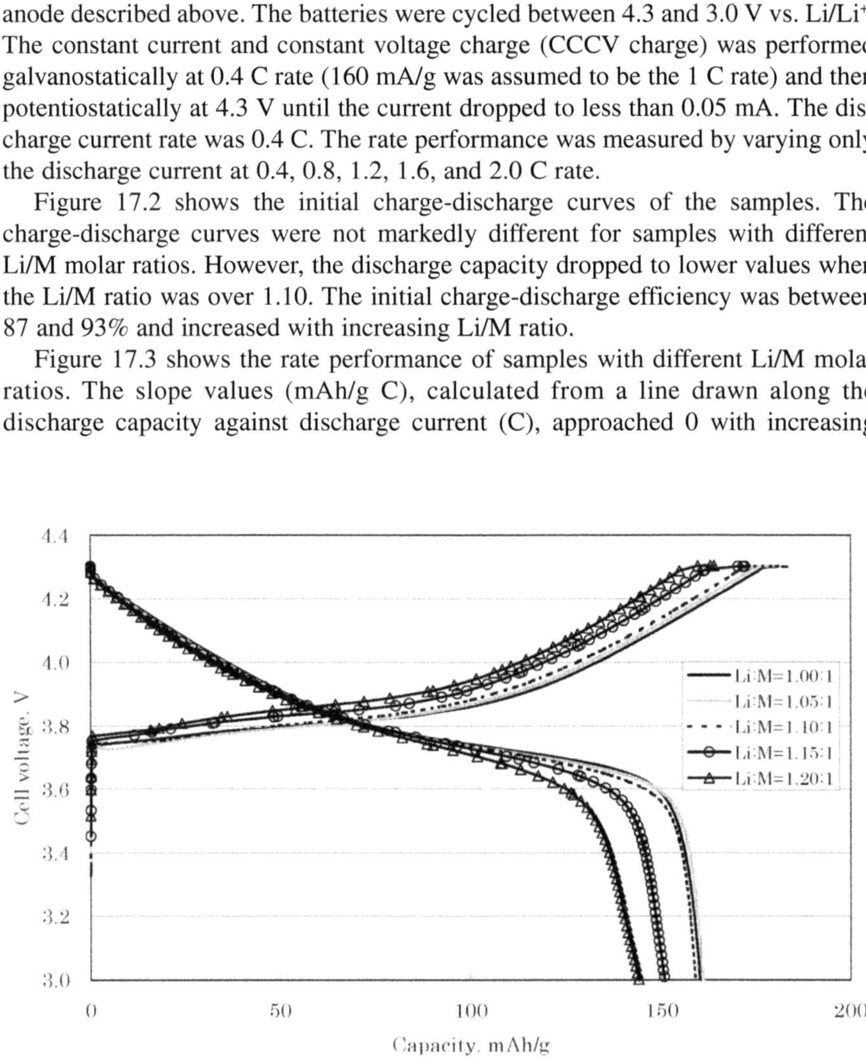

Fig. 17.2 The first charge–discharge curves of lithium–manganese–cobalt–nickel mixed oxide with Li/(Mn + Co + Ni) ratio = 1.00, 1.05, 1.10, 1.15, and 1.20. Only a constant current was applied for discharge

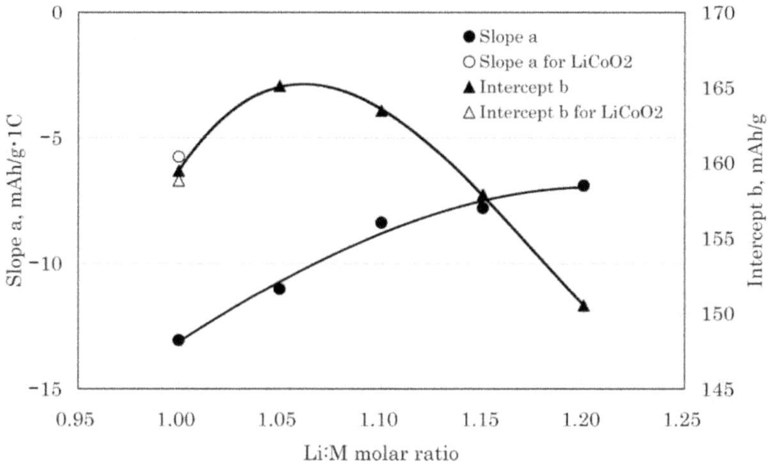

Fig. 17.3 Dependence of rate capability on Li/(Mn + Co + Ni) molar ratio

Table 17.2 Powder characteristics of MMS LSM

Powder characteristics	Values
Mean particle size (μm)	11.7
Tapping density (g/cm³)	2.6
SSA (m²/g)	0.2

Fig. 17.4 SEM of Mitsui Mining and Smelting LSM

lithium content, indicating better rate performance. The rate performance of the samples was better when the Li/M molar ratio was high. Moreover, the sample with Li/M = 1.20 showed rate performance close to that of $LiCoO_2$.

According to these results, the Li/M ratio of our mainstream LSM was around 1.10. The Li/M ratio can be varied depending on the customers' requirements for battery performance.

Table 17.2 shows the typical powder characteristics of our LSM product. The specific surface area of 0.2 m^2/g is small and the tapping density of 2.6 g/cm^3 is very high. Packing density of active materials in the electrode is a very important factor for the energy density of the LIB. These characteristics are derived from the shape and structure of the secondary particles. Figure 17.4 shows an SEM of a particle of our LSM. The secondary particle was spherical in shape and primary particles were packed densely in the secondary particles. These dense particles are expected to be effective for the production of high-energy density batteries. Cathode-active materials containing manganese are very attractive because of their safety and cost-effectiveness. Furthermore, it is expected that LSM will have a larger capacity than $LiCoO_2$ if charged higher than 4.3 V vs. Li/Li^+.

References

1. S.R. Taylor, S.M. McLennan, *Rev. Geophys.*, **33**, 241–265 (1995).
2. G. Pistoia, D. Zone, Y. Zhang, *J. Electrochem. Soc.*, **142** (8), 2551 (1995).
3. J.C. Hunter, *J. Solid State Chem.*, **39**, 142 (1982).
4. M.M. Thackeray, P.J. Johnson, L.A. de Picciotto, P.G. Bruce, J.B. Goodenough, *Mater.Res. Bull.*, **19**, 179 (1984).
5. Y. Gao, J.R. Dahn, *J. Electrochem. Soc.*, **143** (1), 100 (1996).
6. D.H. Jang, Y.J. Shin, S.M. Oh, *J. Electrochem. Soc.*, **143** (7), 2204 (1996).
7. C. Delmas, I. Saadoune, *Solid State Ionics*, **370** (1992).53–56,
8. R. Stoyanova, E. Zhecheva, L. Zarkova, *Solid State Ionics*, **73**, 233 (1994).
9. K. Numata, C. Sakaki, S. Yamanaka, *Chem. Lett.*, 1997, 725 (1997).
10. T. Ohzuku, Y. Makimura, *Chem. Lett.*, 2001, 642 (2001).
11. Z. Wang, Y. Sun, L. Chen, X. Huang, *J. Electrochem. Soc.*, **151** (6), A914–A921 (2004).
12. K.M. Shaju, G.V. Subba Rao, B.V.R. Chowdariz, *J. Electrochem. Soc.*, **151** (9), A1324–A1332 (2004).
13. Y. Koyama, N. Yabuuchi, I. Tanaka, H. Adachi, T. Ohzuku, *J. Electrochem. Soc.*, **151** (10), A1545–A1551 (2004).
14. Y.M. Todorov, K. Numata, *Electrochim. Acta*, **50** (2–3), 493–497 (2004).

Chapter 18
Trends in Carbon Material as an Anode in Lithium-Ion Battery

Tatsuya Nishida

18.1 Introduction

As sales of small mobile equipment such as mobile phones, notebook-sized personal computers, and camcorders have rapidly increased, so too has the demand for the lithium-ion battery used as the main power supply of these devices rapidly increased. Moreover, electric vehicles definitely are expected to be a solution of the worldwide environmental problems, and the development of the lithium-ion (Li-Ion) battery as their power source is now at the stage of practical use.[1]

The Li-Ion battery has advantages compared to the nickel-cadmium (Ni–Cd) battery and the nickel-metal hydride (Ni–MH) battery such as its higher energy densities per volume and per weight, so miniaturization and reduced weight of equipment are possible. Moreover, further developments in the Li-Ion batteries, such as improvement in energy density and high rate charge/discharge capability, are expected due to the requirement of the small mobile devices using them.

Although amorphous carbon[2] and spherical mesophase base graphite have been used mainly as anode materials in Li-Ion batteries, artificial graphitic carbon with lower voltage and the higher energy density[3] is being used mainly to fulfill the difficult requirement for mobile equipment in recent years.

In this chapter, details of the massive artificial graphite (MAG) with excellent anode performance in the Li-Ion battery will be described. MAG is developed for the mass production of a unique graphite for the purpose of battery application, based on various knowledge about conventional production technology of artificial graphite. This material shares about 70% of the anode material market for Li-Ion batteries produced in Japan.

T. Nishida
Hitachi Chemical Co. Ltd., 13-1 Higashi-Cho 4-chome, Hitachi City, 317-8555, Japan
t-nishida@hitachi-chem.co.jp

M. Yoshio et al. (eds.), *Lithium-Ion Batteries*,
DOI 10.1007/978-0-387-34445-4_18 © Springer Science+Business Media LLC 2009

18.2 Powdery Characteristics of MAG

In the development of the high-performance anode material for the Li-Ion battery, studies were carried out in order to produce a unique aggregation structure and crystal structure, which are not found in conventional graphite, such as the spherical graphite or the flaky natural graphite. A new MAG has been developed on above concepts. Its powder properties will be initially described.

18.2.1 Aggregation Structure of MAG

The scanning electron microscopy (SEM) photographs of the apparent shape and the cross section of MAG particles are shown in Fig. 18.1. The MAG particles consist of spherical aggregation of fine, flat crystals. It has an average diameter of 20–30 μm and an aspect ratio of ca. 1. The cross-sectional images of MAG particles shows a pseudoisotropic texture where primary flat particles are randomly aggregated. Moreover, many large pores are distributed around the fine, flat particles in the spherical aggregate. Figure 18.2 shows the data of pore size distribution obtained by the mercury porosimetry measurement. MAG has a wide range of pore size distribution from 10^2 nm to 100 μm. Small pores less than 5 μm is characteristic in MAG, and such small pores contribute higher pore volume (1×10^{-3} m^3/kg) in MAG in comparison with conventional spherical graphite. In general, the charge of graphite without intragranular pores would proceed in the following steps: Li^+ is inserted on the surface of the particle, and then the Li diffuses into the graphite crystal. As for the newly developed anode material, MAG, the insertion and extraction of lithium can be carried out not only on the outer surface of the aggregated MAG particles but also on the inner surface of MAG particles through the intragranular pores. The intragranular pores would be soaked by electrolyte as the same

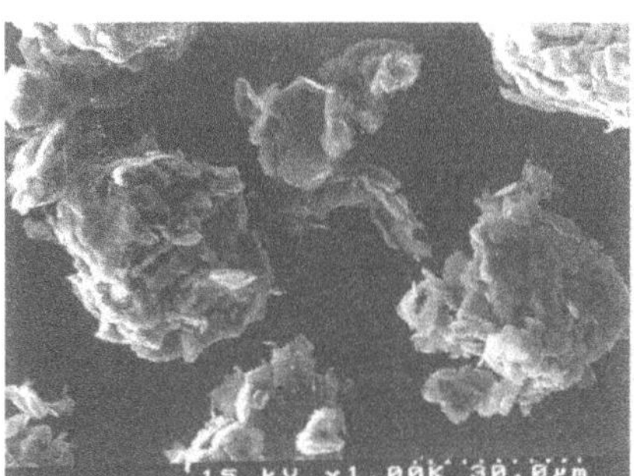

(a) apparence

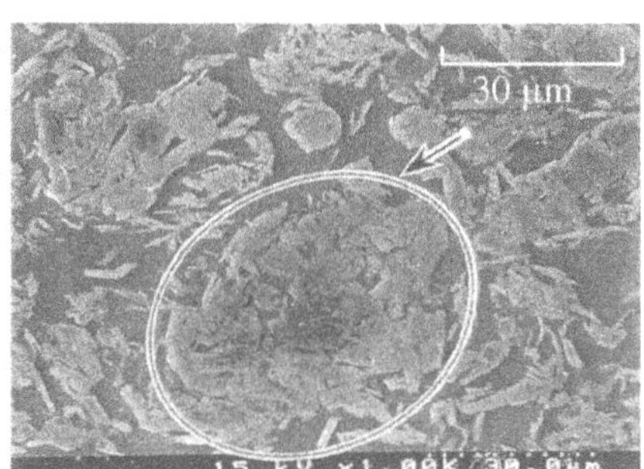

(b) cross section

Fig. 18.1 SEM images of MAG particles

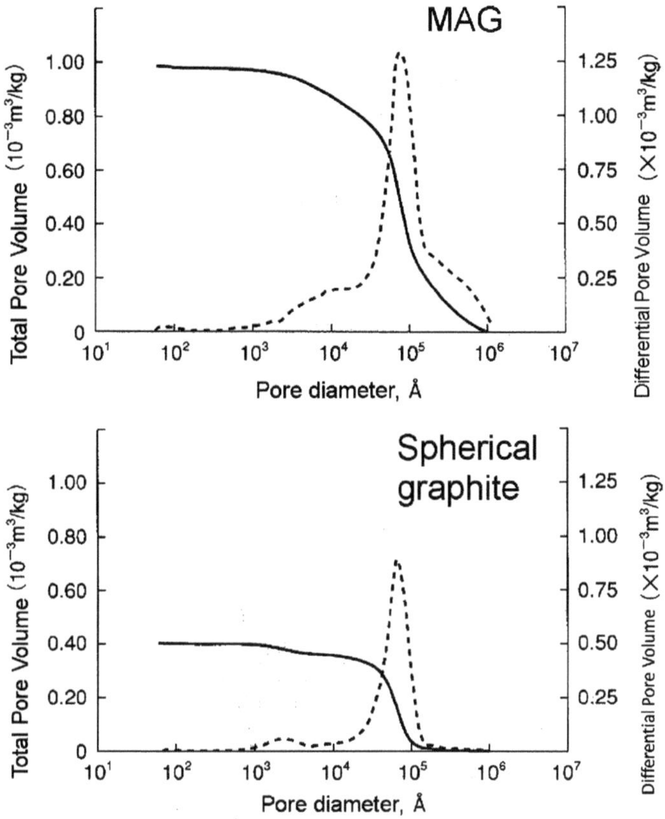

Fig. 18.2 Pore size distributions and accumulated pore volume of MAG and spherical graphite

as liquid mercury. Moreover, although the graphite anode has the property of expanding and shrinking during charge/discharge,[4] the new anode material, MAG, with a pseudoisotropic texture can tolerate this expansion and shrinkage.

18.2.2 Crystal Structure of MAG

Figure 18.3 shows the X-ray diffraction (XRD) patterns of MAG and spherical graphite. The diffraction line at $2\theta = 26.5°$ is assigned to the (002) plane of the graphite crystal.[5] The values of interlayer spacing (d-spacing) of the graphite crystal calculated from the diffraction line of the (002) plane are 0.336 nm for MAG and 0.338 nm for spherical graphite, respectively. As its value of MAG is very close to that of natural graphite, it can be said that the developed anode MAG material has a high crystallinity.

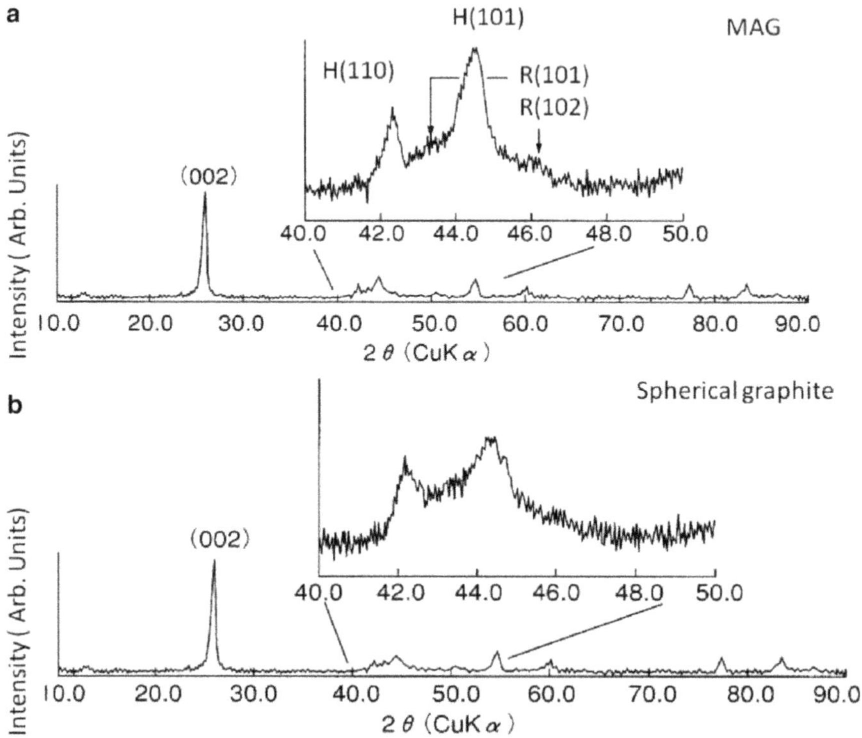

Fig. 18.3 X-ray diffraction pattern of (**a**) MAG and (**b**) spherical graphite

There are two kinds of crystal structures in graphite: the hexagonal structure and the rhombohedral structure. The hexagonal structure consists of graphene sheets stacked in an orderly sequence of ABAB·····; on the other hand, the rhombohedral structure consists of graphene sheets stacked in a sequence of ABCABC·····. These two kinds of crystal structures have differences in their XRD profiles in the range of $2\theta = 40–50°$. MAG has lower peak intensities of rhombohedral (101) and (102) lines [R(101), R(102) in Fig. 18.3] than those of hexagonal (100) and (101) lines [H(100), H(101) in Fig. 18.3], which indicates that the structure of MAG is very close to pure hexagonal.

Figure 18.4 shows the Raman spectra of MAG and spherical graphite. In the hexagonal graphite with D_{6h} symmetry, an E_{2g} vibration mode, assigned to the lattice stretching of the conjugated c–c bond within the graphene sheet, appears around 1,580 cm^{-1} of the Raman shift.[6,7] Moreover, a Raman shift of around 1,350 cm^{-1} is assigned to disordered carbon. While two shifts at 1,568 cm^{-1} and 1,354 cm^{-1} were detected for spherical graphite, only one shift at 1,576 cm^{-1} was observed for MAG; in the other words, it has no shifts in the vicinity of 1,350 cm^{-1}. Therefore, it can be said that MAG is hexagonal crystal with few disordered carbons. Moreover, it is known that the higher the crystallinity, the higher wave number the

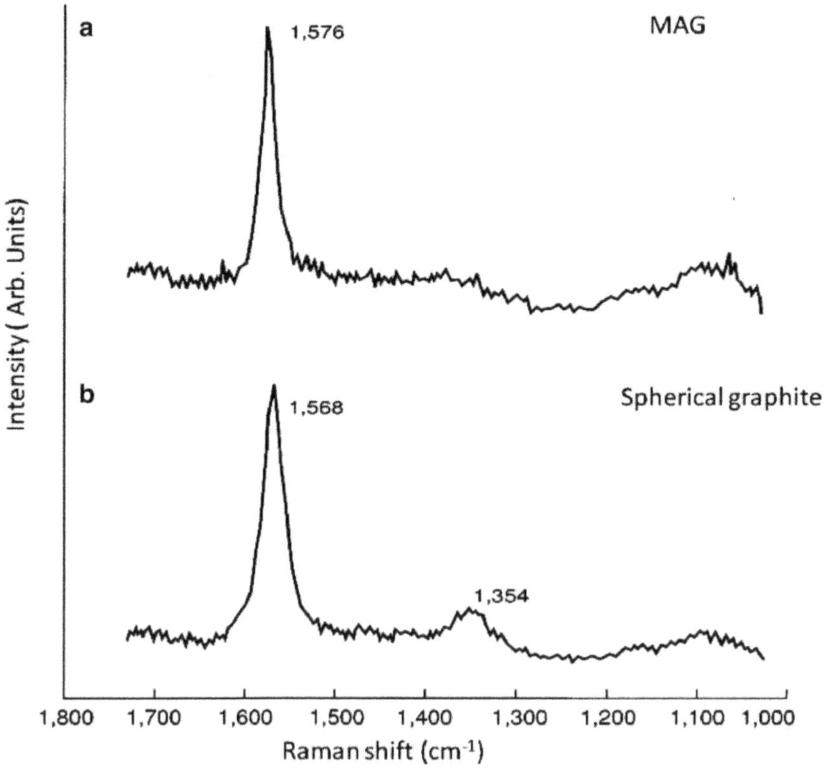

Fig. 18.4 Raman shifts of (**a**) MAG and (**b**) spherical graphite

E_{2g} vibration shifts. The wave number in the shift of the E_{2g} mode of MAG shifts 8 cm^{-1} higher than that of the spherical graphite, so high crystallinity of MAG was confirmed by the Raman spectrum as well as XRD data, such as narrow peak width and strong peaks intensity.

18.3 Charge/Discharge Characteristics of MAG

18.3.1 Crystal Structure at Various Charge/Discharge States

Figure 18.5 shows the initial charge/discharge curve of a MAG anode with a metallic lithium counterelectrode. Several voltage plateaus are observed in a low potential region of both charge and discharge curves, which are peculiar in graphite. Moreover, the plateau potential changes step by step during the charge/discharge; it appears at about 190, 95, and 65 for the Li^{+} insertion and at about 105, 140, and

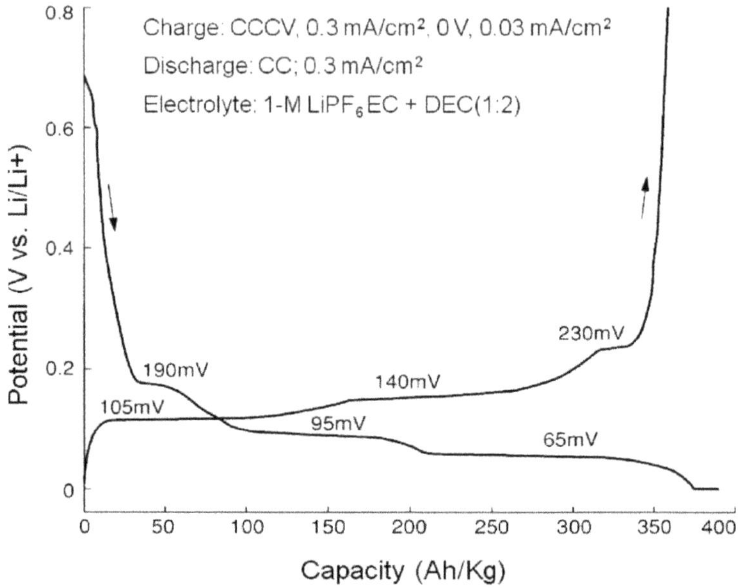

Fig. 18.5 Initial charge/discharge curves of MAG electrode

230 for the extraction. These phenomena are attributed to the changes in the stage structure of lithium-graphite intercalation compound (Li-GIC) lithium insertion or extraction between the graphene layers.

Figure 18.6 shows the XRD patterns of the original MAG and its lithiated one (voltage: 65 mV, charge capacity: 330 Ah/kg). A strong hexagonal (002) diffraction line in the original MAG is observed in the vicinity of $2\theta = 26.5°$. The interlayer spacing of the hexagonal crystal is calculated to be 0.336 nm. The diffraction line around $2\theta = 26.5°$ disappears in the lithiated state and a new diffraction line appears at a lower angle of $2\theta = 24.0°$. This peak corresponds to the first stage of Li-GIC.

The interlayer spacing of the crystal in this stage is calculated to be 0.370 nm. These results indicate that the change from graphite to the first stage of Li-GIC formed by the insertion of Li[8,9] causes expansion in the interlayer spacing of the crystal from 0.336 to 0.370 nm.

Figure 18.7 shows the relation between the interlayer spacing of hexagonal crystal and the discharge capacity, which was determined for low-density electrode (10^4 kg/m^3). The discharge capacity of MAG exhibits a higher value of 362 Ah/kg, which is very close to the theoretical one (372 Ah/kg) of graphite. Graphite with a high crystallinity, which has a smaller interlayer spacing in the crystal, has relatively higher discharge capacity. This could be attributed to the fact that the graphite with the higher crystallinity is mainly composed of regular stacking structure, which is suitable for the formation of Li-GIC, and the content of unsuitable disordered structure would be low.

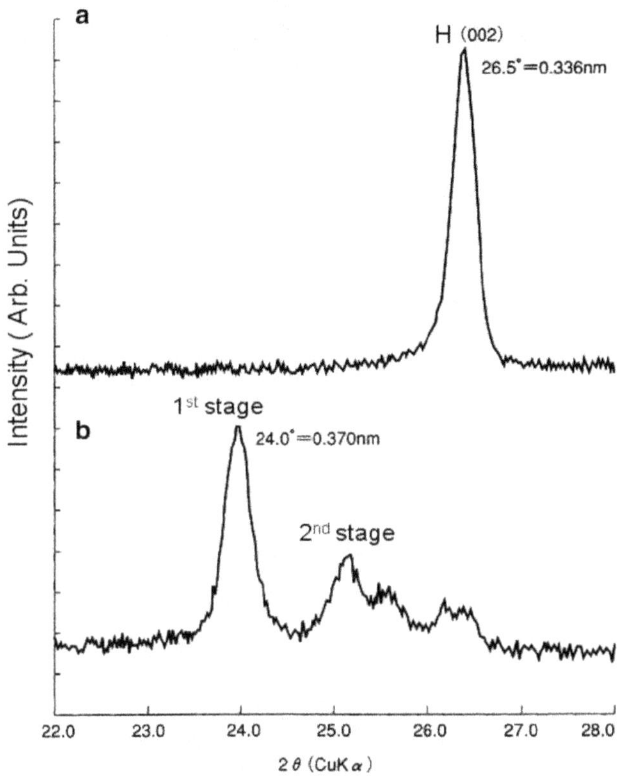

Fig. 18.6 XRD profiles of (**a**) original MAG and (**b**) lithiated one (b: charged capacity: 330 Ah/Kg)

18.3.2 Coulombic Efficiency

Figure 18.8 shows the relation between the number of cycles and the coulombic efficiency for MAG electrode. Here, the coulombic efficiency and the irreversible capacity are defined by (18.1) and (18.2) respectively:

$$\text{Coulombic efficiency (\%)} = \text{(discharge capacity/charge capacity)} \times 100 \quad (18.1)$$

$$\text{Irreversible capacity (Ah/kg)} = \text{charge capacity - discharge capacity} \quad (18.2)$$

The coulombic efficiency of MAG electrode is 91.6% (irreversible capacity 33 Ah/kg) for the first cycle, 99.2% for the second cycle, and it almost reaches an ideal value of 100% after the fifth cycle. In general, the coulombic efficiency of carbon anodes in the first cycle is so low that there exists a high irreversible capacity. Since this irreversible capacity of carbon anode does not contribute to the discharge capacity of lithium-ion battery, it is desirable to make it as low as possible in the viewpoint of increasing the battery capacity. It is known that the irreversible

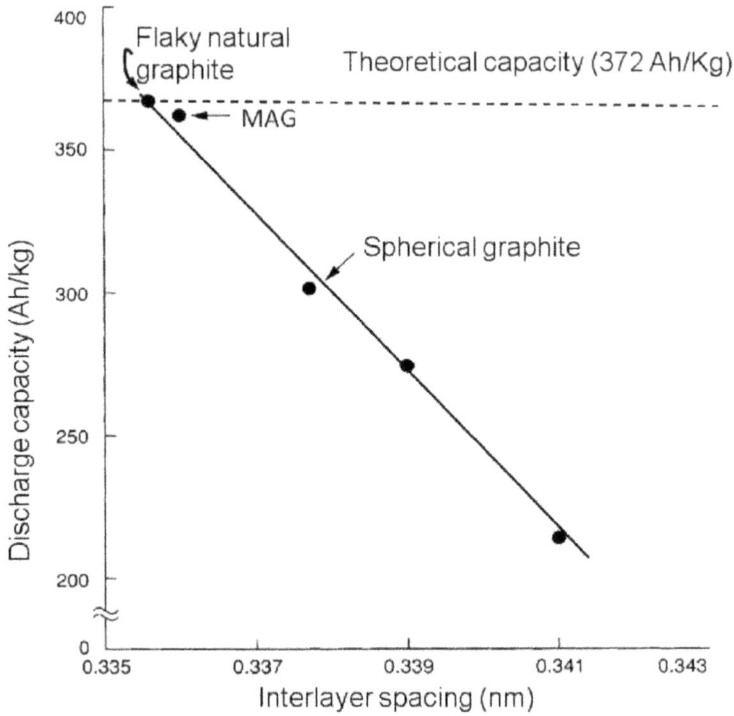

Fig. 18.7 Relation between interlayer spacing of various graphite and its discharge capacity

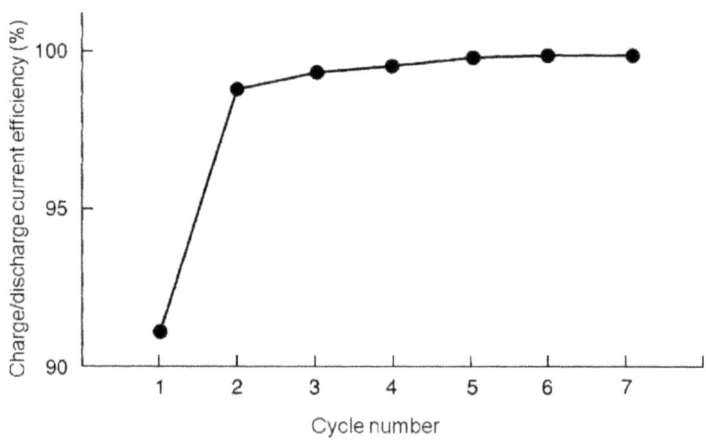

Fig. 18.8 The relation between the number of cycles and the coulombic efficiency for MAG electrode

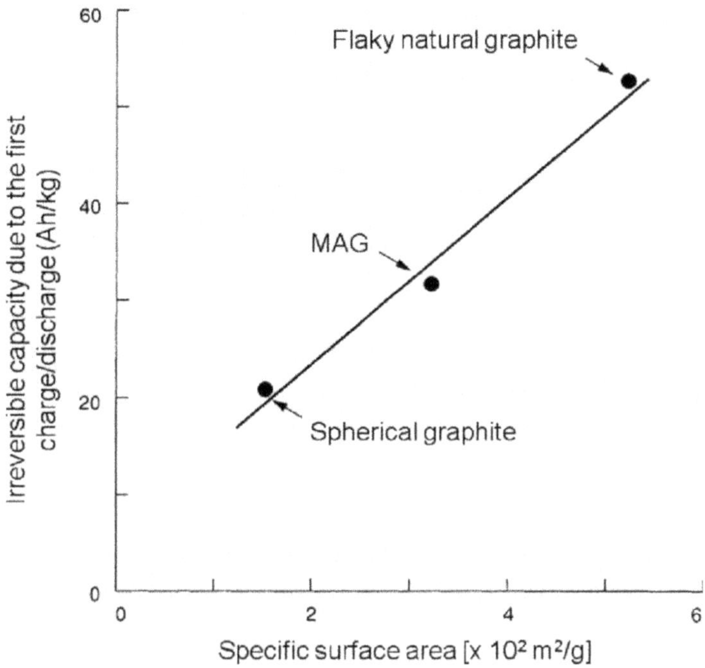

Fig. 18.9 The relation between the irreversible capacity in the first cycle and the specific surface area for various graphitic materials

capacity in the first cycle is attributed to the reductive decomposition of the electrolyte on the anode surface. Such reaction mainly takes place during the initial charge process in the graphitic material.[3] Products on the anode surface generally are called solid electrolyte interface (SEI) film, as this reductive decomposition of the electrolyte is suppressed after the second cycle and the irreversible capacity decreases to zero in elongated cycles.

Figure 18.9 shows the relation between the irreversible capacity in the first cycle and the specific surface area for various graphitic materials. The irreversible capacity in the first cycle shows an almost linear relationship with the specific surface area, and the smaller the specific surface area, the lower the irreversible capacity. It can be postulated that the lower the specific surface area, the smaller the reaction area for the decomposition reaction of the electrolyte becomes.

18.3.3 Cyclicities

Figure 18.10 shows the cyclicity of two kinds of electrode with a density of 1.5×10^3 kg/m^3. Fresh metallic lithium used as an auxiliary electrode is replaced every 15 cycles, so sudden capacity recovery is due to above replacement. MAG

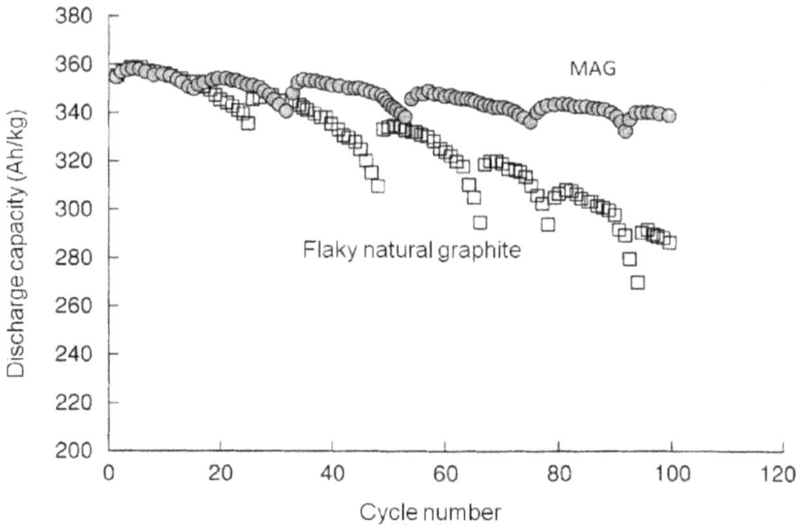

Fig. 18.10 Cycling performance of MAG and flaky natural graphite

exhibits an excellent cyclicity and keeps a discharge capacity of 340 Ah/kg even after 100 cycles (capacity retention: 94%.) On the other hand, flaky natural graphite exhibits a higher capacity fading during the cycling, and its capacity at 100 cycles is only 290 Ah/kg (capacity retention: 81%.)

18.3.4 High-Rate Discharge Performance

When the battery is discharged at a high current drain in a short time, it is desirable that the decrease in the discharge capacity would be low. In recent years, the high-rate discharge performance is one of the most important battery characteristics because the electric devices tend to need high power for its operation.

Figure 18.11 shows the relationship between the discharge capacity and the discharge current density for three graphitic carbon anodes with a constant electrode density of 1.4×10^3 kg/m^3. The constant current method (0.3 mA/cm^2, 0 V cutoff) is used for charge. Higher discharge capacities are maintained for both MAG and the flaky natural graphite at the discharge current density of 2.0 mA/cm^2 or less. However, the discharge capacity of the flaky natural graphite decreased at the discharge current density beyond 2.0 mA/cm^2, and reaches 200 Ah/kg at the current density of 6.0 mA/cm^2. On the other hand, the decrease in the discharge capacity of MAG is much less than that of the flaky natural graphite and MAG shows a higher capacity of 310 Ah/kg even at the higher current density of 6.0 mA/cm^2. Although the discharge capacity of spherical graphite is low, its decrease in the discharge capacity is smaller than that of the flaky natural graphite.

Figure 18.12 shows SEM photographs of cross-sectional images for MAG and the flaky natural graphite anode. Many individual graphite crystals in a natural graphite anode have parallel orientation against the copper collector; on the other hand, the flat crystals in MAG give a random configuration. Movement of Li^+ ion

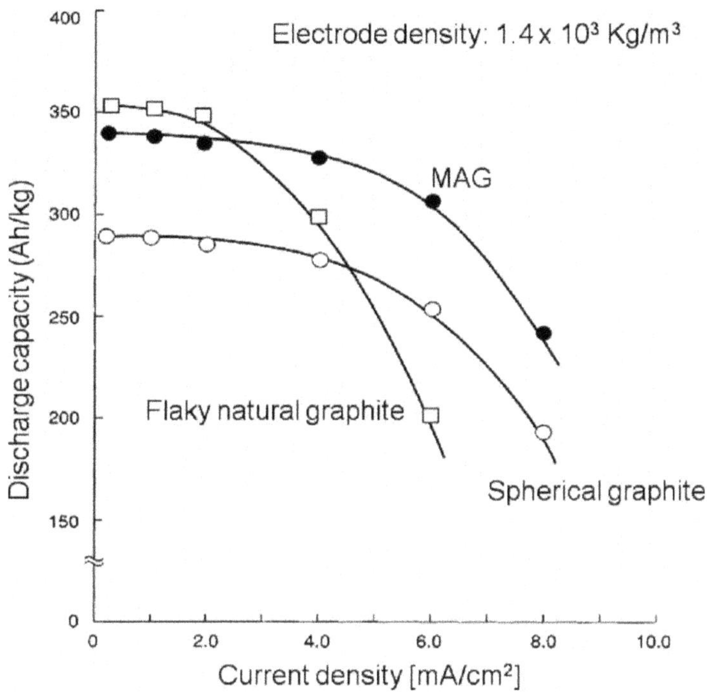

Fig. 18.11 Relationship between discharge capacity and discharge current density for three graphitic carbon anodes

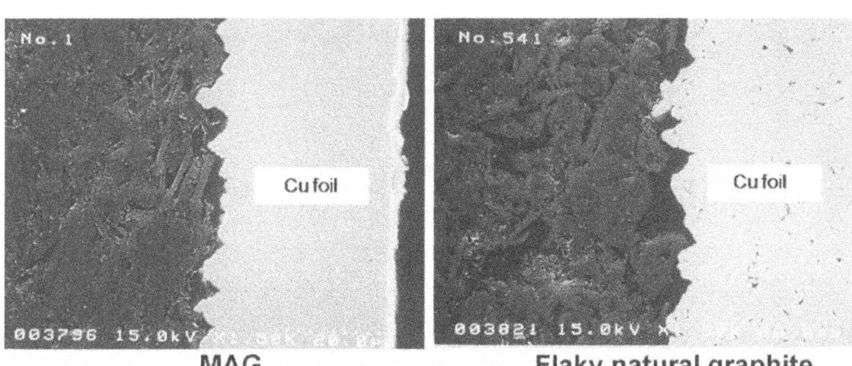

Fig. 18.12 SEM image of a cross section of (*left*) MAG and (*right*) flaky natural graphite anode

occurs in the vertical direction to the collector during the charge/discharge and Li^+ insertion into graphite occurs at the edge of graphite crystal. Therefore, the lithium transfer in MAG during the charge/discharge becomes easier than that in flaky natural graphite because of the randomly arranged texture. We believe that this electrode texture is one of the important factors for the excellent high-rate discharge performance of MAG.

18.3.5 Electrode Density of MAG Anode

Miniaturization is required for the lithium-ion battery; in the other words, an improvement in the volumetric energy density is important. Therefore, it is necessary to develop an anode material with high packing density. Finally, the increase in the electrode density without deterioration of electrochemical property is desirable.

Figure 18.13 shows the relationship between the electrode density and the discharge capacity for three graphitic carbon anodes. The discharge capacity of all three electrodes decreases for every material with the increase in the electrode density.

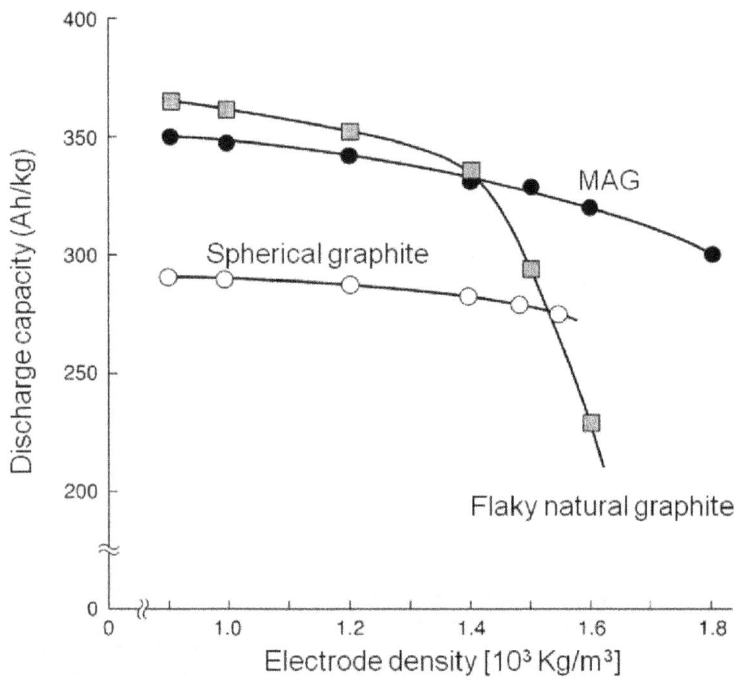

Fig. 18.13 The relation between the electrode density and the discharge capacity for three graphitic carbon anode

It causes the decrease in the volume of the pore, which is formed inside of the electrode by packing the graphitic carbon and the ability to store electrolyte. Lithium diffusion in electrochemical reactions would proceed mainly between the outer electrolyte and anode material because of the lack of electrolyte in the pore. In particular, a rapid decrease in the discharge capacity was observed for the flaky natural graphite with the electrode density of more than 1.4×10^3 kg/m^3. The flaky natural graphite electrode has an orientated texture, as shown in Fig. 18.12. When the electrode density increases, the extent of the orientation further increases, which causes the rapid decrease in the discharge capacity. On the other hand, MAG material and spherical graphite show a slight decrease in the discharge capacity for the increase in the density. The discharge capacity of 320 Ah/kg could be attained for MAG with the density of 1.6×10^3 kg/m^3. This high capacity of MAG indicates that an isotropic texture of graphite crystal becomes an advantage in the high-density electrode.

18.4 Conclusion

Excellent MAG has been developed as an anode material for the lithium-ion battery. The features of this material are described as follows:

- Well-developed primary crystals form massive aggregation with a pseudoisotropic texture, in which pores from 100 nm to 2 μm are included.
- A high discharge capacity of MAG is attained by both its characteristic texture and ideal hexagonal crystal structure. It excels in high-rate discharge performance, cyclicity, and electrode density.

Based on these facts, we believe that the newly developed anode material, MAG, is the most suitable anode material for lithium-ion batteries.

References

1. M. Nagamine, H. Kato, Y. Nishi (eds.). 33rd Battery Conference of Japan, Ab 1C11 (1992).
2. H. Arai, S. Okada, H. Ohtsuka, M. Ichimura, J. Yamaki. Solid State Ionics, **80** (1995) 261.
3. T. Fujieda. J. Jpn. Met. Soc. Mater., **38** (1999) 488.
4. M. Uratani, R. Takagi, K. Sumiya, K. Sekine, T. Takamura. 38th Battery Conference of Japan, Ab 2B12 (1997).
5. Carbon Material Society of Japan (ed.). The Introduction of Carbon Material (in Japanese), 1996, p. 24.
6. L. Nikiel, P. W. Jagodzinski. Carbon, **31** (1993) 1313.
7. F. Tuinstra, J. L. Koenig. J. Chem. Phys., **53** (1970) 1126.
8. J. R. Dahn, R. Fong, M. J. Spoon Phys. Rev., **B42** (1990) 6424.
9. T. Ohzuku, Y. Iwakoshi, K. Sawai. J. Electrochem. Soc., **140** (1993) 2490.

Chapter 19
Functional Electrolytes Specially Designed for Lithium-Ion Batteries

Hideya Yoshitake

19.1 Introduction

Ube Industries Ltd. UBE began commercialization of highly purified solvents for lithium-ion batteries (LIB) in 1992, and in 1997 started to sell a "functional electrolytes," which contained special additives.[1] UBE has a market share of around 65% for LIB in the high-end cellular phone market and around 40% of the total electrolyte market for LIB from 1998. The strategy for developing the functional electrolytes was directed at both novel additives for LIB and the design of electrolytes and development of specialized additives for the different types of anode and/or cathode materials used by different LIB companies. This technology was unique for the industry. UBE significantly changed the market approach for electrolytes.

UBE has several functional electrolytes requested by customers and is now developing more additives for spinel-based LIB to prevent the corrosion of aluminum substrate and so forth. The main reason for the progress of LIB – e.g., increase in energy density, safety, cyclicity, etc. – is fundamentally due to the development of the functional electrolytes. Even when changing the anode or cathode active material or using the different types of graphite and mixed different types of cathode active material, the different additive compounds should be adopted. This type of LIB cell design is only studied and commercialized by UBE.

The reason why the progress of gel-polymer battery is rather slow is the slow technological progress in the area of functional electrolytes. The following discussion presents the history of electrolyte and additive development as well as the process followed in the design of LIB electrolyte.

H. Yoshitake
Specialty Chemicals & Products Company, Ube Industries, Ltd, 1-2-1 Shibaura, Minato-ku, Tokyo, 105-8449 Japan
27452u@ube-ind.co.jp

M. Yoshio et al. (eds.), *Lithium-Ion Batteries*,
DOI 10.1007/978-0-387-34445-4_19

19.2 The Past and Future of Electrolytes for Lithium-Ion Batteries

19.2.1 History of Electrolytes for Lithium-Ion Batteries

Figure 19.1 shows the capacity increase that the18650 size battery has undergone during the period from 1992 to 2005. One can see that the discharge capacity increased by almost three times to 2550 mAh over this period. For this kind of substantial battery capacity increase, high-performance active materials are indispensable. There are several reasons for the increase. The early cells used cokes or hard carbons (~150–200 mAh/g), which were shifted to low-crystallinity graphite (280–250 mAh/g) and finally to high-performance synthetic graphite such as massive artificial graphite (MAG) (320–360 mAh/g). In addition, the charging voltage was increased from 4.1 to 4.4 V, which increased the capacity of LIB as shown in Fig. 19.1 . The invention and use of the functional electrolytes made possible the use of high-performance (high crystalline) graphite, resulting in an increase of charging voltage. The electrolyte composition also changed significantly. The initial cells (1991) used propylene carbonate (PC) and diethyl carbonate (DEC); ethylene carbonate (EC) and γ-butyrolactone (γ-BL) were introduced in 1992. These electrolytes were followed by methyl ethyl carbonate (MEC), methyl propionate (MP), and dimethyl carbonate (DMC) in 1993. The electrolyte solvents remained essentially unchanged until 2000. In the 2000, fluorobenzene (FB) was initiated on a limited production basis and provided an important clue for future changes in electrolyte development for LIB. In fact, fluoroethylene carbonate (FEC) was proposed as a commercial electrolyte in 2005. This is an important

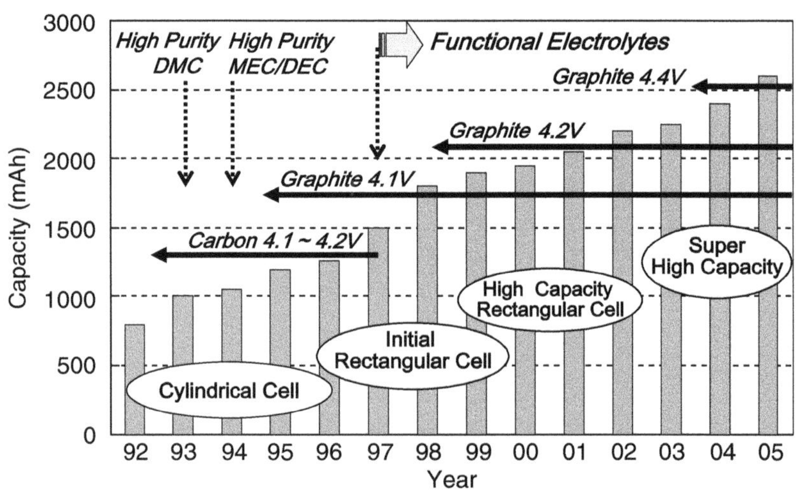

Fig. 19.1 Capacity increase of 18,650 lithium-ion cell based on the improvement of the negative electrode and charging voltage

cornerstone because it was recognized that fluorochemicals have a strong resistance and should be superior solutes for use in lithium-ion cell electrolytes.

The solute lithium salt, $LiPF_6$, was a common choice for carbonate-based electrolytes and $LiBF_4$ for γ-BL-based electrolytes. In 1994, the method for preparing highly purified $LiPF_6$ was developed and cleared the way to the use of highly purified solvents and using additives. This was a very important step in the rapid development of a functional electrolytes. In 2000, novel organic lithium salts of lithium bis-trifluoromethanesulfonyl imide (HQ115) and lithium bis-pentafluoroethanesulfonyl imide (BETI) were introduced in LIB and marked a new direction for electrolyte development.

19.3 Functional Electrolytes

The capacity of the typical 18650 cell had been increasing about 10% per year until 1997. At this point, it became difficult to maintain the growth in capacity. It was discovered that the main cause of the cell's deterioration during cycling was the growth of the solid electrolyte interphase (SEI) on each charge cycle due to the reaction of graphite with the electrolyte and lithium salt. The SEI forms initially during the first charge by reaction of the lithium with the electrolyte. The same reaction occurs on each successive cycle although at a much slower rate. The accumulation SEI produced a thick, highly resistive layer on the anode surface material, eventually blocking the normal cell operation. Its formation was a result of the natural processes of charge and discharge and could not be prevented, given the materials used in cell assembly.

In 1997–1998, Ube Industries Ltd. developed and introduced the first electrolyte composition using highly purified electrolyte with additives specifically designed for different types of graphite anodes. This prevented not only the accumulation of the SEI on the surface of the anode, but made a novel surface film, called "functional film." This breakthrough to overcome the disadvantage of the existence of the accumulative SEI layer was initiated by the new functional electrolytes, which improved the cyclicity of LIB and permitted the use of high-capacity, highly crystalline graphite. The concept of the functional electrolytes is the formation of a protective surface layer formation by the specially designed additives before the SEI is formed. Dr. A.Yoshino said that the author's invention makes it possible to smooth topochemical reactions of lithium ion with graphite material, enabling the carbon electrode material to participate in a smooth topochemical reaction (negative electrode to which lithium ions can be intercalated and deintercalated).

19.3.1 Increase in Oxidation Potential of Solvents Using Highly Purified Solvent

Figure 19.2 shows the oxidation potential vs. purity of the solvents used in 1992. Most solvents with a purity of 99.9% or less had a decomposition potential of around 4.6–4.9 V vs. Li/Li^+, although an oxidation potential of the solvent of around

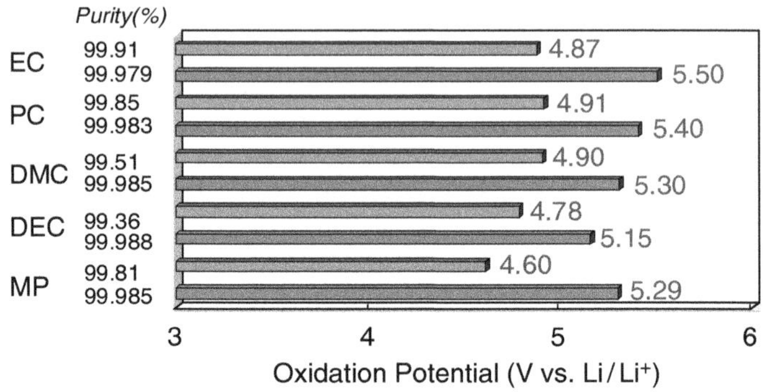

Fig. 19.2 Relation between the oxidation potential and purity of several solvents

5.0 V was commonly accepted. It was a necessary for LIB to have an oxidation voltage larger than 5.0 V.

The typical DMC production method and main impurities are shown in Fig. 19.3. Liquid-phase CO process and phosgene process contain the chlorine-related impurities and are not suitable for LIB solvents. UBE found that the electrochemical impurities in the common solvents produce a small but measurable residual current due to the presence of these electrochemically active impurities. This was the mainfactor causing the short cycle life of LIB. When highly purified solvents were used, the oxidation potential increased to 5.2–5.5 V, as shown in Fig. 19.2. UBE developed a novel production process based on a gas-phase CO process that produce halogen-free materials with incorporated additional purification technology. The new procedure produces very pure materials with wide electrochemical windows. The high purity solvents free from electrochemical impurities produce stable low-resistance SEI layer cells with high electrochemical stability. UBE started the commercial production of DMC in 1992 and started supplying high purity electrolytes in March 1992. High purity DEC and MEC were obtained using high purity DMC as the starting material by the ester-exchange method.

19.3.2 Low-Impedance Electrolytes

Another important factor that affects the SEI formation is the purification of the lithium salt ($LiPF_6$) which is a key component of the electrolyte. A new system for producing a highly purified $LiPF_6$ was developed in 1994. The highly purified salt significantly improves the electrochemical performance. This was an important pre-stage for the development of the functional electrolytes. Figure 19.4 compares the morphology of conventional and highly purified $LiPF_6$ for its use in the functional electrolytes.

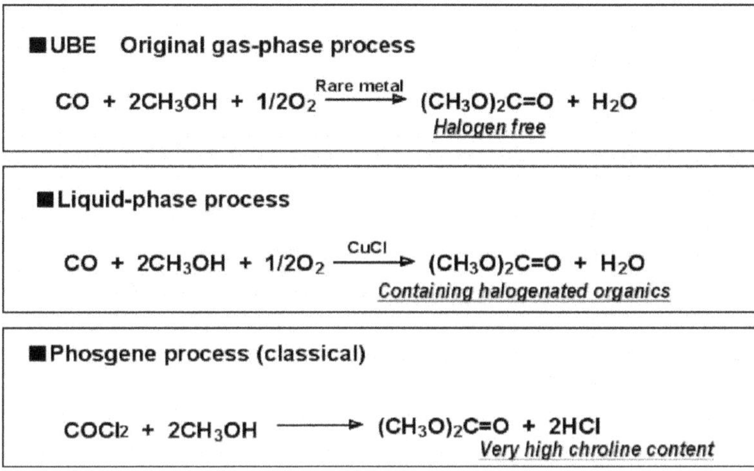

Fig. 19.3 Main impurities in DMC made by different processes

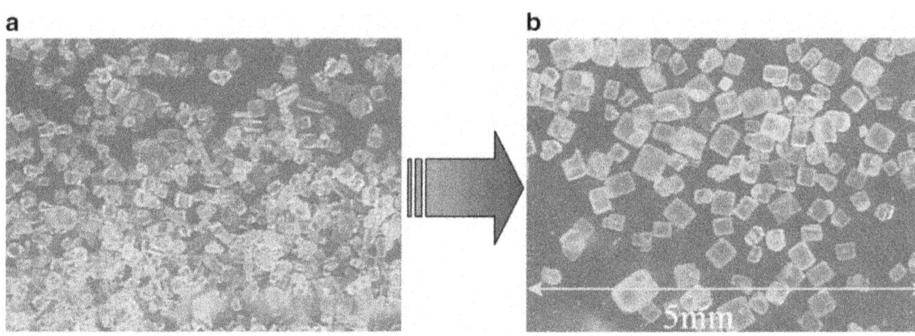

Fig. 19.4 (**a**) Conventional and (**b**) purified $LiPF_6$ for functional electrolytes

The measurement technique used to determine the various impurities of the solvents obtained from different methods was recognized as useful. For example, Fig. 19.5b shows a comparison of the increase of (HF) content of a common electrolyte with the high-purity solvent and highly purified $LiPF_6$ electrolyte. It also compares the storage time at room temperature. When content of impurities are high, the HF content rapidly increases to 62 ppm after 5 days and saturates at 67 ppm after 10 days. The purified UBE's electrolyte saturates at about 18 ppm after 1 day. The impedance of a graphite anode is shown in Fig. 19.5a in a Cole-Cole plot. One should notice how the impedance rapidly increases with storage time for the case of electrolyte with high HF content.

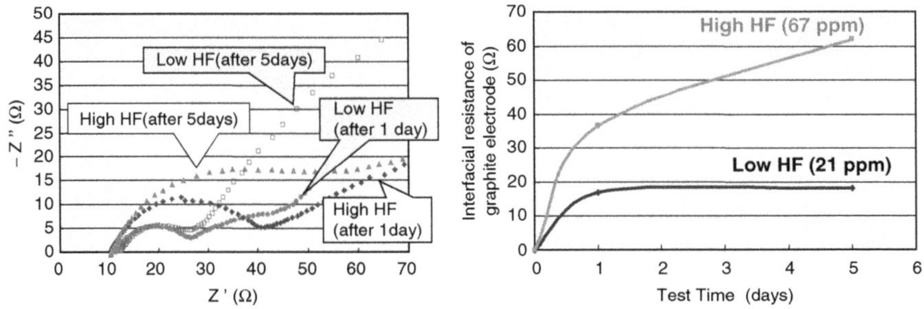

Fig. 19.5 (**a**) Effect of HF content in the electrolyte on the impedance of graphite anodes, and (**b**) the change in HF content with storage time in the electrolyte of Ube electrolyte and conventional electrolyte at room temperature

19.3.3 Functional Electrolytes Creation of a Stable Surface for Topochemical Reactions

The early commercial LIB used amorphous-carbon anode materials (cokes or hard carbon). Graphite anode materials gradually became an important technology for the lithium-ion cells to increase cell capacity. While the hard carbons are stable in a PC-based electrolyte, the intercalation of lithium ion in graphite is impossible for PC-based electrolytes. To avoid this difficulty, EC was used with the graphite electrode cells. However, EC is solid at room temperature (melting point = 36ºC). Even when DMC, MEC, or DEC is mixed with the EC for the electrolyte, the low temperature performance of the cell is poor. The SEI that formed on the graphite surface in the charging process as well as its formation mechanism were widely investigated by Aurbach and other researchers.[2] Their conclusion was that the SEI formation is unavoidable on graphite and it is not possible to change the nature of the SEI as long as an organic electrolyte is used. The common understanding is "the lithium-ion cell electrolyte gradually deteriorates upon the repeated charges and discharges, and a resistive SEI will be produced on the negative and positive electrode surface. Thus, the capacity is gradually reduced on cycling." UBE's novel idea is helping not only with study of the SEI formation mechanism, but also with finding a new functional film using suitable additives.

Since the melting point of PC is -40ºC, an effort was made to develop the PC-based electrolyte. The goal was to produce a more effective surface coating by using additives to modify the surface reactions that occur during the initial charge process, and thereby to produce a stable robust SEI. The first approach was to eliminate the electrochemically active impurities in the solvents and to purify the electrolyte salt, $LiPF_6$. The second approach was the incorporation of novel additives to the electrolyte to change the nature of the SEI and produce a stable surface film (functional film) on graphite that would facilitate lithium-ion intercalation and deintercalation at the graphite anode. The additives should react on negative graphite surface to produce a stable and beneficial film before the electrolyte begins to decompose the SEI formation. Molecular orbital (MO) calculations were used to screen potential additives.[3] This approach proved to be very useful in identifying compounds for practical applications. The simple expression of MO theory is shown in Fig. 19.6. The surface of the negative graphite electrode was modified by providing a suitable electrochemical surface structure (field) of graphite for the electrochemical reaction. This is the basic thinking for the current functional electrolytes.

Referring to Fig. 19.6, compound Y, with a low lowest unoccupied molecular orbital (LUMO) value, is a good electron acceptor and should decompose more easily at the anode than compounds X or Z. Compound Z, with a high highest occupied molecular orbital (HOMO) value, is a good electron donor and should react easily at the positive electrode compared to either X or Y.

The LUMO/HOMO relationship of solvents for 3-V lithium metal battery and 4-V LIB for anode and cathode additives is shown in Fig. 19.7. Solvents for a lithium metal battery – which are mostly ethers – should be stable against lithium metal anode and therefore their LUMO values should be high. The solvents – mainly cyclic and chain carbonates – for 4-V LIB have lower HOMO values, which means they are strong against oxidation. A modified version of the software was used for MO calculation, so the data for additives may differ from values from the usual commercial software. The relation between HOMO and LUMO values for first-generation additives (candidate for anode additives) and second-generation additives for candidate of cathode were calculated using the modified MO theory and are shown in Fig. 19.7. These calculations were used to screen the candidates to identify additives.

Figure 19.8 shows the relationship of HOMO values of selected solvents and their oxidation potential (Eox) measured by cyclic voltammetry using a Pt electrode as working electrode. It was found that Eox is roughly proportional to the HOMO values; this confirms the idea that MO theory is useful to screen the effective additives.

After confirming the effectiveness of MO theory in determining the oxidation and/or reduction potential of chemical compounds, a design diagram to choose

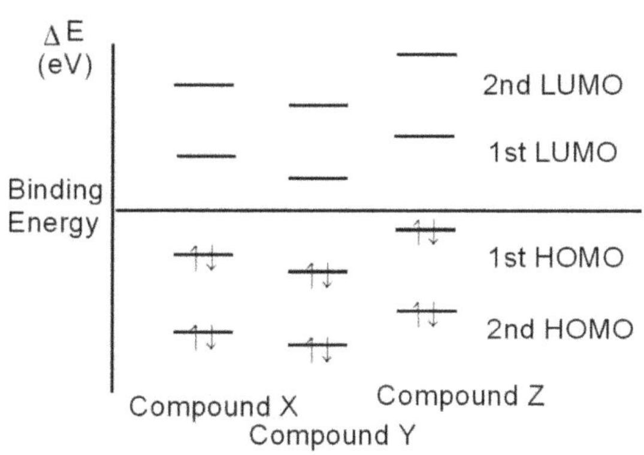

Fig. 19.6 Values of HOMO and LUMO for compounds X, Y, and Z

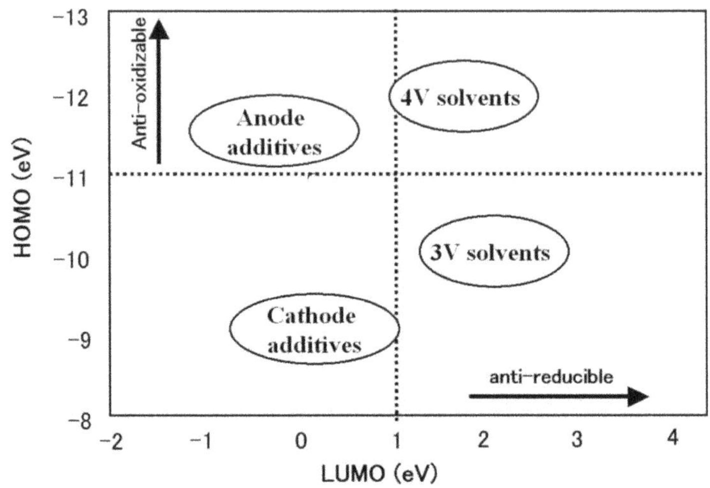

Fig. 19.7 Classification of solvents for lithium-metal battery, for 4-V LIB, for anode and cathode additives by the MO theory

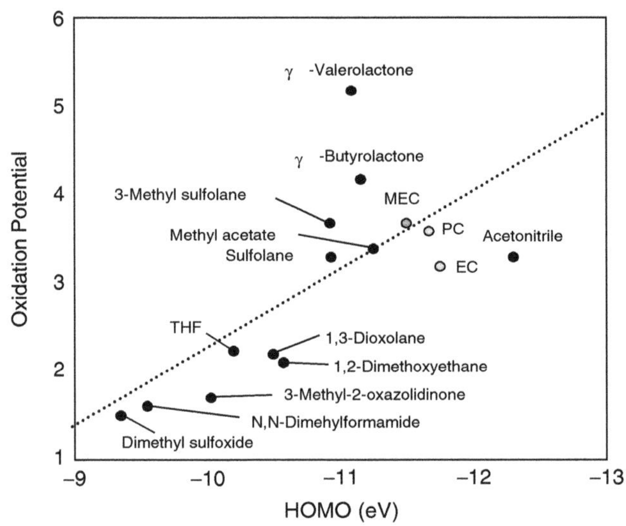

Fig. 19.8 Relationship between the oxidation potential and HOMO values

(screening) the electrolyte additives for functional electrolytes is established, as shown in Fig. 19.9.

A typical result is shown in Fig. 19.10. It shows the effectiveness of propane sulton (PS) as an anode additive to improve the high-temperature stability of the anode graphite. When MCMB (graphitized mesophase carbon) 6–28 is used as the

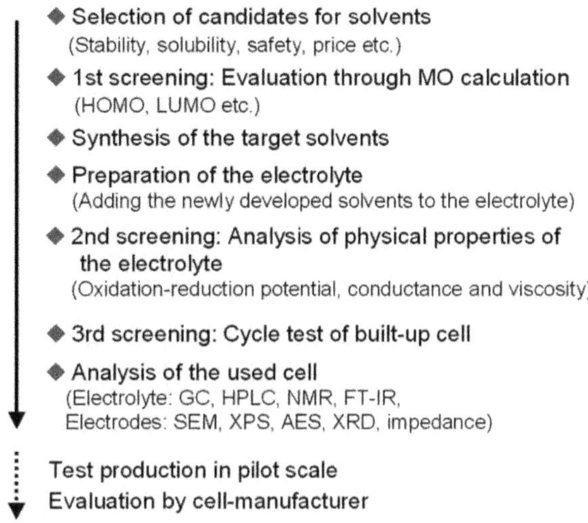

Fig. 19.9 Design diagram for screening electrolyte additives for functional electrolytes

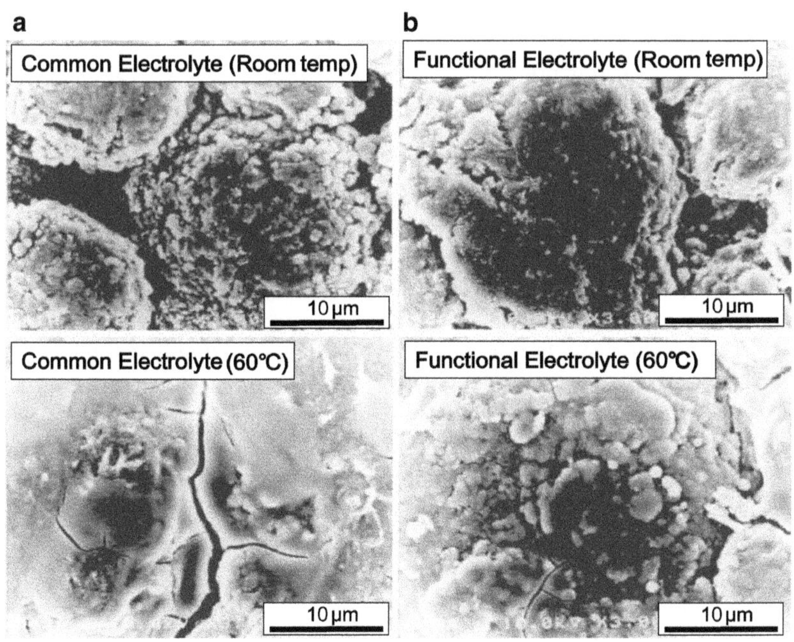

Fig. 19.10 Improvement of surface structure for topochemical reaction of graphite anode. Stored in (**a**) common and (**b**) functional (PS additive, right) electrolytes at room and 60°C temperature

negative electrode material stored for 1 month after charging in a common electrolyte, the surface of graphite is fairly damaged (Fig. 19.10a) and it is impossible to continue its cycling. However, the same graphite stored in a functional electrolyte is not damaged and works very well (Fig. 19.10b). As shown in successive sections, both a better negative electrode and a functional electrolyte improves the performance of cathode and LIB itself.

19.4 Second-Generation Functional Electrolytes

For the first generation of functional electrolytes, this was a period of proposing our new concept and using the highly purified solvents and lithium salt of $LiPF_6$. The second generation is a new period for developing control technology of the surface structure of the anode graphite. The rapid improvement in the electrochemical capacity due to using the high capacity of the negative electrode, accompanied by rapid deterioration of battery cycle life, development of a new precisely designed electrolyte was desirable.

The keyword for the second-generation functional electrolytes is the nanosized film control technology on the anode side. When a common electrolyte is used in the electrochemical cell system, a several hundred nanometer-thick surface film (SEI film) is produced on the negative electrode by successive charge/discharge cycles. This thick surface film inhibits the topochemical reaction of Li ion into graphite anode. The functional electrolytes significantly inhibit the reaction and maintains a low cell impedance. This second-generation technology has moved toward the production of nanoscale films on a 20-μm particle surface.

In this section, the design of several new compounds and the practical aspects of the control function of the second-generation functional electrolytes will be explained. When graphite is used for the negative electrode of the cell, EC has the advantage of low-solvent decomposition; however, EC is a solid at room temperature and the cell performance is poor at low temperature. When PC is used as the solvent, the solvent molecules are decomposed on the graphite and the cell operation becomes impossible. In this type of PC cell, the graphite decomposes the PC and produces a resistive material and exfoliation occurs on the graphite structure. Thus, the topochemical reaction is significantly inhibited.

The battery performance using natural graphite as an anode is often poor due to the reaction with the electrolyte to produce a highly resistive SEI. Special selected compounds such as vinyl acetate (VA), divinyl adipate (ADV), and allyl methyl carbonate (AMC) dissolved in an ordinary electrolyte are presented here to find the effectiveness of the additives.[3–7] Figure 19.11 shows the discharge capacity with cycles of positive active materials with and without the special functional additives compared to well-known ethylene sulfite (ES) additive.

In the cells with a PC-based electrolyte having no functionally active compound destruction of the active material itself was confirmed. The capacity of cells having no additives significantly dropped by 10 cycles making it impossible to continue

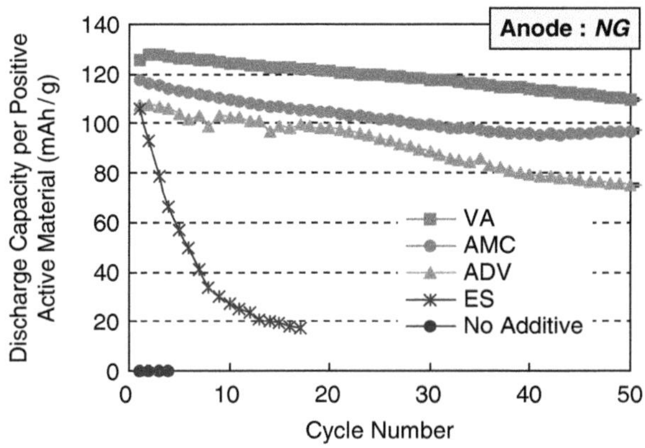

Fig. 19.11 Formation of topochemical reaction field on natural graphite by functional additives [1M $LiPF_6$/PC-DMC (1:2) + Additive 1 wt% at room temperature]

the cycling process. The cells containing VA, ADV, or AMC produced a smooth topochemical reaction structure on the graphite surface as seen in Fig. 19.11. With the EC electrolyte, the cells' discharge and charge cycles were possible without an additive since a better SEI is formed by the decomposition of the EC. Figure 19.12 shows SEM photographs in various solvents with or without the special additive compound. The topochemical charge-discharge operation becomes facilitated when VA is added to the PC-based electrolyte. The special nanosized coating on graphite film, here called the functional film, is not common SEI film but has a functional nature produced by the VA decomposition before PC molecules are decomposed. As shown in Fig. 19.12, the appearance of natural graphite (NG) with VA addition to PC-based electrolyte resembles that of an EC-based electrolyte without additive.

LUMO energy values of various solvents and additives are shown in Fig. 19.13. The electrochemical decomposition voltage on the graphite negative electrode (vs. Li/Li^+) is in the order PC < EC < AMC < VA < ADV < ES. The voltages are proportional to those of LUMO energy values, as shown in Fig. 19.14. Based on these voltage values,

the additive compound is decomposed on graphite anode and produces the functional film at a higher voltage during first charging process before reaching the PC or EC solvent decomposition voltage.

In general, the LUMO energy values are useful guidelines for compound or additive selection for anode additives, but it is necessary to test the additive compound in practical cell experiments in order to examine the details of the coating film for topochemical reaction. These tests provide important information for the development of the functional electrolytes. Figure 19.14 shows the relationship of the LUMO energy values and reduction potentials of additives on NG or Pt electrode. Reduction potential on NG is higher than those on Pt electrode due to the

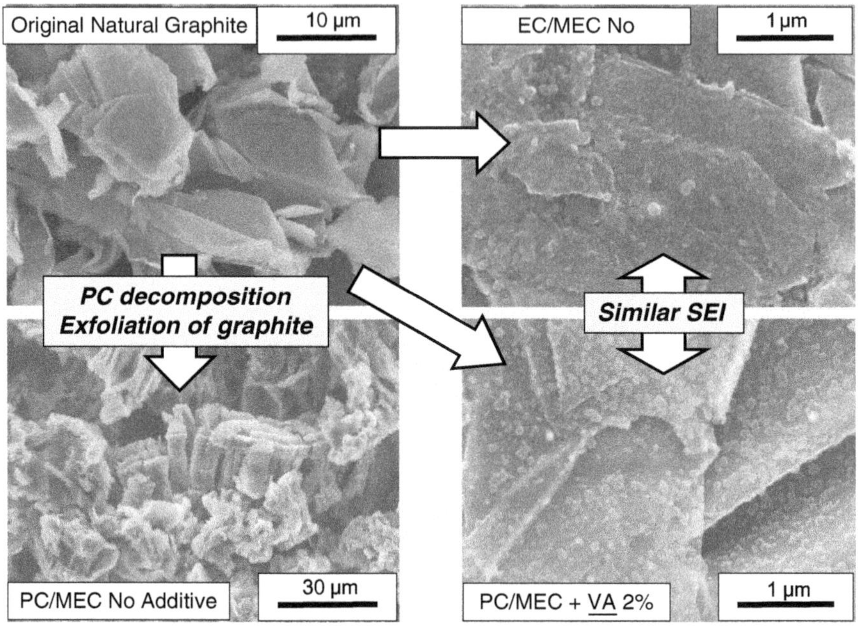

Fig. 19.12 SEM photographs of natural graphite surfaces on (NG). (**a**) Original NG; (**b**) exfoliation occurred NG in PC/MEC (1:2) electrolyte; (**c**) no exfoliation occurs in EC/MEC electrolyte; (**d**) no exfoliation in PC/MEC electrolyte with VA additive

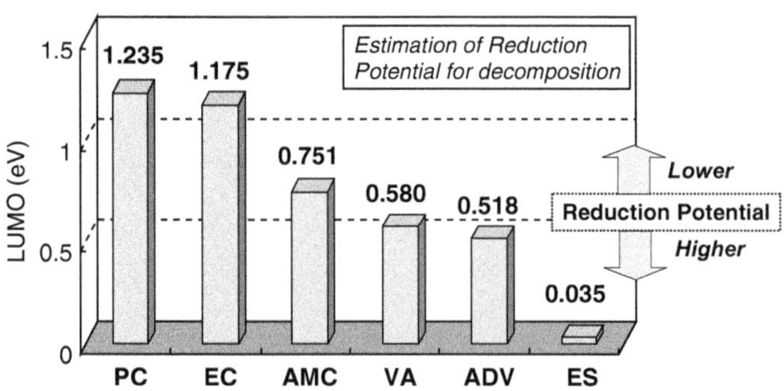

Fig. 19.13 LUMO energy values of various solvents and additives

higher catalytic activity of graphite. Accordingly, the reduction potential of additives is different depending on what kind of graphite is used in the cell. Battery companies sometime use an anode with two or three different types of graphite in their LIB. Therefore, UBE's functional electrolytes are specially designed for a

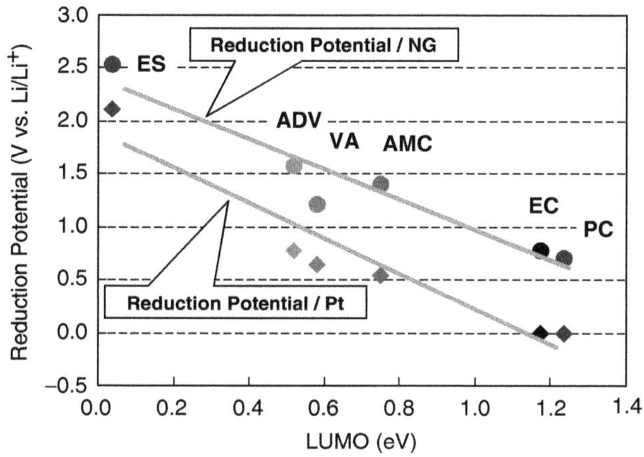

Fig. 19.14 LUMO energy values and reduction potentials of additives and solvents on natural graphite or Pt electrode

Table 19.1 Coating film thickness and impedance of natural graphite electrodes with and without additives

	NG (anode active material)	
Electrolyte	SEI film thickness (Å)	Impedance (Ω)
EC/MEC, no additive	266.7	0.34
PC/MEC no additive	–	–
PC/MEC + VA	211.8	0.15
PC/MEC + ADV	282.2	0.69
PC/MEC + AMC	136.9	0.24
PC/MEC + ES	330.3	1.83

particular kind of graphite anode, which usually means that five or six additives are included depending on the requirements of the customer.

The thickness and the resistivity are important characteristics of the film produced by the additive compound. Table 19.1 shows these characteristics of the film. The thinnest film was produced by AMC and its thickness was 100 nm. The VA additive produced a film with the lowest resistance value. These facts explain the highest capacity retention values when using VA compared to AMC (see Fig. 19.11). The film thickness difference between ES and VA is only 130–330 nm, but the resistance is ten times higher in ES. This small difference of only 100 nm in the film thickness has a large effect on the cell performance: ES cells quickly deteriorate with cycling (Fig. 19.11). Precise control of the film thickness and resistance is necessary for further improvement of the negative electrode performance.

19.5 Third-Generation Functional Electrolytes Designed for Cathode Electrode

The original design concept for the nanosized film control technologies was to separate the topochemical reaction of the positive electrode from that at the graphite anode. This was indispensable for a high-capacity cell for commercialization, especially when charging voltage is increased to improve the capacity of the cathode. The voltage distribution of the cathode surface is not homogeneous due to its very high voltage, as shown in Fig. 19.15. When such a high potential electrode is charged, the electrolyte decomposition takes place on the high voltage areas (active areas) due to the polarization. The functional electrolytes to control the surface of the positive electrode was made by dissolving the additives into the high purity electrolyte.

In this electrolyte, the additive material is designed to decompose before the solvent decomposition takes place at the cathode surface. For this purpose, the additives are designed to have a slightly lower oxidation voltage, but a stable in reduction direction. As an example of this type of design, biphenyl (BP) and o-terphenyl (OTP) were selected as additives for cathode. The HOMO energy and oxidation potential of various compounds measured by CV is shown in Fig. 19.16. It is clearly shown that the higher the oxidation potential, the lower the HOMO energy values.

BP or OTP decomposes at 4.5 V, at which the active area of the cathode surface even if the charging voltage is 4.3 V, and forms a thin film before the solvents are decomposed. The film makes the voltage distribution of the cathode surface homogeneous. We proposed a new mechanism for BP in the functional electrolytes as a voltage stabilizer of the positive electrode.

We confirmed the cell performance using 0.1% of BP in the functional electrolyte as a voltage stabilizer in 18650 size cells containing graphite as the negative

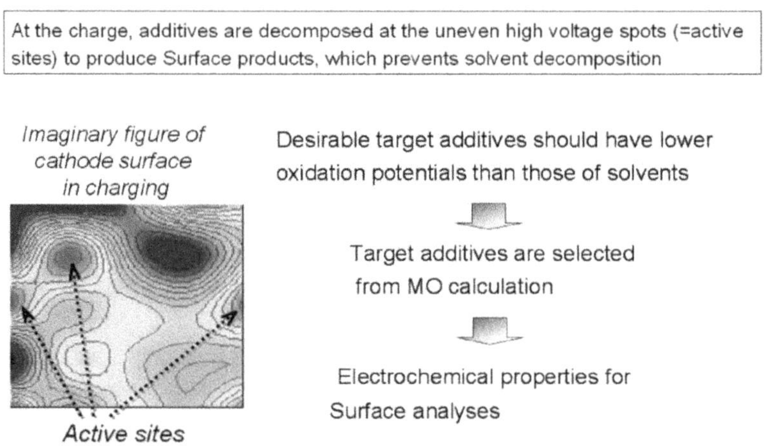

Fig. 19.15 Image of voltage distribution of the cathode surface

	BP	OTP	EC	DMC
compound	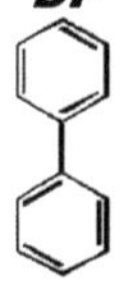	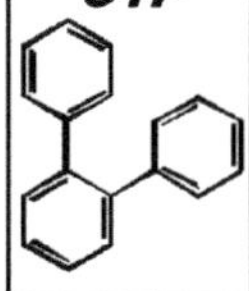	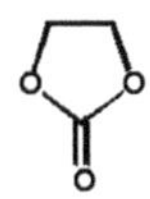	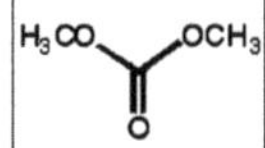
HOMO (eV)	- 8.92	- 9.34	- 11.8	- 11.6
Ox. Pot. (V vs Li/Li^+)	4.5	4.5	5.5	5.3

·*The measured values of BP and OTP are lower in the oxidation potentials than those of EC and DMC*
·*MO calculation has good correlation to the experimental values*

Fig. 19.16 HOMO energy and oxidation potential of several compounds measured by CV

electrode and lithium cobaltate as the positive electrode, as shown in Fig. 19.17. The cycle life tests were performed under hard conditions of 45°C and 4.3 V charge; it was found that the cyclicity of $LiCoO_2$ is rather poor without BP additives. When BP is added to the basic electrolyte, the cycle life is improved by the addition of a very small amount of BP; but if the added BP amount is large, the cycle life becomes shorter. The amount of additive makes a clear difference in the functional electrolytes. Although Moli Energy patented the addition of BP to the electrolyte as an overcharge protection, the amount of addition is too large (2%); hydrogen gas is produced when overcharged, which ruptures the LIB protective film.[8] The mechanisms of BP between overcharge proof and cathode additives are completely different.[9,10]

Before discussing these aspects we have to clarify the state of BP on the surface of the positive electrode material. We measured the depth profile of thecobalt positive electrode after 200 cycles by Auger electron spectroscopy (AES), as shown in Fig. 19.18. Thickness of the electroconductive membrane (ECM) film is estimated by the AES depth profiles atomic concentration of cobalt and oxygen. It reaches 90% with and without BP addition as shown in Fig. 19.18. The observed ECM film thicknesses are as follow: in the basic electrolyte, the ECM film thickness was 45 Å; in the functional electrolyte containing 1% of BP, the ECM film thickness was 68 Å; in the functional electrolyte having 2% of BP, the ECM film thickness was 214 Å. These results clearly show that the ECM film thickness on the positive electrode increased with the amount of BP. Based on these results, the cycle life of the basic electrolyte cell should be better, but the cells with the functional electrolyte containing the small amount of BP (the film thickness of 68 Å) afford the best results.

The special thin film is not a SEI, which is formed on anode, but is formed on the cathode electrode with a conductive nature, shown in a later section, and its thickness is much thinner than the SEI. The author named this very thin film as ECM film. The reason for this can be best determined from cyclic voltammetry

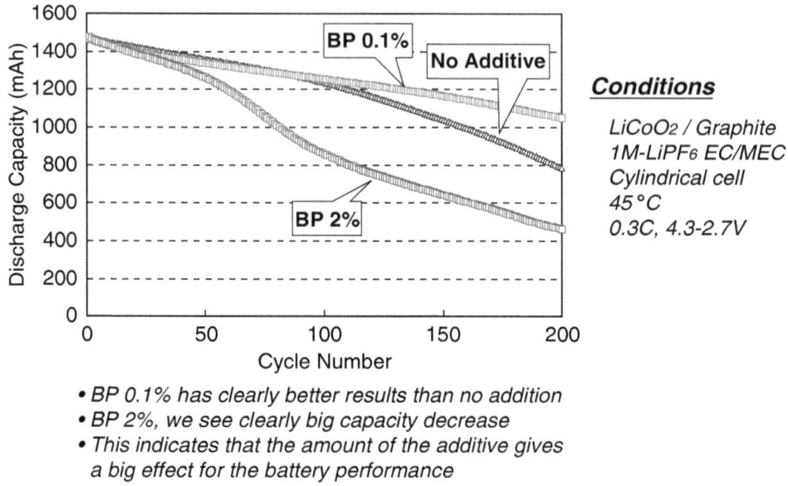

Fig. 19.17 Cyclicity of $LiCoO_2$/graphite cell between 4.3 and 2.7 V at 45°C

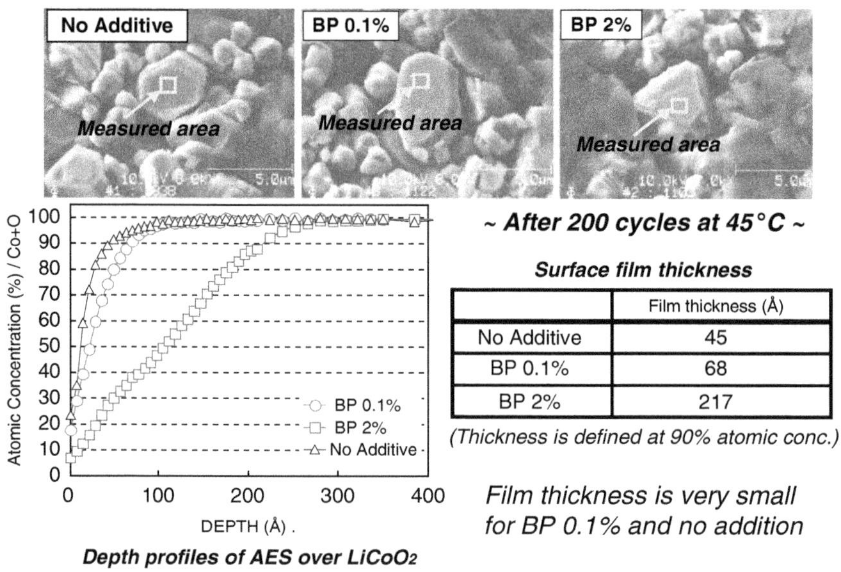

	Film thickness (Å)
No Additive	45
BP 0.1%	68
BP 2%	217

Fig. 19.18 AES measurement of $LiCoO_2$ cathode surface film thickness. $LiCoO_2$ electrodes used for the cylindrical cells after 200 charge–discharge cycles at 45°C in 1 M $LiPF_6$ EC/MEC = 3/7 without additive, with BP 0.1 wt% and with BP 2 wt%

observations. We should note that these three electrolytes have different reduction-oxidation behaviors and the decomposition behavior is difficult to evaluate only the cell function by the depth profile of the ECM film. The important point is the electrochemical characteristics of the functional electrolyte with minute amounts of BP added. The design concept of the functional electrolyte is to deliberately decompose the additive compound before the decomposition of the electrolyte in order to form the best ECM film on the electrode's surface. This decomposition process that takes place on the electrode can be clearly understood by the cyclic voltammetry results. For this electrochemical system using lithium cobaltate as the positive electrode and metallic lithium as the negative electrode, only the profile of the positive electrode can be observed. A very low current drain slowly discharged cells containing 1 M $LiPF_6$ in PC in order to observe the oxidation behavior. By this type of cell testing, it was found that BP added as a small amount to the functional electrolyte decomposed just before the decomposition of the basic electrolyte at about just over 4.5 V. Also, it was confirmed that the impedance increase due to the film produced from BP was negligibly small (Fig. 19.19).

Chemical analyses of the film components by X-ray photoelectron spectroscopy (XPS) are shown in Fig. 19.20. In the C_{1s} spectra, the peaks resulted from the binder polyvinylidene difluoride (PVdF) and carbonate peaks decreased and C–H/C–C peak of organic components increased by the addition of BP. In the F_{1s} spectra, the binder F–C peak is lowered and the Co–O peaks by $LiCoO_2$ are reduced in the Co–2P spectra. These XPS results indicate that the surface of cathode materials is covered with organic compounds and both binder and $LiCoO_2$ are difficult to see when 0.1 wt% of BP exists. Moreover, in the C_{1s} spectra, the peaks of carbonate and

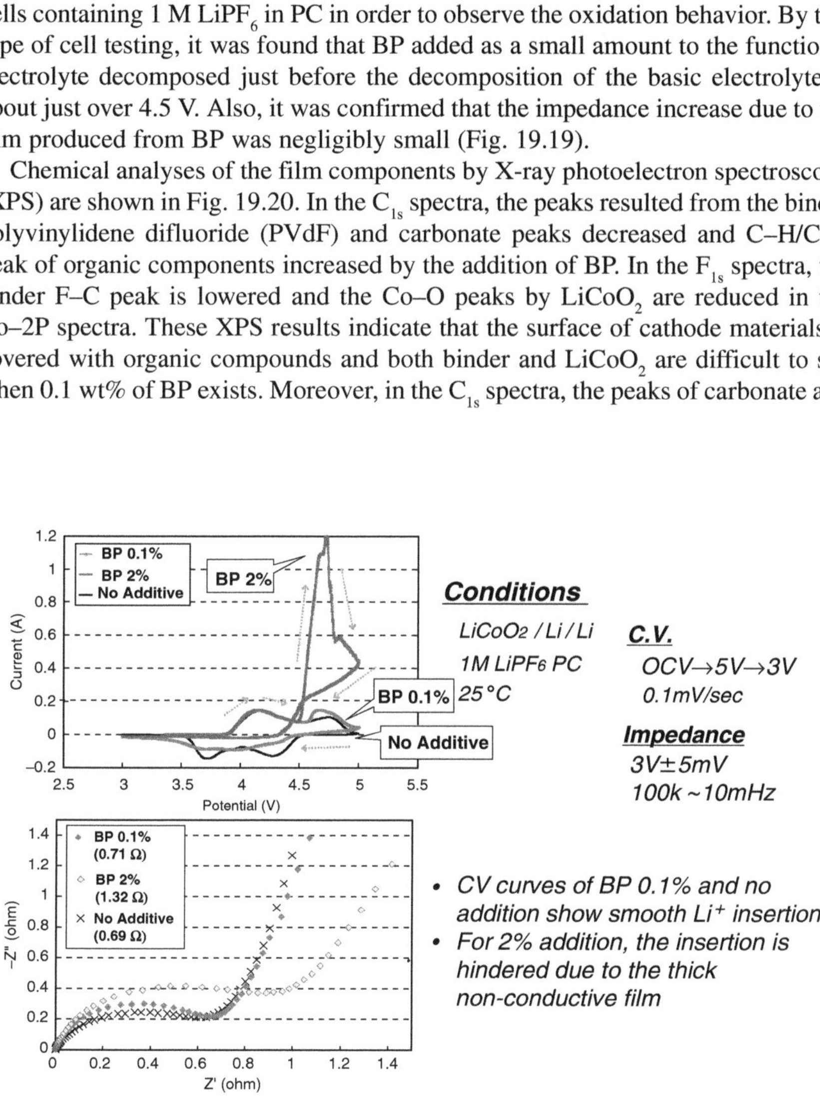

Fig. 19.19 Cyclic voltammograms of $LiCoO_2$ in 1 M$LiPF_6$ PC, without additive, with 0.1 wt% BP, and with 2 wt% BP at 0.1 mV/s and impedance values of the film

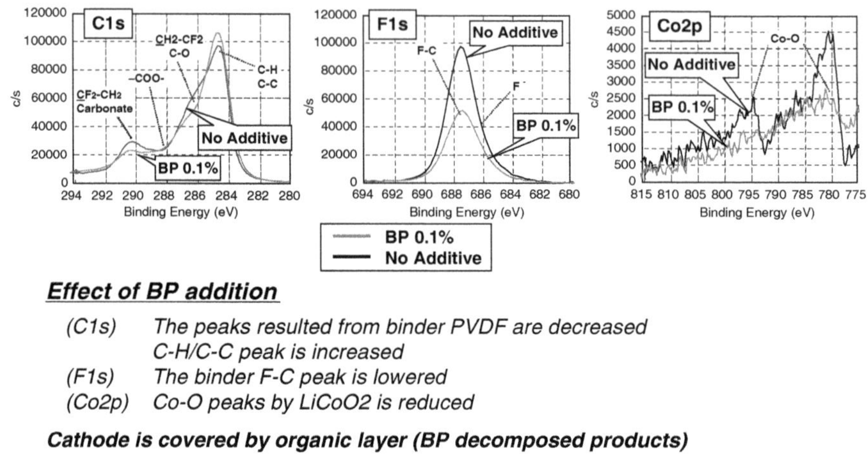

Fig. 19.20 XPS analysis of cathode surface film on $LiCoO_2$ electrodes used for the cylindrical cells after 200 charge–discharge cycles at 45°C in 1 M $LiPF_6$ EC/MEC = 3/7 without additive and with BP 0.1 wt%

esters do not increase. This implies that the film over the cathode is produced from the cathode additives, not from electrolyte component.

It can be seen that in the film that comes from decomposition of BP, cobalt concentration is as low as the lower limit of analysis, which means that the surface of $LiCoO_2$ is very stable in the presence of the BP-derived film. In contrast, cobalt could be obviously found in the film corresponding to the pristine electrolyte. Moreover, the contents of lithium and fluorine in BP-derived film are much smaller than the film in contact with the pristine electrolyte, whereas the carbon content in BP-derived films is higher.

The use of a functional electrolyte including cathode additives makes it possible to increase the energy density of LIB by an increase in charging voltage. For example, magnesium-doped $LiCoO_2$ has been utilized to prolong the $LiCoO_2$ capacity by charging higher than 4.2 V, and layered materials such as $LiMn_{1/3}Co_{1/3}Ni_{1/3}O_2$ and so on can be charged to 4.3 V or more; otherwise, layered material has low capacity when charged to 4.2 V. Based on the results, the existence of a special film was found on the surface of the lithium cobaltate and on the surface of the insulating binder. This indicates that a new theory for the conductive film formation is a reasonable idea. After the precise analysis of the electrochemical decomposition product of BP, we found that the film can be an insulator or it can be a conductor depending on the amount of BP added. Based on these facts, we developed a new concept that the functional electrolytes in which a minute amount of BP is added can produce a topochemical structure on the surface of the positive electrode (Fig. 19.17).

19.6 Other Functional Electrolytes

19.6.1 Functional Electrolytes to Help Stabilize Manganese Spinel Positive Electrodes

What kind of electrolyte can be suitable as the functional electrolytes for large-size, high-power batteries? What is the required performance of such large-size, high-power batteries from the standpoint of the market? The large-size battery must be "inexpensive with a high reliability," and the energy density should be less important. The final design of this large-type battery must be inexpensive, although it may be expensive during its development period. For an inexpensive battery, its components must be selected from low-cost materials. For a high reliability, the battery's components should last a long time. In this study, the manganese spinel was selected as an inexpensive material and its electrochemical structure is quite stable during long-term storage and use.

For the lithium-ion batteries using the manganese spinel as the positive electrode, the design of the functional electrolytes, which can stabilize the positive electrode, centers on how to handle the dissolution of the manganese spinel. In order to examine this concept, the concentration relation between the dissolved manganese and hydrogen fluoride were examined. A special cell design was developed that accelerated the dissolution of the manganese spinel as shown in Fig. 19.21.

The increase in manganese ion concentration during storage is proposed to involve two steps:

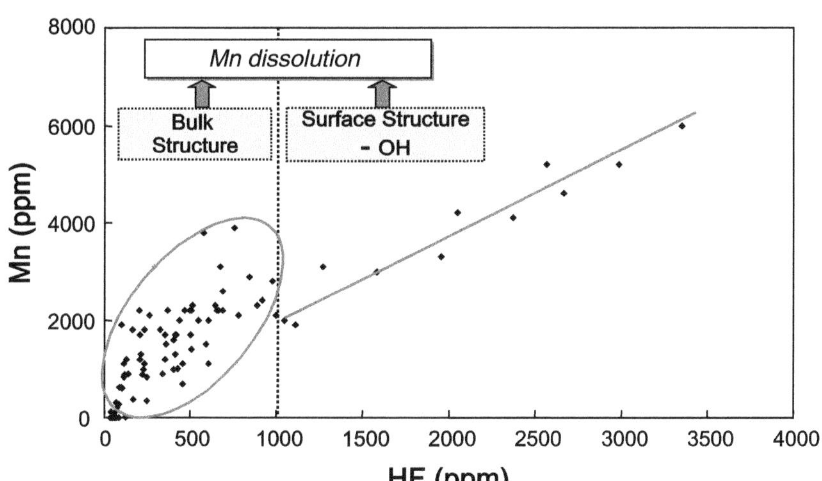

Fig. 19.21 The content of dissolved Mn ion and HF after storage of $LiMn_2O_4$ during 10 days storage at 80°C in 1 M $LiPF_6$ EC/DEC(3/7) solution

- Impurities such as Mn–OH, formed by water included in Mn structure, react with $LiPF_6$ by cation exchange reaction and HF molecules are formed
- Once formed, HF attacks $LiMn_2O_4$ and the spinel is dissolved

The proposed mechanism is shown in Fig. 19.22.

Two reasons for the origin of dissolved manganese ions were proposed. One was from the bulk structure of the manganese spinel and the other was the surface hydroxide. In general, the hydroxides on the manganese spinel are produced from two sources, i.e., water in the spinel and manganese hydroxide derivatives. The water is assumed to be slowly released from the oxides during the charge and discharge operations of spinel cathode. Therefore, the hydroxides in the spinel can be assumed to have the property of base. The hydroxide in the spinel reacts with the $LiPF_6$ to produce HF during the electrochemical reaction and heat energy generated during the charge and discharge processes. HF can react with spinel and dissolve the spinel particle itself and increase the manganese concentration in the electrolyte.

In order to suppress the manganese dissolution, an effective way is to reduce the reaction of the hydroxide and $LiPF_6$ in the electrolyte. Thus, the addition of a compound into the electrolyte which accelerates the reaction between the $LiPF_6$ and the hydroxide in the spinel must be a good quick solution. Namely, the desirable compound is one that reacts with OH in the spinel before the $LiPF_6$ reacts with it. This compound should be designed to produce protons that form a neutral solution in the non-aqueous electrolyte. Such a compound added to the functional electrolyte can easily react with the OH of the spinel in the base electrolyte. The novel functional electrolyte with additive X, especially designed for a spinel-based LIB, is a chemical reaction-controlled solution (Fig. 19.23).

With a cell using the manganese spinel as the positive electrode, the manganese dissolution rate was compared for the base electrolyte and the special functional electrolyte. The comparison tests were carried out at three temperatures (20, 45, and 60°C), first storing the charge cells for 1 month and the cells then repeatedly charged and discharged for 100 cycles. The dissolved manganese in these cells was then measured. The spinel compound in the experimental cells was produced by a

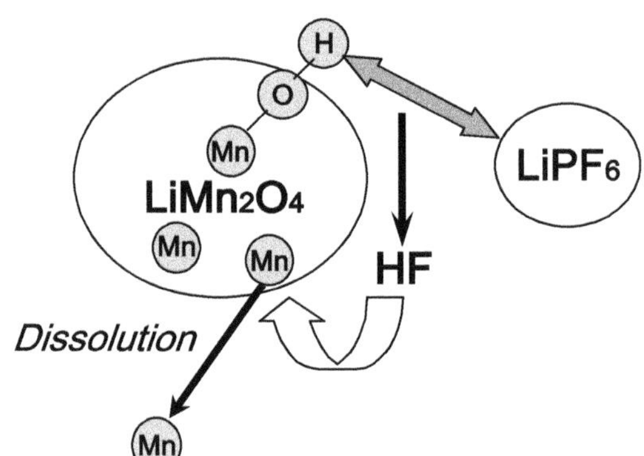

Fig. 19.22 The proposed mechanism of Mn dissolution of spinel structure at 80°C

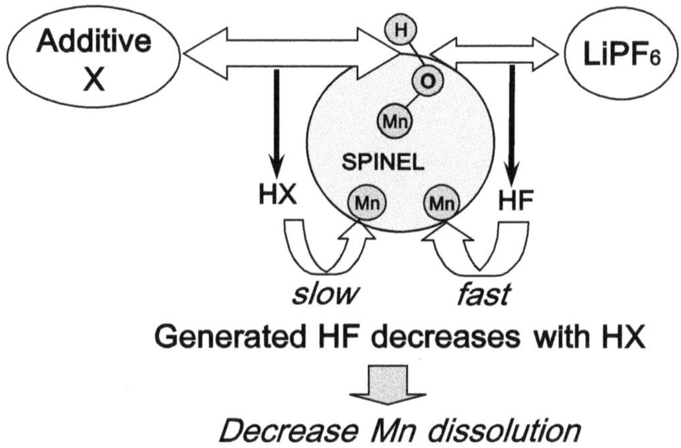

Fig. 19.23 The proposed mechanism explaining the reduced Mn dissolution by the novel functional electrolyte with X additive for spinel-based LIB

highly skilled method. The results clearly show the difference in the amounts of the dissolved manganese. This is one of the more important technological concepts for producing highly reliable cells for long-term stability (Fig. 19.24).

When the manganese dissolution is suppressed from the manganese spinel, the original spinel structure should remain intact without disintegration. In fact, when the cells with the functional electrolyte containing the special compound that stabilizes the manganese spinel are discharged for 100 cycles at the relatively high temperature of 60°C, the spinel structure remains practically unchanged throughout the test (Fig. 19.25).

19.6.2 Functional Electrolytes for Corrosion Prevention of Aluminum Current Collector

To produce a highly reliable battery, the development of only stable active materials is not enough. The prevention of deterioration of the conducting network among the anode or cathode electrodes due to electrolyte decomposition and reaction between the binder and the electrolyte are essential for long life. The electrolyte in the cell is in contact with the active materials, current collector, binder, separator, tab material, and tapes (battery case), and can produce unwanted electrochemical and chemical effects.

In this section, the author describes only the functional electrolytes for corrosion inhibition of the aluminum current collector for use in high-power, large-size batteries. The functional electrolytes not only is improving the active material utilization, but also all other materials, which improves the overall cell performance. If we compare the cells to our human body, the electrolyte can be considered to be

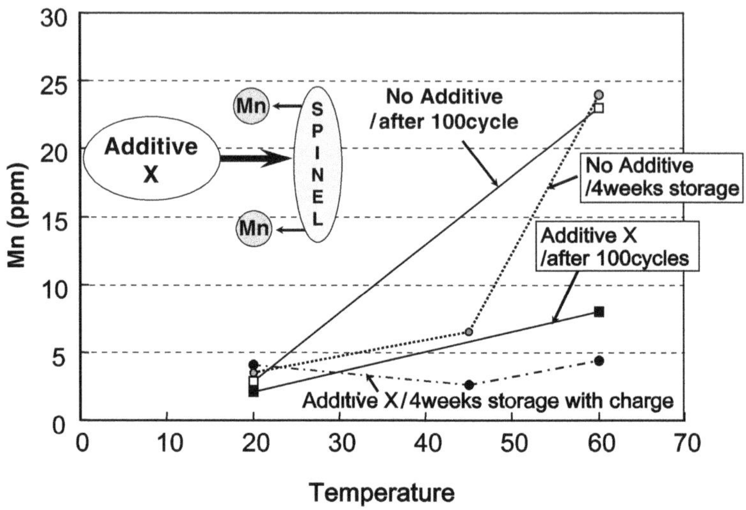

Fig. 19.24 Comparison of dissolved Mn ions from between functional electrolyte with additive X and common electrolyte at various temperatures

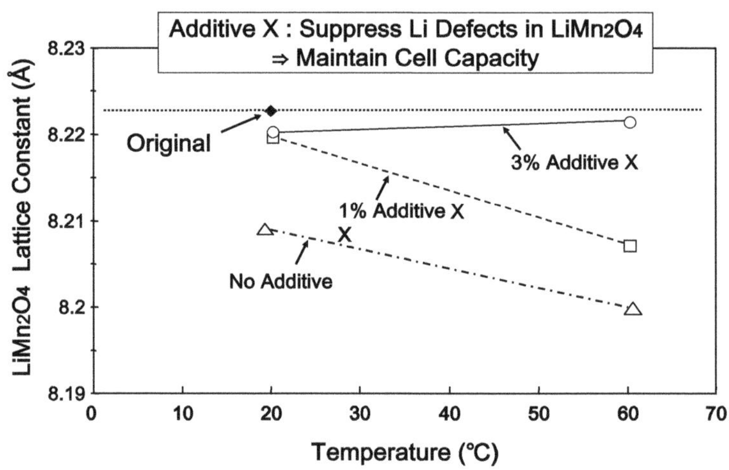

Fig. 19.25 Lattice constant of graphite in spinel/graphite cell with or without additive, X, in the electrolyte after 100 cycles at various temperatures

the blood. The additive material for the cell is a kind of medicine, which can reach the injured part of the cell and remove the problem or cure the disease. The current collector's main function is to collect the current, but it must satisfactorily work for 5–10 years with a high reliability. For the accelerated evaluation of the current collector, the cells were operated at high temperature in order to produce some

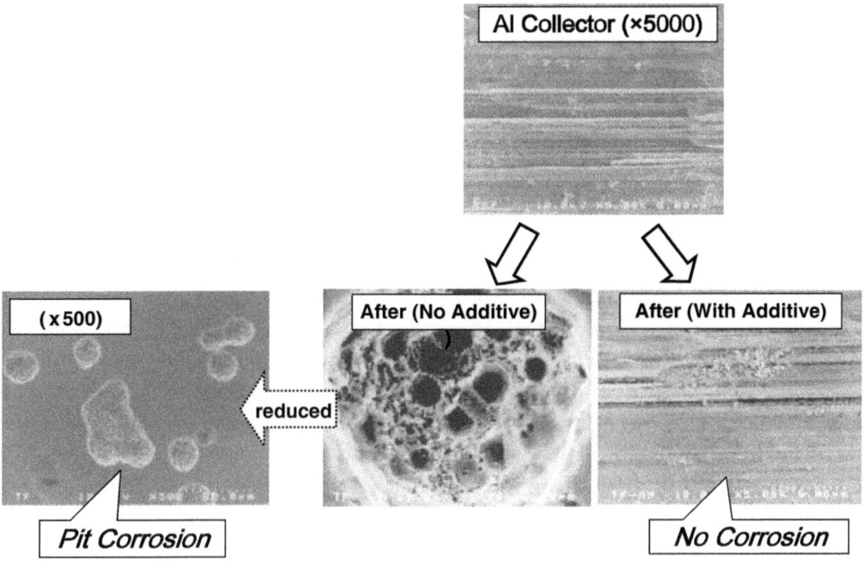

Fig. 19.26 Effect of novel functional electrolyte containing NI as an additive designed to suppress the aluminum substrate corrosion

deterioration in the functional electrolyte containing an aluminum corrosion inhibitor and in the regular basic electrolyte. The aluminum current collectors then were removed from the cells and tested as an electrode undergoing charge-discharge cycles. This type of test is to confirm the electric corrosion of the aluminum. The beneficial effect of the special functional electrolyte containing the aluminum corrosion inhibitor (NI) is compared to the results using the basic electrolyte. The aluminum current collector maintained its initial surface morphology (no pits) after the test in the special functional electrolyte (Fig. 19.26). In contrast, there were many pits on the aluminum. This indicates that the battery that uses our special functional electrolytes has a reliable and long life.

19.7 Additional Comments

In the past, the electrolyte has been basically considered to be a transporter of lithium ions. Now, the various functional electrolytes have been developed in order to produce a specific benefit in order to improve the battery performance. The initiation of the research on functional electrolytes[9] was based on the suggestion of Prof. Masaki Yoshio of Saga University in 1992. Later, the author received the excellent suggestion of a "topochemical theory" by Dr. Akira Yoshino of the Asahi

Chemical Co. The first commercial production was successfully started in 1996. Its trademark was "Purelyte," which was sold only under confidentiality agreement. Mr. Takeshita of IIT (working at NRI at that time) described in an NRI report that commercial production had been initiated and the product then became number one in the world. The author was invited for a lecture on "Battery Power Supply" at the Techno-Frontier Symposium 99. At this lecture, the author presented the "functional electrolytes" as a new technology. In 2001, the product was included as an "electrolyte and additive" in the first Lithium Battery Discussion by LG Chem. In 2003, Mitsubishi Chemical presented this type of product as "Role Assigned Electrolytes in Battery and Power Supply" at Techno-Frontier Symposium 2003.

In the early days, the additive agents in the functional electrolytes were not favorably accepted as it decreased the voltage window of the electrode materials. Today, the functional electrolytes have been accepted as an important aspect of battery development. The author is understandably very happy to have been the initiator of this type of functional electrolytes and feels a large responsibility for future developments in this area. The author hopes to continue battery development work based on this new concept from different points of view for electrochemical batteries based on its chemistry.

References

1. H. Takeshita, NRI Report Market of Advanced Secondary Battery, NRI, 1997.
2. D. Aurbach, *Advances in Lithium Ion Batteries*, Chap. 1, W. A. Schalkwijk, B. Scrosati (eds.).
3. H. Yoshitake, K. Abe, T. Kitakura, J. B. Gong, Y. S. Lee, H. Nakamura and M. Yoshio, *Chem. Lett.*, **32** (2003) 134.
4. P. Ghimire, H. Nakamura, M. Yoshio, H. Yoshitake and K. Abe, *Electrochemistry*, **71** (2003) 1084.
5. K. Abe, H. Yoshitake, T. Kitakura, T. Hattori, C. Wang and M. Yoshio, *Electrochim. Acta*, **49** (2004) 4613.
6. P. Ghimire, H. Nakamura, M. Yoshio, H. Yoshitake and K. Abe, *ITE Letters on Batteries. New Technologies & Medicine*, **6** (2005) 16.
7. P. Ghimire, H. Nakaramura, M. Yoshio, H. Yoshitake and K. Abe, *Chem Left.*, **34** (2005) 1052
8. H. Mao, US Pat. 5,879,834, 9 March 1999.
9. K. Abe, T. Takaya, H. Yoshitake, Y. Ushigoe, M. Yoshio and H. Wang, *Electrochem. and Solid-State Lett.*. **7** (2004) A462.
10. K. Abe, Y. Ushigoe, H. Yoshitake, M. Yoshio, *J. Power Sources*, 153 (2005) 328.

Chapter 20
Lithium-Ion Battery Separators[1]

Zhengming (John) Zhang and Premanand Ramadass

20.1 Introduction

Many advances have been made in battery technology in recent years, both through continued improvement of specific electrochemical systems and through the development and introduction of new battery chemistries. Nevertheless, there still is no one "ideal" battery that gives optimum performance under all operating conditions. Similarly, there is no one separator that can be considered "ideal" for all battery chemistries and geometries.

A separator is a porous membrane placed between electrodes of opposite polarity, permeable to ionic flow but preventing electric contact of the electrodes. A variety of separators have been used in batteries over the years. Starting with cedar shingles and sausage casing, separators have been manufactured from cellulosic papers and cellophane to nonwoven fabrics, foams, ion exchange membranes, and microporous flat-sheet membranes made from polymeric materials. As batteries have become more sophisticated, separator function also has become more demanding and complex.

Separators play a key role in all batteries. Their main function is to keep the positive and negative electrodes apart to prevent electrical short circuits and at the same time allow rapid transport of ionic charge carriers that are needed to complete the circuit during the passage of current in an electrochemical cell. They should be very good electronic insulators and have the capability of conducting ions by either intrinsic ionic conductor or by soaking electrolyte. They should minimize any processes that adversely affect the electrochemical energy efficiency of the batteries.

Very little work (relative to research of electrode materials and electrolytes) is directed toward characterizing and developing new separators. Similarly, not much attention has been given to separators in publications reviewing batteries.[1–10]

[1] Reprinted in part with permission from *Chem. Rev.*[104] (2004) 4419–4462, copyright (2004) American Chemical Society.

Z. Zhang (✉) and P. Ramadass
Celgard, LLC, 13800 South Lakes Dr., Charlotte, NC28273, USA
johnzhang@celgard.com

M. Yoshio et al. (eds.), *Lithium-Ion Batteries*,
DOI 10.1007/978-0-387-34445-4_20 © Springer Science+Business Media LLC 2009

A number of reviews on the cell fabrication, their performance, and application in real life have appeared in recent years but none have discussed separators in detail. Recently a few reviews have been published in both English and Japanese which discuss different types of separators for various batteries.[11–20] A detailed review of lead acid and lithium-ion (Li-Ion) battery separators was published by Boehnstedt[13] and Spotnitz,[14] respectively, in *Handbook of Battery Materials*. Earlier Kinoshita et al. had done a survey of different types of membranes/separators used in different electrochemical systems, including batteries.[11]

The majority of the separators currently used in batteries were typically developed as spin-offs of existing technologies. They usually were not developed specifically for those batteries, and thus are not completely optimized for systems in which they are used. One positive result of adapting existing technologies is that they are produced in high volume at relatively low cost. The availability of low-cost separators is an important consideration in the commercialization of batteries, because the battery industry traditionally operates with thin profit margins and relatively small research budgets.

The purpose of this chapter is to provide a detailed review of separators used in Li-Ion battery applications and their chemical, mechanical, and electrochemical properties. The separator requirements, properties, and characterization techniques are also described with respect to Li-Ion batteries. Despite the widespread use of separators, a great need still exists for improving the performance, increasing its life, and reducing cost. In the following Sections an attempt is made to discuss key issues in various separators with the hope of bringing into focus present and future directions of research and development in separator technologies.

20.2 Battery and Separator Market

The battery industry has seen enormous growth over the past few years in portable, rechargeable battery packs. The majority of this surge can be attributed to the widespread use of cell phones, personal digital assistants (PDAs), laptop computers and other wireless electronics. Batteries remained the mainstream source of power for systems ranging from mobile phones and PDAs to electric and hybrid electric vehicles. The world market for batteries was approximately $41 billion in 2000. This included $16.2 billion primary and $24.9 billion secondary cells.[21]

The rechargeable battery (NiCd, NiMH, and Li-Ion) market for 2003 for portable electronics was around $5.24 billion, around 20% more then 2002. The Li-Ion battery market was around $3.8 billion (~73%). They now are used in more than 90% of cellphones, camcorders, and portable computers worldwide and also have been adopted for use in power tools recently.[22]

A recent market survey shows that the lithium battery is becoming an almighty rechargeable battery and in 2006 the demand is around 2 billion cells with more than 10% annual growth. There is a steady increase in the demand for lithium batteries for notebook PCs and cellular devices, and in addition the demand for another important application, namely power tools is also more than expected even though the nickel–cadmium (Ni–Cd) battery demand for power tools is still increasing. Lithium battery-powered (UPS) already is commercialized and there is no doubt that Li-Ion battery hybrid electric vehicle (LIB-HEV) will be a major breakthrough in 2010 and beyond.[23]

Figure 20.1 presents the details of total sales of all the major rechargeable battery systems (Li–Cd, Ni–MH, Li-Ion battery, and Li-Ion battery-Laminated) from 1991 to 2006.[23] The total market size of rechargeable battery systems keeps over 6 billion US$ and as of now the lithium battery market is about 4.5 billion US$; adding HEV application, the market size of 10 billion US$ and above is possible.

The tremendous progress in Li-Ion cells is clearly visible with as much as a twofold increase in the volumetric and gravimetric energy density for both 18,650 and prismatic cells between 1994 and 2002. In last few years the Li-Ion production has expanded to South Korea (Samsung, LG, etc.) and China (BYD, B&K, Lishen, etc.) from Japan. Several Japanese (Sanyo, Sony, MBI, NEC, etc.) and Korean (LG Chemical) manufacturers also have moved their manufacturing plants to China.[22] Japan, which controlled 94% of the global rechargeable battery market in 2000, has seen its market share drop to about 65% of the global market.[22–25] The continued growth in Li-Ion battery market has led to a strong demand for battery separators. All the major separator manufacturers (Celgard, Asahi, and Tonen) either have increased the capacity in 2003 or were planning to increase in 2004.[26–28]

There is not much information available on battery separator market in the literature. It is estimated that about 30% of the rechargeable lithium battery market, or

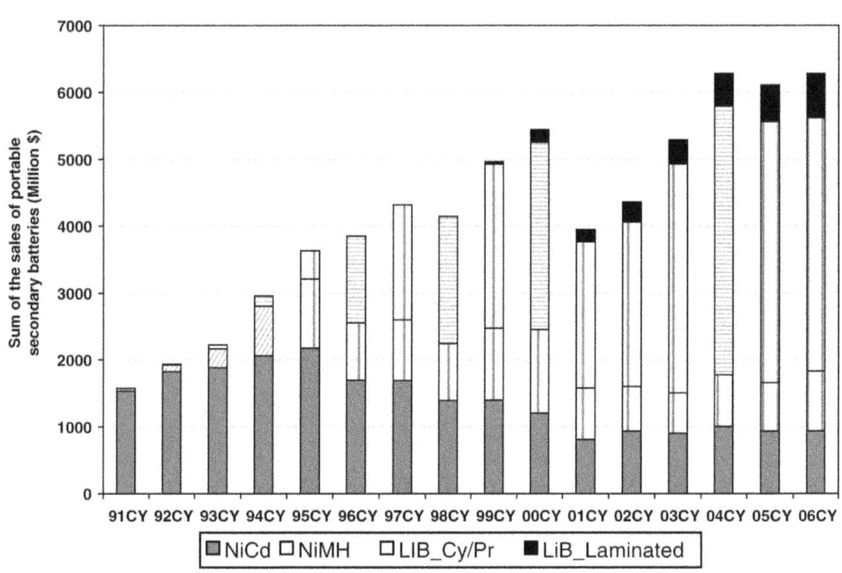

Fig. 20.1 Overview of worldwide rechargeable battery market[23]

$1.5 billion, is the size of the battery materials or components market. Battery separators for lithium batteries are about a $330 million market within the total battery components market.[29,30] Recently, Freedonia Group has reported that the US demand for battery separators will increase to $410 million in 2007 from $237 million in 1977, and $300 million in 2002, respectively.[31,32]

20.3 Separator and Batteries

Batteries are built in many different shapes and configurations—button, flat, prismatic (rectangular), and cylindrical (AA, AAA, C, D, 18,650 etc.). The cell components (including separators) are designed to accommodate a particular cell shape and design. The separators are either stacked between the electrodes or wound together with electrodes to form jellyrolls, as shown in Fig. 20.2. Stacked cells are generally held together by pressure from the cell container. The Li-Ion gel polymer stacked cells are prepared by bonding/laminating layers of electrodes and separators together. The separator properties should not change significantly during the bonding process. In some cases, the separators are coated to help in bonding process, and thus reducing the interfacial resistance.[33–35]

In the conventional way of making spirally wound cells, two layers of separators are wound along with the positive and negative electrodes, resulting in separator-negative/separator-positive configuration. They are wound as tightly as possible to ensure good interfacial contact. This requires the separators to be strong to avoid any contact between the electrodes through the separator. The separator also must not yield and reduce in width, or else the electrodes may contact each other.

Once wound, the jellyroll is inserted into a can and filled with electrolyte. The separator must be wetted quickly by the electrolyte to reduce the electrolyte filling time. A header is then crimped into the cell to cover the can from top. In some prismatic cells, the jellyroll is pressed at high temperatures and pressures and then inserted into thin rectangular cans. A typical 18,650 Li-Ion cell use around 0.07–0.09 m^2 of separator, which is approximately 4–5% of the total cell weight.[36]

20.4 Separator Requirements

A number of factors must be considered in selecting the best separator for a particular battery and application. The characteristics of each available separator must be weighed against the requirements and one selected that best fulfills these needs. A wide variety of properties are required of separators used in batteries. The considerations that are important and influence the selection of the separator include:

- Electronic insulator
- Minimal electrolyte (ionic) resistance
- Mechanical and dimensional stability

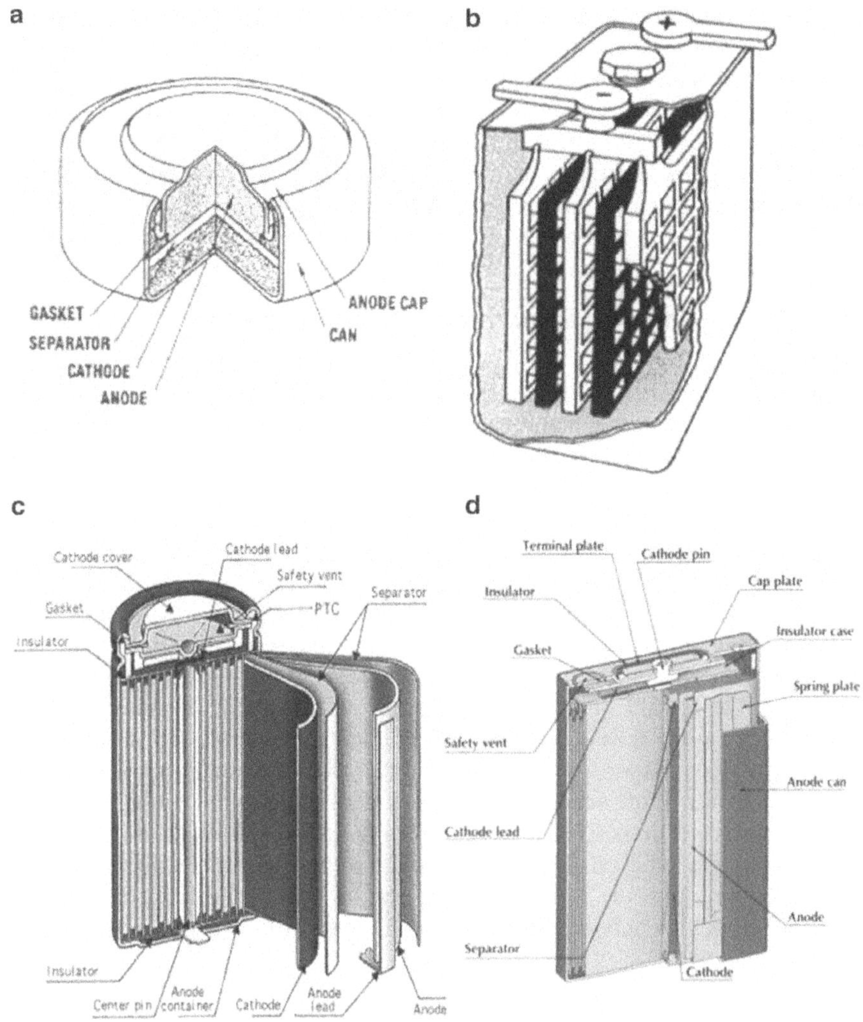

Fig. 20.2 Typical battery configurations (**a**) button cell, (**b**) stack lead acid, (**c**) spiral-wound cylindrical lithium-ion, and (**d**) spiral-wound prismatic lithium ion. Reprinted with permission from *Chem. Rev.*[104] (2004) 4419–4462, copyright (2004), American Chemical Society

- Sufficient physical strength to allow easy handling
- Chemical resistance to degradation by electrolyte, impurities, and electrode reactants and products
- Effective in preventing migration of particles or colloidal or soluble species between the two electrodes
- Readily wetted by electrolyte
- Uniform in thickness and other properties

The order of importance of the various criteria varies, depending on the battery applications. The above list presents a broad spectrum of requirements for separators in batteries. In many applications, a compromise in requirements for the separator generally must be made to optimize performance, safety, cost, and so forth. For example, batteries that are characterized by small internal resistance and consume little power require separators that are highly porous and thin; but the need for adequate physical strength may require that they be thick.

20.5 Separators for Secondary Lithium Batteries

All lithium-based batteries use nonaqueous electrolytes because of the reactivity of lithium in aqueous solution and because of an electrolyte's stability at high voltage. The majority of these cells use microporous membranes made of polyolefins. In some cases, nonwovens made of polyolefins are used either alone or with microporous separators. This section will focus mainly on separators used in secondary lithium batteries followed by a brief summary of separators used in lithium primary batteries.

Lithium secondary batteries can be classified into three types: a liquid-type battery using liquid electrolytes, a gel-type battery using gel electrolytes mixed with polymer and liquid, and a solid-type battery using polymer electrolytes. The types of separators used in different types of secondary lithium batteries are shown in Table 20.1. The liquid Li-Ion cell uses microporous polyolefin separators while the gel polymer Li-Ion cells either use polyvinylidene difluoride (PVdF) separator (e.g., PLION® cells) or PVdF-coated microporous polyolefin separators. The PLION® cells use PVdF loaded with silica and a plasticizer as the separator. The microporous structure is formed by removing the plasticizer and then filling with liquid electrolyte. They also are characterized as plasticized electrolyte. In solid polymer Li-Ion cells, the solid electrolyte acts as both electrolyte and separator. This chapter focuses only on the conventional liquid Li-Ion systems.

Table 20.1 Types of separators used in different types of secondary lithium batteries

Battery system	Type of separator	Composition
Lithium-ion (liquid electrolyte)	Microporous	Polyolefins (PE, PP, PP/PE/PP)
Lithium-ion gel polymer	Microporous	PVdF
	Microporous	Polyolefins (PE, PP, PP/PE/PP) coated with PVdF or other gelling agents
Lithium-polymer (e.g., Li-V6O13)	Polymer electrolyte	Polyethylene oxide with lithium salt

Reprinted with permission from *Chem. Rev.* [104] (2004) 4419–4462, copyright (2004), American Chemical Society.

The last decade has seen significant advances in the ambient temperature lithium battery technology. Lithium-ion batteries are the preferred power source for most portable electronics because of their higher energy density, longer cycle life, and higher operational voltage compared to Ni–Cd and Ni–MH systems. In 2002, 66% of the total rechargeable battery market for mobile IT and communication devices used lithium-based batteries and the rest used nickel-based batteries.[37,38]

A typical Li-Ion cell consists of a positive electrode composed of a thin layer of powdered metal oxide (e.g., $LiCoO_2$) mounted on aluminum foil and a negative electrode formed from a thin layer of powdered graphite, or certain other carbons, mounted on a copper foil. The two electrodes are separated by a porous plastic film soaked typically in $LiPF_6$ dissolved in a mixture of organic solvents such as ethylene carbonate (EC), ethyl methyl carbonate (EMC), or diethyl carbonate (DEC). In the charge/discharge process, lithium ions are inserted or extracted from interstitial space between atomic layers within the active materials.

Sony's introduction of the rechargeable lithium ion battery in the early 1990s precipitated a need for new separators that provided not only good mechanical and electrical properties but also provided added safety through a thermal shutdown mechanism. Although a variety of separators (e.g., cellulose, nonwoven fabric, etc.) have been used in different types of batteries, various studies on separators for Li-Ion batteries have been pursued in the last few years, as separators for Li-Ion batteries require different characteristics than separators used in conventional batteries.

A novel microporous separator using polyolefins has been developed and used extensively in Li-Ion batteries since it is difficult for conventional separator materials to satisfy the characteristics required in Li-Ion batteries. In Li-Ion batteries two layers of separators are sandwiched between positive and negative electrodes and then spirally wound together in cylindrical and prismatic configurations. The pores of the separator are filled with ionically conductive liquid electrolyte.

Microporous polyolefin membranes in current use are thin (< 30 μm) and are made of polyethylene (PE), polypropylene (PP), or laminates[39] of polyethylene and polypropylene. They are made up of polyolefin materials because they provide excellent mechanical properties, chemical stability, and acceptable cost.[40,41] They have been found to be compatible with the cell chemistry and can be cycled for several hundred cycles without significant degradation in chemical or physical properties.

Commercial membranes offer pore sizes in the range of 0.03–0.1 μm, and 30–50% porosity. The low melting point of PE enables their use as a thermal fuse. As the temperature approaches the melting point of the polymer, 135°C for PE and 165°C for PP, porosity is lost. The trilayer material (PP/PE/PP)[42] has been developed by Celgard® where a PP layer is designed to maintain the integrity of the film, while the low melting point of the PE layer is intended to shut down the cell if an overtemperature condition is reached. Asahi Kasai's flat-film membrane "Hipore™" is available in thicknesses ranging from 20 μm to several hundred microns, and with highly uniform pore sizes ranging from 0.05 to 0.5 μm.[43] The major manufacturers of Li-Ion battery separators along with their typical products are listed in Table 20.2.

Table 20.2 Major manufacturers of lithium-Ion battery separators along with their typical products

Manufacturer	Structure	Composition	Process	Trade name
Asahi Kasai	Single layer	PE	Wet	HiPore
Celgard Inc	Single layer	PP, PE	Dry	Celgard
	Multilayer	PP/PE/PP	Dry	Celgard
	PVdF coated	PVdF, PP, PE, PP/PE/PP	Dry	Celgard
Entek Membranes	Single layer	PE	Wet	Teklon
Mitsui Chemical	Single layer	PE	Wet	
Nitto Denko [44]	Single layer	PE	Wet	
DSM	Single layer	PE	Wet	Solupur
Tonen	Single layer	PE	Wet	Setela
Ube Industries	Multilayer	PP/PE/PP	Dry	U-Pore

In recent years there has been a strong demand for higher-capacity Li-Ion cells because of the strong growth in portable electronics. One way to achieve higher capacity is by reducing the thickness of separators. The battery manufacturers have started using 20- and 16-μm separators in higher-capacity (> 2.0 Ah) cylindrical cells, and 9-μm separators in Li-Ion gel polymer cells.

Nonwoven materials also have been developed for Li-Ion cells but have not been widely accepted, in part due to the difficulty in fabricating thin materials with good uniformity and high strength. [14] Nonwoven separators have been used in button cells and bobbin cells when thicker separators and low discharge rates are acceptable.

20.5.1 Separator Development

The process for making Li-Ion battery separators can be broadly divided into dry [45, 46] and wet [47] processes. Both processes usually employ one or more orientation steps to impart porosity and/or increase tensile strength. Dry process involves melting a polyolefin resin, extruding it into a film, thermal annealing to increase the size and amount of lamella crystallites, and precisely stretching to form tightly ordered micropores.[48–52] In this process, a row lamellar crystal structure is generated in the polymer in the initial extrusion step. This non-porous structure is highly oriented as a result of extrusion and annealing conditions. In the next step, film is stretched to form micropores. This microporous structure is continuous throughout the bulk interior of the membrane. [53]

Polypropylene and polyethylene microporous films obtained by this method are available from Celgard, [48, 50, 54, 55] and Ube. [56] The dry process is technologically convenient because no solvents are required. However, only a uniaxial stretching method has been successful to date, and as a result the pores are slitlike in shape and there is an anisotropy of films in mechanical properties. The tensile strength in the lateral direction is relatively low.

Wet process (phase inversion process)[57, 58] involves mixing of hydrocarbon liquid or some other low-molecular-weight substance generally with a polyolefin resin, heating and melting the mixture, extruding the melt into a sheet, orienting the sheet either in the machine direction (MD) or biaxially, and then extracting the liquid with a volatile solvent.[45, 59] Separators made by wet process are available from Asahi Kasei,[60] Tonen,[61–63] Mitsui Chemicals,[64] and more recently from Polypore/ Membrana and Entek.[65] The structure and properties of the membranes can be changed by controlling the composition of the solutions and the evaporation or subtractions of solvents in the gelation and solidification processes. The separators made by wet process use ultra-high-molecular-weight polyethylene (UHMWPE). The use of UHMWPE gives good mechanical properties as well as some degree of melt integrity.

The disadvantage of the solution method is that a large amount of waste solvents have to be dealt with, the reuse of which is usually very difficult. The solution method also is not convenient for polymers such as polypropylene, which do not dissolve in conventional solvents. Ihm et al. have given a nice overview of the wet process by preparing a separator with polymer blends of high-density polyethylene (HDPE) UHMWPE.[58] They showed that the mechanical strength and drawing characteristics are influenced by the content and the molecular weight of the UHMWPE contained in a polyolefin blending solution. The manufacturing process of typical microporous film by dry and wet process is compared in Table 20.3.

Figure 20.3a, b shows polyolefin separators used in Li-Ion batteries and a simplified flowchart for separator manufacturing process, respectively.[66] The virgin polymer is prepared and mixed with processing aids (e.g., antioxidants, plasticizer, etc.) and then extruded. The extruded polymer then goes through different steps, which vary from process to process. For dry process, it can involve film annealing and stretching, while for wet process, it can involve solvent extraction and stretching. The finished film is then slit into required widths and packed into boxes and shipped to the battery manufacturers. With the advent of thinner separators, the film

Table 20.3 Manufacturing Process of Typical Microporous Film

Process	Mechanism	Raw material	Properties	Typical membranes	Manufacturers
Dry process	Drawing	Polymer	Simple process Anisotropic film	PP, PE, PP/PE/PP	Celgard, Ube
Wet process	Phase separation	Polymer + solvent	Isotropic film	PE	Asahi, Tonen
		Polymer + solvent + filler	Large pore size High porosity	PE	Asahi

Reprinted with permission from *Chem. Rev.* **104** (2004) 4419-4462, copyright (2004), American Chemical Society.

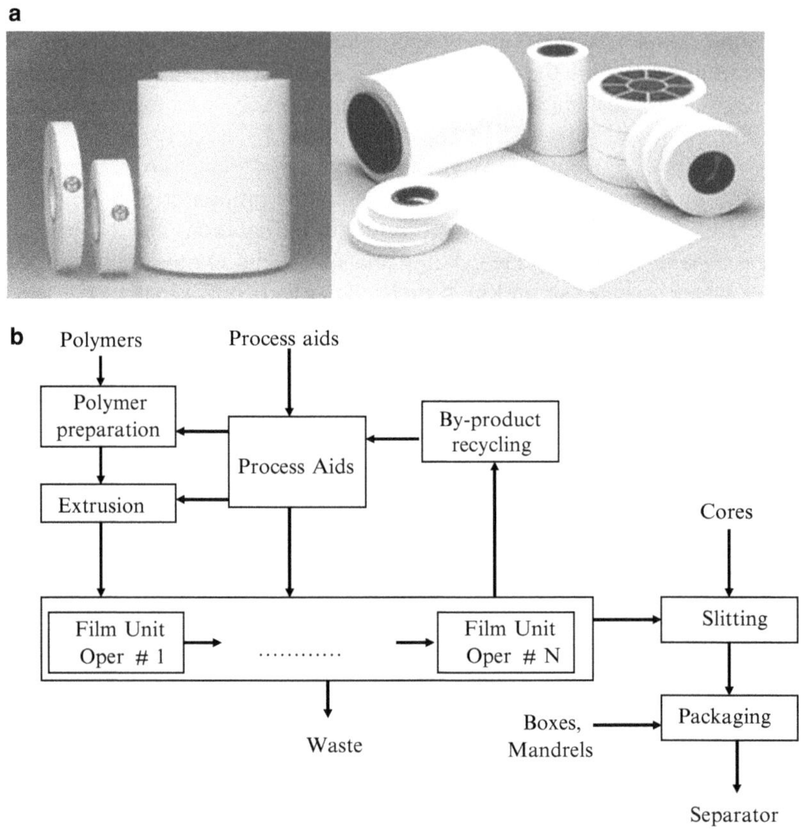

Fig. 20.3 (**a**) Polyolefin separators used in Li-Ion batteries. (**b**) A simplified flowchart for separator manufacturing process.[66] Each step of the separator manufacturing process has online detection systems to monitor the quality of the separator. Reprinted with permission from *Chem. Rev.* **104** (2004) 4419–4462, copyright (2004), American Chemical Society

handling during manufacturing steps has become very important for the final quality of the film. Each step of the separator manufacturing process has online detection systems to monitor the quality of the separators.

Uniaxially oriented films generally have high strength in only one direction, whereas biaxially oriented films are relatively strong in both machine direction (MD) and transverse direction (TD). Although intuitively one might expect biaxially oriented films to be preferred over uniaxially oriented films, in practice biaxial orientation provides no performance advantage. In fact, biaxial orientation tends to introduce TD shrinkage. This shrinkage, at elevated temperatures, can allow electrodes to contact each other. The separator must have sufficient strength in the machine direction so that it does not decrease in width or break under the stress of winding. The strength in the transverse direction is not as important as that in the machine direction during the process of making spirally wound batteries. The minimum generally practical requirement for the mechanical strength of the 25-μm separator is 1,000 kg/mm^2.[58]

The typical properties of some commercial microporous membranes are summarized in Table 20.4. Celgard 2730 and Celgard 2400 are single-layer PE and PP separators, respectively, while Celgard 2320 and 2325 are trilayer separators of 20 and 25 μm thickness. Asahi and Tonen separators are single-layer PE separators made by wet process. Basic properties, such as thickness, gurley, porosity, melt temperature, and ionic resistivity are reported in Table 20.4. These properties are defined in Sect. 20.5.2.

Efforts have been made to find a new route for dry process using biaxial stretching techniques for preparing polypropylene microporous films, which may have submicron pore size and narrow size distribution, high permeability to gases and liquids, combined with good mechanical properties. The biaxially stretched polypropylene microporous films (Micpor®) were made by using nonporous polypropylene films of high beta-crystal content.[67] The porosity of these films can be as high as 30–40%, with an average pore size of approximately 0.05 μm. The pores on the surface were almost circular in shape compared to slitlike pores observed in uniaxial stretched samples and exhibited high permeability to fluids with good mechanical properties and almost circular pore shape with narrow pore size distribution.[68–70]

The PP/PE bilayers[39] and PP/PE/PP trilayer separators were developed by Celgard. Multilayer separators offer advantages of strength and combine the lower melting temperature of PE with the high temperature strength of PP. Nitto Denko also has patented a single-layer separator made from a blend of PE/PP by the dry stretch process.[71] According to the patent, the separator has microporous regions of PE and PP. On heating in an oven, the impedance of the separator increases near the melting point of PE and the impedance remain high until beyond the melting point of PP. However, battery performance data have not been presented.

Microporous polyethylene separator material composed of a combination of randomly oriented thick and thin fibrils of UHMWPE, Solupur®, manufactured by DSM Solutech, also is an interesting separator material for Li-Ion batteries. Solupur® is fabricated in standard grades with base weights ranging from 7 to 16 g/m^2 and mean pore size ranging from 0.1 to 2.0 μm and a porosity of 80–90%.[72] Ooms et al. carried out a study on a series of DSM Solupur materials with different permeability. Rate capability and cycling tests of these materials were compared with commercially available separators in CR2320-type coin cells. Solupur® materials showed low tortuosity, high strength and puncture resistance, excellent wettability, good high-rate capability, and low-temperature performance because of its high porosity and UHMWPE structure.[73]

Recently Nitto Denko developed a battery separator made by a wet process that had high puncture strength and high heat rupture resistance.[74] They used a polyolefin resin with a high-molecular-weight rubber as its main component materials and cross-linked through oxidation in air. The melt rupture temperature, as measured by thermomechanical analysis, was over 200°C in this material. They also tried cross-linking UHMWPE with electron-beam and ultraviolet irradiation, but this had the

Table 20.4 Typical properties of some commercial microporous membranes

Separator/Properties	Celgard 2730	Celgard 2400	Celgard 2320	Celgard 2325	Asahi Hipore	Tonen Setela
Structure	Single layer	Single layer	Trilayer	Trilayer	Single layer	Single layer
Composition	PE	PP	PP/PE/PP	PP/PE/PP	PE	PE
Thickness (μm)	20	25	20	25	25	25
Gurley (s)	22	24	20	23	21	26
Ionic resistivity[a] (Ω cm^2)	2.23	2.55	1.36	1.85	2.66	2.56
Porosity (%)	43	40	42	42	40	41
Melt temperature (°C)	135	165	135/165	135/165	138	137

[a] In 1-M LiPF6 EC:EMC (30:70 by volume). Reprinted with permission from *Chem. Rev.* **104** (2004) 4419-4462, Copyright (2004) American Chemical Society

side effect of causing deterioration in the polyolefin including rupture of the main chains, and therefore resulted in reduced strength.

Figure 20.4 shows the SEM of the Nitto Denko separator that resembles more of a nonwoven texture similar to the SEM of a separator sample made through a wet process. The Nitto Denko separator has more shrinkage in both machine and transverse directions. The separator shows the shutdown behavior when heated close to melting temperature of the polyolefin, but the melt integrity is poor.

Mitsui Chemicals Inc. has developed a microporous film from a high-molecular-weight polyethylene material and they reported that the film is excellent in strength and air-permeability. [64] The microporous film is composed of fibrils comprising extended-chain crystals and lamellar crystals and/or fibrils comprising helicoidal crystals. The high porous film is produced by subjecting an air-impermeable film of a high-molecular-weight polyethylene to thermal treatment followed by, if necessary, a stretching treatment and/or a thermosetting treatment. Figure 20.5 shows the SEM of the Mitsui separator. Mitsui separators also show shutdown behavior but with better melt integrity when compared to Nitti Denko separators.

ENTEK Membranes LLC has developed Teklon,™ a highly porous, UHMWPE separator for Li-Ion batteries. Pekala et al. characterized Celgard™, Setela™, and Teklon™ separators in terms of their physical, mechanical, and electrical properties. [75]Recently, Chen et al. [76] had developed a conducting polymer shunt capable of providing overcharge protection for rechargeable lithium batteries via a reversible, internal, self-actuating mechanism. A microporous polypropylene separator was impregnated with poly(3-butylthiophene) (P3BT), an electrochemically active polymer that becomes electronically conducting when oxidized. Figure 20.6 shows the SEM of the microporous polypropylene separator before and after P3BT deposition. The resulting composite membrane had a thin surface polymer layer, and the polymer filled about 10% of the membrane pores. This conducting polymer composite separator

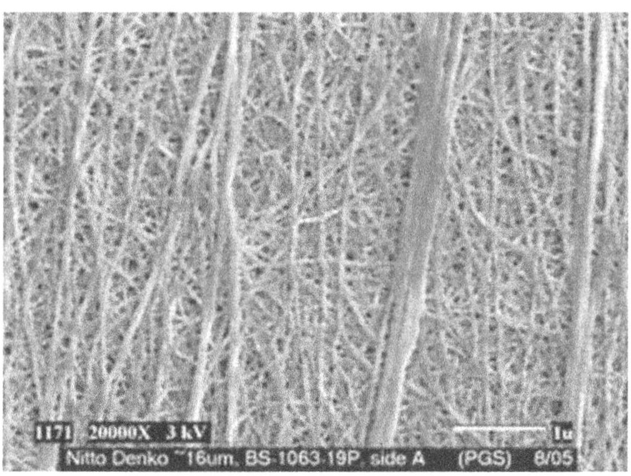

Fig. 20.4 SEM of the Nitto Denko separator

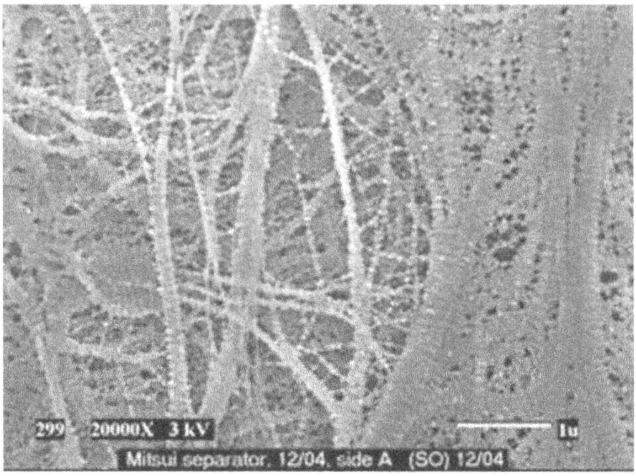

Fig. 20.5 SEM of the Mitsui separator

Fig. 20.6 SEM of the separator (**a**) before and (**b**) after P3BT deposition

membrane was tested with TiS_2-Li cell (3.5-V system) and when the cell was overcharged above 4 V, the cell remained undamaged, since the excess charge current was carried by the polymer, which limited the charging potential to about 3.2 V.

Further research work also is being carried out by Chen et al. at LBNL on polymers with higher oxidation onset potentials, which will be suitable for overcharge protection in 4-V Li-Ion batteries. They developed a bilayer conducting polymer composite separator that provides overcharge protection for high-voltage rechargeable lithium batteries. [77] The polymer with the higher oxidation potential is placed in contact with the cathode and sets the overcharge protection limit, while the lower voltage polymer completes the reversible shunt and protects the high-voltage polymer from degradation due to instability at the anode potential. This approach was successfully adopted by Chen et al. for protection of $LiFePO_4$/Li cells and $LiNi_{0.8}Co_{0.15}Al_{0.05}O_2$/Li cells.

Celgard's separators are by far the best-characterized battery separators in the literature as they have been widely used in numerous battery systems. Bierenbam et al.[45] has described the process, physical and chemical properties, and end-use applications. Fleming and Taskier[78] described the use of Celgard microporous membranes as battery separators. Hoffman et al.[79] presented a comparison of PP and PE Celgard microporous materials. Callahan discussed a number of novel uses of Celgard membranes. Callahan and co-workers[80] also characterized Celgard membranes by SEM image analysis, mercury porosimetry, air permeability, and electrical resistivity, and later characterized the puncture strength and temperature/impedance data for Celgard membranes.[39] Spotnitz et al. reported short-circuit behavior in simulated, spirally wound cells, as well as impedance/temperature behavior and thermomechanical properties.[81] Yu[82] found that a trilayer structure of PP/PE/PP Celgard™ microporous membranes provided exceptional puncture strength.

Nonwoven materials such as cellulosic fibers have never been used successfully in lithium batteries. This lack of interest is related to the hygroscopic nature of cellulosic papers and films, their tendency to degrade in contact with lithium metal, and their susceptibility to pinhole formation at thicknesses of less then 100 μm. For future applications, such as electric vehicles and load leveling systems at electric power plants, cellulosic separators may find a place because of their stability at higher temperatures compared to polyolefins. They may be laminated with polyolefin separators to provide high-temperature melt integrity.

Asahi Chemical Industry carried out an exploratory investigation to determine the requirements for cellulose-based separators for Li-Ion batteries.[83] In an attempt to obtain an acceptable balance of lithium ion conductivity, mechanical strength, and resistance to pinhole formation, they fabricated a composite separator (39–85 μm) that consists of fibrilliform cellulosic fibers (diameter 0.5–5.0 μm) embedded in a microporous cellulosic (pore diameter 10–200 nm) film. The fibers can reduce the possibility of separator meltdown under exposure to heat generated by overcharging or internal short circuiting. The resistance of these films was equal to or lower than the conventional polyolefin-based microporous separators. The long-term cycling performance also was very comparable.

Pasquier et al.[84] used paper-based separators in flat pouch-type Li-Ion batteries and compared the performance with cells made with Celgard-type polyolefin-based separators. The paper separators had good wetting properties and good mechanical properties, but did not provide the shutdown effect essential for large Li-Ion batteries. Their resistance was similar to polyolefin separators and when all water traces were removed from paper, their cycling performance was similar to Celgard separators. The paper-based separators can be used in small flat pouch-type cells where high strength and shutdown behavior is not required. For larger spherically wound cells, which require strong separators with a shutdown feature, can never use paper-based separators.

Recently Degussa announced that they have developed Separion® separators for lithium batteries by combining the characteristics of flexible polymeric separators with the advantages of chemical and thermally resistant and hydrophilic ceramic materials. Separion® is produced in a continuous coating process. Ceramic materials, e.g., alumina, silica, and/or zirconia, are slip-coated and hardened onto a support.[85, 86]

According to Degussa, Separion separators have an excellent high-temperature stability, superior chemical resistance, and good wettability, especially at low temperatures. They tested the performance and safety behavior of Separion separator in 18,650 cells and found the performance to be comparable to polyolefin-based separators. [87]

Figure 20.7 shows the SEM of the Separion separator, where the ceramic materials coated over the surface of the nonwoven bonded polymer can be clearly observed. Separion separators are weak in puncture strength and tensile stress and lower dielectric breakdown, but reasonably high in mix-penetration strength. The separator also is weak in coating adhesion.

The potential use of polymeric ion-exchange membranes in the next generation single-ion secondary lithium-polymer batteries was shown by Sachan et al. [88, 89] Conductivities exceeding 10^{-4} S/cm with transference numbers of unity were achieved for Nafion converted to Li^{+} salt form.

To obtain a thin (less than 15 μm) separator for lithium batteries, Optodot has taken a different approach of high-speed coating of a metal oxide sol-gel coating on a smooth surface followed by a delamination step to provide the freestanding separator. Using this approach, a separator with thicknesses from 6 to 11 μm was made on large-scale production coating equipment. [90] They found that the sol-gel separators with a thickness in the middle of this range of 8–9 μm have the preferred combination of thinness and strength. The metal oxide sol-gel coating is water-based with no organic solvents present. The coating formulations include a polymer and a surfactant. The polymer provides improved coating rheology, mechanical strength, and other properties. The surfactant provides improved wetting properties on the substrate. The films prepared were around 11 μm thick, with 45% porosity, completely wettable in nonaqueous electrolyte and melt temperature greater then 180°C. These films are relatively thin and should help in increasing the capacity but may not be

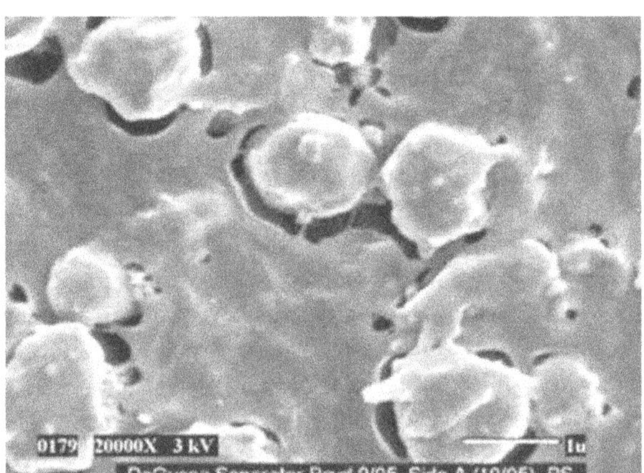

Fig. 20.7 SEM of Separion (Degussa) separator. Coating of the high-temperature-resistant ceramic materials can be observed

strong enough for tightly wound cells. Moreover, the shutdown temperature of the separator seems to be very high, and thus not suitable for Li-Ion batteries.

Gineste et al. carried out the grafting of hydrophilic monomers onto PP or PE separators to improve the wettability of separators used in secondary lithium batteries with a lower content of wetting agents.[91, 92] They used a PP film (Celgard 2505) of 50-μm thickness after irradiating in air by electron beams with a dose ranging from 0.5 to 4 Mrad. The irradiated film was grafted by a monofunctional monomer [acrylic acid (AA)], in the presence of a difunctional, cross-linking agent [diethyleneglycol dimethacrylate (DEGDM)]. The separators start losing mechanical properties, when the grafting ratio is higher than 50%.

20.5.2 Separator Requirements

In lithium-based cells, the essential function of the battery separator is to prevent electronic contact, while enabling ionic transport between the positive and negative electrodes. It should be usable on high-speed winding machines and should possess good shutdown properties. The most commonly used separators for primary lithium batteries are microporous polypropylene membranes. Microporous polyethylene and laminates of polypropylene and polyethylene are widely used in Li-Ion batteries.[93] These materials are chemically and electrochemically stable in secondary lithium batteries. A key requirement for the separators for lithium primary batteries is that their pores be small enough to prevent dendritic lithium penetration through them. The general requirements[94] for lithium-ion battery separators are given below.

20.5.2.1 Thickness

The Li-Ion cells used in consumer applications use thin microporous separators (< 25 μm). The separators being developed for EV/HEV applications will require thicker (~40 μm) separators. The thicker the separator, the greater the mechanical strength and the lower the probability of punctures during cell assembly, but the smaller the amount of active materials that can be placed in the can. The thinner separators take up less space and permit the use of longer electrodes. This increased both capacity and, by increasing the interfacial area, rate capability. The thinness also makes it a low-resistance separator.

20.5.2.2 Permeability

The separators should not limit the electrical performance of the battery under normal conditions. Typically the presence of a separator increases the effective resistivity of the electrolyte by a factor of 6–7. The ratio of the resistivity of the separator filled with electrolyte divided by the resistivity of the electrolyte alone is called the McMullin number. The McMullin number as high as 10–12 has been used in consumer cells.

20.5.2.3 Gurley (Air Permeability)

Air permeability is proportional to electrical resistivity for a given separator morphology. It can be used in place of electrical resistance (ER) measurements once the relationship between Gurley and ER is established. The separator should have low Gurley values for good electrical performance.

20.5.2.4 Porosity

It is implicit in the permeability requirement that typically Li-Ion battery separators have a porosity of 40%. Control of porosity is very important for battery separators. Specification of percent porosity is commonly an integral part of separator acceptance criteria. The porosity of separators used in alkaline zinc MnO_2 cells is typically around 80–90%.

20.5.2.5 Wettability

The separators should wet out quickly and completely in typical battery electrolytes.

20.5.2.6 Electrolyte Absorption and Retention

A separator should be able to absorb and retain electrolyte. Electrolyte absorption is needed for ion transport. The microporous membranes usually do not swell on electrolyte absorption.

20.5.2.7 Chemical Stability

The separators should be stable in battery for a long period of time. It should be inert to both strong reducing and strong oxidizing conditions and should not degrade or lose mechanical strength or produce impurities, which can interfere with the function of the battery. The separator must be able to withstand the strong oxidizing positive electrode and the corrosive nature of the electrolyte at temperatures as high as 75°C. The greater the oxidation resistance, the longer the separator will survive in a cell. Polyolefins exhibit high resistance to most of the conventional chemicals, good mechanical properties, and a moderate temperature range for application, making it one of the ideal polymers for Li-Ion battery separators. Polypropylene films are more oxidation resistant compared to polyethylene films when used in direct contact with the positive electrodes in Li-Ion cells. Thus, the trilayer separators (PP/PE/PP) with PP as outer layers and PE as the central layer show superior oxidation-resistance properties compared to single-layer PE separators. The oxidation of the separators can lead to degradation in mechanical properties, which can ultimately lead to cell failure.

20.5.2.8 Dimensional Stability

The separator should lay flat and should not curl at the edges when unrolled as this can greatly complicate cell assembly. The separator also should not shrink when exposed to the electrolyte. The cell winding should not affect the porous structure in any adverse way.

20.5.2.9 Puncture Strength

The separators used in wound cells require high puncture strength to avoid penetration of electrode material through the separator. If particulate material from the electrodes penetrates the separator, an electrical short will result and the battery will be rejected. The separators used in Li-Ion batteries require more strength than the one used in lithium primary batteries. The primary lithium batteries have only one rough electrode, and thus it requires less strength. As empirically observed, for most applications the puncture strength should be at least 400 g/mil for separators used in Li-Ion cells. The mix penetration strength is a better measure for the strength of separator in batteries.

20.5.2.10 Mix Penetration Strength

The susceptibility of separators to particle penetration is characterized by mix penetration strength.[49] During the winding of the spiral wrap construction, considerable mechanical pressure is applied to the cathode–separator–anode interface. Any loose particle could be forced through the separator and short the cell.

20.5.2.11 Thermal Stability

Lithium-ion batteries can be poisoned by water and so materials going into the cell are typically dried at 80°C under vacuum. Under these conditions, the separator must not shrink significantly and definitely must not wrinkle. Each battery manufacturer has specific drying procedures. The requirement of less then 5% shrinkage after 60 min at 90°C (in vacuum) in both MD and TD direction is a reasonable generalization.

20.5.2.12 Pore Size

A key requirement of separators for lithium batteries is that their pores be small enough to prevent dendritic lithium penetration through them. Membranes with submicron pore sizes have proven adequate for lithium batteries.

20.5.2.13 Tensile Strength

The separator is wound with the electrodes under tension. The separator must not elongate significantly under tension in order to avoid contraction of the width. A tensile strength specification is sometimes given, but the key parameter is Young's modulus in the machine direction. Since Young's modulus is difficult to measure, 2% offset yield is a good measure; less than 2% offset at 1,000 psi is acceptable for most winding machines.

20.5.2.14 Skew

Ideally, when a strip of separator is laid out, the separator should be straight and not bow or skew. In practice, however, some skew often is observed. If sufficiently extreme, the skew can cause misalignment between the electrodes and separator. Skew can be measured by laying the separator flat on a table parallel with a straight meter stick. The skew should be less than 0.2 mm/m of separator.

20.5.2.15 Shutdown

Lithium-ion batteries separators provide some margin of protection against short circuit and overcharge in Li-Ion cells. The separators exhibit a large increase in impedance at a temperature about 130°C that effectively stops ionic transport between the electrodes.[95, 96] The greater the mechanical integrity of the separator above 130°C, the greater the margin of safety the separator can provide. If the separator loses mechanical integrity, then the electrodes can come into direct contact, react chemically, and result in thermal runaway. The shutdown behavior of a separator can be characterized by heating the separator (saturated with electrolyte) to high temperatures and simultaneously monitoring the electrical resistance of the separator.[96, 81]

20.5.2.16 High Temperature Stability

A separator might provide an extra margin of safety if it can prevent the electrodes from contacting one another at high temperatures. Separators with good mechanical integrity at high temperatures can provide a greater margin of safety for Li-Ion cells. Thermal mechanical analysis (TMA) can be used to characterize the high-temperature stability of separators. Utilizing TMA, the separator is held under constant load and the degree of elongation temperature is measured; the temperature at which the separator loses mechanical integrity, the elongation increases dramatically.

20.5.2.17 Electrode Interface

The separator should form a good interface with the electrodes to provide sufficient electrolyte flow. In addition to the above properties, the separator must be essentially free of any type of defects (pinholes, gels, wrinkles, contaminants, etc.). All of the above properties have to be optimized before a membrane qualifies as a separator for a Li-Ion battery. The general requirements for Lithium-Ion battery separators are summarized in Table 20.5.

20.5.3 Separator Properties/Characterization

Separators are characterized by structural and functional properties; the former describes what they are and the latter how they perform. The structural properties include chemical (molecular) and microcrystalline nature, thickness, pore size, pore

Table 20.5 General requirements for lithium-ion battery separator [94]

Parameter	Goal
Thickness [97,98] (μm)	< 25
Electrical resistance (McMullin, [99] dimensionless)	< 8
Electrical resistance (Ω cm²)	< 2
Gurley [100] (s)	~25/mil
Pore size [101] (μm)	< 1
Porosity (%)	~40
Puncture strength [102] (g)	> 300 g/mil
Mix penetration strength (kgf)	> 100 kgf/mil
Shrinkage [103] (%)	< 5% in both MD and TD
Tensile strength [104]	< 2% offset at 1,000 psi
Shutdown temperature (°C)	~130
High temperature melt integrity (°C)	> 150
Wettability	Complete wet out in typical battery electrolytes
Chemical stability	Stable in battery for long period of time
Dimensional stability	Separator should lay flat; be stable in electrolyte
Skew	< 0.2 mm/m

Reprinted with permission from *Chem. Rev.* **104** (2004) 4419–4462, copyright (2004), American Chemical Society size distribution, porosity, and various chemical and physical properties such as chemical stability and electrolyte uptake. The functional properties of interest are electrical resistivity, permeability, and transport number. It is useful to characterize separator materials in terms of their structural and functional properties, and to establish a correlation of these properties with their performance in batteries. A variety of techniques are used to evaluate separators. Some of these techniques are discussed in this section.

20.5.3.1 Gurley

Separator permeability usually is characterized by air permeability. The Gurley number expresses the time required for a specific amount of air to pass through a specific area of separator under a specific pressure. The standard test method is described in ASTM-D726 (B).

The Gurley number is used to characterize separators because the measurement is accurate and easy to make, and deviations from specific values are a good indication of problems. Air permeability (Gurley) is proportional to electrical resistance (ER) for a given separator morphology. [80] Gurley can be used in place of ER measurements once the relationship between Gurley and ER is established. A lower Gurley value means higher porosity, lower tortuosity, and accordingly lower ER.

20.5.3.2 Electrical Resistance

The measurement of separator resistance is very important to the art of battery manufacture because of the influence the separator has on electrical performance. Electrical resistance is a more comprehensive measure of permeability than the Gurley number in that the measurement is carried out in the actual electrolyte solution. The ionic resistivity of the porous membrane is essentially the resistivity of the electrolyte that is embedded in the pores of the separator. Typically, a microporous separator, immersed in an electrolyte, has an electrical resistivity about six to seven times that of a comparable volume of electrolyte, which it displaces. It is a function of the membrane's porosity, tortuosity, the resistivity of the electrolyte, the thickness of the membrane, and the extent to which the electrolyte wets the pores of the membrane. [105] The ER of the separator is the true performance indicator of the cell. It describes a predictable voltage loss within the cell during discharge and allows one to estimate rate limitations.

Classic techniques for measuring electrical resistivity of microporous separators have been described by Falk and Salkind [5] and Robinson and Walker. [106] The resistivity of an electrolyte is more accurately determined by AC methods since DC can polarize the electrodes and cause electrolysis of the solution. Modern AC impedance-measuring systems allow rapid measurements of cell resistance over a wide range or frequencies from which resistance can be calculated free of capacitance effects. Compared to the DC techniques, the equipment required and the theory necessary to interpret the AC techniques are more complex; however, AC measurements yield information about long-range migration of ions and polarization phenomena occurring within the cell. In an AC measurement, a sinusoidal voltage is applied to a cell and the sinusoidal current passing through the cell as a result of this perturbation is determined. A four-electrode cell usually is used for resistivity measurements. The outer two electrodes serve to apply a sinusoidal potential and the resulting current passing through the inner two electrodes is measured. This technique is employed to avoid the complications arising from a nonuniform potential field near the outer two electrodes. An excellent review of experimental techniques for measuring electrical resistivity in aqueous solution is available.[107, 108]

The separator resistance usually is characterized by cutting small pieces of separators from the finished material and then placing them between two blocking electrodes. The separators are completely saturated with the electrolyte. The resistance (Ω) of the separator is measured at a certain frequency by AC impedance techniques. The frequency is chosen so that the separator impedance is equal to the separator resistance. In order to reduce the measurement error, it is best to do multiple measurements by adding extra layers. The average resistance of single layer is determined from multiple measurements. The specific resistivity, ρ_s (Ω-cm), of the separator saturated with electrolyte is given by

$$\rho_s = \frac{R_s A}{l}, \tag{20.1}$$

where R_s is the measured resistance of separator in Ω, A is the electrode area in centimeters squared, and l is the thickness of membrane in centimeters. Similarly, the specific resistivity of the electrolyte, ρ_e (Ω-cm), is given by

$$\rho_e = \frac{R_e A}{l}, \tag{20.2}$$

where R_e is the measured resistance of electrolyte in Ω. The ratio of the resistivity of a separator membrane to that of the electrolyte is called the McMullin number, N_m, which can be used to predict the influence of the separator on battery performance[109]:

$$N_m = \frac{\rho}{\rho_e} = \frac{\tau^2}{\varepsilon}, \tag{20.3}$$

where τ is the tortuosity and ε is the porosity of the separator. The MacMullin number describes the relative contribution of a separator to cell resistance. It is almost independent of electrolyte used and also factors out the thickness of the material. It assumes that the separator wets completely in the electrolyte used for the test. From (20.1) and (20.3) the ER of a microporous membrane is given by the following[5,110]

$$R_m = \rho_e \left(\frac{\tau^2 l}{\varepsilon A} \right). \tag{20.4}$$

It has been shown for Celgard membranes that the membrane resistance can be related to the Gurley number by[80]

$$R_{\mathrm{m}}A = \frac{P_{\mathrm{e}}}{5\times 18\times 10^{-3}}\, t_{\mathrm{gur}}\, d, \tag{20.5}$$

where R_{m} is the membrane resistance (Ω), A is the membrane area (cm^2), ρ_{e} is the specific electrolyte resistance (Ω-cm), t_{gur} is the Gurley number (10 cc air, 2.3 mm Hg), d is the pore size, and $5 \times 18 \times 10^{-3}$ a scaling factor.

The usual procedure for characterizing battery separators is to cut several test samples from the finished material. Thus, only a small portion of the separator is actually examined. Ionov et al. has proposed an alternative technique to measure the resistance of a separator over a large separator area.[111] In this technique the separator material is passed through an electrolyte bath between electrical resistance measuring transducers. The set of transducers installed in the bath transverse to the moving sheet of separator material and examines the entire surface of the material. If the production process ensures good uniformity in the physicochemical properties of the separator material over the entire surface, the transducer outputs clearly will be close to one another. A nonuniform separator will cause significant deviations from the average value at various sections of the material. In this case, the sections having lower or higher resistance compared with the average value should be regarded as flawed.

20.5.3.3 Porosity

The porosity is important for high permeability and also for providing a reservoir of electrolyte in the cell. Higher and more uniform porosity is desirable for unhindered ionic current flow. Nonuniform porosity leads to nonuniform current density, and can lead further to reduced activity of the electrodes. Cell failure can result because during discharge in some areas of the electrodes works harder than others.

Porosity of a separator is defined as the ratio of void volume to apparent geometric volume. It usually is calculated (20.6) from the skeletal density, basis weight, and dimensions of the material and so may not reflect the accessible porosity of the material:

$$\text{Porosity}\,(\%) = 1 - \frac{\left(\text{Sample Weight} \big/ \text{Sample Volume}\right)}{\text{polymer Density}} \times 100\% \tag{20.6}$$

The standard test method is described in ASTM D-2873. The actual or accessible porosity also can be determined by the weight of liquid (e.g., hexadecane) absorbed in the pores of the separator. In this method, the separator weight is measured before and after dipping in hexadecane solvent and the porosity is calculated (20.7) by assuming that volume occupied by hexadecane is equal to the porous volume of the separator.

$$\text{Porosity}(\%) = \frac{\text{Volumeoccupied by Hexadecane}}{\text{Volume of polymer} + \text{Volume occupied by Hexadecane}} \times 100. \quad (20.7)$$

20.5.3.4 Tortuosity

Tortuosity is the ratio of mean effective capillary length to separator thickness. The tortuosity factor, τ, of a separator can be expressed by

$$\tau = \frac{l_s}{d}, \quad (20.8)$$

where l_s is the ion path through the separator and d is the thickness of the separating layer.

Tortuosity is a long-range property of a porous medium, which qualitatively describes the average pore conductivity of the solid. It is usual to define τ by electrical conductivity measurements. With knowledge of the specific resistance of the electrolyte and from a measurement of the sample membrane resistance, thickness, area, and porosity, the membrane tortuosity can be calculated from (20.3).

This parameter is widely used to describe the ionic transport by providing information on the effect of pore blockage. A tortuosity factor $\tau = 1$, therefore, describes an ideal porous body with cylindrical and parallel pores, whereas values of $\tau > 1$ refer to more hindered systems. Higher tortuosity is good for dendrite resistance but can lead to higher separator resistance.

20.5.3.5 Pore Size and Pore Size Distribution

For any battery applications, the separator should have uniform pore distribution to avoid performance losses arising from nonuniform current densities. The submicron pore dimensions are critical for preventing internal shorts between the anode and the cathode of the Li-Ion cell, particularly since these separators tend to be as thin as 25 μm or less. This feature will be increasingly important as battery manufacturers continue to increase the cell capacity with thinner separators. The pore structure usually is influenced by polymer composition and stretching conditions, such as drawing temperature, drawing speed, and draw ratio. In the wet process, the separators produced by the process of drawing after extraction (as claimed by Asahi Chemical and Mitsui Chemical) are found to have much larger pore size (0.24–0.34 μm) and wider pore size distribution than those produced by the process of extraction (0.1–0.13 μm) after drawing (as claimed by Tonen).[58]

The testing of battery separators and control of their pore characteristics are important requirements for proper functioning of batteries. Mercury porosimetry historically has been used to characterize the separators in terms of percentage porosity, mean pore size, and pore size distribution.[112] In this method, the size and volume of pores in a material are measured by determining the quantity of mercury, which can be forced into the pores at increasing pressure. Mercury does not wet most materials and a force must be applied to overcome the surface tension forces opposing entry into the pores.

The hydrophobic (e.g., polyolefins) separators also are characterized using the aquapore (nonmercury porosimetry) technique, where water is used in place of mercury. This is a very useful technique for characterizing polyolefin-based separators used in lithium batteries.[113] Porosimetry gives pore volume, surface area, mean pore diameter, and pore size distribution. In a typical experiment, the sample is placed in the instrument and evacuated. As the pressure increases, the quantity of water forced into the pores increases in proportion to the differential pore volume; the size of the pores corresponding to the instantaneous pressure. Thus, increasing the pressure on a membrane having a given pore size distribution results in a unique volume vs. pressure or pore diameter curve. The pressure required for intrusion of water in to a pore of diameter, D is given by following equation

$$D = \frac{4\gamma \cos\theta}{p}, \tag{20.9}$$

where D is the diameter of the pore assuming the pore to be cylindrical, p is the differential pressure, γ is the surface tension of the nonwetting liquid, water, and $ is the contact angle of water. The pores generally are not of spherical shape of a constant diameter. They usually vary in their form and size. Thus, statements of any pore diameter are always to be viewed with the above in mind.

Another technique, capillary flow porometry has been developed by Porous Materials Inc.[114] to characterize battery separators.[115,116] The instrument can measure a number of characteristics of battery separators such as size of the pore at its most constricted part, the largest pore size, pore size distribution, permeability, and envelope surface area.[117]

Scanning electron microscopy (SEM) also is used to examine separator morphology. SEM pictures of some commercial membranes are shown in Figs. 20.8 – 20.10. The surface SEM of Celgard 2400, 2500, and 2730 are shown in Fig. 20.8. It is clear from the images that the pores are uniformly distributed. Both Celgard 2400 and 2500 are single-layer PP separators, but the pore size of Celgard 2500 is substantially larger than Celgard 2400. Thus it has lower resistance and is more suited for high-rate applications. Figure 20.9 shows the surface SEM and cross-section SEM of Celgard 2325. The surface SEM only shows the PP pores while the PE pores are visible in the cross section. It is clear from the image that all three layers are of equal thickness. The SEM of separators made by wet process is shown in Fig. 20.10. The pore structure of all these membranes is very similar. Hipore-1 (Fig. 20.10b) separator has significantly larger pores compared to the other membranes.

Image analysis has been used to characterize the pore structure of synthetic membrane materials.[118] The Celgard films also have been characterized by scanning tunneling microscopy, atomic force microscopy, and field emission scanning electron microscopy.[53,119] The pore size of the Celgard membranes also can be calculated from (20.5) once the McMullin number and Gurley values are known.

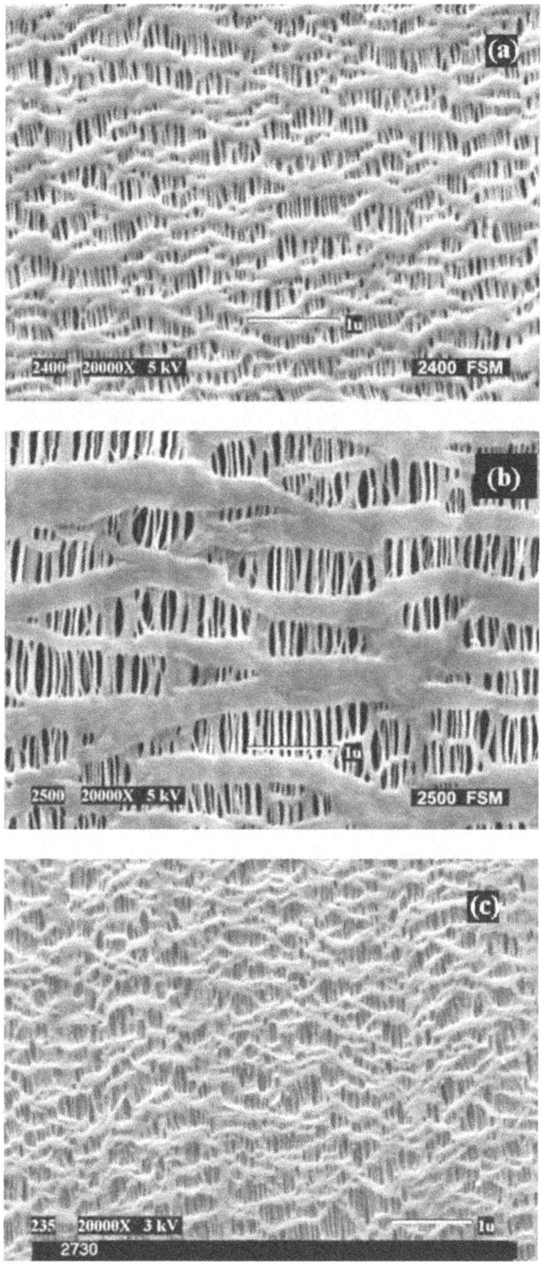

Fig. 20.8 Scanning electron micrographs of surface of single-layer Celgard separators used in lithium batteries (**a**) 2400 (PP) small pore, (**b**) 2500 (PP) large pore, and (**c**) 2730 (PE). Reprinted with permission from *Chem. Rev.* **104** (2004) 4419–4462, copyright (2004), American Chemical Society

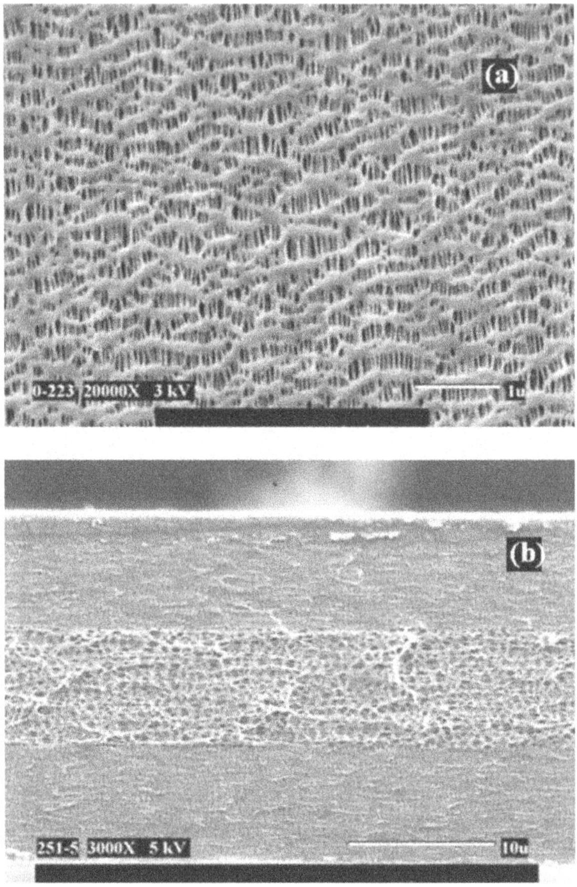

Fig. 20.9 Scanning electron micrographs of Celgard 2325 (PP/PE/PP) separator used in lithium ion batteries (**a**) surface SEM, (**b**) cross-section SEM. Reprinted with permission from *Chem. Rev.* **104** (2004) 4419–4462, copyright (2004), American Chemical Society

20.5.3.6 Puncture Strength

A separator is required to have sufficient physical strength to endure the rigors of cell assembly and day-to-day charge-discharge cycling. Physical strength is required to withstand basic handling, cell blocking/assembly, physical shock, punctures, abrasion, and compression.

The puncture strength (PS) is the weight that must be applied to a needle to force it completely through a separator.[45,120] It has been used to indicate the tendency of separators to allow short-circuits in a cell that may occur due to holes generated in the separator by the rough surface of an electrode during the battery assembly and charge-discharge cycle. The PS requirement for lithium-ion batteries is higher than

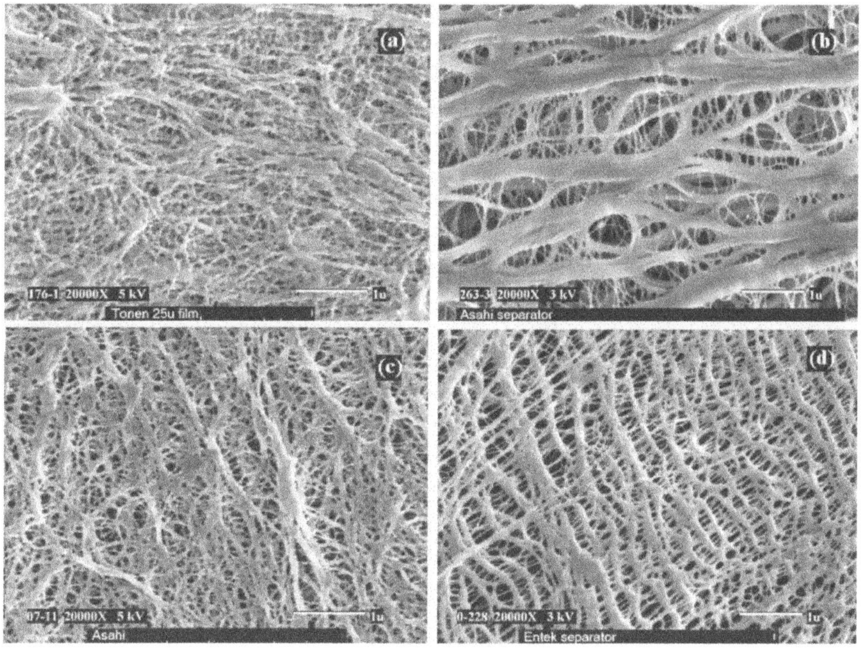

Fig. 20.10 Scanning electron micrographs of separators made by wet process and used in lithium-ion batteries (**a**) Setela (Tonen), (**b**) Hipore–1 (Asahi), (**c**) Hipore-2 (Asahi), and (**d**) Teklon (Entek). Reprinted with permission from *Chem. Rev.***104** (2004) 4419–4462, copyright (2004), American Chemical Society

lithium-foil batteries, because the separator must contend with two rough surfaces. Commercially available puncture strength machines made for textiles tend to give-meaningless results when testing battery separator membranes. More reproducible results can be obtained with a load frame (such as an Instron Machine).

The strength of the separator depends greatly on the materials used and the manufacturing method. The wet-biaxial method simultaneously stretches in the MD and TD directions and thus achieves a material that has tensile modulus and rupture strength in both directions. Both high-polymer entanglement and stretching help increase the physical strength of the separator.

The mix penetration strength is a better measure for the strength of separator in batteries. In case of mix penetration test, electrode mix is pushed through the separator till it creates a short. This test simulates the real situation inside the battery more closely.

20.5.3.7 Mix Penetration Strength

The force required to create a short through a separator due to mix (electrode material) penetration defines mix penetration strength. In this test force (with a half inch diameter ball) is applied on the positive electrode/separator/negative electrode sandwich and the force at which the mix penetrates through the separator and creates an electronic short is called mix penetration force. Mix penetration strength is used to indicate the tendency of separators to allow short-circuits during battery assembly. The mix penetration resistance test is more closely related to particle penetration resistance compared to puncture resistance.[49]

20.5.3.8 Tensile Strength

The tensile strength measurements (e.g., Young's modulus, percent offset strength, elongation at break, stress at break) can be made by utilizing widely known standard procedures. These tests are carried out in both MD and TD directions. The tensile properties are dependent on the manufacturing process. The uniaxially oriented films have high strength in only one direction, whereas biaxially oriented films are more uniformly strong in both MD and TD directions. The ASTM test method D882-00 "standard test method for tensile properties of thin plastic sheeting" is an appropriate test.

The separator should be strong enough to withstand mechanical handling during cell winding and assembly. It should be dimensionally stable and should not neck down during winding. The decrease in width will allow the electrodes to touch each other and create a short. Thus the tensile property of the separator should be very strong in MD direction compared to TD direction.

20.5.3.9 Shrinkage

Shrinkage test is carried out on both MD and TD directions. In this test, the dimensions of separators are measured and then stored at 90°C for a fixed time. The shrinkage is then calculated from the change in dimensions as shown in (20.10):

$$\text{Shrinkage}(\%) = \frac{L_\mathrm{i} - L_\mathrm{f}}{L_\mathrm{i}} \times 100, \quad (20.10)$$

where L_i is the initial length, and L_f is the final length of separator after high temperature storage. The uniaxially stretched separators tend to shrink in MD direction only, while the biaxially stretched separators shrink in both MD and TD directions. The shrinkage of separators also can be compared by carrying out thermal mechanical analysis (TMA) test at a constant load and rate.

20.5.3.10 Shutdown

Separator shutdown is a useful and essential mechanism for limiting temperature and preventing venting in short-circuited cells.[81] It usually takes place close to the melting temperature of the polymer when the pores collapse, turning the porous ionically conductive polymer film into a nonporous insulating layer between the electrodes. At this temperature there is a significant increase in cell impedance and passage of current through the cell is restricted. This prevents further electrochemical activity in the cell, thereby shutting the cell down before an explosion can occur.

The ability of the PE-based separator to shut down the battery is determined by its molecular weight, percent crystallinity (density), and process history. Material properties and processing methods might need to be tailored so that the shutdown response is spontaneous and complete. The optimization needs to be done without affecting the mechanical properties of the material in the temperature range of interest. This is easier to do with the trilayer separators manufactured by Celgard, since one material is utilized for the shutdown response and another for the mechanical properties. Polyethylene containing separators, in particular trilayer laminates of polypropylene, polyethylene, and polypropylene, appear to have the most attractive properties for preventing thermal runaway in Li-Ion cells.[117,121] The shutdown temperature of 130°C usually is enough to control the cell heating and avoid thermal runaway in Li-Ion cells. A lower temperature shut down will be desirable if it does not affect the separator mechanical properties or high-temperature cell performance in any adverse way. The shutdown property of separators is measured by measuring the impedance of a separator while the temperature is linearly increased.[96,81] Figure 20.11 shows actual measurement for Celgard trilayer membrane. The heating rate was around 60°C/min and the impedance was measured at 1 kHz. The rise in impedance corresponds to a collapse in pore structure due to melting of the separator. A 1,000-fold increase in impedance is necessary for the separator to stop thermal runaway in the battery. The drop in impedance corresponds to opening of the separator due to coalescence of the polymer, and/or to penetration of the separator by the electrodes; this phenomenon is referred to as a loss in "melt integrity." This test is fairly reliable in indicating the temperature at which the impedance rises, but shows some variability in characterizing the subsequent drop in impedance.

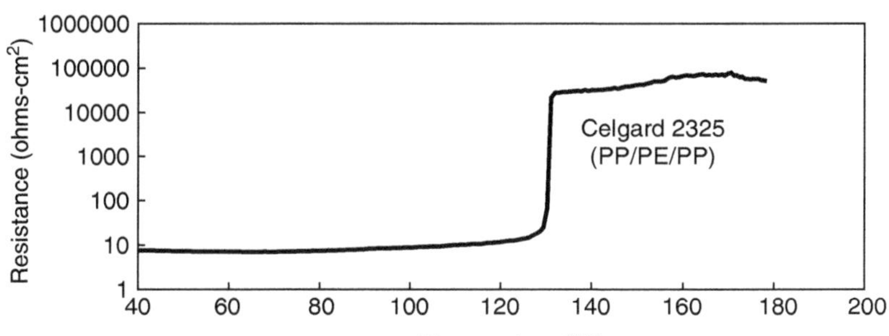

Fig. 20.11 Internal impedance (at 1 KHz) of Celgard 2325 (PP/PE/PP) separator as a function of temperature. Heating rate: 60°C/min. Reprinted with permission from *Chem. Rev.* **104** (2004) 4419–4462, copyright (2004), American Chemical Society

In Fig. 20.11 the shutdown behavior of a Celgard trilayer membrane (PP/PE/PP) is shown. The impedance rise occurred near the melting point of polyethylene (130°C) and remained high until such time as the melting point of polypropylene (165°C) was attained. The shutdown temperature of the separator is governed by the melting point of the separator material. At the melting point the pores in the separator collapse to form a relatively nonporous film between the anode and the cathode. This was confirmed by differential scanning calorimetry (DSC) as shown in Fig. 20.12. The DSC scan in Fig. 20.12 gives a peak melting temperature of 135°C for Celgard 2730, 168°C for Celgard 2400, and 135/165°C for Celgard 2325. The shutdown behavior of thinner separators (<20 μm) is very similar to thicker separators. The battery manufacturers have been very successful in using the thinner separators without compromising on the shutdown behavior of the separators.

Laman et al. introduced the use of impedance measurements as a function of temperature to characterize shutdown separators.[96] Using a temperature scan rate of 1°C/min they found that the impedance increased several orders of magnitude near the melting point of the separator. They verified the patent claims of Lundquist et al.[122] that bilayer separators of PE and PP gave a temperature window of high impedance extending approximately between the melting point of the polymers. The concept of using separators consisting of distinct layers, one of which could act as a fuse, was developed by Lundquist et al.[123,124] Laman's results have been corroborated by Geiger et al.[39] and Spotnitz et al.[81,110] Spotnitz et al. developed a thin-layer cell that allowed temperature scan rates of 5°C/min and higher and obtained results similar to those of Laman et al.

Prior work related to the shutdown separators also involved application of waxes on membranes.[125,126] In these cases, the wax or low-melting polymers were coated on the polyolefin separator. The disadvantage of this technique is that the coating can block the pores of the separator, and thus can affect the performance by increasing separator resistance. Moreover, the coating level has to be very high to get complete shut down.

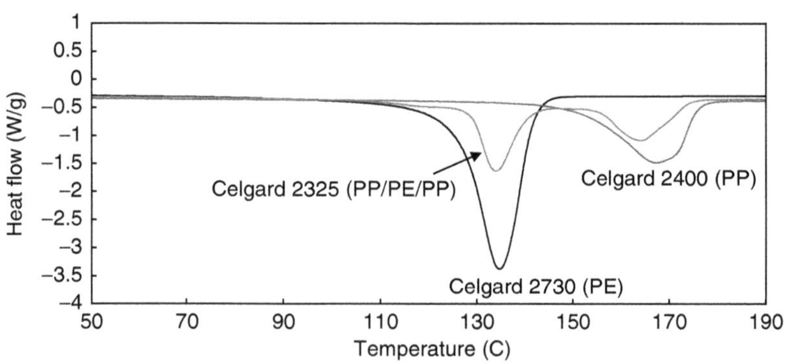

Fig. 20.12 DSC of Celgard 2730 (PE), 2400 (PP), and 2325 (PP/PE/PP). Reprinted with permission from *Chem. Rev.* **104** (2004) 4419–4462, copyright (2004), American Chemical Society

The shutdown characteristic provides protection from external short circuit and during cell overcharge. It provides little protection from internal shorts should they occur. Should the electrodes touch each other or become shorted from a dendritic growth of soluble impurity or other dendrite-forming soluble material, the separator only helps in avoiding delayed failures. In case of an instant failure during internal short circuit, the heating rate is too high and the separator shutdown is not fast enough to control the heating rate.

20.5.3.11 Melt Integrity

The separators used in Li-Ion batteries should have high-temperature melt integrity. The separator should maintain its melt integrity after shut down so that the electrodes do not touch and create a short. This helps to avoid the thermal runaway even when the cell is exposed to high temperatures. Thermal mechanical analysis (TMA) is a very good technique to measure the high-temperature melt integrity of separators.

TMA involves measuring the shape change of a separator under load while the temperature is linearly increased. Typically, separators show some shrinkage and then start to elongate and finally break as shown in Fig. 20.13. This test uses a small separator sample [about 5- to 10-mm length (MD) and about 5-mm width], which is held in mini-instron-type grips. The sample is held with a constant 2-g load while the temperature is ramped at 5°C/min past the melting point until the tension ruptures the film. Three parameters are reported from TMA test: shrinkage onset temperature, melt temperature, and melt rupture temperature. It has proved to be a more reproducible measure of melt integrity of the separator.[81]

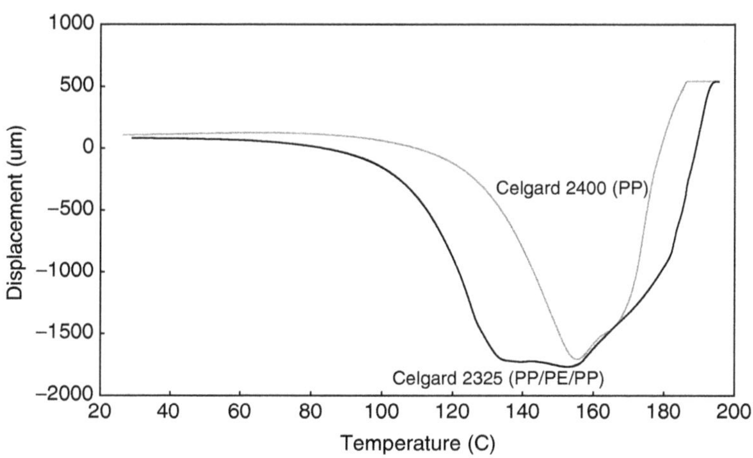

Fig. 20.13 TMA of Celgard 2400 (PP) and 2325 (PP/PE/PP). A constant load (2 g) is applied while the temperature is ramped at 5°C/min. Reprinted with permission from *Chem. Rev.* **104** (2004) 4419–4462, copyright (2004), American Chemical Society

Figure 20.13 shows the TMA data for two different Celgard membranes. The shrinkage onset temperature, deformation temperature, and rupture temperature are summarized in Table 20.6. The single-layer PP membrane (Celgard 2400) showed a higher softening temperature (~121°C), a deformation temperature around 160°C, and a very high rupture temperature around 180°C. The multilayer PP/PE/PP separator (Celgard 2325) combined the low-temperature shutdown property of PE with the high-temperature melt integrity of PP, resulting in a separator with softening (~105°C) and melt temperature (~135°C) very similar to PE and rupture temperature (~190°C) very similar to PP.

Separators with melt integrity greater then 200°C are desirable for Li-Ion cells. The trilayer separators with PP on the outside helps in maintaining the melt integrity of the separators at higher temperatures compared to single-layer PE separators. This is especially important for larger Li-Ion cells being developed for hybrid and electric vehicles.

20.5.3.12 Wettability and Wetting Speed

Two physical properties of separators, which are important to the operating characteristics of a battery, are electrolyte absorption and electrolyte retention. Any good separator should be able to absorb a significant amount of electrolyte and also .retain the absorbed electrolyte when the cell is in operation. These are more important in sealed cells where no free electrolyte is present. A maximum amount of electrolyte in the separator is desirable to achieve minimum cell internal resistance.

The separator wettability can limit the performance of batteries by increasing the separator and cell resistance. Separator wetting speed can be correlated with electrolyte filling time in real cells. The wetting speed is determined by the type of polymer (surface energy), pore size, porosity, and tortuosity of the separators. There generally is no accepted test for separator wettability. However, simply placing a drop of electrolyte on the separator and observing whether or not the droplet quickly wicks into the separator is a good indication of wettability. The contact angle also is a good measure of wettability.

The uptake of electrolyte by many hydrophobic polymer separators can be enhanced either by wetting agents or ionic-functional groups (e.g., ion-exchange membranes).

Table 20.6 TMA data for typical Celgard separators

	Celgard 2730	Celgard 2400	Celgard 2325
Shrinkage onset temperature (°C)	100	121	106
Deformation temperature (°C)	125	156	135, 154
Rupture temperature (°C)	140	183	192

Reprinted with permission from *Chem. Rev.* **104** (2004) 4419-4462, copyright (2004), American Chemical Society

20.5.4 *Effect of Separator on Cell Performance and Safety*

Although the material of a battery separator is inert and does not influence electrical energy storage or output, its physical properties greatly influence performance and safety of the battery. This is especially true for Li-Ion cells, and thus the battery manufacturers have started paying more attention to separators while designing the cells. The cells are designed in such a way that separators do not limit the performance; however, if the separator properties are not uniform or if there are other issues, it can affect the performance and safety of cells. This section will focus on the effect of the separator properties on cell performance and safety. Table 20.7 shows different types of safety and performance tests for Li-Ion batteries and the corresponding important separator property and how it affects performance and/or safety.

In order to achieve good performance of Li-Ion cells, the separators should have low resistance, low shrinkage, and uniform pore structure. Cells with high-resistance separators will perform poorly during high-rate discharge and also will increase the cell-charging time. Low shrinkage is a very important characteristic for separators, especially for higher-capacity cells. These cells are used in high-speed laptop computers, which can experience higher temperatures (~70–75°C) under certain conditions.[127] This can lead to shrinkage of separators and ultimately higher cell resistance and poor long-term cycling. The shrinkage in TD direction can lead to safety issues because of an internal short between the electrodes. Larger pores can lead to shorts during cell manufacturing or can fail during hipot testing. Larger pores will allow more soft shorts and higher self-discharge, especially during high-temperature storage. Very small pore size can lead to higher resistance and poor cycle life during high-temperature cycling and storage. Thus the pore size of the separator should be optimized to achieve good strength and performance.

One of the ways to increase cell capacity is by decreasing the thickness of separators. The newer high-capacity cells (> 2.0 Ah) generally use 20- and 16-μm separators as compared to 25 μm separators used in cells with 1.6- to 1.8-Ah capacity. The thinner separators offer lower resistance and help in increasing the capacity. However, they can hold less electrolyte and their mechanical strength is not as high as thicker separators. Thus appropriate changes should be made in cell design to keep the cell safe. The handling and manufacturing of thinner separators also is a challenge for the separator manufacturers. They are required to maintain the same electrical and mechanical properties and better quality for thinner separators. The separator manufacturers have installed better controls and quality standards and have started offering 16-μm separators. A lot of battery experts are of the opinion that the 16-μm is the thinnest they can use and still maintain the stringent performance and safety requirements of Li-Ion cells.

The separators inside the Li-Ion batteries experience extreme oxidizing environment on the side facing the positive electrode and extreme reducing environment on the side facing the negative electrode. The separators should be stable in these conditions during long-term cycling especially at high temperatures. Separators with

Table 20.7 Safety and performance tests for lithium-ion batteries and the corresponding important separator property and its effect on the cell performance and/or safety

Cell property	Separator property	Comments
Cell capacity	Thickness	Cell capacity can be increased by making the separator thinner
Cell internal resistance	Resistance	Separator resistance is a function of thickness, pore size, porosity, and tortuosity
High-rate performance	Resistance	Separator resistance is a function of thickness, pore size, porosity, and tortuosity
Fast charging	Resistance	Low separator resistance will aid in overall faster charging by allowing higher and/or longer constant current charging
High-temperature storage	Oxidation resistance	Oxidation of separators can lead to poor storage performance and reduce performance life
High-temperature cycling	Oxidation resistance	Oxidation of separators can lead to poor cycling performance
Self-discharge	Weak areas, pinholes	Soft shorts during cell formation and testing can lead to internal current leakage
Long-term cycling	Resistance, shrinkage, pore size	High resistance, high shrinkage, and very small pore size can lead to poor cycling performance
Overcharge	Shutdown behavior; high-temperature melt integrity	Separator should completely shut down and then maintain its melt integrity at high temperatures
External short circuit	Shutdown behavior	Separator shut down stops the cells from overheating
Hotbox	High-temperature melt integrity	Separator should be able to keep the two electrodes apart at high temperatures
Nail crush	Shutdown (to stop delayed failure)	In case of internal shorts, the separator may be the only safety device to stop the cell from overheating
Bar crush	Shutdown (to stop delayed failure)	In case of internal shorts, the separator may be the only safety device to stop the cell from overheating

Reprinted with permission from *Chem. Rev.* **104** (2004) 4419–4462, copyright (2004), American Chemical Society

poor oxidation resistance can lead to poor high-temperature storage performance and poor long-term cycling behavior. Polypropylene separators are more oxidation resistant when used in direct contact with the positive electrodes in Li-Ion cells. Thus, the trilayer separators (PP/PE/PP) with PP as the outer layers and PE as the central layer show superior oxidation-resistance properties compared to single-layer PE separators.

The products formed by the decomposition of the electrolyte also can block the pores of the separator, leading to an increase in cell resistance. The separators with lower resistance also help in better low-temperature performance. At very low temperatures, the resistance of the electrolytes is very high, and thus smaller contributions from the separator help to keep the cell resistance lower.

Zeng et al.[128] has shown that small amounts of active lithium metal could be added to a Li-Ion battery via the separator by using vacuum deposition techniques. The lithium films (4–8 μm) were deposited onto microporous PP film and showed that the lithium electrochemically reacted with either electrode, and thus the intrinsic irreversible capacity of the negative electrode was compensated for using volumetrically efficient lithium metal. This may be a novel idea to allow higher-capacity designs but is likely to be impractical and uneconomical due to issues involved with lithium plating on polymer films and the handling of the resulting films.

The Li-Ion cells have demonstrated power loss when aged and/or cycled at high temperatures. Norin et al. [129] demonstrated that the separator is at least partly responsible for the power loss due to the intrinsic increase in its ionic resistance. They showed that impedance increased significantly on cycling and/or aging of Li-Ion cells at elevated temperatures and that separators accounts for ~15% of the total cell impedance rise. They later reported that the loss in ionic conductivity of the separator was due to blocking of the separator pores by the products formed due to electrolyte decomposition, which was significantly accelerated at elevated temperatures. [130]

The US Department of Transportation (DOT) classifies all Li-Ion batteries as hazardous materials for shipping in the same category as lithium metal primary batteries. [131] It grants exceptions based on the cell capacity and ability of the cells to pass specified tests. There are several groups that regulate, or provide testing, to verify safe operation of Li-Ion cells under abuse conditions. In addition, the UL Laboratories, [132, 133] the International Electrotechnic Commission, [134] and the United Nations [135] have developed standardized safety testing procedures. These tests are designed to assure that cells are safe to ship and are resistant to typical abuse conditions such as internal shorting, overcharge, overdischarge, vibration, shock, and temperature variations that may be encountered in normal transportation environments.

The Underwriters Laboratories (UL) requires that consumer batteries pass a number of safety tests (UL 1642, [136] and UL-2054 [137]). There are similar recommendations from United Nations for transport of dangerous goods, [138] and from International Electrotechnical Commission (IEC), and Japan Battery Association. [139] An abnormal increase in cell temperature can occur from internal heating caused by either electrical abuse – overcharge or short circuit – or mechanical abuse – nail penetration or crush. Higher cell temperature also could be a result of external heating. For this reason, battery packs containing Li-Ion cells are designed with safety control circuits that have redundant safety features (PTC, CID, vent, thermal fuse, etc.). Shutdown separators are one of the safety devices inside the cell and act as a last line of defense. The separator shut down is irreversible, which is fine for polyethylene-based separators, which melt around 130°C.

The impedance of the separator increases by two to three orders of magnitude due to an increase in cell temperature, which results from cell abuse (e.g., short circuit, overcharge). The separator not only should shut down around 130°C, but it also should maintain its mechanical integrity at higher temperatures, preferably at temperatures as high as 200°C. If the separator does not shut down properly, then the cell will continue to heat during an overcharge test and can lead to thermal runaway. The high-temperature melt integrity of separators also is a very important property to keep the cell safe during extended overcharge or during extended exposure to higher temperatures.

Figure 20.14 shows a typical short-circuit curve for an 18,650 Li-Ion cell with a shutdown separator. The cell does not have other safety devices (e.g., CID, PTC), which usually work before separator shut down. As soon as the cell is short circuited externally through a very small shunt resistor, the cell starts heating because of the large current drained through the cell. The shut down of the separator, which occurs around 130°C, stops the cell from heating further. The current decrease is caused by an increase of battery internal resistance due to separator shutdown. The separator shutdown helps to avoid the thermal runaway of the cell.

Cells can be overcharged when the cell voltage is incorrectly detected by the charging control system or when the charger breaks down. When this happens, the lithium ions remaining in the cathode are removed and more lithium ions are inserted into the anode than under standard charging conditions. If the lithium insertion ability of the carbon anode is small, lithium metal in the form of dendrites may be deposited on the carbon and this causes a drastic reduction in thermal stability.

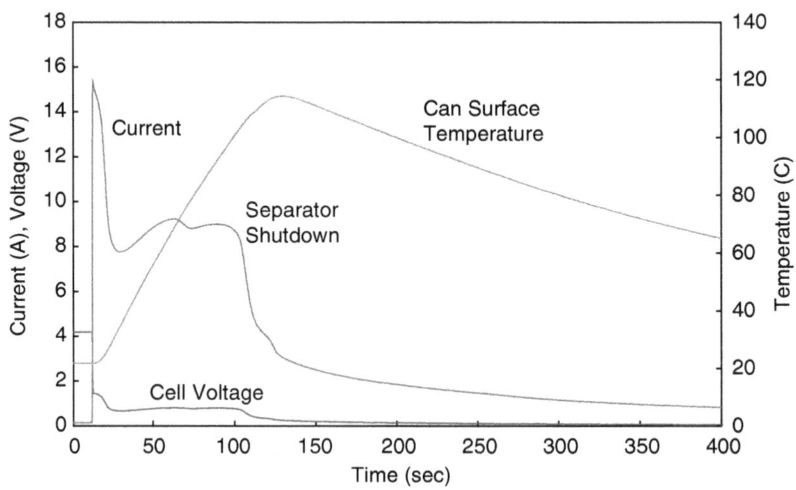

Fig. 20.14 Typical short-circuit behavior of a 18,650 Li-Ion cell with shutdown separator and current interrupt device (CID). This test simulates external short circuit of a cell. Reprinted with permission from *Chem. Rev.* **104** (2004) 4419–4462, copyright (2004), American Chemical Society

At higher charging rates, the heat output increases greatly because the joule heat output is proportional to I^2R. Several exothermic reactions (e.g., reaction between lithium and electrolyte, thermal decomposition of anode and cathode, thermal decomposition of electrolyte, etc.) occur inside the cell as its temperature increases. Separator shut down happens when cell temperature reaches melting point of PE as shown in Fig. 20.15. The CID of the18,650 cell was removed to test the performance of separator alone. The current decrease is caused by an increase of battery internal resistance due to separator shut down. Once the pores of the separator have closed due to softening, the battery cannot continue to be charged or discharged, and thus thermal runaway is prevented. During continued overcharge, the separator should maintain its shutdown feature and should not allow the cell to heat again. It should also maintain its melt integrity and should not allow the two electrodes to touch each other.

The separator also should not allow any dendrite to penetrate through the separator to avoid internal shorts. During an internal short, the separator is the only safety device that can stop the thermal runaway if the failure is not instant. If the heating rate is too high, then the instant failure will occur which cannot be stopped by separator shutdown. If the heating rate is not too high, then the separator shutdown can help in controlling the heating rate and stop thermal runaway.

Generally in a nail penetration test, an instantaneous internal short would result the moment the nail is tucked into the battery. Enormous heat is produced from current flow (double-layer discharge and electrochemical reactions) in the circuit by the metal nail and electrodes. The contact area varies according to the depth of

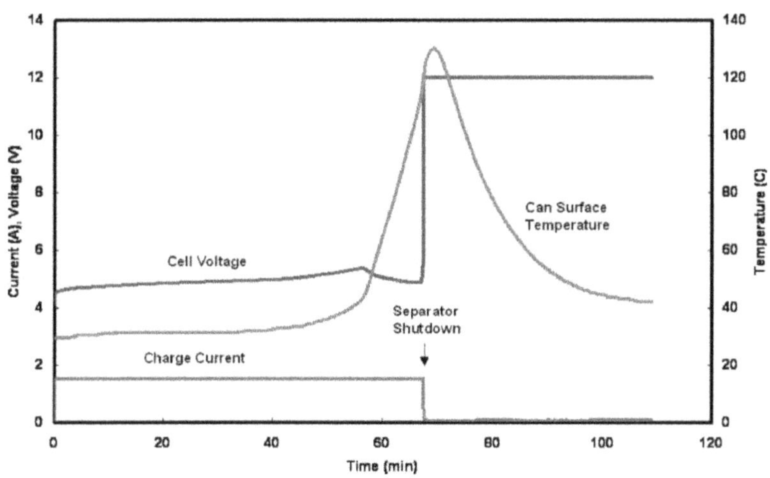

Fig. 20.15 Typical overcharge behavior of a 18,650 lithium-ion cell with shutdown separator. The PTC (Positive Temperature Coefficient) and CID (Current Interrupt Device) were removed from the cell header. Reprinted with permission from *Chem. Rev.* **104** (2004) 4419–4462, Copyright (2004) American Chemical Society

penetration. The shallower the depth, the smaller the contact area and therefore the greater the local current density and heat production. Thermal runaway is likely to take place as local heat generation induces electrolyte and electrode materials to decompose. On the other hand, if the battery is fully penetrated, the increased contact area would lower the current density, and consequently all tests would pass the nail penetration test. Internal short-circuit tests are more difficult to pass than the external short-circuit tests described earlier, because the contact area between metal nail contact is smaller than the contact area between current collectors, where current density therefore would be larger.

Figure 20.16 shows the typical nail penetration behavior of a Li-Ion cell with shutdown separator. Clearly, there was a voltage drop from 4.2 to 0.0 V, instantaneously, as the nail penetrates through (when internal short circuit occur) and temperature rose. When the heating rate is low, the cell stops heating when the temperature is close to separator shutdown temperature as shown in Fig. 20.16a. If the heating rate is very high, then the cell continues to heat and fails the nail penetration test as shown in Fig. 20.16b. In this case, the separator shutdown is not fast enough to stop the cell from thermal runaway. Thus a separator only helps to avoid delayed failures in case of internal short circuit as simulated by nail and bar crush tests. Separators with high-temperature melt integrity and good shutdown feature (to avoid delayed failures) are needed to pass internal short-circuit tests.

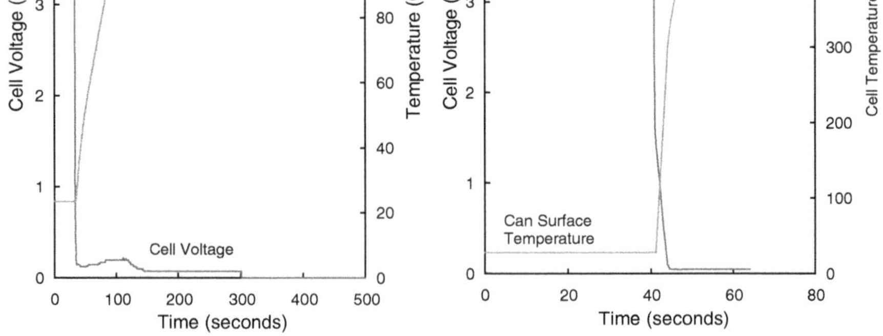

Fig. 20.16 Typical nail penetration behavior of a 18,650 Li-Ion cell with a shutdown separator. This test simulates internal short circuit of cell. (**a**) Cell passed nail penetration test. (**b**) Cell failed nail penetration test. Reprinted with permission from *Chem. Rev.* **104** (2004) 4419–4462, copyright (2004), American Chemical Society

Thinner separators (< 20 μm) used in high-capacity cells should offer similar shutdown and high- emperature melt integrity properties as thicker separators. The decrease in separator strength should be balanced with changes in cell design. The separator properties across the length and width should be very uniform to keep the cell safe during abnormal use.

The mechanism and characteristics of thermal cutoff devices in several prismatic Li-Ion cells was studied by Venugopal, [140] who monitored the impedance at 1 kHz and the open circuit voltage (OCV) of the cells as a function of temperature. All the cells studied contained PE-based separators with a shutdown temperature between 130 and 135°C. Within this narrow temperature range, the shutdown separators caused a sharp and irreversible rise in impedance of the cell. Single-layer PE separators were effective up to around 145°C, above which they demonstrated a meltdown effect. Trilayer separators had meltdown temperatures as high as 160°C because of the presence of additional layers of higher melting PP. It was found that the separators are not able to shut down the cell completely. In case of an overcharged test, the cell could continue to charge at lower currents even after the shutdown event, rendering the cell a potential hazard if not disposed of immediately and safely. This usually does not become an issue in commercial cells because the cell manufacturers have addressed this issue by including multiple cutoff devices within a single cell.

20.6 Summary

The ideal battery separator would be infinitesimally thin, offer no resistance to ionic transport in electrolytes, provide infinite resistance to electronic conductivity for isolation of electrodes, be highly tortuous to prevent dendritic growths, and be inert to chemical reactions. Unfortunately, in the real world the ideal case does not exist. Real-world separators are electronically insulating membranes whose ionic resistivity is brought to the desired range by manipulating the membrane thickness and porosity.

It is clear that no single separator satisfies all the needs of battery designers and that compromises have to be made. It is ultimately the application that decides which separator is most suitable. The information provided is purely technical and does not include other very important parameters, such as cost of production, availability, and long-term stability.

There has been a continued demand for thinner battery separators to increase battery power and capacity. This has been especially true for Li-Ion batteries used in portable electronics. However, it is very important to ensure the continued safety of batteries, and this is where the role of the separator is greatest. Thus, it is essential to optimize all the components of the battery to improve the performance while maintaining the safety of these cells. Separator manufacturers should work along with the battery manufacturers to create the next generation of batteries with increased reliability and performance, but always keeping safety in mind.

20.7 Future Directions

Up until recently, most of the separators and membranes historically used had not been specifically developed for battery applications. Thus, future research should be directed toward developing separators that are specifically tailored for battery applications. The general objectives of separator research should be:

1. To find new and cost-effective separators
2. To understand the separator properties in batteries
3. To optimize separator properties related to specific cell performance, life, and safety

One way to achieve some of these goals will be to develop mathematical models that reflect the effects of separator resistance, thickness, pore size, shrinkage, tortuosity, and mechanical strength on the final performance and safety of batteries. The battery separators for tomorrow will demand more than just good insulation and mechanical filtration; they will require unique electrochemical properties.

Acknowledgments The authors would like to acknowledge numerous colleagues and staff members at Celgard, LLC and Polypore, International Inc. for their help with the preparation and publication of this chapter.

References

1. Linden, D., Reddy, T. B. *Handbook of Batteries*, 3rd ed., McGraw Hill: New York, 2002
2. Besenhard, J. O. Editor, *Handbook of Battery Materials*, Wiley: Weimheim, 1999
3. Berndt, D. *Maintenance Free Batteries*, 3rd ed., Research Studies Press Ltd.: Taunton, Somerset, England, 2003
4. Bode, H. *Lead-Acid Batteries*, Wiley: New York, 1977
5. Falk, S. U., Salkind, A. J. *Alkaline Storage Batteries*, Wiley: New York, 1969
6. Fleischer, A., Lander, J. J. *Zinc-Silver Oxide Batteries*, Wiley: New York, 1971
7. Brodd, R. J., Friend, H. M., Nardi, J. C. Editors, *Lithium Ion Battery Technology*, ITE-JEC: Brunswick, OH, 1995
8. Wakihara, M., Yamamoto, O. Editors, *Lithium Ion Batteries, Fundamentals and Performance*, Wiley: New York, 1998
9. Yoshino, A. *Chem. Ind.*, **146** (1995) 870
10. Schalkwijk, W. A. V. Editor, *Advances in Lithium Ion Batteries*, Kluwer: New York, 2002
11. Kinoshita, K., Yeo, R., *Survey on Separators for Electrochemical Systems*, LBNL, January 1985
12. Benett, J., Choi, W. M. Developments in small cell separators, Proceedings of the Tenth Annual Battery Conference on Application and Advances, IEEE Aerospace and Electronics Systems Society, New York, January 10–13, 1995, 265
13. Boehnstedt, W. *Handbook of Battery Materials*, J. O. Besenhard, Editor, Wiley: Amsterdam and New York, 1999
14. Spotnitz, R. *Handbook of Battery Materials*, J. O. Besenhard, Editor, Wiley: Amsterdam and New York, 1999
15. Shirai, H., Spotnitz, R. *Lithium Ion Secondary Battery – Materials and Applications*, K. Yoshio, Editor, Nikkan Kogyo Shin-bun: Osaka, 1996, p 91 (in Japanese)
16. Shirai, H., Spotnitz, R., Atsushi, A. *Characterization of Separators for Lithium Ion Batteries – A Review, Chemical Industry*, **48** (1997) 47 (in Japanese)

17. Hiroshi, T. *The Latest Technologies of the New Secondary Battery Materials*, Z. Ogumi, Editor, CMC, Tokyo, p. 99, 1997
18. Hiroshi, T. *Advanced Technologies for Polymer Battery*, N. Oyama, Editor, CMC, Tokyo, p. 165, 1998
19. Koichi, K. *Advanced Technologies for Polymer Battery*, N. Oyama, Editor, CMC, Tokyo, p. 174, 1998
20. Kiyoshi, K. *Lithium Secondary Battery Technology for the 21st Century*, K. Kanamura, Editor, CMC, Tokyo, p. 116, 2002
21. Brodd, R. J., Bullock, K. R., Leising, R. A., Middaugh, R. L., Miller, J. R., Takeuchi, E. *J. Electrochem. Soc.*, **151** (2004) K1
22. Takeshita, H. The 21st International Seminar & Exhibit on Primary & Secondary Batteries, Fort Lauderdale, FL, Florida Educational Seminars, Inc., March 8, 2004
23. Takeshita, H. The 23rd International Battery Seminar & Exhibit, Fort Lauderdale, FL, Florida Educational Seminars, Inc., March **13**, 2004
24. *Battery & EV Technology, January*, **28** (2004) 2
25. Pilot, C. *The Worldwide Rechargeable Battery Market, Batteries 2004*, 6th Ed., Paris, June 2nd–4th 2004
26. Celgard Inc. http://www.celgard.com
27. Celgard completes capacity expansion, Global Sources, http://www.globalsources.com, February 04, 2004
28. Asahi Kasei adding battery membrane capacity, *Nikkei Net Interactive*, http://www.nni.nikkei.co.jp/AC/TNKS/Nni20030806D06JFA23.htm, August 7, 2003
29. Advanced Rechargeable Battery Industry, 2001/2002, Nomura Research Institute Limited, 2002
30. About Edison Batteries, Inc, http://www.optodot.com/sys-tmpl/htmlpage/
31. Battery and Fuel Cell Components, The Fredonia Group, August 2003
32. *Advanced Battery Technology*, February, **40** (2004) 22
33. Hamano, K., Yoshida, Y., Shiota, H., Shiraga, S., Aihara, S., Murai, M., Inuzuka, T. U.S. Patent 6,664,007 B2, 2003
34. Sun, L., Chen, G., Xu, D., Abe, T. The Electrochemical Society, Abs 423, 204th Meeting, 2003
35. Sun, L. U.S. Patent 2003/0152828A1, 2003
36. Johnson, B. A., White, R. E. *J. Power Sources*, **70** (1998) 48
37. Bradford, S. M. Battery Power Products & Technology, March 2004, 17
38. Frost & Sullivan Research World Rechargeable Battery Markets for Mobile IT and Communication Devices (A575-27), 2002
39. Geiger, M., Callahan, R. W., Diwiggins, C. F., Fisher, H. M., Hoffman, D. K., Yu, W. C., Abraham, K. M., Jillson, M. H., Nguyen, T. H. The Eleventh International Seminar on Primary and Secondary Battery Technology and Application, Fort Lauderdale, FL, Florida Educational Seminars Inc., February 28–March 3, 1994
40. Tanba, H. *Molding Process*, **11** (1999) 759
41. Adachi, A., Spotnitz, R. M., et al. *Osaka Chemical Marketing Center* (1997), 69–80
42. Yu, W. C., Hux, S. E. US Patent, 5,952,120, 1999
43. Hipore, A. K. http://www.asahi-kasai.co.jp/memrbane/english/tradenm/t07.html
44. Soji, N., Hiroyuk, H., Kiichiro, M., Ryoichi, M. US Patent 5,480,745, 1996
45. Bierenbaum, H. S., Isaacson, R. B., Druin, M. L., Plovan, S. G. *Ind. Eng. Chem. Prod. Res. Dev.*, 13, 2, 1974
46. *Jpn. Ind. News*, Japan Industrial Journal, Tokyo, **91**, 1996
47. Kim, S. S., Lloyd, D. R. *J. Membrane Sci.*, **64** (1991) 13
48. Druin, M. L., Loft, J. T., Plovan, S. G. U.S. Patent 3,801,404. 1974
49. Schell, W. J., Zhang, Z. The Fourteenth Annual Battery Conference on Applications and Advances, Long Beach, CA, IEEE Aerospace and Electronics Systems Society, New York, January 12–15, 1999, 161
50. Isaacson, R. B., Bierenbaum, H. S. U.S. Patent 3,558,764, 1971
51. Kamei, E., Shimomura, Y. U.S. Patent 4,563,317, 1986

52. Yu, T. H. *Processing and Structure-Property Behavior of Microporous Polyethylene – From Resin to final Film*, Ph.D. Dissertation, Virginia Polytechnic Institute and State University, 1996
53. Sarada, T., Sawyer, L. C., Ostler, M. I. *J. Membrane Sci.*, **15** (1983) 97
54. Bierenbaum, H. S., Daley, L. R., Zimmerman, D., Hay, I. L. U.S. Patent, 3,843,761, 1974
55. Hamer, E. A. G. U.S. Patent, 4,620,956, 1986
56. Hiroshi, K., Tetuo, A., Akira, K. U.S. Patent 5,691,047, 1997
57. Kesting, R. E. *Synthetic Polymeric Membranes*, 2nd Ed., Wiley: New York, Ch 2 1985
58. Ihm, D. W., Noh, J. G., Kim, J. Y. *J. Power Sources*, **109** (2002) 388
59. Takita, K., Kono, K., Takashima, T., Okamoto, K. U.S. Patent, 5,051,183, 1991
60. Michiyuki, A.; Jpn. Patent 8064194, 1996
61. Kotaro, T., Koichi, K., Tatsuya, T., Kenkichi, O. U.S. patent 5,051,183, 1991
62. Koichi, K., Kotaro, T., Mamoru, T., Tatsuya, T. Jpn. Patent 8012799, 1996
63. Norimitsu, K., Kotaru, T., Koichi, K., Hidehiko, F. U.S. Patent 6,153,133, 2000
64. Akinao, H., Kazuo, Y., Hitoshi, M. U.S. Patent 6,048,607, 2000
65. Pekala, R. W., Khavari, M. U.S. Patent 6,586,138, 2003
66. Userguide, FreedomCar Separator Costing Document, February 2003
67. Xu, M., Hu, S., Guan, J., Sun, X., Wu, W., Zhu, W., Zhang, X., Ma, Z., Han, Q., Liu, S. U.S. Patent 5,134,174, 1992
68. Fisher, H. M., Wensley, C. G. U.S. Patent 6,368,742, 2002
69. Zhu, W., Zhang, X., Zhao, C., Zu, W., Hou, J., Xu, M. *Polymersr Adv. Technol.s*, 7 (1996) 743
70. Sadamitsu, K., Ikeda, N., Hoki, M., Nagata, K., Ogino, K. World Patent Application 02066233A1, 2002
71. Higuchi, H., Matsushita, K., Ezoe, M., Shinomura, T. U.S. Patent 5,385,777, 1995
72. Calis, G. H. M., Daemen, A. P. M., Gerrits, N. S. J. A., Smedinga, J. T. *J. Power Sources*, **65** (1997) 275
73. Ooms, F. G. B., Kelder, E. M., Schoonman, J., Gerrits, N., Smedinga, J., Calis, G. *J. Power Sources*, **97–98** (2001) 598
74. Yamamura, Y., Ooizumi, S., Yamamoto, K. Separator for rechargeable lithium ion batteries with high puncture strength and high melt rupture temperature, Nitto Denko Technical Report (http://www.nitto.com/rd/rd6_1.html), **39** (2001) 39
75. Pekala, R. W., Khavari, M., Dobbie, G., Lee, D., Fraser-Bell, G. 17th International Seminar & Exhibit on Primary and Secondary Batteries, Fort Lauderdale, Fl, Florida Educational Seminars, March 6–9, 2000
76. Chen, G., Richardson, T. J. *Electrochem. Solid-State Lett.*, **7(2)**, (2004) A23–A26
77. Chen, G., Thomas, J. R. *Electrochem. Solid-State Lett.*, **9(1)**, (2006) A24–A26
78. Fleming, R., Taskier, H. *Prog. Batt. Solar Cells*, **9** (1990) 58
79. Hoffman, D., Fisher, H., Langford, E., Diwiggins, C. *Prog. Batt Solar Cells*, 9 (1990) 48
80. Callahan, R. W., Nguyen, K. V., McLean, J. G., Propost, J., Hoffman, D. K. Proceedings of the 10th International Seminar on Primary and Secondary Battery Technology and Application, Fort Lauderdale, FL, March 1–4, Florida Educational Seminars Inc., 1993
81. Spotnitz, R., Ferebee, M., Callahan, R. W., Nguyen, K., Yu, W. C., Geiger, M., Dwiggens, C., Fischer, H., Hoffman, D. Proceedings of the 12th International Seminar on Primary and Secondary Battery Technology and Applications, Fort Lauderdale, FL,Florida Educational Seminars Inc., 1995March 6–9
82. Yu, W. C., Callahan, R. W., Diwiggins, C. F., Fischer, H. M., Geiger, M. W., Schell, W. J. North America Membrane Society Conference, Breckenridge, CO, North America Membrane Society, 1994
83. Kuribayashi, I. *J. Power Sources*, **63** (1996) 87
84. Pasquier, A. D., Gozdz, A., Plitz, I., Shelburne, J. 201st meeting, The Electrochemical Society, Philadelphia, PA, May 12–17, 2002
85. Augustin, S., Volker, H., Gerhard, H., Christian, H. *Desalination*, **146** (2002) 23
86. http://www.separion.com

87. Hying, C. Separion separators for lithium batteries – safety & performance, Batteries 2004, 6th Ed., Paris, June 2nd–4th 2004
88. Sachan, S., Ray, C. A., Perusich, S. A. *Polymer Engineer. Sci.*, **42** (2002) 1469
89. Sachan, S., Perusich, S. Electrochemical Society Meeting, Seattle, 1999
90. Carlson, S. A.; *Membrane & Separation Technology News*, 2004, 22, 8
91. Abraham, K. M. *Electrochim. Acta*, **38** (1993) 1233
92. Gineste, J. L., Pourcell, G. *J. Membrane Sci*, **107** (1995) 155
93. Hoffman, D. K., Abraham, K. M. Proceedings of the Fifth International Seminar on Lithium Battery Technology and Applications, Deerfield Beach, FL,Florida Educational Seminars Inc., 1991
94. USABC "Development of low cost separators for lithium-ion batteries", RFPI 2001
95. Laman, F. C., Sakutai, Y., Hirai, T., Yamaki, J., Tobishima, S. Ext. Abstr., 6th Int. Meet. Lithium Batteries, Münster, Germany, 10–15 May 1992, p 298–300
96. Laman, F. C., Gee, M. A., Denovan, *J. J. Electrochem. Soc*, **140** (1993) L51
97. ASTM D5947-96, Standard Test Methods for Physical Dimensions of Solid Plastics Specimens, ASTM International, July 2002
98. ASTM D2103, Standard Specification for Polyethylene Film and Sheeting, ASTM International
99. Caldwell, D. L., Poush, K. A. U.S. Patent, 4,464,238, 1984
100. ASTM D726, Standard Test Methods for Identification of Fibers in Textiles, ASTM International
101. ASTM E128-99, Standard test method for Maximum Pore Diameter and Permeability of Rigid Porous Filters for Laboratory Use, ASTM International
102. ASTM D3763, Standard Test Method for High Speed Puncture Properties of Plastics using Load and Displacement Sensors, ASTM International
103. ASTM D1204, Standard Test methods for Linear Dimensional Changes of Nonrigid Thermoplastic Sheeting or Film at Elevated Temperatures, ASTM International
104. ASTM D882, Standard Test Method for Tensile Properties of Thin Plastic Sheeting, ASTM International
105. Abraham, K. M. *Electrochim. Acta*, **38** (1993) 1233
106. Robinson, R. G., Walker, R. L., Batteries, D. H., CollinsEditor, The MacMillan Company: New York, NY, 1963, p15
107. Lander, J. J., Weaver, R. D., Salkind, A. J., Kelley, J. J. in *Characteristics of Separators for Alkaline Siver Oxide Zinc Secondary Batteries. Screening Methods* J. E. Cooper, A. Fleischer, Editors, NASA Technical Report NAS 5-2860 1964
108. Kilroy, W. P., Moynihan, C. T. *J. Electrochem. Soc.*, **125** (1978) 520
109. MacMullin, R. B., Muccini, G. A. *AIChE J.*, **2** (1956) 393
110. Spotnitz, R., Ferebee, M. W. Meeting Abstracts, The Electrochemical Society Inc., Volume 96-2, Fall Meeting, San Antonio, TX, October 6–11, 1996
111. Ionov, V. V., Isakevitch, V. V., Katalevsky, E. E., Chernokoz, A. J. *J. Power Sources*, **30** (1990) 321
112. Lowell, S., Shields, E. *Powder Surface Area and Porosity*, 3rd ed., Chapman and Hall: London, 1991
113. PMI Conference 2000 Proceedings, PMI short course, Ithaca, NY, October 16–19, 2000
114. Porous Materials Inc, http://www.pmiapp.com
115. Jena, A. K., Gupta, K. M. *J. Power Sources*, **80** (1999) 46
116. Jena, A. K., Gupta, K. M. *J. Power Sources*, **96** (2001) 214
117. Venugopal, G., Moore, J., Howard, J., Pendalwar, S. *J. Power Sources*, **77** (1999) 34
118. Zeman, L., Denault, L. *J. Membrane Sci.*, **71** (1992) 221
119. Chen, R. T., Saw, C. K., Jamieson, M. G., Aversa, T. R., Callahan, R. W. *J. Appl. Polym. Sci.*, **53** (1994) 471
120. Fujii, T., Mochizuki, T. U.S. Patent, 5,759,678, 1998

121. Venugopal, G. The role of plastics in lithium-ion batteries, Proceedings of the 3rd Annual Conference on Plastics for Portable and Wireless Electronics, Philadelphia, 11, October 14–15, 1997
122. Lundquist, J. T., Lundsager, C. B., Palmer, N. L., Troffkin, H. J., Howard, J. U.S. Patent 4,731,304, 1998
123. Lundquist, J. T., Lundsager, C. B., Palmer, N. I., Troffkin, H. J. U.S. Patents 4,650,730, 1987
124. Zuckerbrod, D., Giovannoni, R. T., Grossman, K. R. Proceedings of the 34th International Power Sources Symposium, Cherry Hill, NJ, June 25–28, 1990, p. 172
125. Faust, M. A., Suchanski, M. R., Osterhoudt, H. W. U.S. Patent No. 4,741,979, 1988
126. Matthias, U., Dieter, B., Heinrich, R., Thomas, B.; U.S. Patent, 6,511,517, 2003
127. Maleki, H., Shamsuri, A. K. *J. Power Sources*, **115** (2003) 131
128. Zeng, S., Moses, P. R. *J. Power Sources*, **90** (2000) 39
129. Norin, L., Kostecki, R., McLarnon, F. *Electrochem. Solid State Lett.*, **2002**, 5, A67
130. Kostecki, R., Norin, L., Song, X., McLarnon, F. *J. Electrochem. Soc.*, **2004**, 151, A522
131. *Hazardous Materials Regulations*, Code of Federal Regulations, CFR49 173.185
132. UL1640, Lithium Batteries, Underwriters Laboratories, Inc
133. UL2054, Household and Commercial Batteries, Underwriter Laboratories, Inc
134. Secondary Lithium Cells and Batteries for Portable Applications, International Electrotechnic Commission, IEC 61960-1 and IEC 61960-2
135. Recommendations on the Transport of Dangerous Goods, Manual of Tests and Criteria, United Nations, New York, 1999
136. Safety Standard for Lithium Batteries, UL 1642, Underwriters Laboratories Inc, 3rd ed., 1995
137. Standard for Household and Commercial Batteries, UL 2054, Underwriter Laboratories, Inc., 1993
138. UN Recommendations on the Transport of Dangerous Goods, December 2000
139. A Guideline for the Safety Evaluation of Secondary Lithium Cells, Japan Battery Association, 1997
140. Venugopal, G. *J. Power Sources*, **101** (2001) 231

Chapter 21
Polymer Electrolyte and Polymer Battery

Toshiyuki Osawa and Michiyuki Kono

21.1 Introduction

The development of organic or inorganic lithium-ion conductor such as lithium super-ion conducting glass (LISICON), lithium iodide (LiI), solid polymer electrolytes(SPE) were activated with progress of a lithium battery. In these electrolytes, the technology of solid electrolyte is useful to improve the performance of batteries.

Recentry, SPE have attracted much attention because they will be especially advantageous for high-ionic conductivity and high processability. Generally, batteries using SPE for three main elements, such as positive electrode, electrolyte, and negative electrode, are called "polymer batteries".

Polyme battery have been developed by several companies [Sony, Lithium Technology Corporation, Ultralife Batteries Inc., MoltechCorporation, SAFT (Alcatel), TDI Batteries Inc., PolyPlus Battery Company etc.] in order to attain superior characteristics, such as high reliability, nonleakage of liquid electrolyte, ultrathin form, flexibility, and high-energy density.

An aqueous gel electrolyte has been used in dry cells for prevention of leakage of the liquid electrolyte for a long time. By adding a gelatinizer such as dextrin, starch, cross-linked starch, and carboxymethylcellulose to a liquid electrolyte, the electrolyte changes into semisolid state. As well as the gel for the dry cell, solidification technology is indispensable for the advancement of non-aqueous batteries.

SPE will become a key material for the next generation of Lithium batteries.

T. Osawa (✉) and M. Kono
Chemical Technology Division, Kanagawa Industrial Technology Center,
705-1 Shimo-ima-izumi, Ebina-city, Kanagawa, 243-0435 Japan
tohsawa@kanagawa-iri.go.jp

M. Yoshio et al. (eds.), *Lithium-Ion Batteries*,
DOI 10.1007/978-0-387-34445-4_21 © Springer Science+Business Media LLC 2009

21.2 Polymer Electrolyte for Lithium Battery

Features of the polymer battery using SPE are summarized as follows:

(a) The design and size are easy and flexible.
(b) The properties of polymer electrolyte, such as leakage prevention, fire resistance, and automatic extinction can ensure the safety.
(c) The lamination structure of electrode and electrolyte has high reliability for impact and vibration.
(d) The electrode interval can be narrowed by using a polymer electrolyte as a separator. As a result, there will be an increase in capacity.
(e) The improvement of cyclicity is expected because of the decrease in the interfacial impedance of the electrode and the electrolyte.

Generally the polymer electrolyte of the polymer battery is classified into two kinds of the electrolyte: One is a dry-type electrolyte composed of a polymer matrix and the electrolyte salt; the other is a gel-type electrolyte in which polar solvent is added as a plasticizer to appropriate polymer matrix. In addition, the gel-type electrolyte is classified into thermoplastic gel and cross-linked gel. The research flow of SPE is shown in Fig. 21.1.

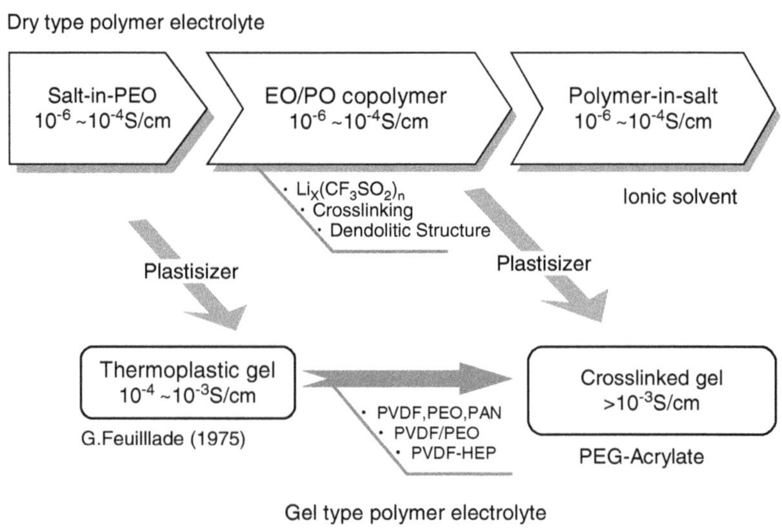

Fig. 21.1 The research flow of SPE

21.3 Dry Polymer Electrolyte

M. B. Armand first proposed the application of a dry polymer electrolyte to a battery[1] since ionic conduction in the EO polymer was discovered by P. V. Wright.[2] The dry polymer electrolyte consisted of a polymer host and alkali metal salts. The polymer hosts have polar groups (repeated function group) that can interact with cation, such as etheric oxygen in oxyethylene or oxypropylene unit. As the function of a polymer host, high solubility of the salt, easy ion dissociation, and easy ion diffusion are required. The ion dissociation units such as ethyleneoxide and ethyleneimine in polymer structures accelerate the dissociation of electrolyte salt into the ions.[3,4] Ion transportation by the polymer host is greatly affected not only by the macrorelaxation which decides the dynamic property of the polymer but also the microrelaxation of the segmental motion of the polymer chain. Thus, the relaxation phenomenon of polymer greatly depends on temperature. Therefore, the ionic conductivity of the polymer electrolyte also is dependent on temperature.

From the behavior of temperature dependence of the ionic conductivity, the ion transport in SPE is explained according to the free volume theory or the configuration entropy model.

Generally, the ionic conductivity σ is approximated by the Williams-Landel-Ferry (WLF) formula[5]

$$\mathrm{Log}[\sigma(\mathrm{T}) / \sigma(\mathrm{T})] = \mathrm{C1}(\mathrm{T}-\mathrm{Tg}) / \{\mathrm{C2}+(\mathrm{T}-\mathrm{Tg})\} \qquad (21.1)$$

where Tg is the glass transition temperature,and C1 and C2 the WLF parameter.

The WLF formula shows that the ionic conductivity of the polymer electrolyte is shown in the temperature range higher than Tg. Ionic conductivity decreases rapidly if its temperature goes below that of Tg. The EO unit is recognized as the most excellent structure from the ionic dissociation viewpoint. The ion is transported coupled with the oxyethylene chain motion in amorphous polymer domain. However, oxyethylene structure easily becomes crystalline. Therefore, in order to accelerate the quick molecular motion of the polymer chain and quick ion diffusion, it is important to lower the crystallization of polymer matrixes. The methods for inhibiting the crystallization of the polymer are, for example, to introduce the polyethylene oxide chain into the low Tg polymer such as polysiloxane and phosphazene,[6,7] or to introduce the asymmetric units such as ethylene oxide/propylene oxide (EO/PO) into polymer main chain.

On the other hand, it also is an effective method to introduce the cross-linked structure into the polymer matrix. This also can solve the problem of the fluidity at the high temperature region. An example of a typical polymer matrix structure is shown in Fig. 21.2.

Recently, dry-type polymer electrolyte having high ionic conductivity has been obtained by introducing branched short-chain polyethyleneoxide.[8] In this system the branched chain acts as an internal plasticizer. Such a polymer shows high ionic conductivity of 10^{-4} S/cm at ca. 30°C.[9]

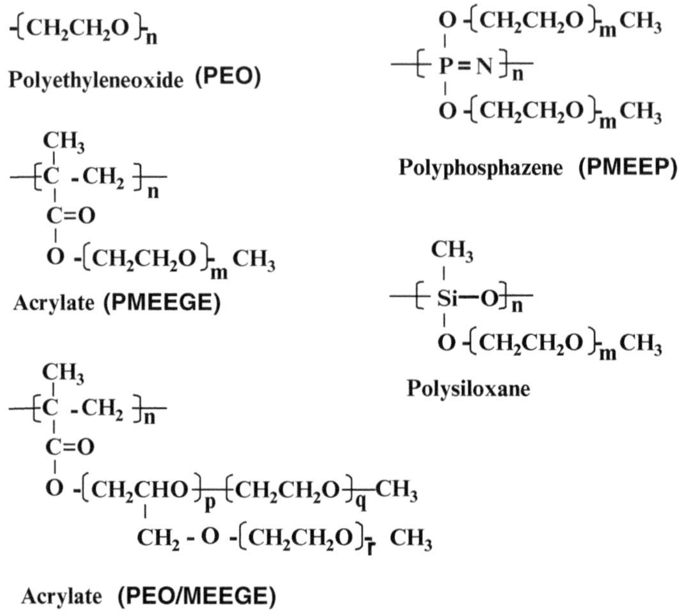

Fig. 21.2 Example of a typical polymer matrix structure

Generally the ionic conductivity of dry-type SPE in ranges above 60°C, which is equal to room-temperature conductivity of the liquid electrolyte. Therefore, the dry-type SPE can be used in high-temperature operation battery.

According to the measurement of differential scanning calorimeter (DSC), exothermic reaction of polymer electrolytes with lithium metal is lower compared with the liquid electrolyte. Polymer electrolytes also are effective for preventing lithium dendrite formation. This fact means that lithium metal can be used as the negative electrode by using polymer electrolytes. Furthermore, the polymer electrolyte is expected to function as separator because it has the sufficient high mechanical strength.

21.4 Gel-Type Polymer Electrolyte

The thermoplastic nonaqueous gel was discovered by Feuillade et al.,[10] around the same period that P. V. Wright discovered the ionic conductivity of dry polymers. Feuillade reported a conductivity of 10^{-4} S/cm for organic liquid electrolytes gelled with polyvinyl acetal. Thermoplastic gel generally is prepared easily by dissolving polymer, such as high-molecular-weight polyethylene oxide (PEO), polyvinylydene difluoride (PVDF), and polyacrylonitrile (PAN), to a heated liquid electrolyte and then by cooling. At room temperature, it becomes an elastic gel and has the same level of ionic conductivity as the liquid electrolyte.

From the safety aspect, flame retardation of the batteries is an important issue. According to a recent report, specific gels using PAN as a polymer matrix have high ionic conductivity and self-extinguishing property. Such phenomenon is observed with PAN-based gel by using $LiPF_6$ as the electrolyte salt.[11] These gels are semi-solid and are "thermoreversible" because the gel becomes a fluid solution again upon heating. The gel electrolyte using polyvinylidene difluoride/hexafluoropropylene copolymer (PVDF/HFP) was proposed by Bellcore Co.[12] This technology attracted much attention from the industrial viewpoint. The point that dramatically differs from the thermoplastic gel is in the high creep and high elasticity in a wide thermal range. The plastic rubber that dissolved liquid electrolyte into Viton® (produced by Du-Pont), which is the copolymer of PVDF and HFP, has been known to show a high ionic conductivity, but by increasing the copolymerization ratio of the HFP (in relation to the PVDF), the processing properties can be further improved. For example, a 4-V lithium-ion-type polymer battery composed of Mn_2O_4 positive electrode, 1-mol $LiPF_6$/EC/DMC gel electrolyte, and carbon negative electrode was reported. Bellcore's gel is made as follows. First, plasticizers such as dioctyl phthalate (DOP) or acetone and fillers such as fumed silica are dissolved in the polymer matrix. Then the DOP is eliminated. Next, fine porosities are formed in the film. Filling the porosities by the liquid electrolyte, the gel-type polymer electrolyte is formed. The filler is added in the polymer matrix in order to ensure necessary degree of swelling, strength as separator membrane, and ionic conductivity. This technology was considered to be the best process for the actual battery manufacturing and many battery manufacturers were licensed from Bellcore.

Unlike the characteristics of the thermoplastic gel, thermosetting gel (cross-linked gel) never becomes a liquid even at a high temperature. Thermosetting gels are prepared by cross-linking reactions of reactive monomers or oligomers such as vinyl polymerization, urethane reaction, polyene–thiol reaction, and the ring-opening reaction of lactone in the liquid electrolyte.[13,14] By choosing appropriate monomers or oligomers, the resulting gel gives high elasticity characteristics.

The most convenient way to prepare thermosetting gel is to add reactive monomers such as acrylate and metacrylate of the alkylene oxide to the liquid electrolyte and cross-link by irradiating heat, ultraviolet, or electron beam. The residual monomer can be detected by methods of infrared absorption measurement (on basis the peak of 1638 cm^{-1}), supercritical fluid chromatography, and photo-DSC (optical difference scanning thermal analysis).

The gel polymer network structure can be varied to that of combs and ladders by changing a kind of acrylate monomer. It also is possible to make an ion-conductive gel with a "polyethylenic" structure by cross-linking a difunctional acrylate compound such as polyethylene-glycol-diacrylate. Valence Inc. was not able to commercialize this battery, although it was proceeding with the development of polymer batteries that combined V_3O_8 positive electrode and lithium metal negative electrode by this electrolyte.[15] Table 21.1 shows the ionic conductivities of various gel polymer electrolytes.

Table 21.1 Ionic conductivity of various gel polymer electrolytes

Electrolyte			Ion conductivity (S/cm)	Temperature (°C)
Polymer matrix	Electrolyte salt	Solvent		
P(VDF-HFP)	$LiPF_6$	EC/PC	3×10^{-3}	22
P(VDF-HFP)	$LiPF_6$	EC/DMC	3×10^{-3}	22
PAN	$LiClO_4$	EC/PC	2.9×10^{-3}	20
PAN	$LiAsF_6$	BL	6.1×0^{-3}	20
PAN	$LiN(CF_3SO_2)_2$	BL/EC	4.0×10^{-3}	20
PAN	$LiCF_3SO_3$	EC/PC	1.4×10^{-3}	20
PEG-DA	$LiCF_3SO_3$	PC	1×10^{-3}	RT
PEG-A/TMPA	$LiBF_4$	PC/DME	3×10^{-3}	25
PEG-A/TMPA	$LiN(CF_3SO_2)_2$	EC-DME	4.6×10^{-3}	25
PEG-A/TMPA	$LiPF_6$	EC type	6.4×10^{-3}	25
P(VDF-HFP)	$LiPF_6$	EC/PC	3×10^{-3}	22
SBR	Li-salt	BL-DME	1.4×10^{-3}	RT

SBR stylene butadiene rubber, *P(VDF–HFP)* polyvinylidenedifluoride/hexafluoropropylene, *PAN* polyacrylonitrile, *PEG-DA* polyethyleneglycoldiacrylate, *PEG-A/TMPA* polyethyleneglycol-monoacrylate/trimethyrolepropaneacrylate

21.5 Commercial Polymer Battery Using Pure Polymer Electrolyte

The Quebec Province electricity company, Hydro Quebec, has been developing a large power supply battery since 1978 that uses a polymer electrolyte. The lithium salt composed of imide anion and carbanion are used as heat-resistant salt with the low lattice energy and bulky anion.

Avestor started the production of the dry polymer battery in September 2002. Figure 21.3 shows the structure of the battery, which AVESTOR has put in the market.

According to the measurement of DSC, exothermic reaction of lithium metals and polymer electrolytes is low in comparison with the electrolytic solution. Polymer electrolytes are effective for preventing lithium dendrite. This fact means that lithium metal can be used as the negative electrode by using polymer electrolytes. Furthermore, the polymer electrolyte is expected as a separator because of the sufficient high mechanical strength.

The goal for this battery is to be a substitution for the valve-regulated lead-acid battery (VRLA) as a backup power supply of for telecommunication. Usually, the relay machines of the telecommunication structure are located in remote areas. The VRLA with its short life is inconvenient from a maintenance aspect. Avestor's lithium metal polymer battery drastically reduced the load and cost of the maintenance, because the life of their battery is over 10 years and is maintenance free. Further, the feature of the battery is that the condition of the battery itself can be monitored in remote settings.

Avestor's manufacturing process also is unique in that they employed an environmentally friendly battery manufacturing process. They manufacture cathodes

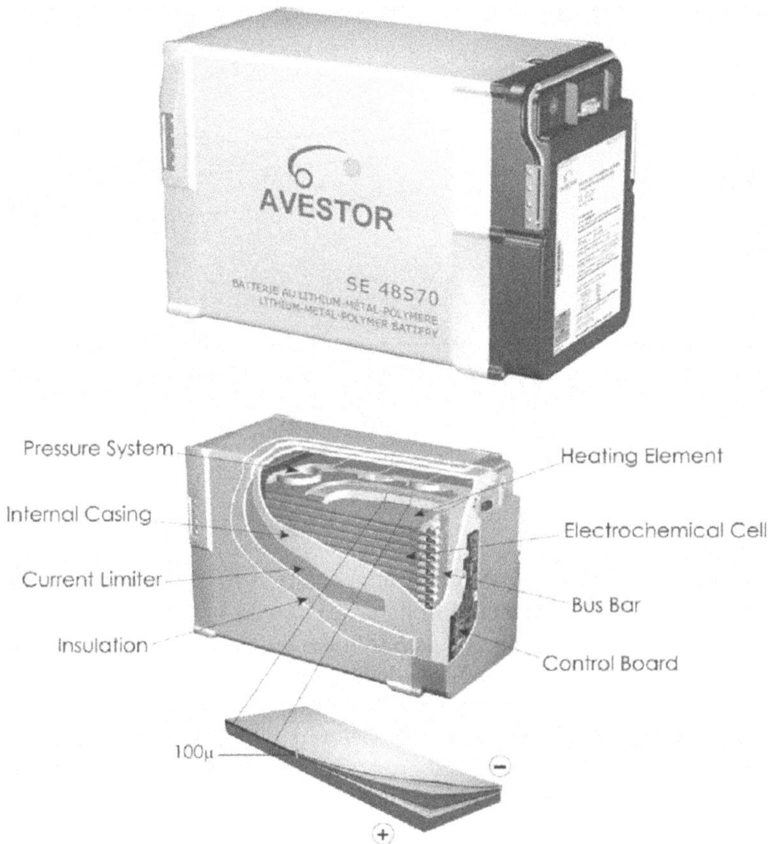

Fig. 21.3 Outlook and internal structure of the lithium pure polymer battery that Avestor has commercialized

and electrolytes by an extrusion process that never uses any volatile organic solvent that usually is used in the existing coating process.

21.6 Commercial Polymer Battery Using Gel Polymer Electrolyte

The practical application of this technology was first attempted by a Japanese battery manufacturer based on Bellcore's technology previously mentioned. However, this first polymer battery was withdrawn from the market because of the liquid electrolyte leakage accident, which occurred through separation of a polymer matrix and liquid electrolyte. After that, the polymer battery using the gel with a

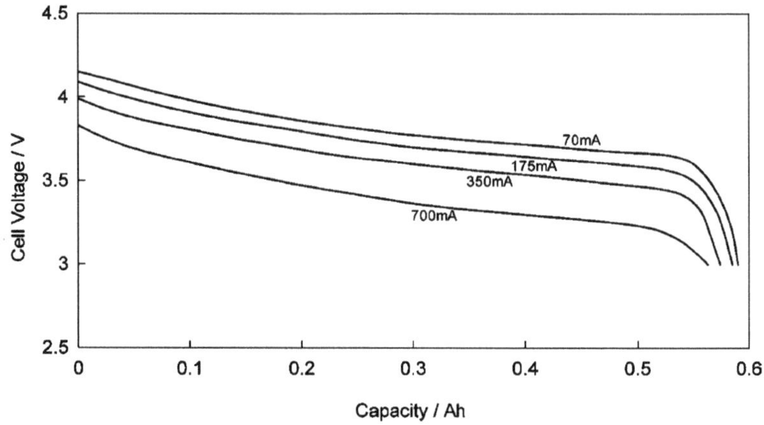

Fig. 21.4 Rate characteristics of lithium-ion polymer battery, UP383562A A2 WIB02H (Sony-made)

similar type of polymer matrix was commercialized by Sony in China. But the technology is different from that of Bellcore. Figure 21.4 shows the rate characteristics of the Sony-made polymer battery. In room temperature, the rate characteristics and cycling characteristics are excellent.

Sony is number one worldwide in production of lithium-ion gel polymer batteries, especially the relatively small-sized (below 1 Ah) battery. Sanyo-GS (Japan) and Samsung SDI (Korea) also are producing this kind of battery. Recently, a Chinese manufacturer started to produce lithium-ion gel polymer battery. ATL, one of the largest battery manufacturers, is producing a polymer battery based on Bellcore technologies. However, those manufacturers still continue their production in the small-sized battery field.

Recent demands of the market is for the mid- or large-sized lithium battery for the power-assisted bicycle, electronic bike, and hybrid vehicle. As the capacity of the battery increases, safety becomes very important. The gel polymer electrolyte contributes to keeping the battery safe even as the capacity of the lithium-ion battery increases.

The performance of acrylate-based gels improves the battery performance remarkably. For example, the gel that had almost equal ionic conductivity to a liquid and that had the elasticity of 10^4 dyne/cm or more by introducing a network polymer structure can be obtained.

Figure 21.5 shows the gel-type polymer electrolyte especially developed for the lithium polymer battery by Dai-Ichi Kogyo Seiyaku Co. Ltd. (DKS). The structure of the acrylate after cross-linking enables it to hold the large quantity of the liquid electrolyte and never leak the liquid electrolyte.[16]

Recently a joint company Shuan Yi Li (SYL) was established in Tianjin, China. SYL is the joint company of Enax Co. Ltd., DKS (both Japan), and Tianjin Yiqing Group (Holding share) Co. Ltd. (Tianjin, China). SYL started to produce midsized lithium-ion polymer batteries. They produce 3.5- to 10-Ah midsized batteries using

Fig. 21.5 Cross-linking-type gel polymer electrolyte

Fig. 21.6 Gel polymer batteries commercialized by SYL

gel polymer electrolyte technologies. Figure 21.6 shows the photograph of midsize polymer battery produced by SYL.

Figure 21.7 shows rate capability, temperature characteristics, cycle life, and overcharge test of the above battery. The battery performances are excellent. The most important point is that this battery is highly safe. Figure 21.7d shows the result of the overcharge test. The temperature of the battery increased with increasing charge capacity, but it showed maximum temperature at 57°C and no further temperature increase was observed. Those batteries will be used for the electric tools, electric bike, hybrid vehicle, and pure electric vehicles (EV).

21.7 New trends

S. J. Visco et al. proposed a new rechargeable battery system at room temperature by using reversible polymerization reaction of organic sulfur compound in place of inorganic sulfur.[17] As for 2,5-dimercapto 1,3,4-thiadiazole (DMcT) chemical

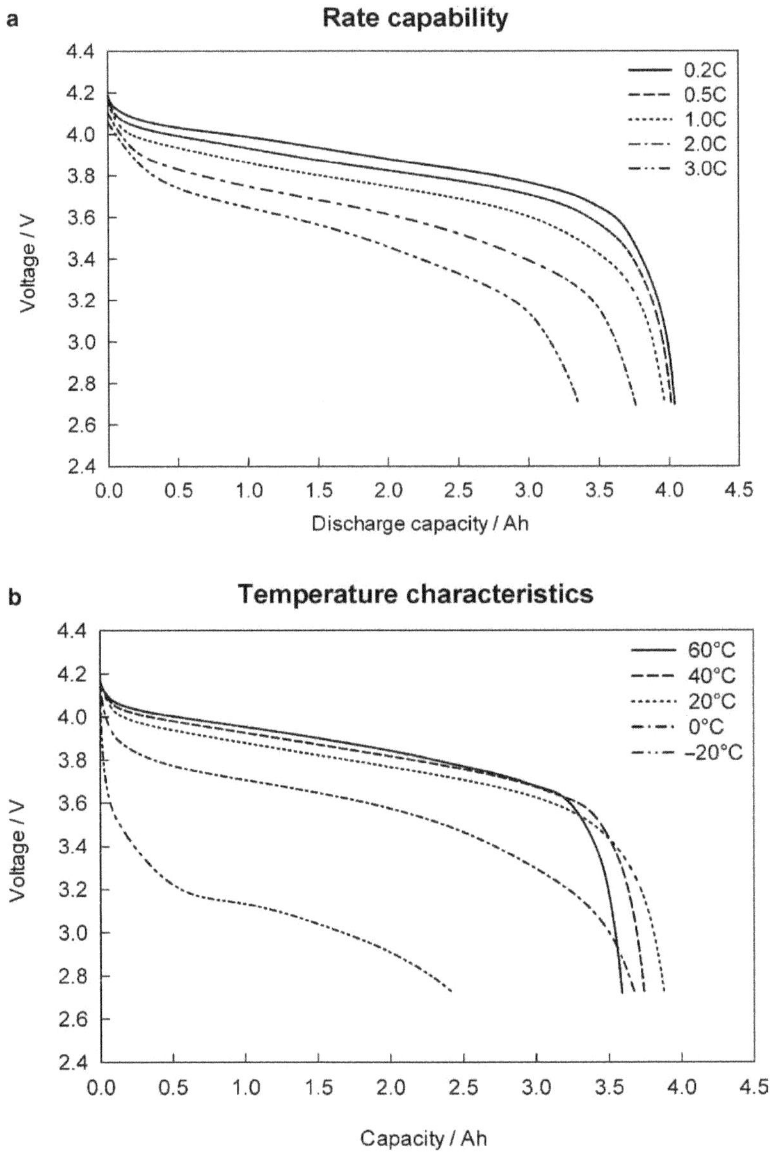

Fig. 21.7 The battery performances of the gel polymer battery made by SYL

compound, which is the organosulfur compound, the reversible change of the monomer and the polymer is possible through the electrochemical redox reaction. The sulfur polymer based on polyacetylene structure deposited by polymerization reaction of carbon disulfide has high-energy density of 2500 mAh/g. As for this battery system, the reversibility of the deposition and dissolution of the polymer is

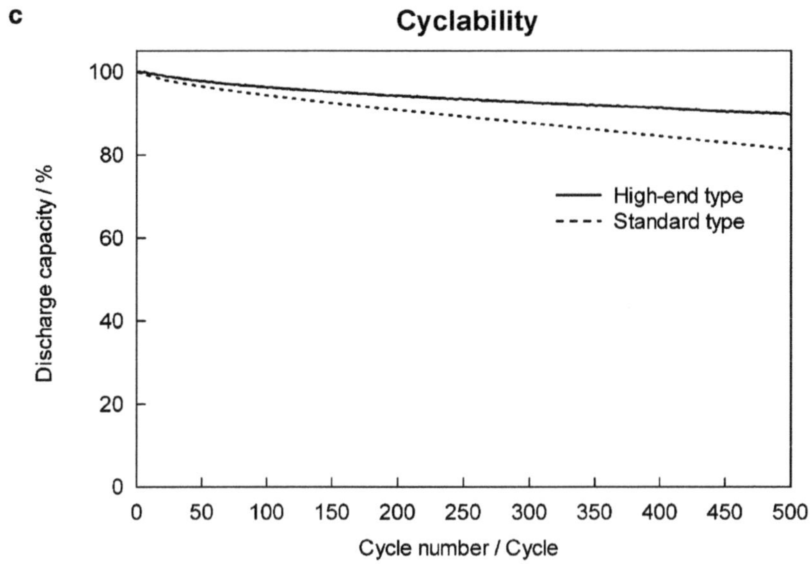

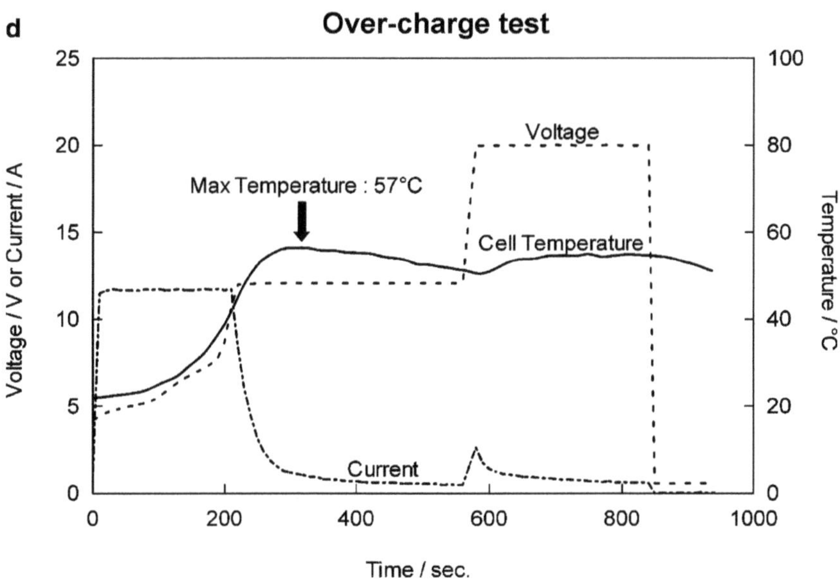

Fig. 21.7 (continued)

required to extend. This gel shows high ionic conductivity of 10^{-2} S/cm. The cycle life was drastically enhanced using polyaniline for the electrode.[18] The energy density of the positive electrode weight reached 180 mAh/g.

The career ion is generated by the interaction of a polymer chain and salt. If the concentration of the electrolytic salt increases, the carrier ion density increases; as a result, the ionic conductivity rises. However, in general when the concentration of a polymer electrolyte increases, the glass transition temperature rises, resulting in the decrease of ion mobility. Therefore, a new concept is necessary to overcome the limit of such electrolytes. The concept of "polymer-in-salt," which is different from the conventional "salt-in-polymer," was proposed by Angell et al.[19] For example, it is possible to obtain the rubberlike SPE by adding polypropylene oxide to ionic liquid, a key material. The examples of the typical ionic liquid structure are shown in Fig. 21.8. The ionic liquid is electrolytic salt based on a cation such as imidazole, pyridinium, and ammonium, which is liquid at room temperature. The ionic liquid is incombustible, and it can be used as an electrolyte in a wide temperature range. The polymer-in-salt type electrolyte draws much attention as polymer electrolyte of the next generation lithium secondary battery. This ionic liquid is prepared by dissolving lithium salt in ionic solution, which is mainly composed of ammonium cation with the wide potential window. Generally, the conductivity of these electrolytes decreases by gelation. However, recently, a new gelling agent, which is a fluorinated type material, was found. There is little change with its conductivity before gelation. The fluorinated gels[20] are prepared through reaction between

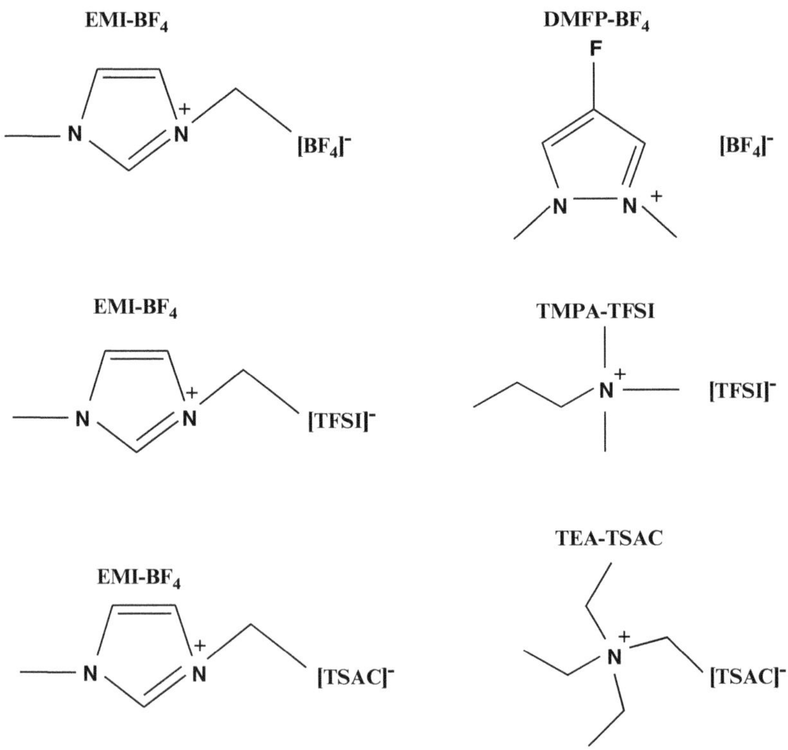

Fig. 21.8 Structure examples of a typical ionic liquid

fluoroalkanoyl peroxides and 2-acrylamid-2-methylpropanesulfonic acid (AMPS) oligomer-containing fluoroalkylated end-capped units {[-Rf-(AMPS)q]p-}. This gel shows high ionic conductivity of 10^{-2} S/cm.

21.8 Future Perspectives

The polymer battery has the advantage of using an electrode for a large area. The wide range of uses of polymer battery from portable electric devices to the electromobiles is expected to contribute to the rechargeable battery market. The polymer battery is creating a new market of the secondary battery for information-processing equipment such as the personal data assistant.

The utilization of the polymer battery has been supported by the technical innovation of polymer electrolyte materials. In order for the polymer battery to establish a strong presence in the battery industry in the near future, the development of the polymer electrolyte material suitable for the manufacturing process of a practical-use battery and development of the process technology which fully demonstrates the performance are much anticipated.

References

1. M. B. Armand, J. M. Chabagno, M. J. Duclot, *First Ion Transport in Solids* P. Vashishta, J. N. Mundy, G. K. Shenoy, Eds., Elsevier, New York (1979), p. 131.
2. P. V. Wright, *Br. Polym. J.*, 7 (1975) 319.
3. S. Claucy, D. F. Shriver, L. A. Ochrymowycz, *Macromolecules*, 19 (1986) 606.
4. C. S. Harris, D. F. Shriver, M. A. Ratner, *Macromolecules*, 19 (1986) 987.
5. M. H. Cohen, D. Turnbull, *J. Chem. Phys.*, 31 (1959) 1164.
6. P. G. Hall, G. R. Davis, J. E. McIntyre, I. M. Ward, D. J. Banister, K. M. F. LeBrocq, *Polym. Commun.*, 27 (1986) 98.
7. P. M. Blonsky, D. F. Shriver, P. Austin, H. R. Allcock, *J. Am. Chem. Soc.*, 106 (1984) 6854.
8. M. Kono, K. Furuta, S. Mori, M. Watanabe, N. Ogata, *Polym. Adv. Technol.* , 4 (1993) 85.
9. M. Kono, E. Hayashi, M. Watanabe, *J. Electrochem. Soc.*, 145 (5) (1998) 1524.
10. G. Feuillade, Ph. Perche, *J. Appl. Electrochem.*, 5 (1975) 63.
11. H. Akashi, K. Tanaka, K. Sekai, *Engineer. Mater.*, 1 (1997) 1.
12. C. Shmutz, J. M. Tarascon, A. S. Gozdz, P. C. Warren, F. K. Shokoohi, *Electrochem. Soc. Proc.*, 28 (1994) 330–335.
13. T. Ohsawa, O. Kimura, T. Kabata, N. Katagiri, T. Fujii, Y. Hayashi, *Electrochem. Soc. Proc.*, 28 (1994) 481–486.
14. M. Kono, E. Hayashi, M. Watanabe, *J. Electrochem. Soc.*, 146 (5) (1999) 1626.
15. US. Patent; US5037712 (1990).
16. M. Kono, E. Hayashi, M. Nishiura, M. Watanabe, *J. Electrochem. Soc.*, 147 (7) (2000) 2517.
17. S. J. Visco, M. Liu, L. C. DeJonghe, *J. Electrochem. Soc.*, 138 (1990) 1191.
18. T. Sotomura, N. Oyama, *Denki Kagaku*, 65 (10) (1997) 796.
19. C. A. Angell, C. Liu, E. Sanchez, *Nature*, 362 (1993) 137.
20. H. Sawada, Y. Murai, M. Kurachi, T. Kawase, T. Minami, J. Kyokane, T. Tomita, *J. Mater. Chem.*, 12 (2002) 188–194.

Chapter 22
A Novel Hard-Carbon Optimized to Large-Size Lithium-Ion Secondary Batteries

Aisaku Nagai, Kazuhiko Shimizu, Mariko Maeda, and Kazuma Gotoh

22.1 Introduction

Kureha Corporation has developed nongraphitizable carbons (so-called hard carbon) prepared from the cross-linked petroleum pitch for almost 20 years as an anode of the lithium-ion battery (LIB).[1] However, in these years graphite is more popularly used as an anode of LIB of small portable equipment, such as cellular phones, digital cameras, and portable personal computers, because high-energy density is much more important in this type of application than long-life durability is. Hard carbon has been used only in the field of professional camcorders, satellites, and electric bikes because it seemed to be difficult to change a new battery at the end of its life.

Recently application to large-size equipments, such as electric tools and hybrid electric vehicles (HEV), gets a lot of attention. The operating voltage window of the battery for HEV should be wide enough to regenerate electric power at a high rate. However, the operating voltage of the battery in which graphite is used as an anode and $LiCoO_2$ as a cathode is too flat and narrow, and the voltage will be immediately cramped to avoid overcharge at more than 4.3V. On the other hand, the operating window of the battery that used hard carbon is wider; it is already evident that the battery that used hard carbon has a much higher input as well as output power at any state of charge than the battery that used graphite. This is the reason why so much of attention is focused on hard carbon.[2]

But the hard carbon that has been developed, Carbotron P (F), is designed for small-size equipment.[3] It has a large charge capacity of more than $500 Ahkg^{-1}$ and long-life durability. In order to improve response, particle size must be small enough to reduce the diffusion path of the lithium ion into carbons. So, there is another type of hard carbon for HEV that is named Carbotron PS (F). The basic electrochemical characteristics of Carbotron P (F) and Carbotron PS (F) are almost

A. Nagai (✉), K. Shimizu, M. Maeda, and K. Gotoh
Research and Technology Division, Kureha Corporation, 3-3-2, Nihonbashi-Hamacho, Chuo-ku, Tokyo 103-8552, Japan
a-nagai@kureha.co.jp

M. Yoshio et al. (eds.), *Lithium-Ion Batteries*,

DOI10.1007/978-0-387-34445-4_22 © Springer Science+Business Media, LLC 2008 the same. However, these hard carbons have low charge-discharge efficiency and these capacities will fade when they are stored in the air before preparation. So, these hard carbons should be redesigned for HEV application.

22.2 Structure and Electrochemical Characteristics of a Novel Hard Carbon

The true density of hard carbon (about 1.5 g cm^{-1}) measured by isopropyl alcohol is smaller than that of graphite (2.26 g cm^{-1}). This fact means self-evidently that there are micropores in hard carbon and that lithium atoms can be stored in those pores. One of the authors first analyzed this new lithium storage mechanism by ^{7}Li-NMR and this mechanism is the reason why the charge capacity of hard carbon is greater than that of graphite.[1] The pores in hard carbon are very different from those in activated carbon because very slow adsorption of N_2 gas into hard carbon for over one week is observed when the nitrogen adsorption isotherm is measured. The O_2 gas and H_2O gas can be adsorbed slowly into the pores, while hard carbon is stored in the air for a long time and those gases may cause the capacity fading. And also, the stored lithium in the ink bottle-type pore may cause the low efficiency of charge-discharge capacity.

One of the unique characteristics of hard carbons as anodes is that the charge-discharge curve shows two regions. In the first region the potential changes gradually (CC region), and in the second region the potential is almost constant (CV region). It was reported that lithium atoms are stored in the pores and form a lithium-cluster like structure at the CV region by the ^{7}Li-NMR measurements.[1, 4–7] So, the origin of the charge capacity at the CV region is the pore structure. On the other hand, the capacity of the CV region cannot be utilized at the real cell because the potential of the CV region is almost the same as that of lithium metal, and lithium-dendrite deposition during charging seems to be inevitable. Also, in the application of HEV the state of charge of LIB is maintained around 50%, and so only the capacity of the CC region can be utilized.

The target of development of a novel hard carbon for HEV is to remove the ink bottle-type pore and to improve the stability and the efficiency. The hard carbon recently developed is called Carbotron P (J),[8] and its particle size is almost the same as Carbotron PS (F), but its electrochemical characteristics are different from Carbotron P (F) and PS (F).

Figure 22.1 shows the surface of Carbotron P (J) by scanning electron microscopy (SEM). X-ray powder diffraction (XRD) patterns of these carbons, measured on Rigaku RAD-C diffractometer with roter-flex and Cu-Kα radiation, are shown in Fig.22.2.[7] The interlayer distances and the crystallite size factors (Lc) calculated from XRD(002) reflection using Scherrer formula[9] are given in Table 22.1. Figure 22.3 also shows the particle size distributions of these three carbons. The average particle sizes estimated from distributions are shown in Table 22.1.

Fig. 22.1 The scanning electron micrograph of the Carbotron P (J). Reprinted from Ref. 7, copyright (2006), with permission from Elsevier Ltd

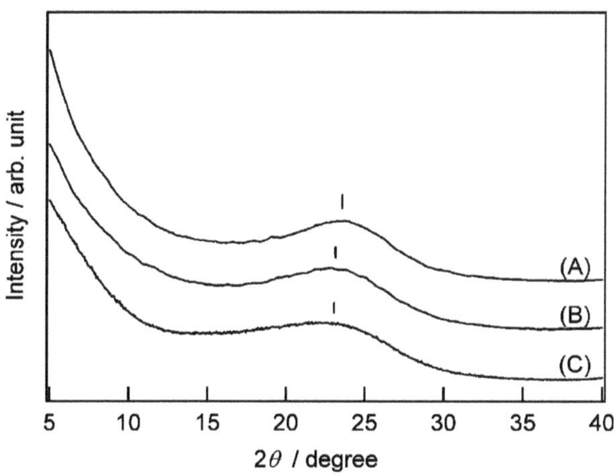

Fig. 22.2 Powder XRD patterns of carbon (A) Carbotron P (J), (B) Carbotron PS (F) and (C) Carbotron P (F). Reprinted from Ref. 7, copyright (2006), with permission from Elsevier Ltd

Table 22.1 The interlayer distances (*d*002), the crystallite size factors (*Lc*002) calculated from XRD(002) reflection using Scherrer formula,[9] and the mean particle sizes of three carbon samples

Sample	d_{002} (nm)	L_{C002} (nm)	Particle size (μm)
Carbotron P (J)	0.374	1.40	9
Carbotron PS (F)	0.379	1.31	9
Carbotron P (F)	0.380	1.10	22

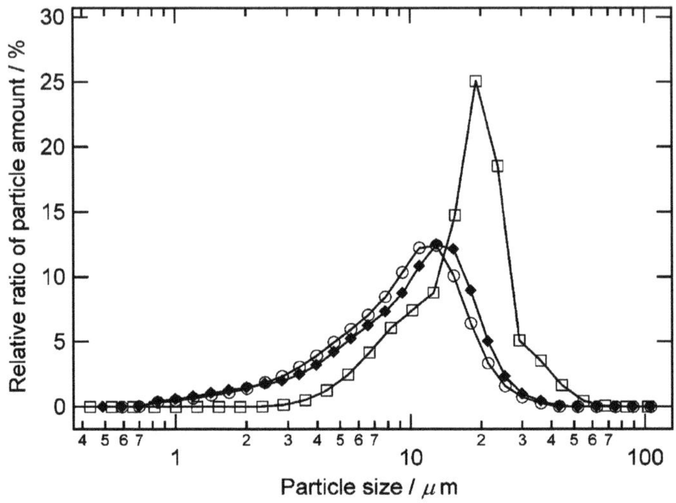

Fig. 22.3 Particle-size distributions of (A) (*closed diamond*) Carbotron P (J), (B) (*open circle*) Carbotron PS (F), and (C) (*open square*) Carbotron P (F). (A) and (B) have almost the same distributions. The mean sizes of (A), (B), and (C) are shown in Table 22.1

According to XRD data there is no significant change in these carbons, but the electrochemical characteristics are quite different, as shown in Fig.22.4 and Table 22.2.[7] These data are obtained by the coin-cell measurements. Lithium metal is employed as a counterelectrode. The electrolyte of 1mol dm^{-3} $LiPF_6$ solution in 1:1 (volume ratio) propylene carbonate (PC):dimethyl carbonate (DMC) is used. The charge is performed with constant current (CC) density of 0.5mA cm^{-2}. After the cell voltage reached 0.0V, the cell is kept at 0.0V until the equilibrium current (the CV lithiation process). Discharge is carried out galvanostatically at 0.5mA cm^{-2}.

The whole charge capacity of Carbotron P (J) apparently is very small compared to that of Carbotron P (F) and PS (F); however, the capacity of Carbotron P (J) at CC region, the effective capacity as the anode for HEV application, is almost the same or even greater than those of the other two products. Another advantage of Carbotron P (J) is that the small irreversible capacity may improve the charge-discharge efficiency of the real cell and increase the battery capacity as a result.

Table 22.3 shows how much water can be adsorbed in carbons when stored in 90%RH at 40°C for 70h. The adsorbed water content in Carbotron P (J) is dramatically decreased to less than half of that in Carbotron PS (F). Further experiments suggested that the adsorbed water content will be able to be decreased to less than 1% in near future.

The true density of Carbotron P (J) is same as that of Carbotron PS (F). So, the pore volume in carbons is not changed. However, the capacity at the CV region and the adsorbed water content apparently are decreased. These phenomena imply that

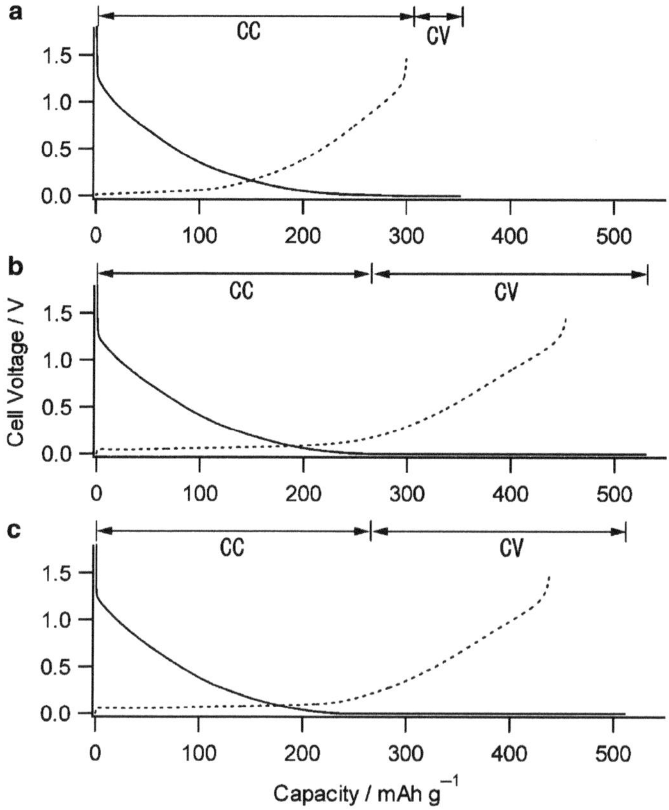

Fig. 22.4 Charge (lithiation process: *solid line*) and discharge (lithium extraction process: *broken line*) curves of carbon: (A) Carbotron P (J), (B) Carbotron PS (F), and (C) Carbotron P (F). Reprinted from Ref. [7], copyright (2006), with permission from Elsevier Ltd

Table 22.2 The results of electrochemical evaluation of cells using samples of carbon: (A) Carbotron P (J); (B) Carbotron PS (F); and (C) Carbotron P (F)

Sample	Whole charge capacity (mAh g^{-1})	CC capacity (mAh g^{-1})	CV capacity (mAh g^{-1})	Initial irreversible capacity (mAh g^{-1})
(A)	352	305	47	53
(B)	530	264	266	77
(C)	511	240	271	73

Table 22.3 The weight% of adsorbed water in carbon: (A) Carbotron P (J), and (B) Carbotron PS (F) Stored in 90%RH at 40°C after 70°h

Sample	wt%
(A)	2.4
(B)	5.8

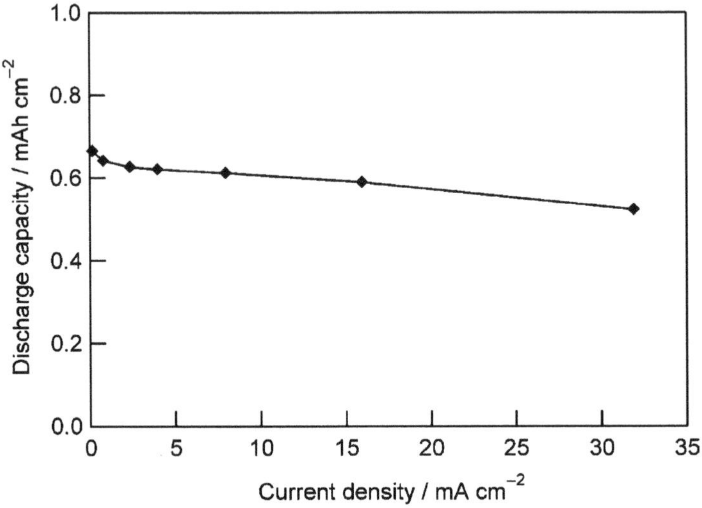

Fig. 22.5 Discharge capacity of Carbotron P (J) measured at various current density. The highest current density corresponds to 60 C

some amount of the pore in Carbotron P (J) may be closed during preparation. This hypothesis was established recently by ^{7}Li-NMR measurements.[7]

Finally the rate capability of Carbotron P (J) is demonstrated in Fig.22.5. The 2016-type coin cells are used for this measurement. Lithium metal is employed as a counter electrode. The electrolyte of 1mol dm^{-3} $LiPF_6$ solution in 1:2:2 (volume ratio) ethylene carbonate (EC): DMC: ethyl methyl carbonate (EMC) is used. Charge is performed with CC density of 0.5mA cm^{-2}. After the cell voltage reached 0.0V, the cell is kept at 0.0V until the current comes down to less than 1/100 C. Discharge is carried out galvanostatically at various current densities and the cut voltage is 1.5V. The highest current density is equivalent to 60 C. It goes without saying that PC is a more suitable solvent for HEV application than EC because of its low viscosity and that there is no continuous electrochemical reaction observed when charging hard carbon in the electrolyte of PC. Figure 22.5 shows that a very high-rate capability can be obtained with this novel hard carbon even though the electrolyte of EC for graphite is chosen.

22.3 Conclusion

A novel hard carbon has been designed for the application of large-size lithium-ion secondary batteries. The carbon's stability in air, its better efficiency of charge capacity, and its high-rate capability have been shown to maintain a suitable charge-discharge voltage profile for HEV application.

References

1. A. Nagai, M. Ishikawa, J. Masuko, N. Sonobe, H. Chuman, T. Iwasaki, *Mater. Res. Soc. Symp. Proc.*, **393** (1995) 339–343.
2. Y. Tanjo, T. Abe, H. Horie, T. Nakagawa, T. Miyamoto, K. Katayama, *Soc. Automot. Eng.*, **SP-1417** (1999) 51–55.
3. Y. Ohashi, Y. Shigaki, Eur. Pat. Appl. (1996) EP 726606.
4. K. Tatsumi, J. Conard, M. Nakahara, S. Menu, P. Lauginie, Y. Sawada, Z. Ogumi, *Chem. Commun.*, (1997) 687–688.
5. K. Tatsumi, T. Kawamura, S. Higuchi, T. Hosotubo, H. Nakajima, Y. Sawada, *J. Power Sources*, **68** (1997) 263–266.
6. K. Tatsumi, J. Conard, M. Nakahara, S. Menu, P. Lauginie, Y. Sawada, Z. Ogumi, *J. Power Sources*, **81–82** (1999) 397–400.
7. K. Gotoh, M. Maeda, A. Nagai, A. Goto, M. Tansho, K. Hashi, T. Shimizu, H. Ishida, *J. Power Sources*, **162** (2006) 1322–1328.
8. K. Shimizu, M. Maeda, S. Morinishi, A. Nagai, A. Hoshi, *PCT Int. Appl.* (2005) WO 2005098999.
9. P. Scherrer, *Nachr. Ges. Wiss. Göttingen*, **2** (1918) 98.

Chapter 23
$LiMn_2O_4$ as a Large-Capacity Positive Material for Lithium-Ion Batteries

Masaki Okada and Masaki Yoshio

23.1 Introduction

The layered oxide materials $LiMO_2$ (M = Co, Ni) have been investigated as the cathode materials for secondary lithium batteries.[1,2] $LiMn_2O_4$ is one of the most promising alternatives to the high-cost cobalt-positive materials including $LiCoO_2$ and $LiNi_{1/3}Co_{1/3}Mn_{1/3}O_2$ because of its low cost, lower toxicity, and material safety in the charge state. $LiMn_2O_4$ shows excellent cycle performance at the 4-V region.[3,4] However, the cycle capacity is lower than other cathode materials such as $LiMO_2$ (M = Co, Ni). If it is possible to reversibly insert the excess lithium into $LiMn_2O_4$ in the 3-V region, this disadvantage should be overcome. However, in previous studies it was reported that a full lithium insertion/extraction process has led to gradual decay of its original structure, and $LiMn_2O_4$ cathode has shown a drastically fast capacity failure when it was used in the 3- to 4-V region.[5]

In this chapter, we report on the relation between the crystal structure of $LiM_XMn_{2-X}O_4$ (M = Mn, Co, Ni) and the electrochemical properties in the 3- to 4-V region. We describe the synthesis of the materials with large rechargeable capacity at wide voltage region.

23.2 Experimental

$LiMn_2O_4$ was prepared by reacting a stoichiometric mixture (Li/Mn = 0.5 by molar ratio) of manganese oxide (CMD, EMD, γ-MnOOH) and $LiNO_3$ as the starting materials. $LiM_XMn_{2-X}O_4$ ($0 < X \leqq 0.5$) were synthesized using γ-MnOOH, $LiNO_3$, and $Ni(OH)_2$ or $Co(OH)_2$ as the starting materials. These mixtures were precalcined

M. Okada (✉) and M. Yoshio
Nanyo Research Laboratory, TOSOH Corporation, 4560, Kaisei-cho, Shin-nanyo, 746-8501, Japan
okada_m@tosoh.co.jp

M. Yoshio et al. (eds.), *Lithium-Ion Batteries*,

DOI 10.1007/987-0-387-34445-4_23

at 450°C for 24 h in air, followed by calcination at 700°C for 24 h in air using the melt-impregnation method.[6] The powder X-ray diffraction using Cu Kα radiation was performed to identify the crystalline phase and the crystalline parameter of resulting products. The chemical composition and the average manganese valence were measured by the inductive coupled plasma (ICP) method and redox titration using $KMnO_4$, respectively. The electrochemical characterization of the compounds was performed on CR2030 coin-type cells. The method of assembling the cell was as follows: the cathode consisted of a mixture of 50 mg of active material and 25 mg of conductive binder (teflonized acetylene black). It was pressed on 2.5 cm^2 stainless steel mesh as a current collector at 1,000 kg cm^{-2} and dried at 200°C for 2 h in a vacuum oven (< 1 mm Hg). The cell was constructed using a cathode, a lithium-metal anode, and a porous polypropylene film as a separator and 1-M $LiPF_6$ propylene carbonate (PC): diethyl carbonate (DEC) (1:4 by volume) as the electrolyte solution. The charge and discharge current density was 1.0 mA cm^{-2} with cutoff voltages of 2.0–4.2 V (cell voltage).

23.3 Results and Discussion

$LiMn_2O_4$ samples were prepared using γ-MnOOH (sample 1), CMD (sample 2) and EMD (sample 3) as the manganese oxide source, respectively. All samples indicated single-phase spinel and show no impurities. Figure 23.1 shows the cycle test results using the prepared samples as a cathode material at 1.0 mA cm^{-2} between 2.0 and 4.2 V. These samples exhibited the different capacity failure on charge-discharge cycling. The sample1 showed the best cycling performance among the samples. In the 3-V region, the crystal structure of $LiMn_2O_4$ changes from cubic

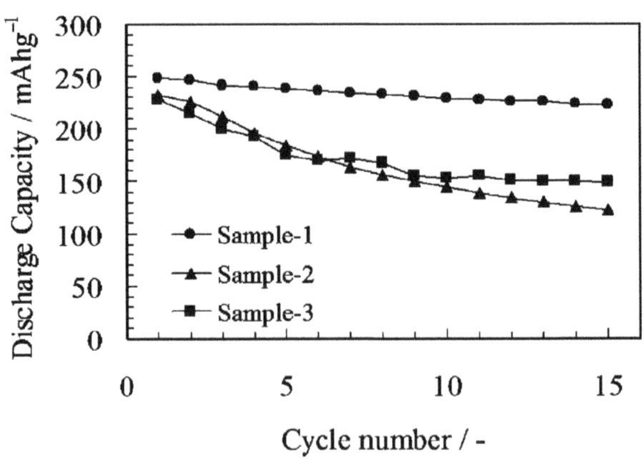

Fig. 23.1 Cycling performance of Li/1 M $LiPF_6$ PC-DEC (1:4)/$LiMn_2O_4$ cells with the various prepared cathodes: (*filled circle*) sample 1 (γ-MnOOH), (*filled triangle*) sample 2 (CMD), and (*filled square*) sample 3 (EMD) at 1.0 mAcm^{-2} between 2.0 and 4.2 V

spinel phase to tetragonal spinel phase.[7] Therefore, to clarify the differences among the samples in the cycling behavior, an analysis of the crystal structure of the samples was performed.

The $\sin\theta/\lambda$–$\beta\cos\theta/\lambda$ plots (= Hall plot)[8] of the samples are shown in Fig.23.2 The following equation can be induced using Scherrer equation[1]:

$$\beta\cos\theta/\lambda = 1/\varepsilon + 2\eta(\sin\theta/\lambda), \tag{23.1}$$

where β is integral width of intensity peak, η indicates strain of crystalline structure, and ε means crystal size. The λ value is 1.5405 Å (CuKα). It was clarified that the discharge capacity failure on cycling in the 3- to 4-V region decreased with an increase of the slope value of the Hall plot (= 2η value) from the results of this crystal analysis. The sample 1 has large η value ($\eta = 1.46 \times 10^{-3}$). It means that the sample 1 has smaller crystallite size and higher strain in the structure. So, we considered that the crystallite size and the strain in the structure were important factors for the crystal change of $LiMn_2O_4$ from cubic spinel phase to tetragonal spinel phase in the charge-discharge process.

To improve the cyclability of $LiMn_2O_4$ in the 4-V region, many studies have been made using doped manganese-substituted spinels such as $LiM_XMn_{2-X}O_4$ (M = Co, Ni, Cr, Al, …). It has been reported that the capacity failure was suppressed by the replacement of a part of manganese with another metal ion because of the doped manganese-substituted spinels inhibited the decay of its original structure in lithium insertion/extraction process. Next we show our investigation of the $LiM_XMn_{2-X}O_4$ compound (M = Co, Ni, X is from 0.0 to 0.5).

The samples were synthesized using γ-MnOOH, $LiNO_3$, and $Ni(OH)_2$ or $Co(OH)_2$ as the starting materials. In the case of $LiCo_XMn_{2-X}O_4$ compounds, all

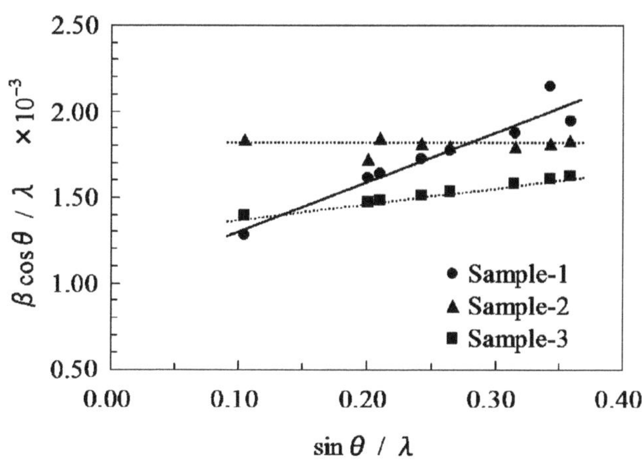

Fig. 23.2 The $\sin\theta/\lambda$–$\cos\theta/\lambda$ plot (Hall plot) of the prepared samples. η values were 1.45×10^{-3} (sample-1), 0.00×10^{-3} (sample-2) and 0.45×10^{-3} (sample-3), respectively

samples were identified as a single-phase spinel and exhibited no impurities in the entire *X* range. All the diffraction peaks of the samples shifted toward a higher angle than the original diffraction position. It seems that the cubic symmetry of spinel structure is maintained. It means that the lattice constant decreased with increase in *X* value without the phase change.

Figure 23.3 shows the X-ray diffraction patterns for the $LiNi_XMn_{2-X}O_4$ compounds. The samples were identified as a single-phase spinel until x was 0.3. When the *X* value was 0.5, there were some other peaks that corresponded to NiO (see Fig. 23.3e). It means that the nickel-doped substituted spinel structure cannot form perfectly solid solutions when *x* is 0.5. It may be very difficult to prepare the nickel-doped substituted spinel by the solid-state reaction. The NiO peak of $LiNi_XMn_{2-X}O_4$ sample, which was prepared by the conventional method, was always observed and it could be removed through several recalcination processes at the high temperature.[9] The NiO peak appeared slightly when the *X* value was only 0.5 in our research and this peak was weak. Our preparation method was more effective than the conventional solid-state method used to synthesize homogeneous metal-doped manganese-substituted spinel.

Figure 23.4 shows the first charge-discharge profiles for $LiCoXMn_{2-}XO_4$ and $LiNi_XMn_{2-X}O_4$ ($X = 0.1, 0.5$). The $LiNi_{0.5}Mn_{1.5}O_4$ cathode showed nearly one plateau in the 3-V region (Fig. 23.4b). The stable valence of nickel is Ni^{2+}.[10] Consequently, the valence of the manganese ion in this compound changes to Mn^{4+} because the spinel matrix must be electrically compensated by the oxidation of Mn^{3+} to Mn^{4+}. The charge-discharge capacity in the 4-V region decreased with an increase in the *X* value. It is suggested that only the Mn^{3+} contributes to the charge-discharge capacity in the 4-V region. This result agrees with that of another group.[10]

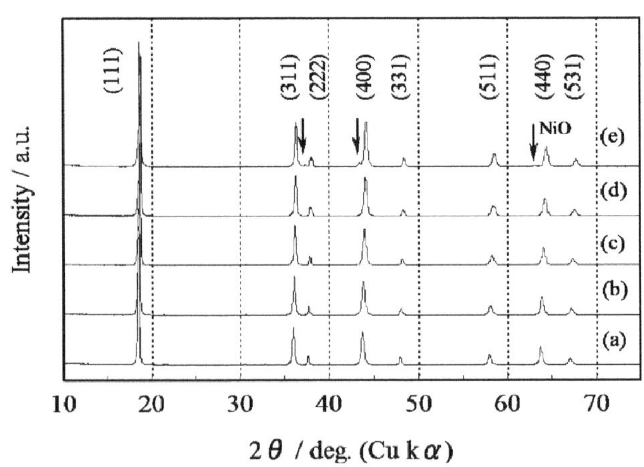

Fig. 23.3 Powder X-ray diffraction patterns for $LiNi_XMn_{2-X}O_4$ samples (**a**) $X = 0.0$, (**b**) $X = 0.1$, (**c**) $X = 0.2$, (**d**) $X = 0.3$, and (**e**) $X = 0.5$, respectively

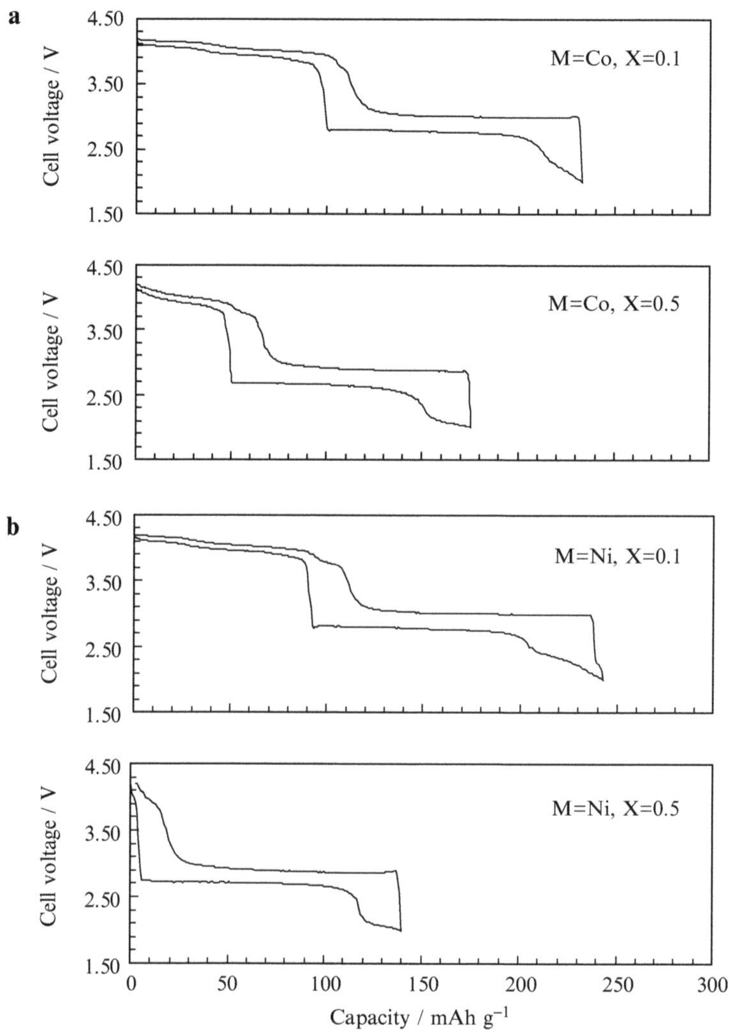

Fig. 23.4 First charge-discharge profiles for $LiM_XMn_{2-X}O_4$ with $X = 0.1$ and 0.5 cycled between 2.0 and 4.2 V at 1.0 $mAcm^{-2}$ (**a**) M = Co and (**b**) M = Ni, respectively

The $LiNi_XMn_{2-X}O_4$ cathode showed lower capacity in the 4-V region than $LiCo_XMn_{2-X}O_4$ cathode at the same X value. Therefore, it is assumed that the valence of cobalt is Co^{3+}.

Now let us focus on the change of structural parameters. The lattice constant decreases with an increase in the X value without the phase change. That is to say, the crystal structure becomes contracted. Wakihara and co-workers reported that the bond length of the 16d–32e in spinel structure referred to the average bond length of M–O bonds (includes both M–O and Mn–O bond for $LiM_XMn_{2-X}O_4$), and the M–O bond becomes shorter with increased doping.[10] Both the nickel ion and cobalt ion were substituted at the manganese 16d site. Therefore, it is reasonable to consider that the M–O and Mn–O bonds become heterogeneously contracted for $LiMXMn_{2-}XO_4$ because the metal ion exhibits the different oxidation state (i.e., Ni^{2+}, Co^{3+}, Mn^{3+}, or Mn^{4+}). Furthermore, it is expected that the increase in the amount of the nickel or cobalt ion contents should enhance the crystal strain. Figure 23.5 shows the Hall plots of $LiCo_XMn_{2-X}O_4$ (Fig. 23.5a) and $LiNi_XMn_{2-X}O_4$ (Fig. 23.5b), respectively. As expected, the η value increased with the increase of the doped content value of X, except for $X = 0.5$ in nickel.

Figure 23.6 shows the cycle test results for $LiMXMn_{2-}XO_4$ cathode at 1.0 mA cm^{-2} between 2.0 and 4.2 V. The initial discharge capacity of $LiMn_2O_4$ was

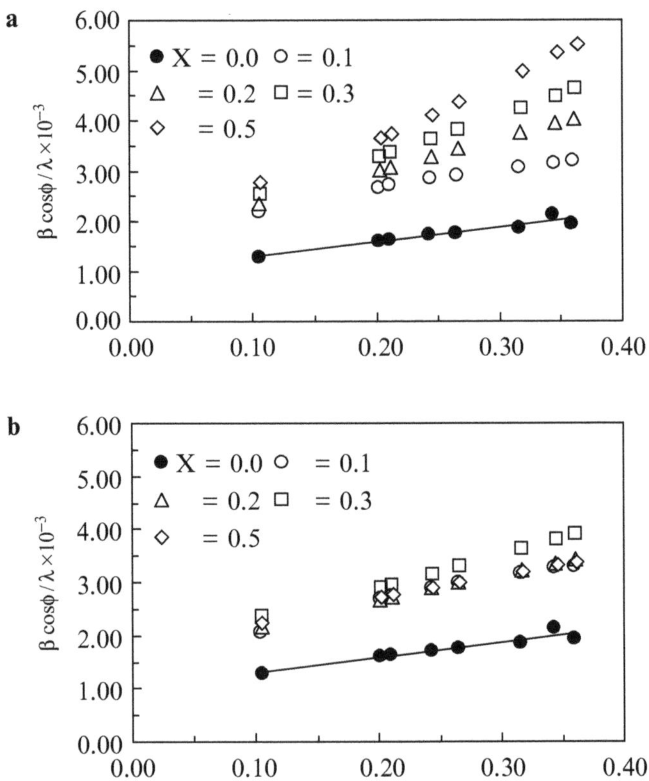

Fig. 23.5 The $\sin\theta/\lambda$–$\cos\theta/\lambda$ plot (Hall plot) of $LiM_XMn_{2-X}O_4$ with several X value (**a**) M = Co and (**b**) M = Ni. The η value was 1.90×10^{-3} ($X = 0.1$), 3.32×10^{-3} ($X = 0.2$), 4.13×10^{-3} ($X = 0.3$) and 5.49×10^{-3} ($X = 0.5$) for M = Co and 2.34×10^{-3} ($X = 0.1$), 2.44×10^{-3} ($X = 0.2$), 3.05×10^{-3} ($X = 0.3$) and 2.19×10^{-3} ($X = 0.5$) for M = Ni, respectively

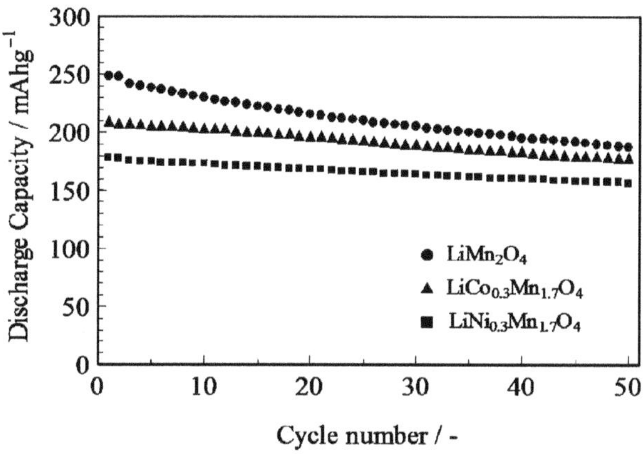

Fig. 23.6 Cycling performance of Li/1 M $LiPF_6$ PC-DEC(1:4)/$LiM_XMn_{2-X}O_4$ cells with various cathodes: (*filled circle*) $LiMn_2O_4$, (*filled square*) $LiCo_{0.3}Mn_{1.7}O_4$, and (*filled triangle*) $LiNi_{0.3}Mn_{1.7}O_4$ at 1.0 mAcm^{-2} between 2.0 and 4.2 V

248.5mAhg^{-1}, and it was reduced to 187.9 mAhg^{-1} after the 50th cycling. In the case of $LiM_{0.3}Mn_{1.7}O_4$, it indicated 209.4 mAhg^{-1} and 178.1 mAhg^{-1} (M = Co), 178.2 mAhg^{-1} and 156.3 mAhg^{-1} (M = Ni), respectively. The cycle retention ratio increased from 76% (X = 0.0) to 85% (Co, X = 0.3) and 88% (Ni, X = 0.3). The stability in charge-discharge cycling increased with increase in η value. We concluded that the crystallite size and the strain in the structure were important factors for the crystal change of $LiMn_2O_4$ from cubic spinel phase to tetragonal spinel phase in charge-discharge process. The replacement of a part of manganese by cobalt or nickel ions in $LiMn_2O_4$ improved the cycling performance and promoted the stability of its structure in the operating 3- to 4-V region.

$LiMn_2O_4$ or $LiM_XMn_{2-X}O_4$ cathodes can be coupled with a carbon anode to construct the lithium-ion battery. When this is the case, the only source of lithium in the cell is the $LiMn_2O_4$ cathode. On the other hand, a main drawback of the present lithium-ion cell is the electrolyte reduction with utilization of lithium on the carbon anode during their initial charging process. This reduction involves an irreversible capacity anywhere from 20 to 60% of lithium intercalation capacity of carbon.[11] Considering these reasons, it is necessary to previously insert the excess lithium into $LiMn_2O_4$ or $LiM_XMn_{2-X}O_4$ in order to use them, as the 4-V or 3- to 4-V cathode. The overlithiated $Li_{1.91}Mn_2O_4$ was prepared by lithiation of $LiMn_2O_4$ (sample 1) in an organic solvent using lithium-ion as a reducing agent.[12] Figure 23.7 shows the initial cycling performance of $Li_{1.91}Mn_2O_4$ at the 3- to 4-V region. As can be seen, this sample exhibited an excellent cycle performance and a large-cycle capacity of about 190 mAh g^{-1}. This can be considered to be an excellent solution to the problem of lithium loss in the lithium-ion cell system; that is, the use of this overlithiated $LiMn_2O_4$ as the cathode material.

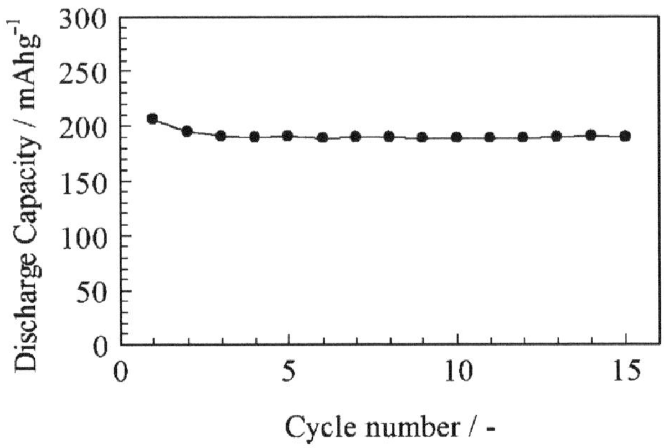

Fig. 23.7 Cycling performance of a Li/1 M $LiPF_6$ PC-DEC(1:4)/$Li_{1.91}Mn_2O_4$ cell at 1.0 mA cm^{-2} between 2.0 and 4.2 V. The overlithiated $Li_{1.91}Mn_2O_4$ was prepared by lithiation of smple-1. The initial charge capacity was 239 mA h g^{-1}

23.4 Conclusion

$LiMn_2O_4$ and $LiM_xMn_{2-x}O_4$ (M = Co, Ni) compounds were synthesized by the melt-impregnation method. We found that the crystallite size and the strain in the structure were important factors for the crystal change of $LiMn_2O_4$ from cubic spinel phase to tetragonal spinel phase during the charge-discharge process. We clarified that the strain in the structure increased with an increase in the doped content. The sample that had smaller crystallite size and higher strain in the structure exhibited excellent cycle behavior in the 3- to 4-V region. The cycle retention ratio of discharge capacity (50th cycle vs. 1st cycle) increased from 76% ($LiMn_2O_4$) to 85% ($LiCo_{0.3}Mn_{1.7}O_4$) and 88% ($LiNi_{0.3}Mn_{1.7}O_4$). The replacement of a part of manganese by cobalt or nickel ions in $LiMn_2O_4$ improved the cycling performance and promoted the stability of its structure for its operation in the 3- to 4-V region. The $Li_{1.91}Mn_2O_4$, which was prepared by lithiation of $LiMn_2O_4$, exhibited an excellent cycle performance and large cycle capacity of about 190 $mAhg^{-1}$.

References

1. K. Mizushima, P. C. Jones, P. J. Wiseman, J. B. Goodenough, *Mater. Res. Bull.*, **15** (1980) 783
2. J. R. Dahn, *J. Electrochem. Soc.*, **138** (1991) 2207
3. Y. Xia, M. Yoshio, *J. Electrochem. Soc.*, **143** (1996) 825
4. Y. Gao, J. R. Dahn, *J. Electrochem. Soc.*, **143** (1996) 1783
5. J. M. Tarascon, E. Wang, F. K. Sehokoohi, *J. Electrochem. Soc.*, **138** (1991) 2859
6. Y. Xia, H. Takeshige, H. Noguchi, M. Yoshio, *J. Power Sources*, **56** (1997) 61
7. T. Ohzuku, M. Kitagawa, T. Hirai, *J. Electrochem. Soc.*, **137** (1995) 769

8. W. H. Hall *J. Inst. Met.*, **75** (1950) 1127
9. K. Amine, H. Tukamoto, H. Yasuda, Y. Fujita, *J. Electrochem. Soc.*, **143** (1996) 1607
10. Li Gvohua, H. Ikuta, T. Uchida, M. Wakihara, *J. Electrochem. Soc.*, **143** (1996) 178
11. D. Peramunage, K. M. Abraham, *J. Electrochem. Soc.*, **145** (1998) 1131
12. J. M. Tarascom, D. Guyomard, *J. Electrochem. Soc.*, **138** (1991) 2864

Index

M

N

www.ingramcontent.com/pod-product-compliance
Ingram Content Group UK Ltd.
Pitfield, Milton Keynes, MK11 3LW, UK
UKHW020226290926
15470UKWH00001B/401
9780387344447